International
REVIEW OF
Neurobiology

Volume 52

Neurobiology
OF THE
Immune System

International
REVIEW OF
Neurobiology
Volume 52

Neurobiology

OF THE

Immune System

EDITED BY

ANGELA CLOW

Department of Psychology
University of Westminster
London W1B 2UW, United Kingdom

FRANK HUCKLEBRIDGE

Department of Biomedical Sciences
University of Westminster
London W1M 8JS, United Kingdom

ACADEMIC PRESS

An imprint of Elsevier Science

Amsterdam   Boston   London   New York   Oxford   Paris
San Diego   San Francisco   Singapore   Sydney   Tokyo

Academic Press
An imprint of Elsevier Science.
525 B Street, Suite 1900, San Diego, California 92101-4495, USA
http://www.academicpress.com

Academic Press
84 Theobalds Road, London WC1X 8RR, UK
http://www.academicpress.com

International Standard Book Number: 0-12-366853-0

PRINTED IN THE UNITED STATES OF AMERICA
02   03   04   05   06   07   MM   9   8   7   6   5   4   3   2   1

CONTENTS

Neuropeptides: Modulators of Immune Responses in Health and Disease

DAVID S. JESSOP

Brain–immune Interactions in Sleep

LISA MARSHALL AND JAN BORN

Neuroendocrinology of Autoimmunity

MICHAEL HARBUZ

Systemic Stress-Induced Th2 Shift and Its Clinical Implications

ILIA J. ELENKOV

Neural Control of Salivary S-IgA Secretion

GORDON B. PROCTOR AND GUY H. CARPENTER

Stress and Secretory Immunity

JOS A. BOSCH, CHRISTOPHER RING, ECO J. C. DE GEUS, ENNO C. I. VEERMAN, AND ARIE V. NIEUW AMERONGEN

Cytokines and Depression

ANGELA CLOW

Immunity and Schizophrenia: Autoimmunity, Cytokines, and Immune Responses

FIONA GAUGHRAN

Cerebral Lateralization and the Immune System

PIERRE J. NEVEU

Behavioral Conditioning of the Immune System

FRANK HUCKLEBRIDGE

Psychological and Neuroendocrine Correlates of Disease Progression

JULIE M. TURNER-COBB

The Role of Psychological Intervention in Modulating Aspects of Immune Function in Relation to Health and Well-Being

J. H. GRUZELIER

CONTRIBUTORS

Numbers in parentheses indicate the pages on which the authors' contributions begin.

Jan Born (93), Department of Clinical Neuroendocrinology, Medical University of Lübeck, 23538 Lübeck, Germany

Jos A. Bosch (213), Department of Oral Biology, College of Dentistry, The Ohio State University, Columbus, Ohio 43218 and Academic Centre for Dentistry Amsterdam Section of Oral Biochemistry, Vrije Universiteit, 1081 BT Amsterdam, The Netherlands

Guy H. Carpenter (187), Salivary Research Group, Guy's, King's and St. Thomas' School of Dentistry, King's College London, The Rayne Institute, London SE5 9NU, United Kingdom

Angela Clow (1, 255), Department of Psychology, University of Westminster, London W1B 2UW, United Kingdom

Eco J. C. de Geus (213), Department of Biological Psychology, Vrije Universiteit, 1081 BT Amsterdam, The Netherlands

Adrian J. Dunn (43), Department of Pharmacology and Therapeutics, Louisiana State University Health Sciences Center, Shreveport, Louisiana 71130

Ilia J. Elenkov (163), Division of Rheumatology, Immunology, and Allergy, Georgetown University Medical Center, Washington, D.C. 20007

Fiona Gaughran (275), Ladywell Unit, University Hospital, Lewisham, London SE13 6LH, United Kingdom

J. H. Gruzelier (383), Department of Cognitive Neuroscience and Behavior, Imperial College London, London W6 8RF, United Kingdom

Michael Harbuz (133), University Research Center for Neuroendocrinology, University of Bristol, Bristol BS2 8HW, United Kingdom

Frank Hucklebridge (1, 325), Department of Biomedical Sciences, University of Westminster, London W1M 8JS, United Kingdom

David S. Jessop (67), University Research Center for Neuroendocrinology, University of Bristol, Bristol BS2 8HW, United Kingdom

Adam P. Kohm (17), Department of Microbiology and Immunology, Northwestern University Medical Center, Chicago, Illinois 60611

Lisa Marshall (93), Department of Clinical Neuroendocrinology, Medical University of Lübeck, 23538 Lübeck, Germany

Pierre J. Neveu (303), Neurobiologie Intégrative, INSERM U394, Institut François Magendie, 33077 Bordeaux, France

Arie V. Nieuw Amerongen (213), Academic Centre for Dentistry Amsterdam, Section of Oral Biochemistry, Vrije Universiteit, 1081 BT Amsterdam, The Netherlands

Gordon B. Proctor (187), Salivary Research Group, Guy's, King' and St. Thomas' School of Dentistry, King's College London, The Rayne Institute, London SE5 9NU, United Kingdom

Christopher Ring (213), School of Sport and Exercise Sciences, University of Birmingham, Edgbaston, Birmingham B15 2TT, United Kingdom

Virginia M. Sanders (17), Department of Molecular Virology, Immunology, and Medical Genetics, The Ohio State University, Columbus, Ohio 43210

Julie M. Turner-Cobb (353), Department of Psychology, University of Kent at Canterbury, Canterbury, Kent CT2 7NP, United Kingdom

Enno C. I. Veerman (213), Academic Centre for Dentistry Amsterdam, Section of Oral Biochemistry, Vrije Universiteit, 1081 BT Amsterdam, The Netherlands

PREFACE

This book addresses the issue of coordination between the brain and the immune system. Although at first it seemed too remarkable to be true it is now widely acknowledged that the immune system and brain "talk" to each other. There are good teleological arguments about why this should be so. The immune system changes its behavior in relation to the light–dark, sleep–wake, cycle. The only way the immune system knows about circadian patterns is that the brain keeps it informed, principally via the neuroendocrine system. Likewise we change our behavior in relation to sickness. The only way the brain knows about sickness is that the immune system can signal directly to it via its own afferent pathways. These relationships and responses are adaptive but a perturbation in either system can have an impact on the functioning of the other with consequences for health and well-being. Relatively speaking it is still early in our understanding of the neurobiology of the immune system. However, what once might have been considered a random collection of phenomenological, disjointed, and uninterpretable findings has now matured into a substantial body of evidence that provides an integrated view of communication pathways between these systems. Our current understanding draws upon the investigative disciplines of neuroscience, immunology, psychology, behavioral and developmental biology, pharmacology, endocrinology, and molecular biology. It is our view that at the current time some of the most exciting developments in immunology and neurobiology are to be found at this interface and are represented in this book.

Early chapters describe both efferent and afferent pathways of communication between the brain and the immune system and their physiological significance. Chapter 5 describes how these pathways are implicated in one very important aspect of normal physiological function, namely the circadian cycle of sleep and wakefulness. Chapter 6 is concerned with abnormal function—the neuroendocrine and immune dysregulation associated with autoimmunity.

One of the most commonly researched but frequently misunderstood aspects of brain–immune system interaction, the association between psychological stress and immune function, is explored in Chapter 7, whereas succeeding chapters deal specifically with neuronal and psychological determinants of mucosal defence. The putative role of the immune system in the major psychological disorders of depression and schizophrenia is considered respectively in Chapters 10 and 11.

Perhaps some of the most important and direct evidence that justifies the concept of the "neurobiology of the immune system" derives from the studies that show that immune function is modulated at the level of the cerebral cortex and various distinct subcortical structures. Chapters concerning cerebral lateralization in relation to immune system functioning and behavioral conditioning of the immune system chart these waters.

Although the clinical implications of our deepening knowledge of brain–immune system interactions are alluded to throughout, the final two chapters deal with these issues specifically.

We acknowledge and thank all of our contributors and hope that the readers will share our enthusiasm and our wonder.

Frank Hucklebridge
Angela Clow

NEUROIMMUNE RELATIONSHIPS IN PERSPECTIVE

Frank Hucklebridge

Department of Biomedical Sciences, University of Westminster,
London W1M 8JS, United Kingdom

Angela Clow

Department of Psychology, University of Westminster,
London W1B 2UW, United Kingdom

I. Introduction
II. Innate and Acquired Immunity
    A. Lymphocytes
    B. Th1/Th2 Cells
III. The Neuroendocrine System and the Balance of the Immune System
    A. Stress
    B. Inflammation
    C. Psychological Stress
IV. Conclusions

I. Introduction

Functional communication between elements of the nervous and immune systems is now so well established that it is appropriate to speak of the *neurobiology* of the immune system. It is equally relevant to talk of the *immunobiology* of the nervous system: signaling of information is bidirectional. Many immunologists have considered the immune system to be autonomous, independent of neuronal modulation. In the same way many neuroscientists had not considered the possibility that the brain may be affected by peripheral cells of the immune system. There were good reasons for these views as the immune system provides protection from infection—a role quite distinct from cognitive, motor, emotional, or even sensory function. Furthermore the predominant view was that the brain is essentially isolated and protected from changes in the circulating milieu by the blood–brain barrier, responding to changes in the internal environment via specialized nerve endings. These conventional views have been radically subverted: the role of the central nervous system (CNS) in monitoring external and internal environmental cues must now be seen to include the vast and diverse world of immune system responses. In essence, the immune system is a dispersed sensory organ, not only handling input signals and generating adaptive responses in its own right, but also informing the CNS of its state of alertness and engagement. At the same time, efferent pathways from the central nervous system bias the activity and sensitivity of the immune system. The recognition that the nervous and immune systems operate as an integrated whole opens up intriguing new areas for investigation that will provide insight into the relationship between psychological variables, immune challenge, and health. The chapters in this book explore the evidence, the mechanisms, and the consequences of these relationships.

II. Innate and Acquired Immunity

The immune system consists of those organs, cells, and secreted molecules whose primary role is "protection from infection," which implies protection from foreign pathogens. A slightly more sophisticated definition would suggest "protection of the cellular integrity of the body" since the immune system is also equipped to recognize damaged or subversive cells, whatever the cause or origin, and terminate their activities.

Immune defenses are generally discriminated into *innate/natural* and *acquired/adaptive*. The distinction lies in the degree to which defensive activity depends upon prior acquaintance with the antigenic stimulus. *Innate* immune defense does not require priming by such initial contact and is fully mobilized regardless. In contrast, *acquired* immune defense is more specific: initially a small number of cells are alerted by contact with the antigen. These cells then undergo a time-consuming period of cellular expansion and differentiation before targeted effector mechanisms are brought into play.

Phagocytic cells are an important first line of defense in the innate branch of the immune system's armory against bacterial infection. Neutrophils and macrophages can destroy bacteria by internalization and enzymatic destruction within a phagolysosome. These cells possess innate receptors that recognize molecular configurations commonly expressed on bacterial surfaces and which distinguish them from mammalian cells (pattern recognition); once bound by the pathogen the phagocytic process is initiated. Natural killer (NK) cells are also part of the innate immune armory, destroying virally infected or transformed (potentially cancerous) cells of the host body. Molecular changes on the surface of such cells are recognized by receptors on the NK cells which target them for cytotoxic activity and hence destruction by induction of apoptotic pathways.

Acquired or adaptive immune responses are the special province of lymphocytes. These cells recognize small details of molecular organization to distinguish self from nonself. Thus molecular recognition by lymphocytes is precise and high resolution. The details of molecular structure, which are the focus for this recognition, are referred to as epitopes. Since the epitopic universe is almost limitless, we require a vast number of lymphocytes to ensure that potentially every different epitope expressed by the microbial world of potentially invading organisms should be duly seen. Complex and unique gene rearrangement events are organized to generate receptor diversity during the development of a lymphocyte to ensure that an enormously diverse repertoire of epitope recognition specificity is expressed by the totality of our lymphocyte population; but generally each individual cell is monospecific, expressing just one receptor specificity, the singular product of gene rearrangement in that cell. To encompass the demands for epitope recognition on this scale and at this degree of resolution the immune system continually generates, from lymphopoietic tissue in the bone marrow, a vast number of cells. The total number of our lymphocyte pool is some $\times 10^{12}$, which is equal to the cellular mass of the brain.

There are a number of distinct lymphocyte populations which perform different roles in the immune system; primarily B lymphocytes and T lymphocytes are distinguished and referred to often simply as B cells and T cells. All lymphocytes (and all blood cells) are generated from stem cells in the bone marrow. Mature B cells are released directly from the bone marrow but T cells migrate to the thymus gland where their development into mature and immunocompetent cells is completed, hence the term T cell—thymic dependent. There are subpopulations of both B cells and T cells; most important, the major population of T cells can be distinguished by the surface expression of either the CD4 molecule or the CD8 molecule. They are therefore said to be either $CD4^+$ or $CD8^+$ T cells. $CD4^+$ cells also diverge to perform either $CD4^+$ Th1 or $CD4^+$ Th2 functions (see later discussion). $CD8^+$ T cells predominantly have cytotoxic (cell killing) functions in the immune system whereas $CD4^+$ cells provide stimulatory signals that activate other effector cells. $CD8^+$ cells are therefore often referred to as cytotoxic T cells (Tcyt.) and $CD4^+$ cells are referred to as helper T cells (Th).

The contribution that lymphocytes make to immunological defense is said to be acquired or adaptive since it alters in relation to repeated stimulation. The number of cells that can specifically recognize an epitope are few, and hence clonal expansion of these cells is required to mount an effector response. Clonal expansion amplifies not only effector cells but also progeny that differentiate into long-lived memory cells that retain the capacity for distinct epitope recognition and can orchestrate more rapid and potent responses on future contact with antigens expressing the original epitope(s) (the basis of immunization). The fundamental role that T cells and B cells play in acquired immunity now will be described.

A. LYMPHOCYTES

1. *B Cells and Antibodies*

B cells express antibody (otherwise referred to as immunoglobulin) on their surface, as their epitope-recognition receptor, the product of rearranged immunoglobulin genes. B cells also have the capacity to divide and differentiate into populations of plasma cells that secrete antibody into body fluids as a soluble defensive effector molecule. Several different kinds of antibody (classes or isotypes) can be secreted depending on the predisposition of the cell, the microenvironment in which it is stimulated, and the presentation of various costimulatory signals. Antibody that protects mucosal surfaces is largely of the IgA isotype. The role of this kind of antibody in immunological defense of mucosal surfaces and the influence of the autonomic nervous system (ANS) and psychological variables on this aspect of immune defense are detailed in Chapters 8 and 9 of this volume.

B cell–derived plasma cells also secrete the isotypes IgM, IgG, and IgE. IgM and IgG are predominant in the vascular circulation although IgG is also transported to extravascular interstitial spaces and indeed crosses the placental maternal/fetal barrier to confer fetal immunological protection. IgE interacts with tissue mast cells and induces atopic, or allergic, responses. Antibody secreting cells can switch isotype from IgM, which is the primary class of antibody, to more specialized isotypes: IgG, or IgA, or IgE.

2. *Helper T Cells and Cytotoxic T Cells*

As already described, T cells fall primarily into two distinct populations: helper T cells (CD4$^+$) and cytotoxic T cells (CD8$^+$). Helper T cells interact with several kinds of antigen-presenting cells (APCs) and in various and different ways activate these cells or other collaborative cells. Three different populations of specialized (professional) APCs are recognized: dendritic cells, macrophages, and B cells (the type of lymphocyte referred to in the previous section). Dendritic cells are thought to belong to the macrophage lineage but, as the name implies, have fine cytoplasmic projections that increase surface area thus maximizing antigen-presentating capacity. Dendritic cells are very potent APCs and are specialized entirely for the presentation of antigen to T cells. By contrast macrophages and B cells are gifted amateurs and play other effector roles in addition to acting as APCs. Similar to dendritic cells, macrophages and B cells can acquire antigen and present it to T cells, but their capacity to contribute to an immunological response, by clearing the antigen, is limited unless they work in collaboration with T cells and receive CD4$^+$ T cell stimulatory signals. Thus macrophage capacity for intracellular phagocyte killing of internalized microorganisms and their production of pro-inflammatory mediators is enhanced as a result of interaction with CD4$^+$ cells. Likewise the capacity of B cells to differentiate into antibody-secreting plasma cells is almost entirely dependent on interaction with CD4$^+$ T cells. Similarly dendritic cell/CD4$^+$ cell interactions promote the cytotoxic effector mechanisms that are protective against viral infection. By contrast cytotoxic T cells (CD8$^+$) recognize different signals expressed on the surface of their interacting cell, which potentially is any infected or otherwise diseased nucleated cell, and respond dramatically differently. These cells are programmed to directly kill the aberrant cell target and thus confer immunological protection.

T cells recognize epitopic detail but only as peptide sequences derived from foreign protein antigens (this is in contrast to B cells which recognize surface features associated with a variety of molecular structures, including protein). Peptide recognition requires the display of peptide on a cellular surface in association with special peptide presentation molecules called class I or class II major histocompatibility complex (MHC) molecules. CD4$^+$ T helper cells recognize peptides in association with class II MHC molecules, whereas CD8$^+$ T cytotoxic cells recognize peptides in association with class I MHC molecules. Since peptides are derived from proteins, cellular processing pathways are required to access the peptide sequences and associate them with the membrane display molecules. There are two distinct processing pathways. Extracellular proteins derived from exogenous (captured) antigen are processed via an endosomal pathway and expressed on the cell surface in association with a class II MHC molecule (this is the particular role of the APCs), whereas proteins synthesized within the cell itself are processed within the cytoplasm and expressed in association with the class I MCH molecule. This marks a fundamental distinction in the way the immune system works. The normal protein synthesis machinery of the cell will tend to be corrupted by viral infection (or cells may otherwise lose their normal internal metabolic regulatory machinery). A manifestation of this subversion is expressed as foreign peptide displayed by the class I MHC molecule on the surface of the diseased cell. Such cells mark themselves as targets for cytotoxic T cell killing since epitope-specific cytotoxic T cells can recognize the peptide–class I MHC signal. In contrast, cells that have captured exogenous antigenic protein as a result of their phagocytic/endocytic activities can process it via the endocytic pathway into peptide sequences for display in association with the class II MHC molecule. All nucleated cells can express foreign peptide together with the class I MHC molecule and are potential targets for CD8$^+$ T cell killing, but only the specialized APCs of the immune system display the peptide–class II MHC molecular signal in such a way as to invite T cell help. The three different types of APC are specialized for presenting peptides from different pathogens. Dendritic cells present peptides derived from viral antigen, macrophages process bacterial antigen, and B cells handle soluble protein antigens such as bacterial toxins. Dendritic cells also activate $CD8^+$ T cells in viral protection via the class I MHC pathway.

   a. MHC Polymorphism. In the human body, the MHC gene locus, which is located on chromosome 6, is known as the human leukocyte antigen (HLA). For class I molecules three genes are expressed: HLA-A, HLA-B, and HLA-C; similarly for class II molecules the designations are HLA-DP, HLA-DQ, and HLA-DR. The MHC is the most highly polymorphic gene system in the human body; that is, numerous alternative alleles exist within the human gene pool for expression within these regions. Each individual, in an outbred population such as the human, inherits a more or less unique combination of alleles. In humans the different alleles are designated by a number system HLA-B15, HLA-DR-4, etc. Certain aspects of disease susceptibility and the manifestation of immunological disorders, such as autoimmune diseases, tend to associate with the expression of particular HLA alleles (see Chapter 6 in this volume). Certain neurological disorders such as schizophrenia also show interesting associations with the inheritance of particular HLA alleles pointing to some immunological involvement in the etiology of the disease (explored in detail in Chapter 11 of this volume). It is not entirely clear why these associations arise but a possibility is that the differing capacities of various HLA alleles to express peptide products of different microorganisms might provoke the immune system in ways that lead to these manifestations of disorder.

B. Th1/Th2 CELLS

1. *The Th1/Th2 Dichotomy*

   There are two distinct populations of $CD4^+$ T helper cells associated with two importantly different avenues for the progress of an immunological response. B cells associate with so-called Th2 cells, whereas macrophages and dendritic cells associate with so-called Th1 cells. Th1 cells drive cellular immunity and are stimulated by intracellular organisms. Cellular defense mechanisms largely provide protection against this kind of infection. By contrast Th2 cells induce antibody production (humoral immunity), which is effective against organisms that invade the body but exploit the extracellular environment. This distinction between alternative Th1 and Th2 response has become enormously important in the conceptualization of how the immune system responds and is controlled and informs much of our understanding of the way that the nervous system and neuroendocrine systems modulate immunological responses.

a. Th2 Immunity. CD4$^+$ Th2 cells interact with B cells expressing peptide derived from foreign soluble protein, originally trapped by specific epitope binding to the B cell surface immunoglobulin (Ig). This interaction involves intimate communication between the Th2 cell and the B cell which takes place within designated regions of secondary lymphoid organs: lymph nodes, spleen, or organized mucosal lymphoid tissue. Cellular contact is established between the peptide recognition receptor of the T cell, the T cell receptor (TcR) and the peptide–class II MHC molecular complex expressed on the surface of the B cell. Interaction between adhesion molecules expressed on the surface of both cells stabilizes the cellular association. Mutual stimulatory signals are provided by ligand receptor interactions. An activation signal to the T cell is modulated through CD4 binding to an extracellular domain on the class II MHC molecule. A T cell costimulatory molecule, CD28, expressed on the T cell surface, interacts with a molecule designated B7 (B7.1 and B7.2) expressed on the B cell surface. This molecular interaction amplifies signaling between the two cells, indeed absence of the B7 signal leads to T cell paralysis (anergy).

Interestingly one of the ways the neuronal signaling influences immune responsiveness is via a β-adrenoreceptor–mediated upregulation of B7 on the B cell triggered by the sympathetic neurotransmitter norepinephrine (NE) (see Chapter 2 herein). Likewise, an activation molecule CD40, expressed on the B cell, is triggered by binding to a T cell CD40 ligand (CD154). Collectively these signals result in mutual activation of both T and B cells. This activation results in the release of cytokines, by the T cells that stimulate B cells, as well as upregulation of cytokine receptors on the part of the B cell. Cytokines are soluble mediators of communication between different cells of the immune system and indeed other cell types. Many are designated as interleukins (IL-1, IL-2, etc.), whereas others belong to the interferon (IFN), tumor necrosis factor (TNF), and transforming growth factor (TGF) families. These areas of cellular contact and signaling specialization between the T and B cells are appropriately referred to as the "immunological synapse." The term *synapse* seems apt since "synaptic" processes in the communication between these immune cells are subject to regulatory plasticity as exemplified by the influence of NE on B7 upregulation and in a sense parallel an important attribute of classical neuronal synapses. Activation results in clonal proliferation of both cell types (B cells and CD4$^+$ T cells) in order to amplify the response. Th2 cytokines also drive the B cells to differentiate into antibody-secreting plasma cells and induce maturation of the immune response in terms of isotype switching and memory B cell differentiation. Important Th2 cytokines that mediate these responses are IL-4 and IL-5. The Th2 cytokine IL-4 is particularly proactive in driving cognate B cell activation and differentiation to antibody-secreting plasma cells. In addition it plays an important role in mediating isotype switch to selective and adaptive isotypes, notably IgG and IgE. IL-5 together with the anti-inflammatory cytokine, TGF-β, promote isotype class switch to IgA within the mucosal immune system.

b. Th1 Immunity. Th1 immunity is driven by intracellular organisms. These are pathogens that can survive and flourish within the cellular domain of the host. Viruses are facultative residents of the host's internal cellular environment. In addition some bacterial and protosoal parasites can subvert the defenses of the macrophage, which is the cell type that will initially internalize them, succeed in avoiding phagocytic killing, and survive and propagate within this cellular sanctuary. Activation of Th1 cells is the adaptive response to this form of microbial challenge. Th1 cells interrelate with antigen-presenting cells (APCs) in much the same way that Th2 cells interact with B cells. Similar cognate recognition and ligand-receptor activation at the "immunological synapse" are involved, although here the cell cooperation is between Th1 CD4$^+$ cells and peptide antigen presented by either macrophages or dendritic cells. Activation of either cell type induces the secretion of the cytokine IL-12 which is strongly Th1 promoting and contributes to the release by these activated Th1 cells of cytokines that promote cellular defenses. The principle Th1 cytokines are IL-2, IFN-γ, and TNF-β. In various ways these Th1 cytokines promote cellular immune activity. This includes macrophage activation and arming, to promote phagocytic killing, and activation of cytotoxic activities of both the innate component of immune defenses NK cells, and the adaptive arm, CD8$^+$ cytotoxic T cells. Whereas Th2 cells remain within the lymphoid environment to maintain close association with their B cell partners, Th1 cells can migrate to the periphery to engage cellular cooperation at the local sites of inflammatory and cellular activation.

2. *The Th1/Th2 Balance*

These two pathways of immunological response are balanced in an immunological equilibrium. Th1 and Th2 immune activities are mutually counterregulatory. Cytokines that promote Th1 activity in relation to intracellular pathogens tend to downregulate Th2 activity and in turn Th2 domination inhibits Th1 activity. The balance between T helper cell production of IFNγ as opposed to IL-4 and IL-10 is particularly important since the Th2 cytokines IL-4 and IL-10 inhibit Th1 activity and the Th-1 cytokine IFNγ inhibits Th2 activity. Th1 and Th2 cytokines also cross-regulate each other's effector mechanisms, for instance, IL-4, IL-10, and IL-13 suppress macrophage activation and are said to be anti-inflammatory. Il-4 is strongly Th2 promoting, stimulating Th2 activity and downregulating Th1 activity.

This dynamic equilibrium within the immune system is fundamentally unstable. Balance is normally maintained as a result of circadian oscillation. Th1 activity is promoted at nighttime, during sleep and rest, whereas the immune system swings toward Th2 domination in relation to awakening and preparation for daytime physical activity. (The relationship between sleep and immune activity is discussed in Chapter 5 of this volume). The relationship between the Th1 and Th2 pathways of the immune system is illustrated in Fig. 1.

Of necessity this dynamic equilibrium leads us to the role that the neuroendocrine system plays in immunomodulation.

III. The Neuroendocrine System and the Balance of the Immune System

The pivotal roles of two neuroendocrine systems in relation to immunomodulation are explored in particular detail in the following chapters. The first of these is the hypothalamic-pituitary-adrenal (HPA) axis. Central control of the HPA axis resides in the paraventricular nucleus (PVN) of the hypothalamus. This system is characterized by a profound circadian cycle such that HPA axis activity, as evidenced by cortisol secretory activity, is at its nadir during nocturnal sleep and increases upon diurnal wakefulness. In contrast activity in the second and related neuroendocrine system, centered on the pineal gland (the product of which is melatonin), shows the reverse cycle being active in darkness and during nocturnal sleep. These cycles are synchronized by input from the hypothalamic suprachiasmatic nucleus (SCN). The secretory products of these neuroendocrine systems serve to balance the immune system: nighttime melatonin promotes Th1 domination whereas daytime cortisol secretory activity, which peaks about 30 min following awakening, is thought to switch the balance toward Th2 domination. Equilibrium is maintained by the circadian cycle.

Although the immune system is balanced in this way, between Th1-mediated type 1 activity and Th2-mediated type 2 immune activity, certain aspects of disease susceptibility and progression are associated with an overall skew in this equilibrium. Chronic inflammatory disorders such as rheumatoid arthritis (RA) and type 1 insulin-dependent diabetes mellitus (IDDM) are associated with an overall shift toward type 1 domination, whereas allergic atopic conditions such as hay fever or asthma are, in part, manifestations of a skew toward type 2 immune domination. The fairly complex associations between the HPA axis and inflammatory autoimmune disorders such as rheumatoid arthritis are explored later in Chapter 6.

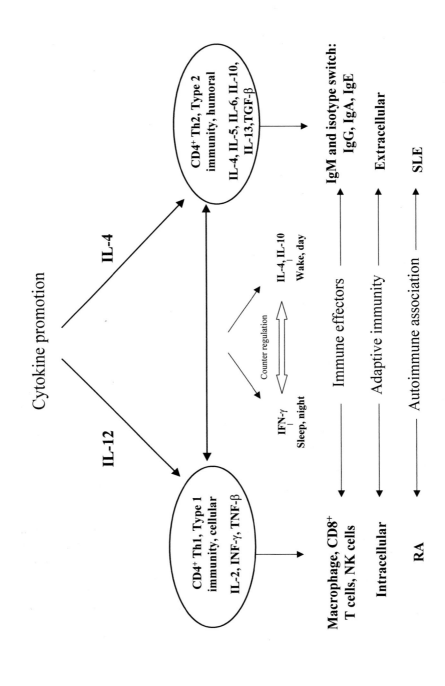

A. Stress

The stress response system consists of two major efferent arms: the HPA axis already described, but outlined in more detail in Chapter 6, and the sympatho-adreno-medullary (SAM) system given focus next in Chapter 2. Central control of the HPA axis resides in the PVN of the hypothalamus and the SAM response is orchestrated in the locus ceruleus (LC). These two centers of the stress response system are coordinated such that noradrenergic neuronal pathways from the LC stimulate PVN activity; likewise the central driver of the HPA axis, corticotropin-releasing factor (CRF) produced by parvocellular cells in the PVN, stimulates LC noradrenergic activity. Serotonergic pathways emanating from the raphe nuclei augment activity in both nuclei. Further details of the central organization of this neuroendocrine system are given later in Chapter 6.

Stressors (various challenges to emotional and physiological equilibrium) mobilize the HPA and SAM stress response systems. Peripheral effectors of this central activation are the adrenocortical hormone cortisol and release of the postganglionic sympathetic neurotransmitter norepinephrine (NE) and additionally the adrenal medullary hormone epinephrine (E) together comprising the peripheral catecholamines. Immunocytes express specific receptors for these soluble stress mediators, as is outlined in this volume in Chapters 2 and 7. Activation of these receptors induces transduction mechanisms that tend to bias the immune balance in favor of Th2 domination. These principle features of these neuroimmune relationship are considered in detail in above-mentioned Chapters 2 and 7.

B. Inflammation

Inflammatory processes mark the acute response to infection. At sites of microbial infection and tissue damage, activated macrophages release the pro-inflammatory cytokines, IL-1, IL-6, and TNF-α. These peripherally

---

FIG. 1. The immune system is balanced between type I (cell-mediated) and type 2 (humoral- or antibody-mediated) responses. This depends on the relative activities of two populations of CD4$^+$ T helper cells—Th1 and Th2—and the cytokines they secrete. Type 1 activity, promoting cytotoxic functions, is a response to intracellular invasion, whereas type 2 immunity protects against extracellular organisms for which antibody is the main effector. Macrophage, CD8$^+$, and NK cell activity is promoted by Th1 cytokines, whereas antibody secretion (with the exception of an IgG subclass referred to as cytotoxic antibody) is driven by Th2 cytokines. Certain autoimmune disorders are associated with a skew toward type 1 activity (e.g., rheumatoid arthritis, RA) or by contrast toward type 2 activity (e.g., systemic lupus erythematosus, SLE). This balance and its implications are referred to throughout.

generated cytokines signal to the brain, importantly influencing mood (affect) and motivation, inducing what is termed malaise or sickness behavior (see Chapters 5 and 12 of this volume). Behavioral changes are energy conserving and considered to be adaptive in infection. Generally there is behavioral withdrawal with a reduction in social and exploratory behavior—loss of appetite and sexual motivation. These cytokines are also pyrogenic (inducing fever) which is an adaptive response to infection since a few degrees in elevation of body temperature favors lymphocyte proliferation and hampers microbial growth and expansion. The energy conservation associated with behavioral changes is necessary to support the metabolic demands of raised body temperature. The behavioral response is referred to as malaise or sickness behavior but is part of what is known as the acute phase response to infection. Evidence has accumulated that disregulation or inappropriate activation of this system may be implicated in the etiology of melancholic depression and perhaps also schizophrenia. These important aspect of the neuroimmune relationship are discussed herein, in Chapters 10 and 11. Physiological aspects of the acute phase response, mediated by pro-inflammatory cytokines are the induction of acute phase, antibacterial proteins, released from the liver, so-called positive acute phase proteins, and chemotactic attraction of phagocytic cells, neutrophils, and monocytes to the site of infection. These cytokines also induce an increase in circulating neutrophils that are summoned from the bone marrow. Although initially the acute phase response is innate, Th1 cells support inflammatory processes; TGF-β and certain Th2 cytokines are anti-inflammatory.

Pro-inflammatory cytokines also stimulate the HPA axis. The pathways by which these peripherally generated signals influence the HPA axis are described in detail in Chapter 3 and also referred to in Chapters 6 and 12 (within this volume) and constitute one of the important avenues by which the immune system signals to the brain. As far as the brain and the stress neuroendocrine system are concerned infections that induce pro-inflammatory cytokines are "stressors." It is generally considered that this linkage has evolved to marshal the anti-inflammatory potential of the steroid hormone cortisol in limiting inflammatory processes. There is little doubt that this is an important pathway in the resolution of the inflammatory process, which if left unchecked can cause severe and irreversible tissue damage and erosion. In the shorter term inflammatory processes can reduce mobility and one role of anti-inflammatory steroids might be to suspend inflammatory processes during periods of vigorous physical activity: Cannon's "fight–flight," and direct these processes toward periods of rest (see discussion in Chapter 5). Chronic inflammatory conditions are aggravated during the cortisol nadir (see later in Chapter 6). It has been argued, with some teleological merit, that the primary role of the stress neuroendocrine system, in all its complexity, is the adaptive response to infection; the cueing of this response to other perceptions of threat (primarily psychological in nature) are a latter acquisition.

Acute inflammation can become chronic in circumstances where resolution, and the various mediators of resolution fail. This may be as a result, at least initially, of persistent and localized infection that continuously provokes the Th1-mediated inflammatory response. The role of the stress neuroendocrine system in the regulation of inflammatory autoimmunity is discussed later in Chapter 6. Adverse immunological responses are implicated not only in what are classically understood to be "autoimmune diseases" but also in major neurological disorders such as depression (see Chapter 10) and schizophrenia (see Chapter 11). In addition to cortisol a number of neuropeptide-secreted products of the stress neuroendocrine system play roles in the regulation of inflammatory processes. Fascinatingly, immune cells themselves can synthesize and secrete various neuropeptides in physiologically significant amounts. The putative roles of these peptides in immuno- and inflammatory regulation are discussed later in Chapter 4.

C. PSYCHOLOGICAL STRESS

Among the stressors that can stimulate the HPA axis and the autonomic nervous system are psychological stressors. Psychological stressors are not easy to define and often are identified in terms of the physiological response: increase in levels of circulating cortisol and catecholamines. There are a number of animal models of stress (see discussion in Chapter 6) but in the human context psychological stressors are usually social stimuli with potentially unpleasant outcomes (negative valence) that are a threat to self-esteem and over which the individual has little behavioral control. Since the process requires cognitive appraisal there is much individual variation in the relationship between the stimulus cues, coping mechanism and psychophysiological response. However, since the response invokes immunoregulatory mechanisms it should be no surprise that the immune system is influenced by psychological stress.

The nature and direction of immune responses to psychosocial stress are commonly mediated by the duration of the stressor. Brief acute stress challenges such as can be engineered under controlled laboratory conditions often induce immunological changes, which differ from those associated with more sustained long-term, enduring life-style, or occupational stress. Acute stress challenges include difficult or paced mental arithmetic, public speaking, and various computer tasks. These generally include the engineering of observed social failure in relation to expected performance criteria. The immune response to these psychological manipulations can often be interpreted as representing mobilization or upregulation of various aspects of immunological defense particularly those arms of the immune system that characterize innate or natural defense. These include increases in peripherally circulating cytotoxic cells, $CD8^+$ and NK cells, and increased NK cell activity; increases in circulating pro-inflammatory cytokines; and, within the mucosal immune system, increases in secretion of S-IgA and other defensive mucosal proteins. These acute responses are generally attributable to autonomic nervous system (ANS) regulatory mechanisms. Detailed discussion of these mechanisms in relation to the mucosal immune defense system is provided herein, in Chapters 8 and 9.

By contrast, enduring chronic stress such as bereavement, marital discord, caring for a relative suffering dementia, or a period of important academic examinations is associated with immune changes which seem to be opposite in nature. These tend to be described as manifestations of immune suppression although evidence is accumulating that these might best be interpreted as representing a shift in Th1/Th2 balance toward a type 2 dominated immune system (see Chapter 7 for detailed discussion). This would be compatible with the influence that elevated cortisol and β_2-adrenergic agonists such as norepinephrine have on the Th1/Th2 balance.

In light of these changes in immune cell function that accompany psychological stress the possibility exists that psychological factors might have an important bearing on disease susceptibility and disease progression. A very large body of evidence has accumulated to suggest that this is indeed the case. This evidence is reviewed in Chapter 14. Of parallel interest is the possibility that various aspects of neurological organization are associated with a biasing of immune responding. In this context hemispheric lateralization of various cortical structures seems of importance. Individuals might be predisposed to respond to various immunological challenges in different ways (with different health outcomes) depending on the degree and direction of hemispheric dominance within the cerebral cortex. Evidence for this view and the various pathways which cortical structures might use to influence the immune system are reviewed in Chapter 12.

IV. Conclusions

The scientific arguments presented in the following chapters provide compelling evidence that the immune system is sensitive to psychological variables and, in turn, exerts its own influence on central neurological processes. One of the earliest and most influential lines of evidence showing that the nervous system might bias immunological responding stems from the pioneering investigations of Cohen and Ader into classical (Pavlovian) conditioning of immunological activity (referred to in Chapter 2 and the subject of Chapter 13). Pairing of an immunologically neutral stimulus (saccharine-flavored water) with an immunosuppressive drug could transfer (condition) immunosuppression to presentation of the neutral stimulus alone. Such conditioning also applied to immuno-enhancing agents and, although first demonstrated in the rat, was shown to be applicable in human studies. This leant persuasive evidence that immune system activity could be modulated by autonomic control.

The possibility therefore exists that psychological intervention might usefully contribute to the armory of therapeutic approaches to disease management. Although still controversial, evidence is beginning to accumulate that psychological intervention strategies are indeed important not just in improving the sense of well-being and quality of life but also the course of infection and pathology itself. Striking examples of beneficial immunomanipulation by classical conditioning paradigms are reported in the literature. Various forms of relaxation therapy and stress management have been shown to have positive outcomes in determining the course of chronic disease. Various aspects of these considerations are discussed in this volume's final chapters, 14 and 15.

Issues raised and experimental evidence discussed in this volume throw important light on the physiology of both the immune system and the nervous system. We hope that for many specialists in either field some knowledge of what their other self is up to might prove illuminating.

SYMPATHETIC NERVOUS SYSTEM INTERACTION WITH THE IMMUNE SYSTEM

Virginia M. Sanders

Department of Molecular Virology, Immunology, and Medical Genetics
The Ohio State University, Columbus, Ohio 43210

Adam P. Kohm

Department of Microbiology and Immunology
Northwestern University Medical Center
Chicago, Illinois 60611

I. Introduction

Some of the first approaches used to indicate that an interaction existed between the nervous and immune systems were those involving behavioral-conditioning paradigms (Ader and Cohen, 1975; Rogers *et al.*, 1976; Wayner *et al.*, 1978; Cohen *et al.*, 1979; Exton *et al.*, 1998). Since then, other research has confirmed and extended these early findings, although a definitive role for such an interaction in the etiology or progression of disease states is inconclusive. This chapter will review the key findings that confirm the existence of a link between these two distinctly different organ systems. Special emphasis will be given to findings that indicate the presence of sympathetic nerve terminals within the parenchyma of lymphoid tissues, the release of norepinephrine following antigen or cytokine administration, the expression of adrenergic receptors on immune cells, and the regulation of immune cell function at both the cellular and the molecular level's by adrenergic receptor stimulation.

17

II. Anatomy and Physiology

The brain communicates with the periphery via two different pathways that include the activation of both the hypothalamo-pituitary-adrenal axis and the sympathetic nervous system (SNS). In this chapter, activation of the SNS, and the effects of this activation on immune cell function, will be reviewed. Sympathetic neurotransmission from the central nervous system (CNS) to the periphery occurs via projections extending from the paraventricular nucleus of the hypothalamus, rostral ventrolateral medulla, ventromedial medulla, and caudal raphe nucleus to preganglionic neurons of the spinal cord where preganglionic cell bodies of sympathetic nerves reside in the intermediolateral cell column of the lateral horn of the spinal cord at T1-L2 (Sawchenko and Swanson, 1982). These cell bodies send myelinated projections that exit from the spinal cord via the ventral roots to synapse on the superior mesenteric ganglia which send projections following the vasculature to innervate target organs. Within the target organ, sympathetic nerves form terminals from which the sympathetic neurotransmitter norepinephrine (NE) is released to bind to adrenergic receptors expressed by various cell populations.

Both primary and secondary lymphoid organs are innervated by sympathetic fibers (Calvo, 1968; Reilly *et al.*, 1979; Williams and Felten, 1981; van Oosterhout and Nijkamp, 1984; Felten *et al.*, 1988). Studies report the presence of sympathetic innervation in the splenic capsule, in trabeculae, and in the white pulp areas containing the T cell–rich periarteriolar lymphoid sheath (PALS), the B cell–rich marginal zone, and the marginal sinus (Livnat *et al.*, 1985; Felten and Olschowka, 1987; Felten *et al.*, 1985, 1987; Ackerman *et al.*, 1987). Sympathetic nerve terminals are in direct apposition to T cells and are adjacent to both interdigitating dendritic cells and B cells (Felten *et al.*, 1987). Therefore, the close proximity of sympathetic nerve terminals to immune cells provides a mechanism for targeting norepinephrine release to immune cells. Human lymphoid tissue is also innervated with sympathetic fibers. Thus, the presence of efferent sympathetic nerve fibers in lymphoid organs provides for the delivery of a message from the brain to immune cells residing within lymphoid organs (Besser and Wank, 1999; Kohm and Sanders, 2000).

However, for norepinephrine to influence immune cell function, it must be released at the immediate site of action because it is either rapidly degraded or taken up into the nerve terminal following release (reviewed in Glowinski and Baldessarini, 1966). Therefore, if norepinephrine is to influence immune cell function in response to antigen, mechanisms must exist for increasing norepinephrine release within the immediate microenvironment of a lymphoid cell responding to antigen. During normal homeostasis, the rate of norepinephrine release is balanced by the rate of norepinephrine synthesis, resulting in a constant tissue level of norepinephrine over a wide range of sympathetic nerve activity. As a result, all studies will need to provide an estimate of the dynamic changes in sympathetic nerve activity instead of making a determination of tissue norepinephrine concentration alone.

Lipopolysaccharide (LPS)-induced activation of immune cell populations increases the rate of norepinephrine release in both the heart and the spleen during the first 12 h of exposure (Pardini *et al.*, 1982), while infection with *Pseudomonas aeruginosa* increases the rate of norepinephrine turnover in both the heart and the bone marrow (Tang *et al.*, 1999). As opposed to changes in the rate of norepinephrine release, immunization of animals with the particulate T cell–dependent antigen sheep red blood cells (sRBCs) appears to decrease the total norepinephrine content of the spleen in comparison to controls (Besedovsky *et al.*, 1979). However, since norepinephrine content may not be reflective of a change in sympathetic nerve activity, the level of the dopamine metabolite 3,4-dihydroxyphenylacetic acid (DOPAC) which correlates with the rate of norepinephrine synthesis, was found to increase in the spleens of mice immunized with sRBCs (Fuchs *et al.*, 1988b). This finding suggests that a particulate antigen precipitates an increase in sympathetic nerve activity and release of norepinephrine within the spleen. Likewise, a cognate-soluble protein antigen increases the rate of norepinephrine release in lymphoid organs in the spleen and bone marrow 18–25 h, but not 1–8 h, following immunization, when measured by norepinephrine turnover analysis in an antigen-specific model system in mice (Kohm *et al.*, 2000). Taken together, these findings suggest that infectious, particulate, and soluble protein antigens precipitate an increase in the rate of norepinephrine release in the spleen.

However, the question remained—What mechanism was used by the immune cells responding to antigen to induce the release of norepinephrine from sympathetic nerve terminals residing with their immediate microenvironment? The hallmark experiments of Besedovsky *et al.* suggest that activated immune cells secrete "soluble factors" into the circulation that ultimately enter the CNS to stimulate neuronal activity in both the hypothalamus and the brainstem (Besedovsky *et al.*, 1983). These studies were some of the first to show that soluble factors produced by cells of the immune system, such as IL-1, were able to alter noradrenergic nerve input into the hypothalamus, affecting both hypothalamic nerve activity and the level of corticotropin-releasing hormone secreted from the hypothalamus (Sapolsky *et al.*, 1987; Akiyoshi *et al.*, 1990; Dunn, 1992; Fleshner *et al.*, 1995), effects that are now known to translate into alterations in efferent sympathetic nerve activity. Peripheral administration of IL-1β increases the rate of norepinephrine turnover in the spleen 15 min to 6 h following exposure (Akiyoshi et al., 1990; Niijima et al., 1991; Takahashi et al., 1992), peaking within 40 min (Ichijo et al., 1992; Shimizu et al., 1994) and raising the basal level of norepinephrine in the spleen from $\sim$1.6–$\sim$3 $\times$ 10^{-6} M (Shimizu et al., 1994). In addition, the effect of IL-1β on sympathetic nerve activity may be specific for sympathetic nerves located in specific organs since it increases the rate of norepinephrine release in the spleen but not in the heart (Akiyoshi et al., 1990). However, in contrast, IL-1β appears to inhibit splenic and atrial sympathetic nerve activity, as measured by microdialysis (Bognar et al., 1994; Abadie et al., 1997). Thus, IL-1β increases sympathetic nerve activity, but the possible mechanisms responsible for this activation remain unclear.

In contrast, another study suggests that macrophage-derived IL-1β may not be the only cytokine responsible for the antigen-induced increase in sympathetic nerve activity. The finding from this study suggests that a soluble mediator produced by a cognate interaction between CD4$^+$ Th cells and B cells may be necessary for an increase in splenic norepinephrine turnover to occur (Kohm et al., 2000). The identity of this cytokine is unknown, but a number of candidates are suggested. For example, IL-6 inhibits [^{3}H]norepinephrine release from sympathetic nerve terminals within 2 h of cytokine exposure in vitro (Ruhl et al., 1994), whereas both IL-2 (Bognar et al., 1994) and TNF-α (Foucart and Abadie, 1996; Abadie et al., 1997) inhibit the rate of splenic norepinephrine release in vivo. Thus, immune cell–derived cytokines other than IL-1β may also affect local sympathetic nerve activity within the microenvironment of immune cells within lymphoid organs.

However, for cytokines to leave the blood and enter the CNS, a major obstacle has to be overcome, the blood–brain barrier. While several mechanisms exist by which blood-borne cytokines cross into the CNS (Banks and Kastin, 1985a,b, 1987), they may not be a primary line of communication from the immune system to the CNS. One alternative mechanism by which immune cell–derived cytokines may signal the CNS is through the stimulation of cytokine receptors expressed on peripheral sensory nerves. By this mechanism, immune responses occurring near sites of sensory innervation might more effectively communicate signals to the CNS.

The interleukin-1 receptor (IL-1R) was the first cytokine receptor reported to be expressed on peripheral sensory nerves, stemming from the findings that the peripheral administration of IL-1β increased CNS activity (Saphier and Ovadia, 1990; Dunn, 1992). Other studies reported that peripheral administration of IL-1β increases vagus nerve activity, suggesting that the IL-1R is expressed on peripheral nerves and that stimulation of these receptors by their specific cytokines induces afferent nerve activity to the CNS (reviewed in Maier *et al.*, 1998). In addition, the effect of peripheral IL-1β on changing hypothalamic levels of norepinephrine is blocked by subdiaphragmatic vagotomy, suggesting a role for vagal afferents in mediating the effect of IL-1β within the CNS (Fleshner *et al.*, 1995). These results were later supported by the finding that vagal paraganglia express the IL-1R, providing a mechanism by which IL-1β directly activates vagal nerve afferent fibers (Goehler *et al.*, 1997). In addition to the IL-1R, sympathetic neurons express IL-2 receptors (Haugen and Letourneau, 1990) and a low level of IL-6 binding subunits that increases after nerve injury (Dinarello, 1998; Marz *et al.*, 1998). Thus, it appears that the expression of a functional IL-1 receptor on the vagus nerve, as well as other cytokine receptors on sympathetic neurons, provides alternative mechanisms by which immune-derived cytokines signal the CNS.

Taken together, the findings summarized in this section suggest that mechanisms are in place for the activated immune system to communicate with the nervous system for inducing the release of norepinephrine from sympathetic nerve terminals residing within the immediate microenvironment of activated, as well as resting, immune cells. At this point, data are needed to show that receptors exist on the surface of immune cells to bind the released norepinephrine.

III. Adrenergic Receptor Expression on Immune Cells

A. CD4$^+$ T Lymphocytes

For local norepinephrine release to influence immune cell function, immune cells must express receptors for the neurotransmitter. Almost every cell associated with the immune system expresses adrenergic receptors that bind norepinephrine, and these findings have been summarized in detail elsewhere (reviewed in Sanders *et al.*, 2001). In this chapter, the findings for CD4$^+$ T cells and B lymphocytes alone will be discussed.

Although few studies have reported the presence of the α-adrenergic receptor (αAR) on T cells, many studies have reported the presence of a functional β-adrenergic receptor (βAR). Early reports show that lymphocyte exposure to a βAR agonist results in adenylyl cyclase activation and increases in cAMP intracellularly, suggesting the presence of a functional βAR to mediate the effects of norepinephrine on signaling intermediates (Makman,

1971; Bourne and Melmon, 1971; Bach, 1975). Williams *et al.* performed the original studies to directly measure the level of βAR expression on human lymphocyte membranes via radioligand saturation binding assays (Williams *et al.*, 1976). Approximately 2000 βAR binding sites per lymphocyte were measured. However, a number of subsequent binding studies report a lower level of βAR expression on purified populations of T cells, as opposed to total lymphocytes, with approximately 200–750 βAR binding sites per T cell and CD8$^+$ cells expressing more binding sites than CD4$^+$ cells (Pochet *et al.*, 1979; Bishopric *et al.*, 1980; Loveland *et al.*, 1981; Krawietz *et al.*, 1982; Bidart *et al.*, 1983; Pochet and Delespesse, 1983a; Khan *et al.*, 1986; Westly and Kelley, 1987; Fuchs *et al.*, 1988a; Van Tits *et al.*, 1990; Radojcic *et al.*, 1991). Many studies also report that a functional β_2AR is expressed on a CD4$^+$ T cell and that stimulation of the receptor activates adenylate cyclase, accumulates cAMP intracellularly, and activates protein kinase A (reviewed in Sanders *et al.*, 2001).

Until the early 1980s, very little was known about specific βAR subtypes expressed on T cells. But subsequently, using competitive binding assays with selective β_1AR and β_2AR antagonists, the primary βAR-subtype expressed on lymphocytes was found to be the β_2AR (Bourne and Melmon, 1971; Williams *et al.*, 1976; Conolly and Greenacre, 1977; Pochet *et al.*, 1979; Loveland *et al.*, 1981a; Meurs *et al.*, 1982; Ramer-Quinn *et al.*, 1997; Sanders *et al.*, 1997). Immature T cells in the thymus express a significantly lower number on their surface in comparison to either circulating peripheral T cells (van de Griend *et al.*, 1983; Pochet and Delespesse, 1983b; Staehelin *et al.*, 1985) or splenic T cells (Fuchs *et al.*, 1988a), suggesting that β_2AR expression may increase on the cell surface during T cell differentiation. The reason for such alterations in β_2AR expression on developing T cells is unclear.

In the mid-1980s, the level of βAR expression on a specific subpopulation of CD4$^+$ T cells was reported. According to these findings, approximately 750 βAR binding sites were expressed on CD4$^+$ effector T cells (Khan *et al.*, 1986), and they were exclusively of the β_2-subtype (Dailey *et al.*, 1988; Robberecht *et al.*, 1989). Importantly, in light of the many current findings in immunology, we know now that these early studies used mixed populations of CD4$^+$ T cells, containing naive, Th1, and Th2 cell populations. The expression of βAR subtypes on CD4$^+$ T cell subsets has been determined at both the protein and the mRNA level. In general, naive CD4$^+$ T cells and Th1 cells, but not Th2 cells, preferentially express the β_2AR as determined by mRNA and functional analysis (Sanders *et al.*, 1997; Swanson *et al.*, 2001; Kohm *et al.*, 2002b), whether or not the Th1 and Th2 cells are cloned cells or cells newly generated *in vitro* from a CD4$^+$ naive cell precursor (Kohm *et al.*, 2002).

Currently, the mechanism regulating the differential expression of the β_2AR on murine Th1 and Th2 cells is unknown, but a few studies have shed light on some possibilities. Since the cytokine microenvironment is the only difference between Th1- and Th2-promoting conditions, it seems reasonable to propose that intracellular signals resulting from cytokine receptor stimulation during CD4$^+$ T cell differentiation may affect β_2AR expression on subsequent generations of effector cells. One mechanism by which cytokine receptor stimulation regulates gene expression is via alterations in both the level of histone acetylation (Ohno et al., 1997; Taplick et al., 1998; Gray et al., 2000; Cheung et al., 2000; Ito et al., 2000; Vanden Berghe et al., 2000) and DNA methylation (Hmadcha et al., 1999; Kang et al., 1999), epigenetic mechanisms that appear to regulate the level of βAR expression in nonimmune cells (Kassis et al., 1988; Buscail et al., 1990). A study shows that exposure of β_2AR-negative Th2 cells to the histone deacetylase inhibitor butyrate results in a dose- and time-dependent induction of β_2AR mRNA expression in these Th2 cells (Kohm et al., 2002). Similarly, exposure of Th2 cells to the methyltransferase inhibitor 5-azacytidine increases β_2AR mRNA expression, but with a longer time of onset in comparison to butyrate. Not surprisingly, when 5-azacytidine DNA hypomethylation is induced prior to butyrate-induced histone hyperacetylation, a synergistic enhancement of β_2AR mRNA expression occurs in Th2 cells. And finally, pretreatment of Th2 cells with either the transcription inhibitor actinomycin-D or the protein synthesis inhibitor cycloheximide shows that the induction of β_2AR mRNA expression following histone hyperacetylation and/or DNA hypomethylation is transcription-dependent, but translation-independent. The latter finding suggests that the basal levels of transcription factor expression are sufficient to induce β_2AR gene transcription once the gene locus is accessible. Thus, epigenetic mechanisms such as histone acetylation and DNA methylation may play a critical role in regulating the differential expression of the β_2AR in murine Th1 and Th2 cells. Unfortunately, due to the difficulty in obtaining mRNA from absolutely pure Th1 and Th2 subpopulations in normal humans, differential expression of the β_2AR in human CD4$^+$ effector T cells remains unknown.

In addition, the state of T cell activation may affect the level of βAR surface expression. For example, an increased level of βAR expression occurs 24 h after mitogen activation, without any effect on the affinity (Kd) of the receptor (Westly and Kelley, 1987; Madden et al., 1989; Radojcic et al., 1991). Also, β_2AR expression upregulates on the surface of activated Th1 cell clones after T cell–receptor stimulation, but remains undetectable on the surface of activated Th2 cell clones (Ramer-Quinn et al., 1997). However, a contrasting finding that T cell activation downregulates β_2AR expression (Cazaux et al., 1995) is supported by a report that mitogen activation of

T cells results in an increase in the expression of the βAR kinase-1 (βARK1) and βARK2 mRNA, with no change in G-protein receptor kinase-5 (GRK5) and GRK6 (De Blasi *et al.,* 1995). Since βARK is a serine–threonine kinase that regulates the level of βAR expression (reviewed in Inglese *et al.,* 1993), βARK activation may downregulate βAR expression following mitogen activation. Thus, while most studies report that T cell activation increases the level of surface βAR expression early after activation, it may decrease levels of βAR expression later after activation.

Taken together, these studies suggest that subsets of CD4^{+} T cells differentially express a functional β_2AR, with detectable expression on CD4^{+} naive T cells and Th1 cells, but not on Th2 cells. Activation of CD4^{+} T cells may result in an initial upregulation of the level of β_2AR surface expression, but then a later downregulation by PKC-dependent mechanisms possibly involving βARK activation.

B. B LYMPHOCYTES

An isoproterenol-induced accumulation of intracellular cAMP was the first finding to suggest the presence of a functional βAR on the B-cell surface (Bach, 1975). A few years later using radioligand binding analysis, an enriched population of murine splenic B cells was shown to express approximately 400–600 βAR binding sites per cell (Pochet *et al.,* 1979). Reports also suggest that purified B cells express a higher level of the βAR on their surface when compared to peripheral T cells (Pochet *et al.,* 1979; Miles *et al.,* 1981; Krawietz *et al.,* 1982; Bidart *et al.,* 1983; Paietta and Schwarzmeier, 1983; Pochet and Delespesse, 1983a,b; Miles *et al.,* 1984a,b, 1985; Korholz *et al.,* 1988; Griese *et al.,* 1988; Fuchs *et al.,* 1988a; Van Tits *et al.,* 1990; Cremaschi *et al.,* 1991). Studies employing salbutamol displacement curves show that the βAR expressed by the B cell is of the β_2-subtype (Korholz *et al.,* 1988; Griese *et al.,* 1988), which is in agreement with the findings of others (Krawietz *et al.,* 1982; Pochet and Delespesse, 1983a; Fuchs *et al.,* 1988a). A study of murine antigen-specific B cells freshly isolated from the spleens of unimmunized mice, using radioligand binding analysis, showed that these B cells express approximately 620 βAR binding sites per cell, while immunofluorescence and mRNA analyses showed the receptor to be β_2-subtype-specific as opposed to β_1- or β_3-specific (Kohm and Sanders, 1999). Functionally, βAR stimulation increases adenylyl cyclase activity and intracellular cAMP accumulation in resting B cells (Bach, 1975; Galant *et al.,* 1978; Bishopric *et al.,* 1980; Pochet and Delespesse, 1983b; Blomhoff *et al.,* 1987; Holte *et al.,* 1988; Kohm and Sanders, 1999). Thus, taken together, these studies suggest that B cells express a functional β_2AR.

IV. Effect of β_2AR Stimulation on CD4$^+$ T Lymphocytes

Early studies suggested that either βAR stimulation or increased cAMP inhibited or enhanced the level of T cell proliferation, depending on the concentration of cAMP generated intracellularly (Smith *et al.*, 1971; Johnson *et al.*, 1981; Glibetic and Baumann, 1986; Feldman *et al.*, 1987; Scordamaglia *et al.*, 1988; Griese *et al.*, 1988; Carlson *et al.*, 1989; Minakuchi *et al.*, 1990; Bartik *et al.*, 1993; Bauman *et al.*, 1994), and that βAR-induced PKA activation correlated with the inhibition of T cell proliferation (Bauman *et al.*, 1994). Although the mechanism responsible for this inhibition is unclear, an early elevation in intracellular cAMP following βAR stimulation at the time of T cell–receptor stimulation inhibits proliferation via disruption of cytoskeletal events leading to cell division (Parsey and Lewis, 1993; Selliah *et al.*, 1995). Another finding suggests a strain-specific effect of norepinephrine-depletion *in vivo* on T cells induced to proliferate *in vitro* (Lyte *et al.*, 1991), with this differential effect *in vivo* depending on the T cell activation stimulus used (Madden *et al.*, 1994). Thus, most studies suggest that norepinephrine and β_2AR stimulation decreases the rate of T cell proliferation *in vitro* and that this effect may be both strain-dependent and activation stimulus–dependent *in vivo*.

In addition to proliferation, T cell differentiation is critical for the development of T cell–effector function. Naive CD4$^+$ T cells differentiate preferentially into Th1 cells when they receive T cell–receptor stimulation in the presence of IL-12 (Seder *et al.*, 1993), whereas they will differentiate preferentially into Th2 effector cells in the presence of IL-4 (Hsieh *et al.*, 1992; Seder *et al.*, 1992). Therefore, in light of the importance of the cytokine microenvironment in naive CD4$^+$ T cell differentiation, any norepinephrine-induced change in this cytokine microenvironment may affect the naive T cell differentiation process. For example, direct addition of db-cAMP, norepinephrine, or a β_2AR-selective agonist to mitogen- or anti-CD40-activated monocytes decreases the level of IL-12 produced, a cytokine critical for naive T cell differentiation into a Th1 cell (Van der Pouw-Kraan *et al.*, 1995; Elenkov *et al.*, 1996; Panina-Bordignon *et al.*, 1997). In contrast, when antigen-presenting cells are activated with peptide antigen, as opposed to LPS or anti-CD40 antibody, norepinephrine and β_2AR stimulation do not affect the level of IL-12 produced (Swanson *et al.*, 2001). Thus, stimulation of the β_2AR on a professional antigen-presenting cell may favor the development of a specific effector cell subset, depending on the stimulus used to activate the antigen-presenting cell.

Norepinephrine and β_2AR stimulation also affect CD4$^+$ T cell differentiation via direct effects on the naive CD4$^+$ T cell. Activation of naive CD4$^+$

T cells by anti-CD3 and anti-CD28 antibodies and IL-12 in the presence of either a β_2AR agonist or norepinephrine (10^{-6} M) generates effector Th1 cells that produce significantly higher levels of IFN-γ per cell on restimulation when compared to effector cells generated without the β_2AR stimulus (Swanson *et al.*, 2001). Furthermore, the presence of IL-12 is essential for the enhancing effect of β_2AR stimulation to be seen on Th1 differentiation, at all concentrations of IL-12 studied. In contrast, direct stimulation of the β_2AR on CD4$^+$ naive cells generated in the presence of IL-4 either increases or decreases the amount of IL-4 produced by restimulated effector cells, depending on the concentration of IL-4 present at the time of naive cell activation, with low concentrations increasing and high concentrations decreasing the level of IL-4 produced (unpublished findings). Thus, although controversial at present, norepinephrine may differentially affect Th1 and Th2 cell function via stimulation of the β_2AR on a naive CD4$^+$ T cell.

The level of IL-2 produced by activated Th1 cells is inhibited by the exogenous addition of a β_2AR-selective agonist or cAMP analog *in vitro* (Sekut *et al.*, 1995; Ramer-Quinn *et al.*, 1997; Holen and Elsayed, 1998). The mechanism for this inhibition may involve a decrease in the rate of calcineurin-dependent IL-2 gene transcription (Paliogianni *et al.*, 1993), a decrease in the binding of nuclear factor of activated T cells (NF-AT) to the IL-2 promoter (Lacour *et al.*, 1994), and/or a decrease in NF-AT and NF-κB nuclear binding (Tsuruta *et al.*, 1995). Similarly, stimulation of the β_2AR on sort-purified naive CD4$^+$ T cells by norepinephrine or a β_2AR agonist at the time of cell activation by anti-CD3 and anti-CD28 antibodies decreases the level of IL-2 produced by these cells 24 h later (Ramer-Quinn *et al.*, 2000), effects that are blocked by the β_2AR-selective antagonist ICI 118,551, but not by the β_1AR-selective antagonist metoprolol (Swanson *et al.*, 2001). Thus, these studies suggest a role for β_2AR stimulation on both naive and effector CD4$^+$ Th1 cells in decreasing the level of IL-2 produced by these cells.

The role of βAR stimulation on the level of IFN-γ produced by effector CD4$^+$ T cells has also been examined. cAMP-elevating agents (10^{-6} M) inhibit the level of IFN-γ produced by Th1-like cells (Betz and Fox, 1991; Van der Pouw-Kraan *et al.*, 1992; Snijdewint *et al.*, 1993). High concentrations of a β_2AR-selective agonist also decrease the level of IFN-γ produced by mitogen-activated T cells (Paul-Eugene *et al.*, 1992; Borger *et al.*, 1998). However, the timing between T cell activation and β_2AR stimulation may be an important factor that determines the effect induced on the level of cytokine produced. For example, the β_2AR-selective agonist terbutaline decreases the level of IFN-γ produced by Th1 clones when the β_2AR is stimulated *before* the Th1 cells are activated (Sanders *et al.*, 1997), whereas the level of IFN-γ produced increases when the β_2AR is stimulated at the time of cell activation (Ramer-Quinn *et al.*, 1997). And finally, norepinephrine may affect the level of effector T cell cytokine production by stimulating the β_2AR expressed on naive CD4$^+$ T cells as discussed earlier. Thus, the majority of findings suggest that β_2AR stimulation affects the level of cytokine produced by Th1-like cells but that the direction of the effect is dependent on both when the β_2AR is stimulated in relation to T cell activation and which CD4$^+$ T cell subset is stimulated.

In contrast to effects on naive and Th1 cells, cAMP-elevating agents or the β_2AR-selective agonist salbutamol fail to affect the level of IL-4 production by either mature T cells or Th2 cell clones (Novak and Rothenberg, 1990; Betz and Fox, 1991; Paul-Eugene *et al.*, 1992; Van der Pouw-Kraan *et al.*, 1992; Snijdewint *et al.*, 1993). Since Th2 cells do not express the β_2AR (Sanders *et al.*, 1997), it is not surprising that salbutamol does not influence IL-4 production by Th2 cell clones. However, in contrast to the previous findings that cAMP appears to cause no effect on the ability of Th2 cells to produce IL-4, other findings suggest that Th2 cells may respond to cAMP elevations (Lacour *et al.*, 1994; Teschendorf *et al.*, 1996; Wirth *et al.*, 1996). However, high concentrations of cAMP may inhibit the level of IL-4 mRNA expression (Borger *et al.*, 1996). Therefore, the type of stimulus used to raise the intracellular concentration of cAMP may determine if the level of IL-4 produced by Th2 cells will be affected. If Th2 cells do not express the β_2AR, then norepinephrine may not affect Th2 cell cytokine production directly, but rather via an indirect effect on another cell population that influences Th2 function or stimulation of another adrenergic-receptor subtype that may be expressed on the resting or activated Th2 cell.

Unfortunately, the effect of norepinephrine stimulation of the β_2AR on cytokine production *in vivo* is unclear. Findings suggest that norepinephrine and β_2AR stimulation differentially affect CD4$^+$ T cell cytokine production *in vivo*, depending on the strain of mouse, the mode of T cell activation, and the specific cytokine being measured (Madden *et al.*, 1994; Kruszewska *et al.*, 1995). In addition, since the β_2AR may be differentially expressed by subpopulations of CD4$^+$ T cells, norepinephrine may selectively influence the level of cytokine produced *in vivo* by naive and Th1 cells but not by Th2 cells.

Surprisingly, both Th1 and Th2 cytokine responses are affected in dopamine β-hydroxylase deficient mice lacking the enzyme that converts dopamine to norepinephrine (Alaniz *et al.*, 1999). When mice are infected with *Listeria monocytogenes*, T cells produce less IFN-γ, TNF-α, and IL-10 in comparison to cells isolated from wild-type mice. However, when mice are infected with *Mycobacterium tuberculosis*, T cells produce less IFN-γ and TNF-α but increased levels of IL-10 in comparison to wild-type mice. This study suggests that norepinephrine differentially affects both Th1- and Th2-like cytokine production and that this effect may be exerted on the naive and/or the effector T cells located within lymphoid organs at the time of antigen administration.

V. Effect of β_2AR Stimulation on B Lymphocytes

Since few B cells in the body are capable of responding to a specific antigen, it is critical that the antigen-specific B cell population expand in number in response to antigen so that a suitable number of B cells will differentiate into both antigen-specific antibody-secreting cells and memory B cells. Many studies emphasize the fact that the effect of cAMP on B cell proliferation *in vitro* depends on the activation stimulus used and the cytokines present within the activated B cell microenvironment (Blomhoff *et al.*, 1987; Hoffmann, 1988; Cohen and Rothstein, 1989; Vazquez *et al.*, 1991). Early studies found that cAMP elevation in B cells either enhances or inhibits the number of antibody-secreting cells produced upon B cell activation, depending on the concentration of cAMP and the timing of cAMP exposure in relation to B cell activation. These findings suggest that cAMP exerts concentration- and time-dependent effects on B cell function, such that early elevations increase the number of antibody-secreting cells, whereas later elevations inhibit this number (Ishizuka *et al.*, 1971; Robison and Sutherland, 1971; Watson *et al.*, 1973; Melmon *et al.*, 1974; Montgomery *et al.*, 1975; Marchalonis and Smith, 1976; Teh and Paetkau, 1974; Kishimoto and Ishizaka, 1976; Kishimoto *et al.*, 1977; Cook *et al.*, 1978; Sanford *et al.*, 1979; Koh *et al.*, 1995). Other studies showed that norepinephrine and isoproterenol both inhibit the number of antibody-secreting cells in response to the particulate antigen sheep red blood cells (sRBCs) when added at times well after B cell activation (Melmon *et al.*, 1974). In contrast, early addition of either norepinephrine or terbutaline to sRBC-exposed spleen cells enhances the number of antibody-secreting cells, with signals generated during the first 6 h of cell activation being the most critical (Sanders and Munson, 1984). Of importance, the previously discussed studies used whole lymphocyte cultures containing various cell populations other than B cells. Thus, cAMP-elevating agents may have affected either B cells directly or other accessory cells that affect the level of antibody produced by a B cell.

Therefore, since the frequency of antigen-specific B cells is relatively low in the spleen, and since other cell types are present in these cultures that also express adrenergic receptors, later studies use enriched populations of antigen-specific B cells and T cells to determine the effects of norepinephrine and β_2AR stimulation. A hapten-carrier protocol can be used to tease out the activity of B cells compared to $CD4^+$ Th2 cells. In this paradigm the B cell recognizes the hapten determinant trinitrophenol (TNP) expressed on the surface of a large carrier protein, keyhole-limpet hemocyanin (KLH). In this way specific B cell–T cell interactions can be engineered and analyzed. When TNP-specific B cells are cocultured with KLH-specific $CD4^+$ Th2 cell clones in the presence of TNP-KLH and the β_2AR agonist terbutaline, the number of anti-TNP IgM-secreting cells increases in comparison to cells exposed to antigen alone (Sanders and Powell-Oliver, 1992). A later study using the same antigen-specific model system and limiting dilution analysis reported that terbutaline also increases the total amount of IgG1 and IgE secreted by B cells, without affecting the number of secreting cells (Kasprowicz et al., 2000). In a similar manner, exposure of human peripheral blood mononuclear cells to the β_2AR-selective agonists salbutamol or fenoterol increases the level of IL-4-dependent IgE produced (Paul-Eugene et al., 1992). Other findings also show that elevations in cAMP increase the level of IgG_1 and IgE produced by lipopolysaccharide (LPS)-activated B cells in the presence of varying concentrations of IL-4 (Roper et al., 1990; Coqueret et al., 1996), possibly via effects on the level of germline $\gamma 1$ transcript produced (Lycke et al., 1990; Roper et al., 1995). Also, exposure of IFN-γ-pulsed B cells to cAMP-elevating agents enhances both the number of IgG_{2a}-producing B cells and the total amount of IgG_{2a} produced following LPS-induced activation (Stein and Phipps, 1991). Thus, elevations in cAMP within a B cell may influence the effects that cytokines exert on the level of a particular antibody isotype produced by the B cell.

β_2AR stimulation on only the B cell augments the level of IL-4-dependent IgG_1 and IgE produced by either the Th2 cell- or the CD40L-activated B cell, and this effect did not occur if B cells were isolated from the spleens of β_2AR $-/-$ mice (Kasprowicz et al., 2000). Importantly, the β_2AR agonist terbutaline appears to exert this effect on the B cell by increasing the level of B cell responsiveness to IL-4, as well as by increasing the level of B7-2 (CD86) expressed on the B cell surface (Kasprowicz et al., 2000). The relationship between the β_2AR-induced increase in B7-2 expression and the level of IgG1 and IgE produced by a B cell is an area of intense investigation currently.

B7-2 is a protein molecule expressed on the surface of a B cell that plays an important role in B cell function. B7-2 is a costimulatory molecule that serves two functions. First, B7-2 stimulates CD28 that is expressed on T cells to increase the level of both the cytokine produced and the CD40L expressed, effects which then affect B cell function. Second, stimulation of B7-2 appears to directly signal the B cell to regulate the level of antibody produced (Jeannin et al., 1997; Kasprowicz et al., 2000). The latter finding that B7-2 signals directly to a B cell may be dependent on stimulation of the B cell receptor for antigen. However, findings suggest that the requirement for B cell–receptor stimulation to attain B7-2 competency may be an effect that depends on when the B cell receives the activation signal through CD40 in relation to B7-2 stimulation (Podojil and Sanders, submitted for publication). Thus, changes in the level of B7-2 expression on the B cell may affect B cell function directly and/or indirectly, and much more investigation needs to be conducted before a definitive mechanism can be ascertained.

While the B7-2 protein and mRNA are expressed at very low levels in resting B cells (Lenschow *et al.*, 1993), cell activation upregulates expression (Freeman *et al.*, 1993; Lenschow *et al.*, 1993, 1994). Stimulation of the β_2AR alone on resting B cells increases the level of B7-2 expression, and concomitant stimulation of both the B cell–antigen receptor and the β_2AR result in an additive increase in the level of B7-2 expression on B cells (Kasprowicz *et al.*, 2000), via effects on mRNA stability and NF-κB-dependent gene transcription (Kohm *et al.*, 2002a). Thus, the change in B7-2 expression by β_2AR stimulation on a B cell may affect not only the level of B7-2 on a B cell, but also, if B7-2 directly signals the B cell, the level of antibody produced during a T cell–dependent antibody response. Taken together, these findings suggest that stimulation of the β_2AR on a B cell either prior to or at the time of cell activation increases the amount of antibody produced by the B cell via a mechanism in which the stimulation of the β_2AR on a B cell may increase signals generated in either the IL-4R or the B7-2 signaling pathways.

In vivo, the effect of norepinephrine and β_2AR stimulation on B cell differentiation into antibody-secreting cells has been studied extensively. Using a low dose of 6-hydroxydopamine (6-OHDA) to selectively destroy peripheral sympathetic nerve terminals, both the primary hemagglutinin titer and the number of plaque-forming cells in response to primary immunization with sheep red blood cells decreases (Kasahara *et al.*, 1977b; Williams *et al.*, 1981; Hall *et al.*, 1982), without affecting the secondary response to antigen (Kasahara *et al.*, 1977a), However, if 6-OHDA is administered concurrently with the secondary exposure to antigen, the level of antibody produced decreases, suggesting that the level of norepinephrine at the time of antigen administration may be important. The level of serum TNP-specific antibody produced by dopamine β-hydroxylase deficient mice (i.e., norepinephrine-deficient mice) was lower than the level of antibody produced by wild-type mice (Alaniz *et al.*, 1999), suggesting that norepinephrine affects the magnitude of an antibody response to antigen. In contrast, one study reports a strain-specific increase in antibody production in norepinephrine-depleted C57Bl/6J (Th1-slanted strain) and Balb/c (Th2-slanted strain) mice (Kruszewska *et al.*, 1995). However, the problem with 6-OHDA model systems is that resident lymphocytes may be exposed to a burst of nor-epinephrine initially, since the mechanism of action of 6-OHDA is to dis-place norepinephrine before destruction of the sympathetic nerve terminal.

To address this possibility, a model system was developed in which norepinephrine-depleted T cell– and B cell–deficient *scid* mice are reconsti-tuted with KLH-specific Th2 cell clones (β_2AR-negative) and TNP-specific B cells (β_2AR-positive). When these mice are immunized, the serum level of TNP-specific IgM and IgG$_1$ decreases, splenic follicles do not expand, and germinal centers do not form in comparison to norepinephrine-intact mice (Kohm and Sanders, 1999). Importantly, the effect of norepinephrine de-pletion is reversed by the administration of either the β_2AR-selective agonist terbutaline or metaproterenol, suggesting that the effect of norepinephrine-depletion on the *in vivo* antibody response is mediated via a lack of β_2AR stimulation on the B cell. In addition, although the level of TNP-specific IgM returns to control levels following secondary immunization in norepine-phrine-depleted mice, serum levels of TNP-specific IgG$_1$ remain signifi-cantly lower, but seem to increase gradually in a delayed manner. Thus, these data suggest that stimulation of the B cell β_2AR by endogenous nor-epinephrine released during the course of a T-dependent immune response (Kohm *et al.*, 2000) is necessary to maintain an optimal level of antibody production *in vivo*.

VI. Conclusion

The relevance of the reviewed findings to the clinical community is under intense investigation but remains unclear. Collectively, these findings indi-cate that both T cells and B cells express the β_2AR and bind norepinephrine that is released within lymphoid organs, and that norepinephrine plays a role in modulating the activity of CD4$^+$ T cells and B cells participating in an immune response against antigen. This role of norepinephrine in reg-ulating T and B cell activity needs to be fully understood since antibodies preserve our well-being by defending us against bacteria, viruses, and aller-gens and cytokines provide help to B cells, allowing the B cell to differentiate and thus secrete antibodies of particular isotypes. Given the importance to the host of having T cells and B cells that function optimally, it is likely that the mechanisms regulating and modulating these functions will be varied and interrelated. For example, it will be necessary to understand how prod-ucts released from the hypothalamic-pituitary-adrenal axis, at the same time norepinephrine is released by sympathetic nerve terminals within lymphoid organs, will affect immune cell function.

By understanding the mechanism by which the SNS modulates the level of cytokines and antibody produced, we will be able to develop therapeutic approaches for treating and preventing changes in the immunocompetent state of persons experiencing any disease that involves an alteration in either nervous or immune system function. For example, the level of immunocompetence may vary as a result of changes in (1) the level of locally secreted norepinephrine in lymphoid organs, (2) the level of expression of the β_2AR on lymphocytes, or (3) the ratio of T cell subsets participating in a particular immune response. Such changes in immunocompetence may not be immediately life-threatening to an individual, but could alter long-term health status and quality of life.

References

Abadie, C., Foucart, S., Page, P., and Nadeau, R. (1997). Interleukin-1 β and tumor necrosis factor-α inhibit the release of [^{3}H]-noradrenaline from isolated human atrial appendages. *Naunyn-Schmiedeberg's Arch. Pharmacol.* **355**, 384–389.

Ackerman, K. D., Felten, S. Y., Bellinger, D. L., and Felten, D. L. (1987). Noradrenergic sympathetic innervation of the spleen. III. Development of innervation in the rat spleen. *J. Neurosci. Res.* **18**, 49–54, 123–125.

Ader, R., and Cohen, N. (1975). Behaviorally conditioned immunosuppression. *Psychosom. Med.* **37**, 333–340.

Akiyoshi, M., Shimizu, Y., and Saito, M. (1990). Interleukin-1 increases norepinephrine turnover in the spleen and lung in rats. *Biochem. Biophys. Res. Commun.* **173**, 1266–1270.

Alaniz, R. C., Thomas, S. A., Perez-Melgosa, M., Mueller, K., Farr, A. G., Palmiter, R. D., and Wilson, C. B. (1999). Dopamine β-hydroxylase deficiency impairs cellular immunity. *Proc. Natl. Acad. Sci. USA* **96**, 2274–2278.

Bach, M.-A. (1975). Differences in cyclic AMP changes after stimulation by prostaglandins and isoproterenol in lymphocyte subpopulations. *J. Clin. Invest.* **55**, 1074–1081.

Banks, W. A., and Kastin, A. J. (1985a). Peptides and the blood–brain barrier: Lipophilicity as a predictor of permeability. *Brain Res. Bull.* **15**, 287–292.

Banks, W. A., and Kastin, A. J. (1985b). Permeability of the blood–brain barrier to neuropeptides: The case for penetration. *Psychoneuroendocrinology* **10**, 385–399.

Banks, W. A., and Kastin, A. J. (1987). Saturable transport of peptides across the blood–brain barrier. *Life Sci.* **41**, 1319–1338.

Bartik, M. M., Brooks, W. H., and Roszman, T. L. (1993). Modulation of T cell proliferation by stimulation of the β-adrenergic receptor: Lack of correlation between inhibition of T cell proliferation and cAMP accumulation. *Cell. Immunol.* **148**, 408–421.

Bauman, G. P., Bartik, M. M., Brooks, W. H., and Roszman, T. L. (1994). Induction of cAMP-dependent protein kinase (PKA) activity in T cells after stimulation of the prostaglandin E2 or the β-adrenergic receptors: Relationship between PKA activity and inhibition of anti-CD3 monoclonal antibody-induced T cell proliferation. *Cell. Immunol.* **158**, 182–194.

Besedovsky, H., Del Rey, A., Sorkin, E., Da Prada, M., Burri, R., and Honegger, C. (1983). The immune response evokes changes in brain noradrenergic neurons. *Science* **221**, 564–566.

Besedovsky, H. O., Del Rey, A., Sorkin, E., Da Prada, M., and Keller, H. H. (1979). Immunoregulation mediated by the sympathetic nervous system. *Cell. Immunol.* **48,** 346–355.

Besser, M., and Wank, R. (1999). Cutting edge: Clonally restricted production of the neurotrophins brain-derived neurotrophic factor and neurotrophin-3 mRNA by human immune cells and Th1/Th2-polarized expression of their receptors. *J. Immunol.* **162,** 6303–6306.

Betz, M., and Fox, B. S. (1991). Prostaglandin E2 inhibits production of Th1 lymphokines but not of Th2 lymphokines. *J. Immunol.* **146,** 108–113.

Bidart, J. M., Motte, P., Assicot, M., Bohuon, C., and Bellet, D. (1983). Catechol-O-methyltransferase activity and aminergic binding sites distribution in human peripheral blood lymphocyte subpopulations. *Clin. Immunol. Immunopathol.* **26,** 1–9.

Bishopric, N. H., Cohen, H. J., and Lefkowitz, R. J. (1980). β-Adrenergic receptors in lymphocyte subpopulations. *J. Allergy Clin. Immunol.* **65,** 29–33.

Blomhoff, H. K., Smeland, E. B., Beiske, K., Blomhoff, R., Ruud, E., Bjoro, T., Pfeifer-Ohlsson, S., Watt, R., Funderud, S., Godal, T., and Ohlsson, R. (1987). Cyclic AMP-mediated suppression of normal and neoplastic B cell proliferation is associated with regulation of *myc* and Ha-*ras* protooncogenes. *J. Cell. Physiol.* **131,** 426–433.

Bognar, I. T., Albrecht, S. A., Farasaty, M., Schmitt, E., Seidel, G., and Fuder, H. (1994). Effects of human recombinant interleukins on stimulation-evoked noradrenaline overflow from the rat perfused spleen. *Naunyn-Schmiedeberg's Arch. Pharmacol.* **349,** 497–502.

Borger, P., Kauffman, H. F., Postma, D. S., and Vellenga, E. (1996). Interleukin-4 gene expression in activated human T lymphocytes is regulated by the cyclic adenosine monophosphate-dependent signaling pathway. *Blood* **87,** 691–698.

Borger, P., Hoekstra, Y., Esselink, M. T., Postma, D. S., Zaagsma, J., Vellenga, E., and Kauffman, H. F. (1998). β-Adrenoceptor-mediated inhibition of IFN-γ, IL-3, and GM-CSF mRNA accumulation in activated human T lymphocytes is solely mediated by the β_2-adrenoceptor subtype. *Am. J. Respir. Cell Mol. Biol.* **19,** 400–407.

Bourne, H. R., and Melmon, K. L. (1971). Adenyl cyclase in human leukocytes: Evidence for activation by separate β adrenergic and prostaglandin receptors. *J. Pharmacol. Exp. Ther.* **178,** 1–7.

Buscail, L., Robberecht, P., DeNeef, P., Bui, D. N., Hooghe, R., and Christophe, J. (1990). Divergent regulation of β_2-adrenoceptors and adenylate cyclase in the Cyc-mouse T lymphoma cell line TL2–9. *Immunobiology* **181,** 51–63.

Calvo, W. (1968). The innervation of the bone marrow in laboratory animals. *J. Anat.* **123,** 315–328.

Carlson, S. L., Brooks, W. H., and Roszman, T. L. (1989). Neurotransmitter-lymphocyte interactions: Dual receptor modulation of lymphocyte proliferation and cAMP production. *J. Neuroimmunol.* **24,** 155–162.

Cazaux, C. A., Sterin-Borda, L., Gorelik, G., and Cremaschi, G. A. (1995). Down-regulation of β-adrenergic receptors induced by mitogen activation of intracellular signaling events in lymphocytes. *FEBS Lett.* **364,** 120–124.

Cheung, P., Tanner, K. G., Cheung, W. L., Sassone-Corsi, P., Denu, J. M., and Allis, C. D. (2000). Synergistic coupling of histone H3 phosphorylation and acetylation in response to epidermal growth factor stimulation. *Molecular Cell* **5,** 905–915.

Cohen, D. P., and Rothstein, T. L. (1989). Adenosine 3', 5'-Cyclic monophosphate modulates the mitogenic responses of murine B lymphocytes. *Cell. Immunol.* **121,** 113–124.

Cohen, N., Ader, R., Green, N., and Bovbjerg, D. (1979). Conditioned suppression of a thymus-independent antibody response. *Psychosom. Med.* **41,** 487–491.

Conolly, M. E., and Greenacre, J. K. (1977). The β-adrenoceptor of the human lymphocyte and human lung parenchyma. *Br. J. Pharmacol.* **59,** 17–23.

Cook, R. G., Stavitsky, A. B., and Harold, W. W. (1978). Regulation of the *in vitro* anamnestic antibody response by cyclic AMP. II. Antigen-dependent enhancement by exogenous prostaglandins of the E series. *Cell. Immunol.* **40,** 128–140.

Coqueret, O., Demarquay, D., and Lagente, V. (1996). Role of cyclic AMP in the modulation of IgE production by the β_2-adrenoceptor agonist, fenoterol. *Eur. Respir. J.* **9,** 220–225.

Cremaschi, G. A., Fisher, P., and Boege, F. (1991). β-Adrenoceptor distribution in murine lymphoid cell lines. *Immunopharmacology* **22,** 195–206.

Dailey, M. O., Schreurs, J., and Schulman, H. (1988). Hormone receptors on cloned T lymphocytes. Increased responsiveness to histamine, prostaglandins, and β-adrenergic agents as a late stage event in T cell activation. *J. Immunol.* **140,** 2931–2936.

De Blasi, A., Parruti, G., and Sallese, M. (1995). Regulation of G-protein-coupled receptor kinase subtypes in activated T lymphocytes. Selective increase of β-adrenergic receptor kinase 1 and 2. *J. Clin. Invest.* **95,** 203–210.

Dinarello, C. A. (1998). Interleukin-1, interleukin-1 receptors and interleukin-1 receptor antagonist. *Int. Rev. Immunol.* **16,** 457–499.

Dunn, A. J. (1992). Endotoxin-induced activation of cerebral catecholamine and serotonin metabolism: Comparison with interleukin-1. *J. Pharmacol. Exp. Ther.* **261,** 964–969.

Elenkov, I. J., Papanicolaou, D. A., Wilder, R. L., and Chrousos, G. P. (1996). Modulatory effects of glucocorticoids and catecholamines on human interleukin-12 and interleukin-10 production: Clinical implications. *Proc. Assoc. Am. Physicians* **108,** 374–381.

Exton, M. S., von Horsten, S., Schult, M., Voge, J., Strubel, T., Donath, S., Steinmuller, C., Seeliger, H., Nagel, E., Westermann, J., and Schedlowski, M. (1998). Behaviorally conditioned immunosuppression using cyclosporine A: Central nervous system reduces IL-2 production via splenic innervation. *J. Neuroimmunol.* **88,** 182–191.

Feldman, R. D., Hunninghake, G. W., and McArdle, W. L. (1987). β-Adrenergic-receptor-mediated suppression of interleukin 2 receptors in human lymphocytes. *J. Immunol.* **139,** 3355–3359.

Felten, D. L., Felten, S. Y., Carlson, S. L., Olschowka, J. A., and Livnat, S. (1985). Noradrenergic and peptidergic innervation of lymphoid tissue. *J. Immunol.* **135,** 755s–765s.

Felten, D. L., Ackerman, K. D., Wiegand, S. J., and Felten, S. Y. (1987). Noradrenergic sympathetic innervation of the spleen. I. Nerve fibers associate with lymphocytes and macrophages in specific compartments of the splenic white pulp. *J. Neurosci. Res.* **18,** 28–36.

Felten, S. Y., and Olschowka, J. (1987). Noradrenergic sympathetic innervation of the spleen. II. Tyrosine hydroxylase (TH)-positive nerve terminals form synaptic-like contacts on lymphocytes in the splenic white pulp. *J. Neurosci. Res.* **18,** 37–48.

Felten, S. Y., Felten, D. L., Bellinger, D. L., and Livnat, S. (1988). Noradrenergic sympathetic innervation of lymphoid organs. *Prog. Allergy* **43,** 14–36.

Fleshner, M., Goehler, L. E., Hermann, J., Relton, J. K., Maier, S. F., and Watkins, L. R. (1995). Interleukin-1 β-induced corticosterone elevation and hypothalamic NE depletion is vagally mediated. *Brain Res. Bull.* **37,** 605–610.

Foucart, S., and Abadie, C. (1996). Interleukin-1 β and tumor necrosis factor-α inhibit the release of [^{3}H]-noradrenaline from mice isolated atria. *Naunyn-Schmiedeberg's Arch. Pharmacol.* **354,** 1–6.

Freeman, G. J., Gribben, J. G., Boussiotis, V. A., Ng, J. W., Restivo, V. A. J., Lombard, L. A., Gray, G. S., and Nadler, L. M. (1993). Cloning of B7-2: A CTLA-4 counter-receptor that costimulates human T cell proliferation. *Science* **262,** 909–911.

Fuchs, B. A., Albright, J. W., and Albright, J. F. (1988a). β-Adrenergic receptor on murine lymphocytes: Density varies with cell maturity and lymphocyte subtype and is decreased after antigen administration. *Cell. Immunol.* **114,** 231–245.

Fuchs, B. A., Campbell, K. S., and Munson, A. E. (1988b). Norepinephrine and serotonin content of the murine spleen: Its relationship to lymphocyte β-adrenergic receptor density and the humoral immune response *in vivo* and *in vitro*. *Cell. Immunol.* **117,** 339–351.

Galant, S. P., Underwood, S. B., Lundak, T. C., Groncy, C. C., and Mouratides, D. I. (1978). Heterogeneity of lymphocyte subpopulations to pharmacologic stimulation. I. Lymphocyte responsiveness to β-adrenergic agents. *J. Allergy Clin. Immunol.* **62,** 349–356.

Glibetic, M. D., and Baumann, H. (1986). Influence of chronic inflammation on the level of mRNA for acute-phase reactants in the mouse liver. *J. Immunol.* **137**(5), 1616–1622.

Glowinski, J., and Baldessarini, R. J. (1966). Metabolism of norepinephrine in the central nervous system. *Pharmacol. Rev.* **18,** 1201–1238.

Goehler, L. E., Relton, J. K., Dripps, D., Kiechle, R., Tartaglia, N., Maier, S. F., and Watkins, L. R. (1997). Vagal paraganglia bind biotinylated interleukin-1 receptor antagonist: A possible mechanism for immune-to-brain communication. *Brain Res. Bull.* **43,** 357–364.

Gray, S. G., Svechnikova, I., Hartmann, W., O'Connor, L., Aguilar-Santelises, M., and Ekstrom, T. J. (2000). IGF-II and IL-2 act synergistically to alter HDAC1 expression following treatments with trichostatin a. *Cytokine* **12,** 1104–1109.

Griese, M., Korholz, U., Korholz, D., Seeger, K., Wahn, V., and Reinhardt, D. (1988). Density and agonist-promoted high and low affinity states of the β-adrenoceptor on human B- and T-cells. *Eur. J. Clin. Invest.* **18,** 213–217.

Hall, N. R., McClure, J. E., Hu, S., Tare, S., Seals, C. M., and Goldstein, A. L. (1982). Effects of 6-hydroxydopamine upon primary and secondary thymus dependent immune responses. *Immunopharmacology* **5,** 39–48.

Haugen, P. K., and Letourneau, P. C. (1990). Interleukin-2 enhances chick and rat sympathetic, but not sensory, neurite outgrowth. *J. Neurosci. Res.* **25,** 443–452.

Hmadcha, A., Bedoya, F. J., Sobrino, F., and Pintado, E. (1999). Methylation-dependent gene silencing induced by interleukin 1β via nitric oxide production. *J. Exp. Med.* **190,** 1595–1604.

Hoffmann, M. K. (1988). The requirement for high intracellular cyclic adenosine monophosphate concentrations distinguishes two pathways of B cell activation induced with lymphokines and antibody to immunoglobulin. *J. Immunol.* **140,** 580–582.

Holen, E., and Elsayed, S. (1998). Effects of β_2-adrenoceptor agonists on T-cell subpopulations. *APMIS* **106,** 849–857.

Holte, H., Torjesen, P., Blomhoff, H. K., Ruud, E., Funderud, S., and Smeland, E. B. (1988). Cyclic AMP has the ability to influence multiple events during B cell stimulation. *Eur. J. Immunol.* **18,** 1359–1366.

Hsieh, C. S., Heimberger, A. B., Gold, J. S., O'Garra, A., and Murphy, K. M. (1992). Differential regulation of T helper phenotype by interleukins 4 and 10 in an $\alpha\beta$ T cell receptor transgenic system. *Proc. Natl. Acad. Sci. USA* **89,** 6065–6060.

Ichijo, T., Katafuchi, T., and Hori, T. (1992). Enhancement of splenic sympathetic nerve activity induced by central administration of interleukin-1-β in rats. *Neurosci. Res.* **Suppl. 17,** S150–S157.

Inglese, J., Freedman, N. J., Koch, W. J., and Lefkowitz, R. J. (1993). Structure and mechanism of the G protein-coupled receptor kinases. *J. Biol. Chem.* **268,** 23735–23738.

Ishizuka, M., Braun, W., and Matsumoto, T. (1971). Cyclic AMP and immune responses. I. Influence of poly A : U and cAMP on antibody formation *in vitro*. *J. Immunol.* **107,** 1027–1035.

Ito, K., Barnes, P. J., and Adcock, I. M. (2000). Glucocorticoid receptor recruitment of histone deacetylase 2 inhibits interleukin-1β-induced histone H4 acetylation on lysines 8 and 12. *Mol. Cell. Biol.* **20,** 6891–6903.

Jeannin, P., Delneste, Y., Lecoanet-Henchoz, S., Gauchat, J.-F., Ellis, J., and Bonnefoy, J.-Y. (1997). CD86 (B7-2) on Human B cells: A functional role in proliferation and selective differentiation into IgE- and IgG4-producing cells. *J. Biol. Chem.* **272**, 15613–15619.

Johnson, D. L., Ashmore, R. C., and Gordon, M. A. (1981). Effects of β-adrenergic agents on the murine lymphocyte response to mitogen stimulation. *J. Immunopharmacol.* **3**, 205–219.

Kang, S. H., Bang, Y. J., Im, Y. H., Yang, H. K., Lee, D. A., Lee, H. Y., Lee, H. S., Kim, N. K., and Kim, S. J. (1999). Transcriptional repression of the transforming growth factor-β type I receptor gene by DNA methylation results in the development of TGF-β resistance in human gastric cancer. *Oncogene* **18**, 7280–7286.

Kasahara, K., Tanaka, S., and Hamashima, Y. (1977a). Suppressed immune response to T-cell dependent antigen in chemically sympathectomized mice. *Res. Commun. Chem. Pathol. Pharmacol.* **18**, 533–542.

Kasahara, K., Tanaka, S., Ito, T., and Hamashima, Y. (1977b). Suppression of the primary immune response by chemical sympathectomy. *Res. Comm. Chem. Pathol. Pharmacol.* **16**, 687–694.

Kasprowicz, D. J., Kohm, A. P., Berton, M. T., Chruscinski, A. J., Sharpe, A. H., and Sanders, V. M. (2000). Stimulation of the B cell receptor, CD86 (B7-2) and the β_2-adrenergic receptor intrinsically modulates the level of IgG1 produced per B cell. *J. Immunol.* **165**, 680–690.

Kassis, S., Sullivan, M., and Fishman, P. H. (1988). Modulation of the β-adrenergic receptor-coupled adenylate cyclase by chemical inducers of differentiation: Effects on β receptors and the inhibitory regulatory protein Gi. *J. Recept. Res.* **8**, 627–644.

Khan, M. M., Sansoni, P., Silverman, E. D., Engleman, E. G., and Melmon, K. L. (1986). β-Adrenergic receptors on human suppressor, helper, and cytolytic lymphocytes. *Biol. Psychiatry* **35**, 1137–1142.

Kishimoto, T., and Ishizaka, K. (1976). Regulation of antibody response *in vitro*. X. Biphasic effect of cyclic AMP on the secondary anti-hapten antibody response to anti-immunoglobulin and enhancing soluble factor. *J. Immunol.* **116**, 534–541.

Kishimoto, T., Nishizawa, Y., Kikutani, H., and Yamamura, Y. (1977). Biphasic effect of cyclic AMP on IgG production and on the changes of non-histone nuclear proteins induced with anti-immunoglobulin and enhancing soluble factor. *J. Immunol.* **118**, 2027–2033.

Koh, W. S., Yang, K. H., and Kaminski, N. E. (1995). Cyclic AMP is an essential factor in immune responses. *Biochem. Biophys. Res. Commun.* **206**, 703–709.

Kohm, A. P., and Sanders, V. M. (1999). Suppression of antigen-specific Th2 cell-dependent IgM and IgG1 production following norepinephrine depletion *in vivo*. *J. Immunol.* **162**, 5299–5308.

Kohm, A. P., and Sanders, V. M. (2000). Norepinephrine: A messenger from the brain to the immune system. *Immunol. Today* **21**, 539–542.

Kohm, A. P., Tang, Y., Sanders, V. M., and Jones, S. B. (2000). Activation of antigen-specific CD4$^+$ Th2 cells *in vivo* increases norepinephrine release in the spleen and bone marrow. *J. Immunol.* **165**, 725–733.

Kohm, A. P., Mozaffarian, A., and Sanders, V. M. (2002a). B cell receptor- and beta-2-adrenergic receptor-induced regulation of B7-2(CD86) expression in B cells. *J. Immunol.* **168**, 6314–6322.

Kohm, A. P., Huber, N., Swanson, M. A., and Sanders, V. M. (2002b). Epigenetic regulation of beta-2-adrenergic receptor expression in CD4$^+$ effector T cell subsets. (submitted for publication).

Korholz, D., Seeger, K., Griese, M., Wahn, V., Reifenhauser, A., and Reinhardt, D. (1988). β-Adrenoceptor density and resolution of high and low affinity state on B- and T-cells in asthmatic and non-asthmatic children. *Eur. J. Pediatr.* **147**, 116–120.

Krawietz, W., Werdan, K., Schober, M., Erdmann, E., Rindfleisch, G. E., and Hannig, K. (1982).

Different numbers of β-receptors in human lymphocyte subpopulations. *Biochem. Pharmacol.* **31**, 133–136.

Kruszewska, B., Felten, S. Y., and Moynihan, J. A. (1995). Alterations in cytokine and antibody production following chemical sympathectomy in two strains of mice. *J. Immunol.* **155**, 4613–4620.

Lacour, M., Arrighi, J. F., Muller, K. M., Carlberg, C., Saurat, J. H., and Hauser, C. (1994). cAMP up-regulates IL-4 and IL-5 production from activated CD4$^+$ T cells while decreasing IL-2 release and NF-AT induction. *Int. Immunol.* **6**, 1333–1343.

Lenschow, D. J., Su, G. H. T., Zuckerman, L. A., Nabavi, N., Jellis, C. L., Gray, G. S., Miller, J., and Bluestone, J. A. (1993). Expression and functional significance of an additional ligand for CTLA-4. *Proc. Natl. Acad. Sci. USA* **90**, 11054–11058.

Lenschow, D. J., Sperling, A. I., Cooke, M. P., Freeman, G., Rhee, L., Decker, D. C., Gray, G., Nadler, L. M., Goodnow, C. C., and Bluestone, J. A. (1994). Differential up-regulation of B7-1 and B7-2 costimulatory molecules after Ig receptor engagement by antigen. *J. Immunol.* **153**, 1990–1997.

Livnat, S., Felten, S. Y., Carlson, S. L., Bellinger, D. L., and Felten, D. L. (1985). Involvement of peripheral and central catecholamine systems in neural–immune interactions. *J. Neuroimmunol.* **10**, 5–30.

Loveland, B. E., Jarrott, B., and McKenzie, I. F. (1981a). The detection of β-adrenoceptors on murine lymphocytes. *Int. J. Immunopharmacol.* **3**, 45–55.

Lycke, N., Severinson, E., and Strober, W. (1990). Cholera toxin acts synergistically with IL-4 to promote IgG1 switch differentiation. *J. Immunol.* **145**, 3316–3324.

Lyte, M., Ernst, S., Driemeyer, J., and Baissa, B. (1991). Strain-specific enhancement of splenic T cell mitogenesis and macrophage phagocytosis following peripheral axotomy. *J. Neuroimmunol.* **31**, 1–8.

Madden, K. S., Felten, S. Y., Felten, D. L., Sundaresan, P. R., and Livnat, S. (1989). Sympathetic neural modulation of the immune system. I. Depression of T cell immunity *in vivo* and *in vitro* following chemical sympathectomy. *Brain Behav. Immun.* **3**, 72–89.

Madden, K. S., Moynihan, J. A., Brenner, G. J., Felten, S. Y., Felten, D. L., and Livnat, S. (1994). Sympathetic nervous system modulation of the immune system. III. Alterations in T and B cell proliferation and differentiation *in vitro* following chemical sympathectomy. *J. Neuroimmunol.* **49**, 77–87.

Maier, S. F., Goehler, L. E., Fleshner, M., and Watkins, L. R. (1998). The role of the vagus nerve in cytokine-to-brain communication. *Ann. N. Y. Acad. Sci.* **840**, 289–300.

Makman, M. H. (1971). Properties of adenylate cyclase of lymphoid cells. *Proc. Natl. Acad. Sci. USA* **68**, 885–889.

Marchalonis, J. J., and Smith, P. (1976). Effects of dibutyrylcyclic AMP on the *in vitro* primary response of mouse spleen cells to sheep erythrocytes. *Aust. J. Exp. Biol. Med. Sci.* **54**, 1–10.

Marz, P., Cheng, J. G., Gadient, R. A., Patterson, P. H., Stoyan, T., Otten, U., and Rose-John, S. (1998). Sympathetic neurons can produce and respond to interleukin 6. *Proc. Natl. Acad. Sci. USA* **95**, 3251–3256.

Melmon, K. L., Bourne, H. R., Weinstein, Y., Shearer, G. M., Kram, J., and Bauminger, S. (1974). Hemolytic plaque formation by leukocytes *in vitro*: Control by vasoactive amines. *J. Clin. Invest.* **53**, 13–21.

Meurs, H., Van Den Bogaard, W., Kauffman, H. F., and Bruynzeel, P. L. B. (1982). Characterization of (−)-[^{3}H]Dihydroalprenolol binding to intact and broken cell preparations of human peripheral blood lymphocytes. *Eur. J. Pharmacol.* **85**, 185–194.

Miles, K., Quintans, J., Chelmicka-Schorr, E., and Arnason, B. G. W. (1981). The sympathetic nervous system modulates antibody response to thymus-independent antigens. *J. Neuroimmunol.* **1**, 101–105.

Miles, K., Atweh, S., Otten, G., Arnason, B. G., and Chelmicka-Schorr, E. (1984a). β-Adrenergic receptors on splenic lymphocytes from axotomized mice. *Int. J. Immunopharmacol.* **6,** 171–177.

Miles, K., Atweh, S., Otten, G., Arnason, B. G. W., and Chelmicka-Schorr, E. (1984b). β-Adrenergic receptors on splenic lymphocytes from axotomized mice. *Int. J. Immunopharmacol.* **6**(3), 171–177.

Miles, K., Chelmicka-Schorr, E., Atweh, S., Otten, G., and Arnason, B. G. W. (1985). Sympathetic ablation alters lymphocyte membrane properties. *J. Immunol.* **135,** 797s–801s.

Minakuchi, R., Wacholtz, M. C., Davis, L. S., and Lipsky, P. E. (1990). Delineation of the mechanism of inhibition of human T cell activation by PGE2. *J. Immunol.* **145,** 2616–2625.

Montogomery, P. C., Skandera, C. A., and Kahn, R. L. (1975). Differential effects of cyclic AMP on the *in vitro* induction of antibody synthesis. *Nature (London)* **256,** 137–138.

Niijima, A., Hori, T., Aou, S., and Oomura, Y. (1991). The effects of interleukin-1 β on the activity of adrenal, splenic and renal sympathetic nerves in the rat. *J. Auton. Nerv. Syst.* **36,** 183–192.

Novak, T. J., and Rothenberg, E. V. (1990). cAMP inhibits induction of interleukin 2 but not of interleukin 4 in T cells. *Proc. Natl. Acad. Sci. USA* **87,** 9353–9357.

Ohno, Y., Lee, J., Fusunyan, R. D., MacDermott, R. P., and Sanderson, I. R. (1997). Macrophage inflammatory protein-2: Chromosomal regulation in rat small intestinal epithelial cells. *Proc. Natl. Acad. Sci. USA* **94,** 10279–10284.

Paietta, E., and Schwarzmeier, J. D. (1983). Differences in β-adrenergic receptor density and adenylate cyclase activity between normal and leukaemic leukocytes. *Eur. J. Clin. Invest.* **13,** 339–346.

Paliogianni, F., Kincaid, R. L., and Boumpas, D. T. (1993). Prostaglandin E2 and other cyclic AMP elevating agents inhibit interleukin 2 gene transcription by counteracting calcineurin-dependent pathways. *J. Exp. Med.* **178,** 1813–1817.

Panina-Bordignon, P., Mazzeo, D., Lucia, P. D., D'Ambrosio, D., Lang, R., Fabbri, L., Self, C., and Sinigaglia, F. (1997). β2-Agonists prevent Th1 development by selective inhibition of interleukin 12. *J. Clin. Invest.* **100,** 1513–1519.

Pardini, B. J., Jones, S. B., and Filkins, J. P. (1982). Contribution of depressed reuptake to the depletion of norepinephrine from rat heart and spleen during endotoxin shock. *Circ. Shock* **9,** 129–143.

Parsey, M. V., and Lewis, G. K. (1993). Actin polymerization and pseudopod reorganization accompany anti-CD3-induced growth arrest in Jurkat T cells. *J. Immunol.* **151,** 1881–1893.

Paul-Eugene, N., Kolb, J. P., Calenda, A., Gordon, J., Kikutani, H., Kishimoto, T., Mencia-Huerta, J. M., Braquet, P., and Dugas, B. (1992). Functional interaction between β2-adrenoceptor agonists and interleukin-4 in the regulation of CD23 expression and release and IgE production in human. *Mol. Immunol.* **30,** 157–164.

Pochet, R., and Delespesse, G. (1983a). β-Adrenoceptors display different efficiency on lymphocyte subpopulations. *Biochem. Pharmacol.* **32,** 1651–1655.

Pochet, R., and Delespesse, G. (1983b). β-Adrenoreceptors display different efficiency on lymphocyte subpopulations. *Biochem. Pharmacol.* **32,** 1651–1655.

Pochet, R., Delespesse, G., Gausset, P. W., and Collet, H. (1979). Distribution of β-adrenergic receptors on human lymphocyte subpopulations. *Clin. Exp. Immunol.* **38,** 578–584.

Podojil, J., and Sanders, V. M. (2002). CD86 (B7-2) and beta-2-adrenergic receptor stimulation on a CD40L/IL-4-activated B cell increases mature IgG1 transcription, without affecting germline IgG1 transcription. (submitted for publication).

Radojcic, T., Baird, S., Darko, D., Smith, D., and Bulloch, K. (1991). Changes in β-adrenergic receptor distribution on immunocytes during differentiation: An analysis of T cells and macrophages. *J. Neurosci. Res.* **30,** 328–335.

Ramer-Quinn, D. S., Baker, R. A., and Sanders, V. M. (1997). Activated Th1 and Th2 cells differentially express the β_2-adrenergic receptor: A mechanism for selective modulation of Th1 cell cytokine production. *J. Immunol.* **159,** 4857–4867.

Ramer-Quinn, D. S., Swanson, M. A., Lee, W. T., and Sanders, V. M. (2000). Cytokine production by naive and primary effector CD4$^+$ T cells exposed to norepinephrine. *Brain Behav. Immun.* **14,** 239–255.

Reilly, F. D., McCuskey, P. A., Miller, M. L., McCuskey, R. S., and Meineke, H. A. (1979). Innervation of the periarteriolar lymphatic sheath of the spleen. *Tissue Cell* **11,** 121–126.

Robberecht, P., Abello, J., Damien, C., de Neef, P., Vervisch, E., Hooghe, R., and Christophe, J. (1989). Variable stimulation of adenylate cyclase activity by vasoactive intestinal-like peptides and β-adrenergic agonists in murine T cell lymphomas of immature, helper, and cytotoxic types. *Immunobiology* **179,** 422–431.

Robison, G. A., and Sutherland, E. W. (1971). Cyclic AMP and the function of eukaryotic cells: An introduction. *Ann. N. Y. Acad. Sci.* **185,** 5–9.

Rogers, M. P., Reich, P., Strom, T. B., and Carpenter, C. B. (1976). Behaviorally conditioned immunosuppression: Replication of a recent study. *Psychosom. Med.* **38,** 447–451.

Roper, R. L., Conrad, D. H., Brown, D. M., Warner, G. L., and Phipps, R. P. (1990). Prostaglandin E2 promotes IL-4-induced IgE and IgG1 synthesis. *J. Immunol.* **145,** 2644–2651.

Roper, R. L., Brown, D. M., and Phipps, R. P. (1995). Prostaglandin E2 promotes B lymphocyte Ig isotype switching to IgE. *J. Immunol.* **154,** 162–170.

Ruhl, A., Hurst, S., and Collins, S. M. (1994). Synergism between interleukins 1 β and 6 on noradrenergic nerves in rat myenteric plexus. *Gastroenterology* **107,** 993–1001.

Sanders, V. M., and Munson, A. E. (1984). β-Adrenoceptor mediation of the enhancing effect of norepinephrine on the murine primary antibody response *in vitro. J. Pharmacol. Exp. Ther.* **230**(1), 183–192.

Sanders, V. M., and Powell-Oliver, F. E. (1992). β_2-Adrenoceptor stimulation increases the number of antigen-specific precursor B lymphocytes that differentiate into IgM-secreting cells without affecting burst size. *J. Immunol.* **148,** 1822–1828.

Sanders, V. M., Baker, R. A., Ramer-Quinn, D. S., Kasprowicz, D. J., Fuchs, B. A., and Street, N. E. (1997). Differential expression of the β_2-adrenergic receptor by Th1 and Th2 clones: Implications for cytokine production and B cell help. *J. Immunol.* **158,** 4200–4210.

Sanders, V. M., Kasprowicz, D. J., Kohm, A. P., and Swanson, M. A. (2001). Neurotransmitter receptors on lymphocytes and other lymphoid cells. *In* "Psychoneuroimmunology" 3rd ed., (Ader, Cohen, and Felten, eds.), Academic Press, San Diego, CA.

Sanford, L. P., Stavitsky, A. B., and Cook, R. G. (1979). Regulation of the *in vitro* anamnestic antibody response by cyclic AMP. III. Cholera enterotoxin induces lymph node cells to release soluble factor(s) which enhance(s) antibody synthesis by antigen-treated lymph node cells. *Cell. Immunol.* **48,** 182–194.

Saphier, D., and Ovadia, H. (1990). Selective facilitation of putative corticotropin-releasing factor-secreting neurones by interleukin-1. *Neurosci. Lett.* **114,** 283–288.

Sapolsky, R., Rivier, C., Yamamoto, G., Plotsky, P., and Vale, W. (1987). Interleukin-1 stimulates the secretion of hypothalamic corticotropin-releasing factor. *Science* **238,** 522–524.

Sawchenko, P. E., and Swanson, L. W. (1982). The organization of noradrenergic pathways from the brainstem to the paraventricular and supraoptic nuclei in the rat. *Brain Res.* **257,** 275–325.

Scordamaglia, A., Ciprandi, G., Ruffoni, S., Caria, M., Paolieri, F., Venuti, D., and Canonica, G. W. (1988). Theophylline and the immune response: *In vitro* and *in vivo* effects. *Clin. Immunol. Immunopathol.* **48,** 238–246.

Seder, R. A., Paul, W. E., Davis, M. M., and de St. Groth, B. F. (1992). The presence of interleukin 4 during *in vitro* priming determines the lymphokine-producing potential of CD4[+] T cells from T cell receptor transgenic mice. *J. Exp. Med.* **176,** 1091–1098.

Seder, R. A., Gazzinelli, R., Sher, A., and Paul, W. E. (1993). Interleukin 12 acts directly on CD4[+] T cells to enhance priming for interferon-γ production and diminishes interleukin 4 inhibition of such priming. *Proc. Natl. Acad. Sci. USA* **90,** 10188–10192.

Sekut, L., Champion, B. R., Page, K., Menius, J. A. Jr., and Connolly, K. M. (1995). Anti-inflammatory activity of salmeterol: Down-regulation of cytokine production. *Clin. Exp. Immunol.* **99,** 461–466.

Selliah, N., Bartik, M. M., Carlson, S. L., Brooks, W. H., and Roszman, T. L. (1995). cAMP accumulation in T-cells inhibits anti-CD3 monoclonal antibody-induced actin polymerization. *J. Neuroimmunol.* **56,** 107–112.

Shimizu, N., Hori, T., and Nakane, H. (1994). An interleukin-1-β-induced noradrenaline release in the spleen is mediated by brain corticotropin-releasing factor: An *in vivo* microdialysis study in conscious rats. *Brain Behav. Immun.* **7,** 14–23.

Smith, J. W., Steiner, A. L., and Parker, C. W. (1971). Human lymphocyte metabolism. Effects of cyclic and noncyclic nucleotides on stimulation by phytohemagglutinin. *J. Clin. Invest.* **50,** 442–448.

Snijdewint, F. G., Kalinski, P., Wierenga, E. A., Bos, J. D., and Kapsenberg, M. L. (1993). Prostaglandin E2 differentially modulates cytokine secretion profiles of human T helper lymphocytes. *J. Immunol.* **150,** 5321–5329.

Staehelin, M., Muller, P., Portenier, M., and Harris, A. W. (1985). β-Adrenergic receptors and adenylate cyclase activity in murine lymphoid cell lines. *J. Cyclic Nucleotide Protein Phosphor. Res.* **10,** 55–64.

Stein, S. H., and Phipps, R. P. (1991). Antigen-specific IgG2a production in response to prostaglandin E2, immune complexes, and IFN-γ. *J. Immunol.* **147,** 2500–2506.

Swanson, M. A., Lee, W. T., and Sanders, V. M. (2001). IFN-γ production by Th1 cells generated from naive CD4[+] T cells exposed to norepinephrine. *J. Immunol.* **166,** 232–240.

Takahashi, H., Nishimura, M., Sakamoto, M., Ikegaki, I., Nakanishi, T., and Yoshimura, M. (1992). Effects of interleukin-1 β on blood pressure, sympathetic nerve activity, and pituitary endocrine functions in anesthetized rats. *Am. J. Hypertens.* **5,** 224–229.

Tang, Y., Shankar, R., Gamelli, R., and Jones, S. (1999). Dynamic norepinephrine alterations in bone marrow: Evidence of functional innervation. *J. Neuroimmunol.* **96,** 182–189.

Taplick, J., Kurtev, V., Lagger, G., and Seiser, C. (1998). Histone H4 acetylation during interleukin-2 stimulation of mouse T cells. *FEBS Lett.* **436,** 349–352.

Teh, H.-S., and Paetkau, V. (1974). Biphasic effect of cyclic AMP on an immune response. *Nature (London)* **250,** 505–507.

Teschendorf, C., Trenn, G., Hoffkes, H. G., and Brittinger, G. (1996). Differential effect of the activation of protein kinase A on the protein synthesis and secretion in the T-helper 2 cell line D10. G4.1. *Scand. J. Immunol.* **44,** 150–156.

Tsuruta, L., Lee, H. J., Masuda, E. S., Koyano-Nakagawa, N., Arai, N., Arai, K., and Yokota, T. (1995). Cyclic AMP inhibits expression of the IL-2 gene through the nuclear factor of activated T cells (NF-AT) site, and transfection of NF-AT cDNAs abrogates the sensitivity of EL-4 cells to cyclic AMP. *J. Immunol.* **154,** 5255–5264.

van de Griend, R. J., Astaldi, A., Wijermans, P., van Doorn, R., and Roos, D. (1983). Low β-adrenergic receptor concentration on human thymocytes. *Clin. Exp. Immunol.* **51,** 53–60.

Vanden Berghe, W., Vermeulen, L., De Wilde, G., De Bosscher, K., Boone, E., and Haegeman, G. (2000). Signal transduction by tumor necrosis factor and gene regulation of the inflammatory cytokine interleukin-6. *Biochem. Pharmacol.* **60,** 1185–1195.

Van der Pouw-Kraan, T., Van Kooten, C., Rensink, I., and Aarden, L. (1992). Interleukin (IL)-4 production by human T cells: Differential regulation of IL-4 vs. IL-2 production. *Eur. J. Immunol.* **22,** 1237–1241.

Van der Pouw-Kraan, T. C. T. M., Boeije, L. C. M., Smeenk, R. J. T., Wijdenes, J., and Aarden, L. A. (1995). Prostaglandin-E2 is a potent inhibitor of human interleukin 12 production. *J. Exp. Med.* **181,** 775–779.

van Oosterhout, A. J. M., and Nijkamp, F. P. (1984). Anterior hypothalamic lesions prevent the endotoxin-induced reduction of β-adrenoceptor number in the guinea pig lung. *Brain Res.* **302,** 277–280.

Van Tits, L. J. H., Michel, M. C., Grosse-Wilde, H., Happel, M., Eigler, F.-W., Soliman, A., and Brodde, O.-E. (1990). Catecholamines increase lymphocyte β_2-adrenergic receptors via a β_2-adrenergic, spleen-dependent process. *Am. J. Physiol.* **258,** E191–E202.

Vazquez, A., Auffredou, M. T., Galanaud, P., and Leca, G. (1991). Modulation of IL-2- and IL-4-dependent human B cell proliferation by cyclic AMP. *J. Immunol.* **146,** 4222–4227.

Watson, J., Epstein, R., and Cohn, M. (1973). Cyclic nucleotides as intracellular mediators of the expression of antigen-sensitive cells. *Nature (London)* **246,** 405–409.

Wayner, E. A., Flannery, G. R., and Singer, G. (1978). Effects of taste aversion conditioning on the primary antibody response to sheep red blood cells and *Brucella abortus* in the albino rat. *Physiol. Behav.* **21,** 995–1000.

Westly, H. J., and Kelley, K. W. (1987). Down-regulation of glucocorticoid and β-adrenergic receptors on lectin-stimulated splenocytes. *Proc. Soc. Exp. Biol. Med.* **185,** 211–218.

Williams, J. M., and Felten, D. L. (1981). Sympathetic innervation of murine thymus and spleen: A comparative histofluorescence study. *Anat. Rec.* **199,** 531–542.

Williams, J. M., Peterson, R. G., Shea, P. A., Schmedtje, J. F., Bauer, D. C., and Felten, D. L. (1981). Sympathetic innervation of murine thymus and spleen: Evidence for a functional link between the nervous and immune systems. *Brain Res. Bull.* **6**(1), 83–94.

Williams, L. T., Snyderman, R., and Lefkowitz, R. J. (1976). Identification of β-adrenergic receptors in human lymphocytes by $(-)^3$H-alprenolol binding. *J. Clin. Invest.* **57,** 149–155.

Wirth, S., Lacour, M., Jaunin, F., and Hauser, C. (1996). Cyclic adenosine monophosphate (cAMP) differentially regulates IL-4 in thymocyte subsets. *Thymus* **24,** 101–109.

MECHANISMS BY WHICH CYTOKINES SIGNAL THE BRAIN

Adrian J. Dunn

Department of Pharmacology and Therapeutics
Louisiana State University Health Sciences Center
Shreveport, Louisiana 71130

I. Introduction

A. The Messengers

Cytokines are proteins or glycoproteins that are synthesized and secreted by immune cells and that act as messengers regulating the immune response. Thus they are the hormones of the immune system. Although the traditional view of the immune system was that it received little input from other parts of the body, it is now widely accepted that there are extensive interactions between the immune system and other organ systems, and especially the central nervous system (CNS). Not surprisingly, cytokines play a role in this communication process, and they are the principal known messengers from the immune system to the brain. However, the mechanisms by which this communication occurs are unresolved. Nevertheless, we now know that there are multiple possibilities, and that more than one different mechanism may be used, even for the same or similar responses. Thus the question "how do cytokines signal the brain?" does not have a simple answer, and it needs to be refined to ask "which cytokine?" and for "what specific message?" In

43

practical terms this requires defining the cytokine and the specific response to the signal.

Many of the experiments relevant to this topic have been performed using endotoxin (lipopolysaccharide, LPS). LPS is a degradation product of the cell walls of gram-negative bacteria, which has the ability to induce fever in animals. LPS is a class of molecules with similar structures, rather than a uniquely defined molecular entity, thus different preparations of LPS from different sources may have differing activities. LPS is a potent inducer of the three major pro-inflammatory cytokines, interleukin-1 (IL-1), interleukin-6 (IL-6), and tumor necrosis factor-α (TNF-α). Many of the effects of administered LPS closely resemble those of these three cytokines (especially IL-1), but it should not be assumed that LPS works only via IL-1 (see Section II.D).

B. THE BLOOD–BRAIN BARRIER

Because cytokines are relatively large molecules, it is unlikely that they enter the brain owing to the existence of the blood–brain barrier. The barrier is not an impenetrable shield surrounding the brain. It is a selective, not an absolute, barrier. Some substances have ready access to the brain, and certain parts of the brain do not exhibit a barrier. To understand how cytokines might affect the brain, it is important to understand the nature of the blood–brain barrier and the structures that comprise it. Molecules that can easily traverse cell membranes are not hampered by the barrier. Thus, lipids and lipid solvents pass the barrier easily, whereas large or highly charged molecules are impeded. For example, anesthetics, such as chloroform, cross readily into the brain as do gases, such as nitrous oxide as well as oxygen and carbon dioxide. It should be obvious that ethanol also has access. Nevertheless, the brain also needs a ready supply of charged molecules as precursors for the synthesis of many important brain chemicals. This is accomplished by selective brain uptake systems for important fuels and synthetic materials, such as glucose and the amino acids, as well as for cations such as Na^+ and Ca^{2+}. In general, lipids and small molecules can traverse the blood–brain barrier, whereas larger and highly charged molecules cannot, unless their structures match the needs of one of the selective uptake systems.

Certain structures in the brain lack a barrier. The blood–brain barrier was discovered in the nineteenth century when it was observed that dyes injected systemically into live animals stained all the tissues in the body except the brain and spinal cord. Subsequent careful examination revealed that a small number of brain structures do stain. These structures include the pituitary (both anterior and posterior) and the pineal, as well as five structures known as the circumventricular organs (CVOs) because they all border the cerebral ventricles: the hypothalamic median eminence, the organum vasculosum laminae terminalis (OVLT), the area postrema, the subfornical organ, and the subcommissural organ. Each of these areas is involved in regulating functions that require feedback from the periphery. Thus cytokines could act directly on neurons within any of these areas.

The physical basis of the barrier is reasonably clear. The major component is the endothelial cells that form the blood vessels within the brain. In contrast with the endothelial cells in most bodily tissue, the junctions between brain endothelia are tight and apparently impenetrable to large molecules. The brain endothelial cells also lack the fenestrations (windows) present in endothelia in other organs which permit passage of macro-molecules. However, the barrier to a molecule originating in the blood and destined for a neuron may well include crossing several other membranes, such as the plasma membranes of the astrocytes making up the glia limitans which surrounds the capillaries as well as those of other neurons and glia en route, and/or long sojourns through extracellular space. The endothelia of the CVOs lack tight junctions and are fenestrated, largely explaining the easy accessibility. However, because these openings would breach the barrier, in most cases the glia bordering the CVOs form tight barriers between the CVO and the brain parenchyma.

C. THE MECHANISMS

The potential mechanisms by which cytokines might signal the brain are listed in Table I. Banks *et al.* (1995) have provided evidence for selective uptake systems that are capable of transporting a range of different cytokines into the brain. The existence of such systems has been criticized but not refuted. Their significance is not known, but their capacity appears to be quite limited (see later discussion). The second avenue is to exploit the absence of a blood–brain barrier in the CVOs. There is no doubt that circulating cytokines can access the CVOs. It is believed that this may be an important pathway for the pyrogenic activities of IL-1 and LPS and specifically involves the OVLT (see Section II.B). It is also likely that this may be a mechanism for the rapid activation of the hypothalamic-pituitary-adrenocortical (HPA) axis, but whether the OVLT, the median eminence, or the area postrema is the important CVO is unresolved (see Section II.C). The third possible mechanism is the activation of peripheral neurons that can send afferent signals to the brain. The principal known pathway is the vagus nerve. This radical hypothesis originated with the discovery that lesions of the vagus nerve could prevent the appearance of the immediate-early gene protein

TABLE I

THE PRINCIPAL KNOWN MECHANISMS FOR CYTOKINE SIGNALING OF THE BRAIN

1. Cytokines can be transported into the brain to a limited extent using selective uptake systems (transporters) that bypass the blood–brain barrier.

2. Cytokines can act on brain tissue at sites where the blood–brain barrier is weak or nonexistent, e.g., the CVOs.

3. Cytokines may act directly or indirectly on peripheral nerves that send afferent signals to the brain.

4. Cytokines can act on peripheral tissues inducing the synthesis of molecules whose ability to penetrate the brain is not limited by the barrier. A major target appears to be endothelial cells which bear receptors for IL-1 (and endotoxin).

5. Cytokines can be synthesized by immune cells that infiltrate the brain.

Fos in hypothalamic neurons in response to peripheral injection of LPS (Wan *et al.*, 1994). The lesions were made subdiaphragmatically (below the diaphragm), because vagal lesions above this level are life-threatening. Subsequently, Watkins *et al.* (1995a) showed that subdiaphragmatic lesions of the vagus attenuated the febrile response of rats to intraperitoneal (ip) IL-1β, and Bret-Dibat *et al.* (1995) showed that subdiaphragmatic vagotomy in mice attenuated the IL-1- and LPS-induced depression of food-motivated behavior. Subdiaphragmatic vagotomy also prevented the IL-1β-induced reduction in hypothalamic norepinephrine (NE) and attenuated the increase in plasma corticosterone (Fleshner *et al.*, 1995). To date, many of the responses to peripherally injected IL-1 have been shown to be attenuated by subdiaphragmatic lesions of the vagus (Goehler *et al.*, 2000; Watkins *et al.*, 1995b). However, rarely are the responses blocked completely. The detailed mechanism of this effect is not yet known, but it is believed that administered IL-1, or IL-1 produced in the periphery, binds to paraganglion cells in the vagal ganglia, and that this induces a signal in the vagal afferents that is transmitted to the nucleus tractus solitarius (NTS; Goehler *et al.*, 1999).

The existence of the blood–brain barrier does not preclude the possibility that cytokines may act on cells outside the barrier to produce molecules for which there is no barrier. Thus the fourth important possibility is that cytokines could act on peripheral tissues inducing the synthesis of molecules whose ability to penetrate the brain is not limited by the barrier. This possibility may take many forms. A major target is thought to be brain endothelial cells which can be induced to produce prostaglandins. Endothelial cells bear receptors for both IL-1 and LPS, which appear to be coupled to cyclooxygenase (COX). COX is an enzyme essential for the synthesis of prostaglandins as well as prostacyclins and thromboxanes. The principal known pathway is activation by IL-1/LPS of the synthesis of prostaglandin (PG)E$_2$, known to be an important inducer of fever. This pathway may be involved in IL-1-induced fever as well as activation of the HPA axis and perhaps some of the behavioral effects of IL-1. The fifth possibility involves infiltration of the brain by immune cells, principally macrophages. It is well established that peripheral treatment with LPS results in the appearance of IL-1 within the brain. The unresolved question is the nature of the cells involved. It is clear that following LPS injection, microglia immunoreactive for IL-1 appear within the brain (Quan *et al.*, 1998b; Van Dam *et al.*, 1992). The study by Quan *et al.* (1998b) suggested a slow migration of IL-1 immunoreactive microglia, which appeared to be migrating from the external surface of the brain inward. However, others have reported IL-1 immunoreactivity in astroglia (Lee *et al.*, 1993), and even in neurons (Breder *et al.*, 1988), but this is still highly controversial. More evidence suggests that stimulation of vagal afferents can induce IL-1 synthesis within the brain (Dantzer *et al.*, 1999). Because the only established presence of IL-1 in the brain is in microglia, the mechanism of this is unclear (see Section II.D).

D. The Responses Measured

From an empirical perspective, one must define the CNS response to be measured. Such responses could be internal to the brain, for example, measuring changes in electrophysiological activity or in various biochemicals or chemical processes. *In vivo* studies can be conducted with standard electrophysiological techniques. Chemical measures can be made with *in vivo* microdialysis or *in vivo* voltammetry or chronoamperometry. We can now include imaging techniques such as magnetic resonance imaging (MRI) or positron emission tomography (PET) scanning, although we are unaware of any such studies with cytokines. Yet another possibility is to assess brain output in terms of physiological and behavioral responses. Those most commonly studied have been changes in body temperature; endocrine measures, especially those of the HPA axis; and behavioral measures, especially sleep and so-called sickness behaviors.

E. The Hypothalamic-Pituitary-Adrenocortical (HPA) Axis

The HPA axis is a hierarchy in which activation of neurons in the paraventricular nucleus of the hypothalamus (PVN) containing corticotropin-releasing factor (CRF) results in the secretion of CRF in the median eminence region of the hypothalamus (Fig. 1). The CRF is secreted into the portal blood supply which carries it to the anterior pituitary where it acts

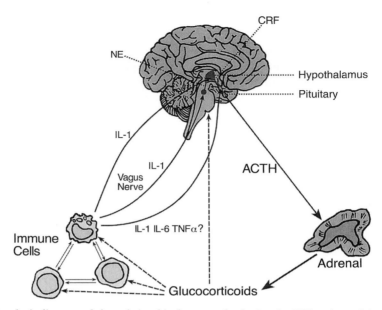

FIG. 1. A diagram of the relationship between the brain, the HPA axis, and immune cells. Interleukin-1 (IL-1) produced peripherally during immune responses activates the hypothalamic-pituitary-adrenocortical (HPA) axis. Release of corticotropin-releasing factor (CRF) occurs in the median eminence region of the hypothalamus and is secreted into the portal blood system. CRF then stimulates the secretion of ACTH from the anterior lobe of the pituitary. The ACTH is carried in peripheral blood to the adrenal cortex where it activates the synthesis and secretion of glucocorticoid hormones. The glucocorticoids provide a negative feedback on cytokine production by lymphocytes. IL-1 may act on circumventricular organs (CVOs) such as the median eminence or the OVLT, or it may activate vagal afferents which in turn stimulate noradrenergic neurons in the brain stem (nucleus tractus solitarius) which innervate the hypothalamus, specifically paraventricular neurons containing CRF. Interleukin-6 (IL-6) and tumor necrosis factor–α (TNE-α) can also activate the HPA axis, but the mechanism(s) are not established.

on specific receptors on corticotrophs resulting in the secretion of adrenocorticotropin (ACTH) into the general circulation. ACTH is transported throughout the body, but as it passes through the adrenal cortex it binds to specific receptors that stimulate the synthesis and subsequent secretion of adrenal glucocorticoids (corticosterone in rodents, cortisol in man and many other species). The glucocorticoids circulate and are ultimately responsible for a host of metabolic and regulatory responses, including the mobilization of glucose stores, the generation of glucose from amino acids, the disposition and distribution of fat in the body, the inhibition of visceral functions, and the modulation of immune system activity.

II. Interleukin-1 (IL-1)

The literature on cytokine signaling of the brain is dominated by studies of interleukin-1 (IL-1). When the structure of IL-1 was determined and the molecule characterized, it was considered to be an endogenous pyrogen, that is, a substance that induces fever (Dinarello, 1984). Another cytokine, interleukin-6 (IL-6) also qualifies as a pyrogen. In fact, IL-6 may be responsible for the pyrogenic activity of IL-1 (Chai *et al.*, 1996; Kozak *et al.*, 1998). TNF-α has been proposed as an endogenous pyrogen, but most of the experimental data suggest that it may be antipyrogenic in rats (Kozak *et al.*, 1995a), and it has little or no effect on body temperature in mice (Wang *et al.*, 1997). It rapidly became apparent that IL-1 had additional physiological activities. McCarthy *et al.* (1985) speculated that IL-1 might account for the reduced appetite that is frequently associated with infections, and went on to show that IL-1 did indeed possess potent anorexic activity (McCarthy *et al.*, 1986). This was the first demonstration of the behavioral activity of IL-1, but we now know that IL-1 is potently active across the behavioral spectrum (see Dantzer *et al.*, 2001). Adding to this in 1986, Besedovsky *et al.* reported that purified recombinant human IL-1β administered ip to rats potently activated the HPA axis. This activity was rapidly confirmed by other researchers and in other species (Berkenbosch *et al.*, 1987; Dunn, 1988; Sapolsky *et al.*, 1987; Uehara and Arimura, 1987). The phenomenon rapidly captured the imagination of many researchers and it is now clear that the ability of IL-1 to activate this stress axis is a critical physiological function of this cytokine. Thus IL-1 has a spectrum of physiological activities (perhaps unique among the cytokines) and can account for many of the changes observed in animals sick from infections or other causes. These responses include fever, endocrine responses, sleep, anorexia, and a variety of other behavioral activities.

IL-1 exists in two forms, IL-1α and IL-1β, which are closely related but have significant differences in their amino acid sequences. There are two receptors for IL-1, type I and type II, but only the type I receptor is thought to be involved in physiological responses. Both IL-1α and IL-1β act on type I receptors, and although IL-1α is somewhat less potent than IL-1β, both induce similar responses.

The available data suggest that there is more than one mechanism by which IL-1 can signal the brain to induce these effects. These may include all the mechanisms listed in Table I. It may be the case that all the possible mechanisms by which IL-1 might signal the brain are functional. Thus the actions of IL-1 on the brain may be a model for the actions of cytokines and other large molecular messengers.

A. Intracerebral and Peripheral Administration of Cytokines

Intracerebral administration of cytokines has been shown to elicit a variety of physiological responses. Because in many cases these effects are very similar to those observed following peripheral administration of cytokines (e.g., fever, HPA activation, behavioral responses), and because these responses could well be initiated by the brain, many researchers leaped to the conclusion that the cytokines somehow penetrated the brain and acted directly on various brain structures, especially in the hypothalamus. This conclusion was bolstered by the observation that in many cases the doses of the cytokines that were effective intracerebroventricularly (icv) were considerably lower than those effective peripherally. However, the effects of intracerebral injection of cytokines are complex. The injection procedure itself causes tissue damage and subsequent inflammation which may result in the generation of inflammatory mediators, including cytokines. In addition the inflammation may cause breakdown of the blood–brain barrier permitting entry of many circulating factors. The local injection of pro-inflammatory cytokines such as IL-1 adds to the inflammation, the net result of which is that many pro- and anti-inflammatory mediators, including cytokines, are produced. Thus it cannot be assumed that intracerebral administration of cytokines simply mimics their local production. It is relevant that most responses to icv IL-1 are delayed in comparison to those after systemic administration, again suggesting that the usual physiological signaling is from the periphery to the brain.

Another problem is the paucity of receptors for IL-1 in the CNS. Although an early report indicated a high density of IL-1 binding to slices of rat brain (Farrar et al., 1987), subsequent studies have failed to support this. Careful experimentation by Haour et al. (1992) and (Takao et al., 1993) failed to find significant binding of radiolabeled IL-1 to rat brain tissue, except on endothelial cells and the choroid plexus. Interestingly, both groups did find significant binding in the mouse brain, but this was limited and almost completely confined to the hippocampus (Ban et al., 1991; Takao et al., 1993). It has been suggested that the number of IL-1 receptors necessary to induce a response is very small (perhaps as little as one per cell) and that this small number might not be detectable in binding studies. In fact, mRNA for IL-1 receptors has been detected in brain tissue, but the significance of this is difficult to assess without expression of the receptors, especially because a small number of necrotic cells could readily explain the presence of this mRNA.

B. IL-1-Induced Changes in Body Temperature

As indicated earlier, an early identified physiological function of IL-1 was its ability to induce fever. Interestingly, this response is relatively slow, and in mice, for example, the initial response to peripherally injected IL-1 is a decrease in body temperature, but this is followed by an increase which is significantly delayed (Long *et al.*, 1992; Wang *et al.*, 1997). Careful analysis indicates that there are multiple phases to the fever response, and it is possible or even likely that each phase is mediated by a different mechanism (Blatteis and Sehic, 1997; Szekely *et al.*, 2000). Changes in body temperature are controlled by the anterior hypothalamus, specifically the preoptic area (POA). The question then is how does IL-1 signal the POA. The potential mechanisms for this response to IL-1 were reviewed by Blatteis and Sehic (1997). According to the calculations of these authors, the amount of IL-1 that could be accumulated in the POA by the cytokine transporter systems would not be sufficient to explain the early febrile response. Because the POA borders the OVLT, the latter is an obvious target. However, studies using OVLT lesions have produced conflicting results, ablating the febrile response to LPS in sheep and guinea pigs (Blatteis and Sehic, 1998), but enhancing it in rats and rabbits (Stitt, 1985). Radiolabeled IL-1 injected into the carotid labeled endothelial cells in the OVLT, but no radioactivity appeared in the POA, suggesting that IL-1 itself did not gain access to the POA, but instead induced secondary mediators within the OVLT. IL-1 and LPS both induce Fos protein in the POA immediately surrounding the OVLT.

It has long been suspected that PGE_2 is the mediator of febrile responses, but it was originally thought that PGE_2 acted directly on the POA. The production of PGE_2 requires cleavage of arachidonic acid from membrane phospholipids and subsequent cyclization by COX. IL-1 and LPS have been shown to increase PGE_2 in the OVLT (Komaki *et al.*, 1992) and this is inhibited by local application of COX inhibitors. PGE_2 is lipophilic and can readily cross cell membranes and thus could diffuse through the tissue to reach the POA, but it is unresolved whether that is the mechanism, or whether there is a neural link from the OVLT to the POA (Blatteis and Sehic, 1997). It is also not known which cells (endothelia, glia, or neurons) provide the COX that makes the PGE_2. COX2 is induced in endothelia in response to IL-1 and LPS (Matsumura *et al.*, 1997; Quan *et al.*, 1998a), and this occurs in the OVLT. This is consistent with the sensitivity of the major component of IL-1-induced fever to selective COX2 inhibitors (Blatteis and Li, 2000) and its absence in COX2 knockout mice (Li *et al.*, 1999). However, the induction of COX2 is too slow to account for the early phases of the fever response, and other mechanisms must be involved. There has been speculation for a role of NE and complement in the more rapid responses (Blatteis *et al.*, 2000).

A number of studies has clearly shown that afferents of the vagus nerve constitute another important mechanism by which IL-1 can elicit fever (Blatteis and Sehic, 1998; Romanovsky *et al.*, 1997; Watkins *et al.*, 1995a).

The hepatic branch of the vagus appears to be the critical one involved in eliciting fever (Romanovsky *et al.*, 1997).

Thus the effects of IL-1 on body temperature may be mediated by mechanisms 2, 3, and 4 in Table I. Whether mechanisms 1 or 5 are involved is not known.

C. IL-1-Induced Activation of the HPA Axis

As previously indicated, strong evidence that peripheral cytokines could affect the central nervous system was derived from the seminal discovery by Besedovsky's group that ip administration of microgram amounts of IL-1β to rats activated the HPA axis (Besedovsky *et al.*, 1986). Although this effect could have been peripheral (i.e., by action on the pituitary or the adrenal glands), as we shall see later, subsequent studies strongly suggested a central mechanism of action. The precise mechanism of this action of IL-1 has been the subject of considerable research and some aspects are still controversial.

A priori, IL-1 could act directly on the adrenal cortex or the anterior pituitary or indirectly via the hypothalamus (Fig. 1). Interestingly, there is evidence for actions of IL-1 at all three levels. The evidence for direct actions on the pituitary and adrenal was derived largely from *in vitro* experiments and is therefore susceptible to artifact. However, the *in vivo* evidence strongly favors a role for hypothalamic CRF as the major mechanism in normal healthy animals (Dunn, 1990; Rivier, 1995; Turnbull and Rivier, 1995). A brief review of this evidence follows.

IL-1 has been shown to stimulate the secretion of corticosterone from the adrenal cortex *in vitro* in some studies (Andreis *et al.*, 1991; Roh *et al.*, 1987; Winter *et al.*, 1990) but not others (Cambronero *et al.*, 1992; Gwosdow *et al.*, 1990; Harlin and Parker, 1991). A direct effect on the adrenal cortex is unlikely to explain the *in vivo* response because IL-1 administration elevates both plasma ACTH and corticosterone in rats (Besedovsky *et al.*, 1986) and mice (Dunn, 2000b). IL-1β elicited increases in plasma ACTH (already elevated by adrenalectomy alone) in rats (Rivier and Vale, 1991) and mice (A. J. Dunn, unpublished observations, 1993). Also, IL-1 failed to induce an increase in plasma corticosterone in hypophysectomized rats (Gwosdow *et al.*, 1990) and mice (Dunn, 1993), although one study did find some response (Andreis *et al.*, 1991). Moreover, the corticosterone response to IL-1 was largely inhibited by treatment with an antibody to CRF in rats (Berkenbosch *et al.*, 1987; Sapolsky *et al.*, 1987; Uehara *et al.*, 1987) and mice (Dunn, 1993).

Many reports have indicated that IL-1 stimulates ACTH release from the anterior pituitary *in vitro* (Beach *et al.*, 1989; Bernton *et al.*, 1987; Kehrer *et al.*, 1988; Malarkey and Zvara, 1989). However, others have failed to find such an effect (Berkenbosch *et al.*, 1987; McGillis *et al.*, 1988; Suda *et al.*, 1989; Tsagarakis *et al.*, 1989). There are earlier reports of IL-1-induced ACTH secretion from AtT20 cells, but studies of abnormal cells as with all *in vitro* studies cannot answer questions concerning *in vivo* mechanisms. Interestingly, the conflicting results may be explained by the observation that prolonged incubation appears to increase the sensitivity to IL-1 while decreasing the response to CRF (Suda *et al.*, 1989). A direct pituitary effect appears to be excluded as the normal physiological mechanism because hypothalamic lesions (Rivest and Rivier, 1991) and CRF-antibody treatment prevented the ACTH and glucocorticoid responses to IL-1 (Berkenbosch *et al.*, 1987; Dunn, 1993; Sapolsky *et al.*, 1987; Uehara *et al.*, 1987).

That the hypothalamus is involved in the HPA response to peripherally administered IL-1 is suggested because lesions of the PVN (Rivest and Rivier, 1991) prevented the ACTH and corticosterone responses. CRF is implicated by the observation that ip IL-1 elevates concentrations of CRF in portal blood (Sapolsky *et al.*, 1987), as well as the apparent release of CRF from the median eminence (Berkenbosch *et al.*, 1987; Watanobe *et al.*, 1991). Also, IL-1 increases the electrophysiological activity of CRF neurons (Saphier and Ovadia, 1990). Moreover, immunoneutralization of CRF prevented the increases in plasma ACTH and corticosterone induced by IL-1 (Berkenbosch *et al.*, 1987; Dunn, 1993; Sapolsky *et al.*, 1987; Uehara *et al.*, 1987). Also, CRF-knockout mice showed only a very small increase in plasma corticosterone after IL-1 administration (Dunn and Swiergiel, 1999). In support of this, IL-1 *in vitro* stimulates CRF release from hypothalamic slabs (Bernardini *et al.*, 1990; Tsagarakis *et al.*, 1989), although such *in vitro* studies cannot provide definitive conclusions.

These studies suggest that the major mechanism of action of IL-1 involves hypothalamic CRF. However, in our studies in mice treated with antibody to CRF (Dunn, 1993), there was a small increase in plasma corticosterone following ip IL-1, and a similar small increase was also observed in CRF-knockout mice (Dunn and Swiergiel, 1999). This suggests that when the HPA axis is impaired, the pituitary or adrenal cortex may gain the ability to mount a small glucocorticoid response.

The normal mechanism by which the HPA axis is activated involves various inputs to the CRF-containing neurons in the PVN. There may be a variety of such inputs, but a major one involves noradrenergic (and possibly adrenergic) projections from the NTS and the locus coeruleus (LC) (see Al-Damluji, 1988). Intraperitoneal administration of IL-1β increased the brain content of the NE catabolite 3-methoxy, 4-hydroxyphenylethyleneglycol (MHPG) in mouse brain, especially in the medial hypothalamus (Dunn, 1988; Zalcman *et al.*, 1994). Dopamine and its catabolites were not significantly affected. This finding was confirmed in rats (Kabiersch *et al.*, 1988), and concordant data were obtained studying NE turnover using metyrosine to block NE synthesis (Terao *et al.*, 1993). Subsequent microdialysis studies have shown increased extracellular concentrations of NE in the medial hypothalamus following intravenous (iv) (Smagin *et al.*, 1996) or ip administration of IL-1 (Ishizuka *et al.*, 1997; Kaur *et al.*, 1998; Merali *et al.*, 1997; Smagin *et al.*, 1996), suggesting that the increases in NE catabolism indeed reflect increased synaptic release of NE.

The IL-1-induced changes in NE closely parallel the increases in plasma corticosterone (Smagin *et al.*, 1996). Surprisingly, adrenoreceptor antagonists did not prevent the IL-1-induced increase in plasma corticosterone. Combined treatment with α- and β-adrenoreceptor antagonists failed to alter the increase in plasma ACTH in response to IL-1 in rats (Rivier, 1995). In mice, the β-adrenoreceptor antagonist, propranolol, did not affect the plasma corticosterone response to IL-1, whereas the α_1-adrenoreceptor antagonist, prazosin, inhibited the response, but not by more than 50%, even at high doses (Dunn, 2001). Nevertheless, lesions of the ventral noradrenergic ascending bundle (VNAB) with 6-hydroxydopamine (6-OHDA) and 6-OHDA injected into the PVN depleted PVN NE by approximately 75% and prevented the increase in plasma corticosterone in response to ip injection of IL-1 in rats (Chuluyan *et al.*, 1992). A similar sensitivity to VNAB 6-OHDA was observed when the IL-1 was administered icv (Weidenfeld *et al.*, 1989). These results suggest that cerebral NE systems are necessary for the HPA response to ip (and icv) IL-1. Consistent with this, the IL-1-induced appearance of Fos in the PVN (which normally accompanies HPA activation) is largely prevented when whole brain NE was depleted with 6-OHDA (Swiergiel *et al.*, 1996). However, when mice were injected icv with 6-OHDA depleting whole brain NE by more than 90%, there was little or no reduction in the plasma corticosterone in response to ip IL-1 (Swiergiel *et al.*, 1996). The reason for the discrepancy between the rat and the mouse results is not known. It could be the species difference, but it is also possible that the direct injection of 6-OHDA into brain tissue causes more nonspecific damage (i.e., to non-noradrenergic systems), and that some of the peptide contransmitters present in NE neurons are involved. Interestingly, the noradrenergic response to IL-1 is more or less prevented by the cyclooxygenase inhibitor, indomethacin, which, however, attenuates but does not prevent the increase in plasma corticosterone (Dunn and Chuluyan, 1992; V. Fillipov and A. J. Dunn, unpublished observations, 2001).

Although these studies strongly suggest the involvement of hypothalamic CRF in the HPA response to IL-1, they do not reveal its site of action. The site of action may well depend on the route of injection. Thus, the HPA response to iv IL-1 occurs very rapidly after injection, dissipates rapidly, and is selectively sensitive to COX inhibitors (Chuluyan *et al.*, 1992). Thus it seems most likely that iv-injected IL-1 acts primarily on a circumventricular organ such as the median eminence or the OVLT. An OVLT site was favored by Katsuura *et al.* (1990) because IL-1 in the OVLT stimulated the synthesis of PGE_2, which in turn may elicit CRF release by stimulating PVN neurons. This hypothesis is consistent with the observations that iv IL-1β administration increased hypothalamic concentrations of PGE_2 as determined by microdialysis, and that hypothalamic infusions of PGE_2 can stimulate the HPA axis (Komaki *et al.*, 1992). On the other hand, the median eminence could be considered a more attractive target because of its proximity to CRF-containing nerve terminals (Rivier, 1995). Sharp's group has proposed that IL-1 may act directly on CRF-containing terminals in the median eminence (Matta *et al.*, 1990). IL-1 injected directly into this region elicits CRF release, as indicated by increases of plasma ACTH, and these effects can be prevented by local administration of the IL-1 receptor antagonist (IL-1ra) (Matta *et al.*, 1993). Interestingly this effect appears to be indirect because icv administration of 6-OHDA prevents this response to IL-1, suggesting that noradrenergic neurons in the median eminence normally regulate the release of CRF. Consistent with this, injection of phenoxybenzamine or propranolol into the median eminence prevented the IL-1-induced elevation of plasma ACTH (Matta *et al.*, 1990).

Yet, another CVO, the area postrema, has been implicated in this response. The area postrema is located quite close to the nucleus tractus solitarius. Lee *et al.* (1998) found that lesions of this structure prevented the induction of *c-fos* mRNA in the PVN and attenuated the increases in plasma ACTH and corticosterone. However, Ericsson *et al.* (1997) found no changes in IL-1-induced *c-fos* expression following area postrema lesions.

The sensitivity to COX inhibitors could be interpreted to mean that COX in endothelial cells is activated, and that PGE_2 (or other COX products) travels through the brain to activate CRF secretion. In either case, this would explain the sensitivity of the HPA response to IL-1 injected intravenously. On the other hand, ip-injected IL-1 probably acts by multiple mechanisms, a rapid response like that to iv IL-1 from the IL-1 that rapidly enters the circulation, and slower responses that involve hypothalamic CRF and perhaps vagal afferents (Dunn and Chuluyan, 1992; Watkins *et al.*, 1995b). Moreover, there may be some slower pituitary effects caused by IL-6 production and some small direct effects on the adrenal cortex.

Thus the effects of IL-1 on the HPA axis may be mediated by mechanisms 2, 3, and 4 in Table I. The roles of mechanisms 1 and 5 are uncertain.

IL-6 and TNF have also been shown to activate the HPA axis, although both are considerably less potent than IL-1—the maximum response is smaller, and the duration is shorter (Ando and Dunn, 1999; Wang and Dunn,

1998). The mechanisms are not known, but there is some evidence that IL-6 may act directly on the pituitary (Dunn, 2000a).

D. IL-1-INDUCED EFFECTS ON BEHAVIOR

Peripheral administration of IL-1 has been shown to affect a variety of different behaviors. For the most part these include what are known as classic sickness behaviors, such as decreased general activity, decreased feeding, decreased exploration of the environment, decreased sexual activity, and increased sleep (for reviews, see Dantzer *et al.*, 2001; Larson and Dunn, 2001). In addition, there may be effects on social behavior (Dantzer *et al.*, 2001), memory (Pugh *et al.*, 2001), and motivation (Larson *et al.*, 2002). Although in most cases IL-1 and LPS have very similar effects, the effects of the latter tend to be more profound, and the two agents should be clearly separated in considering cytokine actions on behavior, because it is clear that IL-1 is not responsible for all of the effects of LPS (Bluthé *et al.*, 2000; Fantuzzi *et al.*, 1996; Kent *et al.*, 1992; Kozak *et al.*, 1998; Swiergiel *et al.*, 1997b). In general the behavioral effects of IL-1 are slower than those discussed earlier. Thus the onset of inhibition of social investigation of a juvenile rat, for example, tends to occur after 2 h or so (Dantzer *et al.*, 2001). Nevertheless, using a model in which mice are allowed to drink sweetened milk for a limited period each day, we have determined that a response to IL-1 can be observed within 30 min (Swiergiel and Dunn, 2002). Such a rapid response would tend to implicate IL-1 uptake or action on a CVO as the mechanism, but the IL-1-induced reduction in milk intake is sensitive to COX inhibitors (Swiergiel *et al.*, 1997a), and more specifically to COX2-selective inhibitors (Dunn and Swiergiel, 2000). Data from COX2 knockout mice are consistent with that mechanism (Swiergiel and Dunn, 2001). However, the induction of COX2 by IL-1 is delayed for at least 1–1.5 h and occurs largely in brain endothelial cells (Matsumura *et al.*, 1998). Additionally, we have determined that the early response of milk drinking to IL-1 is sensitive to COX inhibitors, but that the selectivity is for COX1 (based on selective inhibitors and knockout mice), whereas at later times COX2 appears to be more important (Swiergiel and Dunn, 2002).

The behavioral responses appear to be particularly sensitive to vagotomy. The first indication that the vagus nerve might be involved in a behavioral response was made by Bluthé *et al.* (1994) who observed that the effects of LPS on social investigation of a juvenile rat were impaired by vagotomy. Subsequently, Bret-Dibat *et al.* (1995) found that vagotomy attenuated the reduction in food-motivated behavior induced in mice by IL-1 and LPS, and Bluthé *et al.* (1996) found that the reductions in social behavior in rats induced by IL-1 were similarly attenuated. This was followed by a number of other observations from several laboratories (see reviews by Dantzer *et al.*, 1999; Kent *et al.*, 1996; Watkins *et al.*, 1995b). It is important that in most cases subdiaphragmatic vagotomy only attenuated the responses. This has been attributed to the experimental necessity not to lesion the entire vagus. However, some investigators have found that subdiaphragmatic vagotomy did not alter the hypophagic responses to IL-1 (or LPS) in rats (Laviano *et al.*, 1995; Schwartz *et al.*, 1997). Thus vagal afferents are clearly not the only mechanism for the behavioral responses to IL-1.

The Bordeaux group has proposed that the induction of sickness behavior in response to peripheral IL-1 and LPS involves induction of IL-1 in the CNS mediated by vagal stimulation (Dantzer *et al.*, 1999). This hypothesis was based on the observation that mRNA for IL-1 was induced in the brain following LPS stimulation and that this appearance was prevented by subdiaphragmatic vagotomy (Layé *et al.*, 1995). This was bolstered by the observation that central administration of IL-1ra can prevent certain sickness behavioral responses to peripheral IL-1 and LPS (Dantzer *et al.*, 1999). Specifically, this has been reported for sleep (Imeri *et al.*, 1993), appetitive behavior (Kent *et al.*, 1996), and food and water intake (Linthorst *et al.*, 1995). Interestingly, such effects were also observed with the febrile response (Luheshi *et al.*, 1997). However, fever was induced by LPS in IL-1 knockout (Kozak *et al.*, 1995b) and IL-1-receptor knockout mice (Leon *et al.*, 1996), indicating either that IL-1 is not involved or that there are alternative mechanisms. This is an extraordinarily interesting concept, but one which raises some troubling issues.

Whether or not IL-1 exists in the normal brain is still controversial. Although there have been many reports (e.g., Quan *et al.*, 1996), none of them has attained the standard biochemical criteria for universal acceptance. That would require purification of IL-1 from normal brain involving at least two different chromatographic procedures, followed by at least one good bioassay and one well characterized immunoassay. No purification has been reported, and proponents have relied on immunoassays and immunohistochemistry both of which are notoriously prone to error. Surprisingly, to date the presence of IL-1 in normal brain has not even been demonstrated using Western blots. Demonstration of the existence of mRNA for IL-1 is not definitive, because there are many instances in which mRNA is synthesized but not translated (e.g., for Fos). In the absence of this evidence, it must be concluded that there is very little if any IL-1 in the normal healthy brain.

Nevertheless, when brain tissue is damaged by injury or infection, local production of cytokines can occur, although the relative contributions of invading immune cells and brain cells remain to be determined. Most of the experimental studies have employed LPS. Following peripheral treatment with LPS, there is evidence that IL-1 does appear in discrete cells within the CNS (although even this has not been established by the previous criteria). But, as discussed earlier, the presumed IL-1 immunoreactivity that appears after peripheral administration of LPS or IL-1 is found primarily in microglia and not in neurons (Quan *et al.*, 1998b; Van Dam *et al.*, 1992). Moreover, the IL-1-containing microglia appear first on the surface of the brain, especially in the CVOs, and the IL-1 immunoreactivity detected appears to reflect migration of IL-1-containing microglia, presumably derived from invasion of peripheral macrophages, through the brain parenchyma (Quan *et al.*, 1998b). This raises a number of questions. How does vagal stimulation cause the microglial invasion? Does this occur because of some neural or hormonal signal to the peripheral macrophages? Or, is the effect transmitted to the brain endothelia allowing it to recognize and grant admission to circulating macrophages? How do the microglia know where to go and when to secrete the IL-1? What is the mechanism of the IL-1 secretion? How can the presumed IL-1-containing microglia carry a specific message for brain cells? The idea that glia might talk to neurons is not new, but this mechanism is cumbersome, and it seems to lack any obvious means for specificity. Why, when the neural messages have already reached the brain, do macrophages become involved?

There has been no direct demonstration that it is the IL-1 contained in the migrating microglia that is affected by the injection of IL-1ra into the ventricles. It should be noted that in most cases the intracerebral antagonists only attenuate the responses, and that the presumed effect of the endogenously produced IL-1 is always to depress behavioral activity. Could this indicate a nonspecific (pathological) depression of cerebral activity? Other questions concern the temporal aspects of the responses. The microglial migration is relatively slow, too slow to account for many of the behavioral responses. The maximal appearance of IL-1 in the brain occurs long after the maximal expression of the behavioral responses. Until some of these questions are answered, and plausible mechanisms are proposed, this hypothesized mechanism should be regarded as tenuous.

Thus the effects of IL-1 on behavior may be mediated by mechanisms, 2, 3, 4, and 5 in Table I. The role of cytokine transporters is unknown.

III. Conclusions

The evidence reviewed earlier clearly indicates that several different mechanisms exist for cytokines to signal the brain. Interestingly, several different mechanisms may exist for the same cytokine to induce the same or similar responses. Obviously which of these mechanisms is relevant depends on the cytokine and the response under consideration. The existing evidence almost exclusively relates to effects of IL-1. This is largely because brain effects of IL-6 and TNF and other cytokines are weaker and have been harder to demonstrate than those of IL-1. Despite the evidence from clinical studies of the effects of the interferons, the responses to interferons in animals have proven to be inconsistent and/or difficult to study. Mechanisms 2 (CVOs), 3 (neural afferents), and 4 (peripheral relays) appear to be relevant to the fever-inducing effects of IL-1 as well as to the HPA axis activation. The behavioral effects of IL-1 appear to involve mechanisms 2, 3, 4, and perhaps 5 (immune cell migration). There is currently little evidence directly implicating the uptake of cytokines into the brain by cytokine transporters. Current and future studies will undoubtedly reveal an increase in the documented effects of cytokines on the brain, and these findings may bring with them the recognition of new mechanisms for cytokine signaling of the brain.

References

Al-Damluji, S. (1988). Adrenergic mechanisms in the control of corticotrophin secretion. *J. Endocrinol.* **119,** 5–14.

Ando, T., and Dunn, A. J. (1999). Mouse tumor necrosis factor–α increases brain tryptophan concentrations and norepinephrine metabolism while activating the HPA axis in mice. *Neuroimmunomodulation* **6,** 319–329.

Andreis, P. G., Neri, G., Belloni, A. S., Mazzocchi, G., Kasprzak, A., and Nussdorfer, G. G. (1991). Interleukin-1β enhances corticosterone secretion by acting directly on the rat adrenal gland. *Endocrinology* **129,** 53–57.

Ban, E., Milon, G., Prudhomme, N., Fillion, G., and Haour, F. (1991). Receptors for interleukin-1 (α and β) in mouse brain: Mapping and neuronal localization in hippocampus. *Neuroscience* **43,** 21–30.

Banks, W. A., Kastin, A. J., and Broadwell, R. D. (1995). Passage of cytokines across the blood–brain barrier. *Neuroimmunomodulation* **2,** 241–248.

Beach, J. E., Smallridge, R. C., Kinzer, C. A., Bernton, E. W., Holaday, J. W., and Fein, H. G. (1989). Rapid release of multiple hormones from rat pituitaries perifused with recombinant interleukin-1. *Life Sci.* **44,** 1–7.

Berkenbosch, F., van Oers, J., del Rey, A., Tilders, F., and Besedovsky, H. (1987). Corticotropin-releasing factor–producing neurons in the rat activated by interleukin-1. *Science* **238,** 524–526.

Bernardini, R., Calogero, A. E., Mauceri, G., and Chrousos, G. P. (1990). Rat hypothalamic corticotropin-releasing hormone secretion *in vitro* is stimulated by interleukin-1 in an eicosanoid-dependent manner. *Life Sci.* **47,** 1601–1607.

Bernton, E. W., Beach, J. E., Holaday, J. W., Smallridge, R. C., and Fein, H. G. (1987). Release of multiple hormones by a direct action of interleukin-1 on pituitary cells. *Science* **238,** 519–521.

Besedovsky, H. O., del Rey, A., Sorkin, E., and Dinarello, C. A. (1986). Immunoregulatory feedback between interleukin-1 and glucocorticoid hormones. *Science* **233**, 652–654.

Blatteis, C. M., and Li, S. (2000). Pyrogenic signaling via vagal afferents: What stimulates their receptors? *Autonom. Neurosci.* **85**, 66–71.

Blatteis, C. M., and Sehic, E. (1997). Fever: How may circulating pyrogens signal the brain. *News Physiol. Sci.* **12**, 1–9.

Blatteis, C. M., and Sehic, E. (1998). Cytokines and fever. *Ann. N.Y. Acad. Sci.* **840**, 608–618.

Blatteis, C. M., Sehic, E., and Li, S. (2000). Pyrogen sensing and signaling: Old views and new concepts. *Clin. Infect. Disease* **31**, S168–S177.

Bluthé, R.-M., Walter, V., Parnet, P., Layé, S., Lestage, J., Verrier, D., Poole, S., Stenning, B. E., Kelley, K. W., and Dantzer, R. (1994). Lipopolysaccharide induces sickness behaviour in rats by a vagal mediated mechanism. *C. R. Acad. Sci. Paris* **317**, 499–503.

Bluthé, R.-M., Michaud, B., Kelley, K. W., and Dantzer, R. (1996). Vagotomy attenuates behavioural effects of interleukin-1 injected peripherally but not centrally. *NeuroReport* **7**, 1485–1488.

Bluthé, R.-M., Layé, S., Michaud, B., Combe, C., Dantzer, R., and Parnet, P. (2000). Role of interleukin-1β and tumor necrosis factor-α in lipopolysaccharide-induced sickness behaviour: A study with interleukin-1 type I receptor-deficient mice. *Eur. J. Neurosci.* **12**, 4447–4456.

Breder, C. D., Dinarello, C. A., and Saper, C. B. (1988). Interleukin-1 immunoreactive innervation of the human hypothalamus. *Science* **240**, 321–324.

Bret-Dibat, J.-L., Bluthé, R.-M., Kent, S., Kelley, K. W., and Dantzer, R. (1995). Lipopolysaccharide and interleukin-1 depress food-motivated behavior in mice by a vagal-mediated mechanism. *Brain Behav. Immun.* **9**, 242–246.

Cambronero, J. C., Rivas, F. J., Borrell, J., and Guaza, C. (1992). Is the adrenal cortex a putative site for the action of interleukin-1? *Horm. Metab. Res.* **24**, 48–49.

Chai, Z., Gatti, S., Toniatti, C., Poli, V., and Bartfai, T. (1996). Interleukin (IL)-6 gene expression in the central nervous system is necessary for fever response to lipopolysaccharide or IL-1β: A study on IL-6-deficient mice. *J. Exp. Med.* **183**, 311–316.

Chuluyan, H., Saphier, D., Rohn, W. M., and Dunn, A. J. (1992). Noradrenergic innervation of the hypothalamus participates in the adrenocortical responses to interleukin-1. *Neuroendocrinology* **56**, 106–111.

Dantzer, R., Aubert, A., Bluthé, R.-M., Gheusi, G., Cremona, S., Layé, S., Konsman, J.-P., Parnet P., and Kelley, K. W. (1999). Mechanisms of the behavioural effects of cytokines. *In* "Cytokines, Stress, and Depression" (R. Dantzer, E. E. Wollman, and R. Yirmiya, eds.), pp. 83–105. Kluwer Academic/Plenum, New York.

Dantzer, R., Bluthé, R.-M., Castanon, N., Chauvet, N., Capuron, L., Goodall, G., Kelley, K. W., Konsman, J.-P., Layé, S., Parnet, P., and Pousset, F. (2001). Cytokine effects on behavior. *In* "Psychoneuroimmunology" (R. Ader, D. Felten, and N. Cohen, eds.), pp. 703–727. Academic Press, San Diego.

Dinarello, C. A. (1984). Interleukin-1. *Rev. Infect. Dis.* **6**, 51–95.

Dunn, A. J. (1988). Systemic interleukin-1 administration stimulates hypothalamic norepinephrine metabolism parallelling the increased plasma corticosterone. *Life Sci.* **43**, 429–435.

Dunn, A. J. (1990). Interleukin-1 as a stimulator of hormone secretion. *Prog. NeuroEndocrinImmunol.* **3**, 26–34.

Dunn, A. J. (1993). Role of cytokines in infection-induced stress. *Ann. N.Y. Acad. Sci.* **697**, 189–202.

Dunn, A. J. (2000a). Cytokine activation of the HPA axis. *Ann. N.Y. Acad. Sci.* **917**, 608–617.

Dunn, A. J. (2000b). Effects of the interleukin-1 (IL-1) receptor antagonist on the IL-1- and endotoxin-induced activation of the HPA axis and cerebral biogenic amines in mice. *Neuroimmunomodulation* **7,** 36–45.

Dunn, A. J. (2001). Effects of cytokines and infections on brain neurochemistry. *In* "Psychoneuroimmunology" (R. Ader, D. L. Felten, and N. Cohen, eds.), pp. 649–666. Academic Press, New York.

Dunn, A. J., and Chuluyan, H. (1992). The role of cyclo-oxygenase and lipoxygenase in the interleukin-1-induced activation of the HPA axis: Dependence on the route of injection. *Life Sci.* **51,** 219–225.

Dunn, A. J., and Swiergiel, A. H. (1999). Behavioral responses to stress are intact in CRF-deficient mice. *Brain Res.* **845,** 14–20.

Dunn, A. J., and Swiergiel, A. H. (2000). The role of cyclooxygenases in endotoxin- and interleukin-1-induced hypophagia. *Brain Behav. Immun.* **14,** 141–152.

Ericsson, A., Arias, C., and Sawchenko, P. E. (1997). Evidence for an intramedullary prostaglandin-dependent mechanism in the activation of stress-related neuroendocrine circuitry by intravenous interleukin-1. *J. Neurosci.* **17,** 7166–7179.

Fantuzzi, G., Zheng, H., Faggioni, R., Benigni, F., Ghezzi, P., Sipe, J. D., Shaw, A. R., and Dinarello, C. A. (1996). Effect of endotoxin in IL-1β-deficient mice. *J. Immunol.* **157,** 291–296.

Farrar, W. L., Kilian, P. L., Ruff, M. R., Hill, J. M., and Pert, C. B. (1987). Visualization and characterization of interleukin 1 receptor in brain. *J. Immunol.* **139,** 459–463.

Fleshner, M., Goehler, L. E., Hermann, J., Relton, J. K., Maier, S. F., and Watkins, L. R. (1995). Interleukin-1β induced corticosterone elevation and hypothalamic NE depletion is vagally mediated. *Brain Res. Bull.* **37,** 605–610.

Goehler, L. E., Gaykema, R. P. A., Nguyen, K. T., Lee, J. E., Tilders, F. J. H., Maier, S. F., and Watkins, L. R. (1999). Interleukin-1β in immune cells of the abdominal vagus nerve: A link between the immune and nervous systems? *J. Neurosci.* **19,** 2799–2806.

Goehler, L. E., Gaykema, R. P. A., Hansen, M. K., Anderson, K., Maier, S. F., and Watkins, L. R. (2000). Vagal immune-to-brain communication: A visceral chemosensory pathway. *Autonom. Neurosci.* **85,** 49–59.

Gwosdow, A. R., Kumar, M. S. A., and Bode, H. H. (1990). Interleukin 1 stimulation of the hypothalamic-pituitary-adrenal axis. *Am. J. Physiol.* **258,** E65–E70.

Haour, F., Ban, E., Marquette, C., Milon, G., and Fillion, G. (1992). Brain Interleukin-1 receptors: Mapping, characterization and modulation. *In* "Interleukin-1 in the Brain" (N. J. Rothwell and R. D. Dantzer, eds.), pp. 13–25. Pergamon Press, Oxford.

Harlin, C. A., and Parker, C. R. (1991). Investigation of the effect of interleukin-1β on steroidogenesis in the human fetal adrenal gland. *Steroids* **56,** 72–76.

Imeri, L., Opp, M. R., and Krueger, J. M. (1993). An IL-1 receptor and an IL-1 receptor antagonist attenuate muramyl dipeptide- and IL-1-induced sleep and fever. *Am. J. Physiol.* **265,** R907–R913.

Ishizuka, Y., Ishida, Y., Kunitake, T., Kato, K., Hanamori, T., Mitsuyama, Y., and Kannan, H. (1997). Effects of area postrema lesion and vagotomy on interleukin-1β-induced norepinephrine release in the hypothalamic paraventricular nucleus region in the rat. *Neurosci. Lett.* **223,** 57–60.

Kabiersch, A., del Rey, A., Honegger, C. G., and Besedovsky, H. O. (1988). Interleukin-1 induces changes in norepinephrine metabolism in the rat brain. *Brain Behav. Immun.* **2,** 267–274.

Katsuura, G., Arimura, A., Koves, K., and Gottschall, P. E. (1990). Involvement of organum vasculosum of lamina terminalis and preoptic area in interleukin 1β-induced ACTH release. *Am. J. Physiol.* **258,** E163–E171.

Kaur, D., Cruess, D. F., and Potter, W. Z. (1998). Effect of IL-1α on the release of norepinephrine in rat hypothalamus. *J. Neuroimmunol.* **90**, 122–127.

Kehrer, P., Turnill, D., Dayer, J.-M., Muller, A. F., and Gaillard, R. C. (1988). Human recombinant interleukin-1β and -α, but not recombinant tumor necrosis factor α stimulate ACTH release from rat anterior pituitary cells *in vitro* in a prostaglandin E2 and cAMP independent manner. *Neuroendocrinology* **48**, 160–166.

Kent, S., Kelley, K. W., and Dantzer, R. (1992). Effects of lipopolysaccharide on food-motivated behavior in the rat are not blocked by an interleukin-1 receptor antagonist. *Neurosci. Lett.* **145**, 83–86.

Kent, S., Bret-Dibat, J. L., Kelley, K. W., and Dantzer, R. (1996). Mechanisms of sickness-induced decreases in food-motivated behavior. *Neurosci. Biobehav. Rev.* **20**, 171–175.

Komaki, G., Arimura, A., and Koves, K. (1992). Effect of intravenous injection of IL-1β on PGE2 levels in several brain areas as determined by microdialysis. *Am. J. Physiol.* **262**, E246–E251.

Kozak, W., Conn, C. A., Klir, J. J., Wong, G. H. W., and Kluger, M. J. (1995a). TNF soluble receptor and antiserum against TNF enhance lipopolysaccharide fever in mice. *Am. J. Physiol.* **269**, R23–R29.

Kozak, W., Zheng, H., Conn, C. A., Soszynski, D., Van Der Ploeg, L. H. T., and Kluger, M. J. (1995b). Thermal and behavioral effects of lipopolysaccharide and influenza in interleukin-1β-deficient mice. *Am. J. Physiol.* **38**, R969–R977.

Kozak, W., Kluger, M. J., Soszynski, D., Conn, C. A., Rudolph, K., Leon, L. R., and Zheng, H. (1998). IL-6 and IL-1β in fever. Studies using cytokine-deficient (knockout) mice. *Ann. N.Y. Acad. Sci.* **856**, 33–47.

Larson, S. J., and Dunn, A. J. (2001). Behavioral effects of cytokines. *Brain Behav. Immun.* **15**, 371–387.

Larson, S. J., Romanoff, R. L., Dunn, A. J., and Glowa, J. R. (2002). Effects of interleukin-1β on food-maintained behavior in the mouse. *Brain Behav. Immun.* **16**, 398–410.

Laviano, A., Yang, Z. J., Meguid, M. M., Koseki, M., and Beverly, J. L. (1995). Hepatic vagus does not mediate IL-1α induced anorexia. *NeuroReport* **6**, 1394–1396.

Layé, S., Bluthé, R. M., Kent, S., Combe, C., Médina, C., Parnet, P., Kelley, K., and Dantzer, R. (1995). Subdiaphragmatic vagotomy blocks induction of IL-1β mRNA in mice brain in response to peripheral LPS. *Am. J. Physiol.* **268**, R1327–R1331.

Lee, H. Y., Whiteside, M. B., and Herkenham, M. (1998). Area postrema removal abolishes stimulatory effects of intravenous interleukin-1β on hypothalamic-pituitary-adrenal axis activity and c-*fos* mRNA in the hypothalamic paraventricular nucleus. *Brain Res. Bull.* **46**, 495–503.

Lee, S. C., Liu, W., Dickson, D. W., Brosnan, C. F., and Berman, J. W. (1993). Cytokine production by human fetal microglial and astrocytes—Differential induction by lipopolysaccharide and IL-1β. *J. Immunol.* **150**, 2659–2667.

Leon, L. R., Conn, C. A., Glaccum, M., and Kluger, M. J. (1996). IL-1 type I receptor mediates acute-phase response to turpentine, but not lipopolysaccharide, in mice. *Am. J. Physiol.* **40**, R1668–R1675.

Li, S., Wang, Y., Matsumura, K., Ballou, L. R., Morham, S. G., and Blatteis, C. M. (1999). The febrile response to lipopolysaccharide is blocked in cyclooxygenase-2-/-, but not in cyclooxygenase-1-/- mice. *Brain Res.* **825**, 86–94.

Linthorst, A. C. E., Flachskamm, C., Müller-Preuss, P., Holsboer, F., and Reul, J. M. H. M. (1995). Effect of bacterial endotoxin and interleukin-1β on hippocampal serotonergic neurotransmission, behavioral activity, and free corticosterone levels: An *in vivo* microdialysis study. *J. Neurosci.* **15**, 2920–2934.

Long, N. C., Morimoto, A., Nakamori, T., and Murakami, N. (1992). Systemic injection of TNF-α attenuates fever due to IL-1β LPS in rats. *Am. J. Physiol.* **263**, R987–R991.

Luheshi, G., Stefferl, A., Turnbull, A. V., Dascombe, M. J., Brouwer, S., Hopkins, S. J., and Rothwell, N. J. (1997). Febrile response to tissue inflammation involves both peripheral and brain IL-1 and TNF-α in the rat. *Am. J. Physiol.* **272,** R862–R868.

Malarkey, W. B., and Zvara, B. J. (1989). Interleukin-1β and other cytokines stimulate adreno-corticotropin release from cultured pituitary cells of patients with Cushing's disease. *J. Clin. Endocrinol. Metab.* **69,** 196–199.

Matsumura, K., Cao, C., and Watanabe, Y. (1997). Possible role of cyclooxygenase-2 in the brain vasculature in febrile response. *Ann. N.Y. Acad. Sci.* **813,** 302–306.

Matsumura, K., Cao, C., Ozaki, M., Morii, H., Nakadate, K., and Watanabe, Y. (1998). Electron microscopic evidence for induction of cyclooxygenase-2 in brain endothelial cells. *Ann. N.Y. Acad. Sci.* **856,** 278–285.

Matta, S. G., Singh, J., Newton, R., and Sharp, B. M. (1990). The adrenocorticotropin response to interleukin-1β instilled into the rat median eminence depends on the local release of catecholamines. *Endocrinology* **127,** 2175–2182.

Matta, S. G., Linner, K. M., and Sharp, B. M. (1993). Interleukin-1-α and interleukin-1-β stimu-late adrenocorticotropin secretion in the rat through a similar hypothalamic receptor(s): Effects of interleukin-1 receptor antagonist protein. *Neuroendocrinology* **57,** 14–22.

McCarthy, D. O., Kluger, M. J., and Vander, A. J. (1985). Suppression of food intake during infections: is interleukin-1 involved? *Amer. J. Clin. Nutr.* **42,** 1179–1182.

McCarthy, D. O., Kluger, M. J., and Vander, A. J. (1986). Effect of centrally administered interleukin-1 and endotoxin on food intake of fasted rats. *Physiol. Behav.* **36,** 745–749.

McGillis, J. P., Hall, N. R., and Goldstein, A. L. (1988). Thymosin fraction 5 (TF5) stimulates secretion of adrenocorticotropic hormone (ACTH) from cultured rat pituitaries. *Life Sci.* **42,** 2259–2268.

Merali, Z., Lacosta, S., and Anisman, H. (1997). Effects of interleukin-1β and mild stress on alterations of norepinephrine, dopamine and serotonin neurotransmission: A regional microdialysis study. *Brain Res.* **761,** 225–235.

Pugh, C. R., Fleshner, M., Watkins, L. R., Mater, S. F., and Rudy, J. W. (2001). The immune system and memory consolidation: A role for cytokine IL-1β. *Neurosci. Biobehav. Rev.* **25,** 29–41.

Quan, N., Zhang, Z., Emery, M., Bonsall, R., and Weiss, J. M. (1996). Detection of interleukin-1 bioactivity in various brain regions of normal healthy rats. *Neuroimmunomodulation* **3,** 47–55.

Quan, N., Whiteside, M., and Herkenham, M. (1998a). Cyclooxygenase 2 mRNA expression in rat brain after peripheral injection of lipopolysaccharide. *Brain Res.* **802,** 189–197.

Quan, N., Whiteside, M., and Herkenham, M. (1998b). Time course and localization patterns of interleukin-1β messenger RNA expression in brain and pituitary after peripheral ad-ministration of lipopolysaccharide. *Neuroscience* **83,** 281–293.

Rivest, S., and Rivier, C. (1991). Influence of the paraventricular nucleus of the hypothala-mus in the alteration of neuroendocrine functions induced by intermittent footshock or interleukin. *Endocrinology* **129,** 2049–2057.

Rivier, C. (1995). Influence of immune signals on the hypothalamic-pituitary axis of the rodent. *Front. Neuroendocrinol.* **16,** 151–182.

Rivier, C., and Vale, W. (1991). Stimulatory effect of interleukin-1 on adrenocorticotropin secretion in the rat: Is it modulated by prostaglandins? *Endocrinology* **129,** 384–388.

Roh, M. S., Drazenovich, K. A., Barbose, J. J., Dinarello, C. A., and Cobb, C. F. (1987). Direct stimulation of the adrenal cortex by interleukin-1. *Surgery* **102,** 140–146.

Romanovsky, A. A., Simons, C. T., Székely, M., and Kulchitsky, V. A. (1997). The vagus nerve in the thermoregulatory response to systemic inflammation. *Am. J. Physiol.* **273,** R407–R413.

Saphier, D., and Ovadia, H. (1990). Selective facilitation of putative corticotropin-releasing factor–secreting neurones by interleukin-1. *Neurosci. Lett.* **114**, 283–288.

Sapolsky, R., Rivier, C., Yamamoto, G., Plotsky, P., and Vale, W. (1987). Interleukin-1 stimulates the secretion of hypothalamic corticotropin-releasing factor. *Science* **238**, 522–524.

Schwartz, G. J., Plata-Salaman, C. R., and Langhans, W. (1997). Subdiaphragmatic vagal deafferentation fails to block feeding-suppressive effects of LPS and IL-1β in rats. *Am. J. Physiol.* **273**, R1193–R1198.

Smagin, G. N., Swiergiel, A. H., and Dunn, A. J. (1996). Peripheral administration of interleukin-1 increases extracellular concentrations of norepinephrine in rat hypothalamus: Comparison with plasma corticosterone. *Psychoneuroendocrinology* **21**, 83–93.

Stitt, J. T. (1985). Evidence for the involvement of the organum vasculosum laminae terminalis in the febrile response of rabbits and rats. *J. Physiol.* **368**, 501–511.

Suda, T., Tozawa, F., Ushiyama, T., Tomori, N., Sumitomo, T., Nakagami, Y., Yamada, M., Demura, H., and Shizume, K. (1989). Effects of protein-kinase-C-related adrenocorticotropin secretagogues and interleukin-1 on proopiomelanocortin gene expression in rat anterior pituitary cells. *Endocrinology* **124**, 1444–1449.

Swiergiel, A. H., and Dunn, A. J. (2001). Cyclooxygenase 1 is not essential for hypophagic responses to interleukin-1 and endotoxin in mice. *Pharmacol. Biochem. Behav.* **69**, 659–663.

Swiergiel, A. H., and Dunn, A. J. (2002). Distinct roles for cyclooxygenases 1 and 2 in interleukin-1-induced hypophagia. *J. Pharmacol. Exptl. Therap.*, in press.

Swiergiel, A. H., Dunn, A. J., and Stone, E. A. (1996). The role of cerebral noradrenergic systems in the Fos response to interleukin-1. *Brain Res. Bull.* **41**, 61–64.

Swiergiel, A. H., Smagin, G. N., and Dunn, A. J. (1997a). Influenza virus infection of mice induces anorexia: Comparison with endotoxin and interleukin-1 and the effects of indomethacin. *Pharmacol. Biochem. Behav.* **57**, 389–396.

Swiergiel, A. H., Smagin, G. N., Johnson, L. J., and Dunn, A. J. (1997b). The role of cytokines in the behavioral responses to endotoxin and influenza virus infection in mice: Effects of acute and chronic administration of the interleukin-1-receptor antagonist (IL-1ra). *Brain Res.* **776**, 96–104.

Szekely, M., Balasko, M., Kulchitsky, V. A., Simons, C. T., Ivanov, A. I., and Romanovsky, A. A. (2000). Multiple neural mechanisms of fever. *Autonom. Neurosci.* **85**, 78–82.

Takao, T., Newton, R. C., and De Souza, E. B. (1993). Species differences in [^{125}I]interleukin-1 binding in brain, endocrine and immune tissues. *Brain Res.* **623**, 172–176.

Terao, A., Oikawa, M., and Saito, M. (1993). Cytokine-induced change in hypothalamic norepinephrine turnover: Involvement of corticotropin-releasing hormone and prostaglandins. *Brain Res.* **622**, 257–261.

Tsagarakis, S., Gillies, G., Rees, L. H., Besser, M., and Grossman, A. (1989). Interleukin-1 directly stimulates the release of corticotrophin-releasing factor from rat hypothalamus. *Neuroendocrinology* **49**, 98–101.

Turnbull, A. V., and Rivier, C. (1995). Regulation of the HPA axis by cytokines. *Brain Behav. Immun.* **9**, 253–275.

Uehara, A., and Arimura, A. (1987). Involvement of interleukin-1 in the hypothalamic/pituitary/adrenocortical axis. *Clin. Res.* **35**, 37A.

Uehara, A., Gottschall, P. E., Dahl, R. R., and Arimura, A. (1987). Interleukin-1 stimulates ACTH release by an indirect action which requires endogenous corticotropin-releasing factor. *Endocrinology* **121**, 1580–1582.

Van Dam, A. M., Brouns, M., Louisse, S., and Berkenbosch, F. (1992). Appearance of interleukin 1 in macrophages and in ramified microglia in the brain of endotoxin-treated rats: A pathway for the induction of non-specific symptoms of sickness? *Brain Res.* **588**, 291–296.

Wan, W., Wetmore, L., Sorensen, C. M., Greenberg, A. H., and Nance, D. M. (1994). Neural and biochemical mediators of endotoxin and stress-induced *c-fos* expression in the rat brain. *Brain Res. Bull.* **34,** 7–14.

Wang, J. P., and Dunn, A. J. (1998). Mouse interleukin-6 stimulates the HPA axis and increases brain tryptophan and serotonin metabolism. *Neurochem. Int.* **33,** 143–154.

Wang, J. P., Ando, T., and Dunn, A. J. (1997). Effect of homologous interleukin-1, interleukin-6 and tumor necrosis factor α on the core body temperature of mice. *Neuroimmunomodulation* **4,** 230–236.

Watanobe, H., Sasaki, S., and Takabe, K. (1991). Evidence that intravenous administration of interleukin-1 stimulates corticotropin-releasing hormone secretion in the median eminence of freely moving rats: Estimation by push–pull perfusion. *Neurosci. Lett.* **133,** 7–10.

Watkins, L. R., Goehler, L. E., Relton, J. K., Tartaglia, N., Silbert, L., Martin, D., and Maier, S. F. (1995a). Blockade of interleukin-1 induced hyperthermia by subdiaphragmatic vagotomy: Evidence for vagal mediation of immune–brain communication. *Neurosci. Lett.* **183,** 27–31.

Watkins, L. R., Maier, S. F., and Goehler, L. E. (1995b). Cytokine-to-brain communication: A review and analysis of alternative mechanisms. *Life Sci.* **57,** 1011–1026.

Weidenfeld, J., Abramsky, O., and Ovadia, H. (1989). Evidence for the involvement of the central adrenergic system in interleukin 1-induced adrenocortical response. *Neuropharmacology* **28,** 1411–1414.

Winter, J. S. D., Gow, K. W., Perry, Y. S., and Greenberg, A. H. (1990). A stimulatory effect of interleukin-1 on adrenocortical cortisol secretion mediated by prostaglandins. *Endocrinology* **127,** 1904–1909.

Zalcman, S., Green-Johnson, J. M., Murray, L., Nance, D. M., Dyck, D., Anisman, H., and Greenberg, A. H. (1994). Cytokine-specific central monoamine alterations induced by interleukin-1, -2 and -6. *Brain Res.* **643,** 40–49.

NEUROPEPTIDES: MODULATORS OF IMMUNE RESPONSES IN HEALTH AND DISEASE

David S. Jessop

University Research Center for Neuroendocrinology, University of Bristol
Bristol BS2 8HW, United Kingdom

I. Introduction

Activation of the immune system is a critical defense in response to life-threatening foreign pathogens. Just as critical are the regulatory mechanisms which tune an immune response according to the magnitude of the antigenic challenge and prevent excessive and potentially damaging release of inflammatory agents. This control system is coordinated through circulating cytokines, glucocorticoids, and catecholamines, which compose a bidirectional homeostatic network of brain–immune interactions. In addition to systemic and neuronal control, immune responses are also under paracrine regulation by neuropeptides at sites of inflammation. Expression and release of neuropeptides from neurons or leukocytes within inflamed tissue is increased in many autoimmune diseases, and neuropeptides can alter immune functions *in vivo* and *in vitro*.

Despite overwhelming evidence for key physiological roles, immunoneuropeptides have not always enjoyed a good press, and modern textbooks of immunology either give scant attention to these compounds or ignore them entirely. This chapter will highlight the principal contentious issues which have arisen in this area of research and offer explanations to harmonize apparently inconsistent data from different laboratories. The intention is not to provide an inventory of all neuropeptides in the immune system but to critically analyze some functional roles which selected immunoneuropeptides might play and to explore potential mechanisms for facilitation of these actions. Evidence for important immunoregulatory functions will be presented for neuropeptides in (1) control of inflammation in autoimmune diseases, and (2) local modulation of immune functions during stress.

II. CRH and Related Peptides

In addition to its widespread distribution in the central nervous system (CNS), corticotropin-releasing hormone (CRH) is present in peripheral lymphocytes (Stephanou *et al.*, 1990; Aird *et al.*, 1993), principally in macrophages (Brouxhon *et al.*, 1998) although also in T and B cells. CRH expression in the immune system is responsive to acute and chronic stressors, notably in rodent models of inflammation such as adjuvant arthritis (AA) (Jessop *et al.*, 1995a), experimental allergic encephalomyelitis (EAE), and systemic lupus erythematosis (SLE) (Jessop, 2001). CRH bioactivity (Hargreaves *et al.*, 1989), mRNA, and peptide levels (Crofford *et al.*, 1992) are elevated in inflamed rat synovial tissues and in synovium from patients with rheumatoid arthritis (RA) (Crofford *et al.*, 1993). CRH in synovial tissue is probably derived both from infiltrating macrophages and from neuronal input. Dramatic upregulation of CRH-R1 receptors in mouse splenocytes was observed following injection of lipopolysaccharide (Radulovic *et al.*, 1999), and CRH-R1 receptors within inflamed synovial tissue were increased in arthritic rats (Mousa *et al.*, 1996). Levels of CRH in synovial fluid from patients with RA correlated positively with the acute-phase protein serum amyloid A (Nishioka *et al.*, 1996). This evidence for an early involvement of CRH in the onset of inflammation is corroborated by reports of increased expression of CRH-R1 (McEvoy *et al.*, 2001) and CRH mRNA (Murphy *et al.*, 2001) in synovial cells from recently diagnosed RA patients.

Other than sensitivity to stress and inflammation, there is little information about mechanisms which control immune CRH expression. The CRH gene contains glucocorticoid and estrogen response elements, but few studies have been undertaken into immune CRH sensitivity to glucocorticoids and estrogens. Interleukin-(IL)-1β stimulation of CRH expression in the thymus, but not in the spleen, is dependent on the presence of circulating glucocorticoids (Jessop *et al.*, 1997b). Rat spleen and thymic macrophages positive for CRH immunoreactivity exist in proximity to noradrenergic fibers (Brouxhon *et al.*, 1998), but the degree of sympathetic control over immune CRH expression is unknown.

In contrast to the anti-inflammatory effects of hypothalamic CRH through stimulation of pituitary ACTH and adrenal glucocorticoid secretion, evidence has emerged that immune CRH can act as a pro-inflammatory peptide (Karalis *et al.*, 1997). CRH stimulates lymphocyte proliferation and cytokine release (Singh, 1989; Singh and Leu, 1990), immunoneutralization of peripheral CRH reduced inflammation associated with carrageenan-induced arthritis in rats (Karalis *et al.*, 1991), and oligonucleotide probe blockade of CRH mRNA translation inhibited splenocyte proliferation (Jessop *et al.*, 1997a). In a mouse transgenic model, overexpression of hypothalamic CRH was associated with a reduction in B cells (Stenzel-Poore *et al.*, 1996) and an increase in T lymphocytes (Stenzel-Poore *et al.*, 1996; Boehme *et al.*, 1997), while CRH deficiency was accompanied by increased circulating IL-6 (Venihaki *et al.*, 2001). It is difficult, however, to assign a direct role to immune-derived CRH in these models since manipulation of central CRH expression will alter catecholamine and glucocorticoid release, both of which are immunomodulatory.

Although there is substantial evidence for pro-inflammatory effects of CRH in animal models of arthritis, CRH infusion did not alter the severity of inflammation in AA (Harbuz *et al.*, 1996). Furthermore, peripheral injection of CRH prevented inflammation in a rat model of EAE (Poliak *et al.*, 1997). Since CRH was protective in adrenalectomized rats in this study, its effects may be exerted through direct influence on the immune system, or via pituitary ACTH, rather than through corticosterone release. To reconcile these conflicting results, CRH may exert pro- or anti-inflammatory properties depending on the dose of CRH, the type (and severity) of inflammation, and the timing of administration relative to disease onset. It should also be considered that the effects of CRH injection on inflammation may not be specific but may be due to stress. CRH is a potent activator of the pituitary-adrenal axis, and application of a stressor, in the form of an acute immunological challenge, prior to onset of disease can prevent inflammation (Harbuz *et al.*, 2002).

CRH may act to modulate inflammatory pathways in synergy with substance P (SP). CRH and its receptors are located in sympathetic efferent neurons and primary afferent nociceptors (PAN) (Elenkov and Chrousos, 1999; Udelsman *et al.*, 1986) where CRH is colocalized with SP in capsaicin-sensitive fibers (Skofitsch *et al.*, 1984). SP, released from both PAN and immune cells, is an important mediator of the acute inflammatory response through stimulation of histamine secretion from mast cells and cytokines from monocytes. CRH also activates mast cells and monocytes (Elenkov and Chrousos, 1999), and there is potential for SP stimulation of CRH release from macrophages, thereby potentiating release of histamine and cytokines.

In addition to mediating inflammation, both SP (Chancellor-Freeland *et al.,* 1995) and CRH (Jessop *et al.,*1997b) in immune cells are responsive to stressors and may interact to mediate the multiple effects of stress on immune function (Jessop *et al.,* 2001). Establishing a functional relationship between CRH and SP in the onset and severity of inflammation may lead to the development of SP and CRH antagonists as anti-inflammatory agents, perhaps acting in synergy. The challenge will be to design a CRH antagonist which is effective at the site of inflammation without passing into the blood where it has the potential to exert pro-inflammatory effects by inhibiting HPA axis activity at the anterior pituitary.

Urocortin (Ucn), another member of the peptide CRH family, binds with high affinity to type 1 and type 2 CRH receptors, both classes of which are present in the immune system (Baigent, 2001). Ucn mRNA has been detected in rat spleen and thymus by riboprobe hybridization (Kageyama *et al.,* 1999) but not by polymerase chain reaction using Ucn cDNA-specific PCR primers (Park *et al.,* 2000). Upregulation of Ucn mRNA and immunoreactivity in synovial tissues from RA patients has been reported (Kohno *et al.,* 2001; Uzuki *et al.,* 2001) but these studies did not determine whether synovial Ucn is of immune or neural origin. There is still a question about whether immune cells produce authentic Ucn, since some groups have failed to locate either Ucn immunoreactivity or mRNA in the immune system. Chromatographic studies have revealed multiple CRH-like peptides in immune tissues which may be novel products of the CRH or Ucn genes or even undiscovered genes of the CRH family. Assumptions that all immunomodulatory actions of the CRH family can be ascribed to CRH and Ucn alone may not be sustained. Further characterization of the physico- and biochemical properties of CRH and Ucn isoforms, and their affinities for CRH receptor subgroups, is necessary if this research is to lead to development of small CRH/Ucn analogs as modulators of inflammation.

III. Pro-opiomelanocortin (POMC) Peptides

Pioneering work by Blalock (1999) and colleagues established that POMC peptides ACTH and β-endorphin are expressed within the immune system. Research in this area has not progressed without controversy, with widely variable amounts and forms of immunoreactivity reported between laboratories. Such discrepancies may be due in part to differing degrees of lymphocyte activation in experimental systems. Only activated T and B lymphocytes contain significant amounts of ACTH, while the expression of ACTH in macrophages is constitutive and independent of cell activation

(Lyons and Blalock, 1995). β-Endorphin contents in a range of lymphocyte subtypes also vary widely throughout the course of inflammation (Rittner *et al.*, 2001). Discrepancies have also arisen from differential POMC precursor processing in immune tissues into forms other than those which exist within the pituitary. Although it has now been convincingly demonstrated that the immune system is capable of producing full-length POMC mRNA and ACTH (1–39) peptide (Lyons and Blalock, 1997), immune tissues also contain a wide range of high molecular weight forms and small fragments of ACTH and β-endorphin (Lolait *et al.*, 1986; Buzzetti *et al.*, 1989; Jessop *et al.*, 1994). Due to differing methods of measurement, levels of total immunoreactivity and chromatographic profiles are not always comparable between laboratories.

Anti-inflammatory effects of ACTH, through stimulation of glucocorticoids, have been recognized for decades but ACTH and other POMC peptides can be anti-inflammatory in their own right. A considerable range of immunomodulatory effects have been reported in studies using synthetic ACTH, β-endorphin, or α-melanocyte-stimulating hormone (α-MSH) (Jessop, 1998). In most of these reports, POMC peptides attenuate immune cell proliferation or block the actions of inflammatory cytokines. The literature, however, is inconsistent, and reports of biological effects of POMC peptides should be interpreted according to whether the concentration employed is in the range reported within immune tissues, which is frequently not the case with *in vitro* studies. Immune cell status is also critical. The biological effects of ACTH are dependent on the degree of cell activation, since activated lymphocytes in culture display a substantially increased number of high affinity ACTH binding sites (Clarke and Bost, 1989; Johnson *et al.*, 2001). *In vivo* determination of the degree of POMC expression in immune tissues, and cellular responses to synthetic POMC peptides, can depend on many factors such as the immunological history of the animal, transportation from breeding facilities, breeding and housing conditions, and prior exposure to stress. Data from *in vitro* cell culture studies can be influenced by the degree of cell activation, the type of antigen employed to activate cells, and the culture milieu, there being a particular requirement for a critical number of CD4$^+$ cells (Gaillard *et al.*, 1998). Discrepant reports in the literature on the immunomodulatory effects of β-endorphin may also be explained by immune status. Activation of T cells by cytokines secreted during inflammatory autoimmune diseases and during infection dramatically increases the affinity and number of opioid binding sites (Roy *et al.*, 1991). The well-recognized biphasic effects of β-endorphin (Heijnen *et al.*, 1987; Williamson *et al.*, 1988; Pasnik *et al.*, 1999) may also lead to empirical confusion, since, depending on the concentration and *in vitro* conditions, β-endorphin may inhibit, stimulate, or leave unaffected cell proliferation and activity. Although most reports highlight the anti-inflammatory effects of POMC peptides, one study which used oligonucleotide antisense cDNA probes to inhibit total POMC peptide translation in splenocytes *in vitro* resulted in increased mitosis (Fulford *et al.*, 2000), which suggests a net proliferative synergy of POMC peptides in this experimental system.

Despite these caveats, there have been promising reports that POMC peptides may form the basis of anti-inflammatory strategies, acting through melanocortin (MC) or opioid receptors. MC-1 and MC-3 receptors have been reported only in macrophages and monocytes and not in resting T and B lymphocytes (Smith *et al.*, 1987; Star *et al.*, 1995; Bhardwaj *et al.*, 1997; Getting *et al.*, 1999). MC-2 receptors are present in some, but not all, T and B cells and macrophages (Johnson *et al.*, 2001). However, it is not possible to be categorical about whether or not a cell subtype expresses a class of receptors, since receptors which are not expressed on resting leukocytes may be upregulated during antigen presentation or inflammation (Johnson *et al.*, 2001). ACTH (4–10) has an antiproliferative effect on mouse macrophages that can be blocked by an MC-3 receptor antagonist (Getting *et al.*, 1999), and MC-3 agonists can attenuate neutrophil migration and macrophage secretory activity in a mouse model of inflammation (Getting *et al.*, 2001). α-MSH is a potent antagonist of IL-β activity (Lipton and Catania, 1998) and can induce expression of the Th2 anti-inflammatory cytokine IL-10 through α-MSH-specific MC-1 receptors on monocytes (Bhardwaj *et al.*, 1996). Use of MC-1 and MC-3 agonists as anti-inflammatory agents has the advantage that the pituitary MC-2 receptor involved in steroidogenesis will not be activated. Therefore systemic application of these compounds in the treatment of chronic inflammatory disorders need not result in hypercortisolemia and associated problems such as obesity, hypertension, and depression.

μ-Opioid receptors are located in all major leukocyte subtypes and there are many reports of immunomodulatory effects of opioid peptides (Jessop, 1998). In one study, β-endorphin prolonged skin allograft survival in mice whereas naloxone accelerated rejection (Sacerdote *et al.*, 1998), demonstrating that endogenous β-endorphin can tonically inhibit the immune response in tissue transplantation. Although many of the effects of opioid peptides are naloxone-reversible and therefore mediated through δ-, κ-, or μ-opioid receptors, there is a body of evidence in support of opioid effects which cannot be explained through classical receptors. Naloxone specifically blocks N-terminal β-endorphin receptor binding, but β-endorphin may also act through C-terminal binding to a naloxone-insensitive class of opioid receptors (McCain *et al.*, 1982; Shahabi *et al.*, 1990; Owen *et al.*, 1998). Morphine has been shown to bind to activated T lymphocytes but no trace of μ-opioid receptor mRNA was detected, suggesting a novel morphine binding site on T cells (Madden *et al.*, 2001). Evidence for novel receptors in immune cells highlights the difficulties in investigating functional mechanisms mediating neuropeptide effects on immune function using pharmacological tools developed for investigation of neuropeptides within the CNS.

Many studies have demonstrated changes in POMC expression in immune tissues during inflammation. POMC mRNA and peptide products are elevated in rat immune tissues in inflammatory autoimmune diseases such as AA, EAE, and SLE (Jessop, 2001). In a rat model of AA, thymic contents of ACTH were elevated in response to inflammation whereas ACTH in the spleen was increased well before clinical signs of inflammation (Jessop *et al.*, 1995a), suggesting tissue-specific control of POMC expression. In contrast to the early increase in splenic ACTH, β-endorphin contents in the spleen were only elevated around the time of inflammation, which is consistent with differential POMC processing in the spleen and possible separate functions for β-endorphin and ACTH in modulating the course of AA. Increased β-endorphin immunoreactivity has been observed in inflamed synovial tissue (Przewlocki *et al.*, 1992). Although β-endorphin contents are increased in inflamed tissues, lymph node contents of β-endorphin are decreased which may reflect differential immunocyte trafficking in response to inflammation (Cabot *et al.*, 1997).

There have been few reports on immune ACTH expression in human pathologies, but biologically active ACTH is produced by HIV-infected lymphocytes in culture (Hashemi *et al.*, 1998), and β-endorphin levels in mononuclear leukocytes are reduced in patients with chronic fatigue syndrome (Conti *et al.*, 1998). Alterations in peripheral human blood lymphocyte contents of β-endorphin in the predominantly Th1-mediated diseases multiple sclerosis, RA, and Crohn's disease have led to the hypothesis that β-endorphin may be influential in mediating the switch from the Th1 to the Th2 profile of predominantly anti-inflammatory cytokines (Panerai and Sacerdote, 1997). Naloxone reduced production of IL-4 and increased secretion of IL-2 and IFN-γ, a cytokine secretory profile which is consistent with an endogenous opioid-induced Th2 switch (Sacerdote *et al.*, 2000). β-Endorphin inhibited secretion of IL-6 from mouse spleen slices (Straub *et al.*, 1998) and IL-8 from synovial fibroblasts taken from RA patients (Raap *et al.*, 2000), demonstrating that β-endorphin can attenuate important immune responses such as antibody production, neutrophil activation, and T-cell chemotaxis. Leukocyte production of β-endorphin is stimulated by the pro-inflammatory cytokine IL-1 (Kavelaars *et al.*, 1989) and also by the Th2 cytokine IL-4 (Kraus *et al.*, 2001), indicating that β-endorphin can respond to changes in the Th1/Th2 cytokine balance, a balance which is highly sensitive to modulation by catecholamines and glucocorticoids released by stress.

An alternative anti-inflammatory mode of action for opioids is the induction of apoptosis in immune cells. Morphine was able to induce DNA fragmentation in mouse thymocytes *in vivo* but not *in vitro* (Fuchs and Pruett, 1993), suggesting an indirect effect, whereas other *in vitro* studies have shown that β-endorphin enhances spermidine transport (Ientile *et al.*, 1997) and morphine increases Fas expression (Yin *et al.*, 1999). Induction of cell death by apoptosis through the Fas/Fas ligand pathway would be a potent anti-inflammatory mechanism for immune-based opioid peptides, but this mechanism remains controversial since a number of groups have failed to demonstrate any effects of opioids on apoptosis (Jessop *et al.*, 2002). Research into possible interactions of opioids with TNF-α in the NF-κB–mediated antiapoptotic pathway might prove fruitful.

POMC peptides in the immune system, as in the anterior pituitary, are responsive to acute stressors such as restraint (Jessop *et al.*, 1997b), hyperglycemia (D. S. Jessop and R. Kvetnansky, unpublished observations, 1999), or hypoglycemia (Gaillard *et al.*, 1998). ACTH and β-endorphin contents in splenocytes and thymocytes are increased following central injection of IL-1β (Jessop *et al.*, 1997b; Panerai *et al.*, 1997), and ACTH and β-endorphin are secreted in response to CRH (Smith *et al.*, 1986). Stress-induced increases in immune ACTH and β-endorphin may be secondary to local secretion of immune CRH. Synthetic CRH upregulates POMC gene expression in mouse lymphocytes (Galin *et al.*, 1991) and stimulates the secretion of ACTH and β-endorphin from leukocytes (Smith *et al.*, 1986; Kavelaars *et al.*, 1990). Although some groups have been unable to observe an effect of CRH on immune POMC expression or ACTH secretion, failure to elicit a POMC response to CRH may be due to differential activation of lymphocytes and the consequent variable numbers of CRH receptors. A functional relationship between CRH and POMC expression has been demonstrated in inflammation whereby CRH injected into inflamed tissue produced an analgesic effect which was reversed by antisera to β-endorphin (Cabot *et al.*, 1997; Schafer *et al.*, 1997) or to the adhesion molecule selectin (Machelska *et al.*, 1998), indicating complex interactions between CRH, β-endorphin, and selectin in mediating peripheral pain transmission. There was a strong correlation, implying interdependence, between splenic contents of ACTH, β-endorphin, and CRH in the acute response to central injection of IL-1β (Jessop *et al.*, 1997b). CRH receptors are present in mouse splenic macrophages and neutrophils and in human lymphocytes (Baigent, 2001). Thus the components are present for a functional interaction of CRH with POMC expression in the immune system. However, in AA, splenic ACTH contents are elevated many days before CRH (Jessop *et al.*, 1995a), a time course inconsistent with stimulation of POMC expression by CRH in this disease model.

Immune POMC expression can be regulated by glucocorticoids as in the anterior pituitary (Smith *et al.*, 1986) but not under all circumstances, since the stimulatory effects of IL-1β on POMC expression are glucocorticoid-dependent in the thymus but not in the spleen (Jessop *et al.*, 1997b). There is also potential for SNS regulation of immune POMC, since sympatho-adrenergic neurons abut neuropeptide-positive splenocytes (Felton *et al.*, 1987) and β-endorphin secretion from human mononuclear cells was stimulated by the β-adrenergic agonist isoprenaline (Kavelaars *et al.*, 1990). Much, however, remains to be learned about the mechanisms which control POMC expression in immune cells. Glucocorticoid and catecholaminergic regulation of POMC expression in immune tissues is a potentially fertile area of investigation into mechanisms whereby immune functions can be influenced by CNS responses to acute or chronic stress. Neurotransmitters such as GABA, serotonin, and dopamine can also exert multiple influences on immune functions. These neurotransmitters are responsive to stress and are located in immune cells as well as peripheral neurons, but little is known about their effects on POMC or expression of any other immunoneuropeptides. Selective peripheral lesioning of these neurotransmitter systems with neurotoxic agents should in the future provide valuable information about which neuronal systems exert tonic control over immunopeptide expression, under normal conditions or during stress.

IV. Opioid Peptides Other Than β-Endorphin

Enkephalins, dynorphins, nociceptins, and endomorphins (EM)—and their respective δ-, κ-, nociceptin-, and μ-opioid receptors—have all been located within immune tissues (Peluso *et al.*, 1998; Sharp *et al.*, 1998). Studies using antisense probes to inhibit enkephalin mRNA translation have demonstrated increased human T-cell proliferation and cytokine secretion (Kamphuis *et al.*, 1998) and inhibition of mouse thymocyte (Linner *et al.*, 1995) and rat splenocyte (Fulford *et al.*, 2000) proliferation, results which are consistent with anti-inflammatory roles for enkephalins in several species and cell types. Preproenkephalin A mRNA and Met-enkephalin are expressed in activated T cells (Linner *et al.*, 1995; Kamphuis *et al.*, 1997) and are abundant in inflamed tissue surrounding the joints of arthritic rats (Przewlocki *et al.*, 1992). Cells containing opioid immunoreactivity were identified in this study as T and B lymphocytes, macrophages, and monocytes located in tissue adjacent to primary afferent nociceptive neurons. Both Met- and Leu-enkephalin (and β-endorphin) inhibited release of TNF-α, IL-1, and matrix metalloproteinases from synovial cells in patients with RA through a naxolone-reversible mechanism (Takeba *et al.*, 2001). Met-Enkephalin and β-endorphin levels are elevated in synovial tissues from RA patients (Stein *et al.*, 1993) who reported a lower pain threshold following

naloxone injection into the inflamed joint compared to intravenous administration, suggesting that endogenous opioids in synovial tissue are capable
of modulating neurogenic pain as well as inflammation in RA.

Dynorphin, present in rat lymphocytes (Hassan *et al.*, 1992), is an en-
hancer of IL-2 secretion and lymphocyte proliferation (Ni *et al.*, 1999) and
at ultralow concentrations inhibits TNF-α secretion from glial cells (Kong
et al., 1997). Nociceptin (orphanin FQ) mRNA has been reported in the
porcine immune system (Pampusch *et al.*, 2000) and nociceptin can modulate immune functions (Peluso *et al.*, 2001; Serhan *et al.*, 2001). Nociceptin receptor mRNA, but not functional receptors, has been reported
in porcine lymphocytes (Pampusch *et al.*, 2000), and it is probable that lymphocyte activation is a prerequisite for receptor expression. Although there
is no doubt that dynorphin and nociceptin of neural origin play important
roles in mechanisms mediating neurogenic pain, these peptides have not
been investigated in any detail in the immune system, and there is little
information about the responses of these peptides, or their receptors, to
onset of inflammation either in animal models or in human autoimmune
diseases.

The opioid tetrapeptide endomorphins, EM-1 and EM-2, which have
very high selectivity and affinity for μ-receptors, are located within rat and
human immune tissues as well as in the CNS (Jessop *et al.*, 2000), and their
synovial tissue levels are increased during inflammation (Jessop *et al.*, 2002).
Endomorphins exerted opposite effects on superoxide anion production by
neutrophils depending on the degree of cell activation (Azuma *et al.*, 2000)
and, in a rat model of acute inflammation, EM-1 was effective in attenuating
the inflammatory response to SP, an effect reversed by naloxone (Khalil *et al.*,
1999). However, one careful study found no evidence for EM-1 involvement
over a range of immune function tests *in vivo* in normal rats (Carrigan
et al., 2000). These apparently discrepant observations suggest that immune
responses to endomorphins may only be evident during inflammation when
opioid receptors are upregulated in activated cells.

Although pharmacological studies strongly associate EM-1 and EM-2
with μ-opioid receptors through which are mediated the analgesic, respiratory, and tolerance effects of morphine, endomorphins may exert some
immunomodulatory effects through novel receptors. Centrally injecting endomorphins induces phenomena which cannot be explained by a common
receptor for endomorphins and morphine. EM-1 and EM-2 differentially induced analgesia in mice where similar responses were predicted on the basis
of action through the μ-receptor (Ohsawa *et al.*, 2001; Sakurada *et al.*, 2001).
The central effects of morphine on corticosterone release could not be
blocked by preadministration of EM-1 or EM-2 (Coventry *et al.*, 2001). EM-1
acted as a potent analgesic without demonstrating any of the immunological

effects normally exerted by morphine (Carrigan *et al.*, 2000). This cumulative evidence suggests that endomorphins can exert effects independent of the morphine receptor. At least two μ-opioid receptor subtypes exist through which EM-1 and EM-2 may act, as well as through other novel μ-opioid receptor variants. Any preferential association of endomorphins with selective μ-opioid receptor subtypes remains to be investigated. Endomorphins, such as dynorphin (Kong *et al.*, 1997) and β-endorphin (Williamson *et al.*, 1988), have been reported to modulate immune functions *in vitro* at concentrations lower than 10^{-15} M. Since ligand occupancy of the classical opioid receptors at these doses is negligible, this pharmacology supports the existence of novel opioid receptors in the immune system or heterodimerization of existing receptors in immune cells to form complexes with extremely high affinity for opioid peptides (Jordan and Devi, 1999). Therefore the potential exists for endomorphins to exert physiological actions at low concentrations through novel opioid receptors.

Intact undegraded EM-1 and EM-2 have been characterized within circulating human lymphocytes (Fig. 1), whereas multiple forms of immunoreactivity have been reported in spleen, thymus, and plasma (Jessop *et al.*, 2000), suggesting extensive degradation. Thus endomorphins can be transported encapsulated and protected within lymphocytes in biologically active forms to target inflamed tissue. This is consistent with the mechanism proposed for directing β-endorphin to inflamed tissue through lymphocyte trafficking (Cabot *et al.*, 1997) and selectin-dependent sequestration (Machelska *et al.*, 1998). In this way opioid peptides can be released from circulating lymphocytes directly at sites of inflammation to act as anti-inflammatory and analgesic agents through inhibition of cytokine and SP release (Fig. 2). Such targeted delivery would enable low concentrations of opioids to exert maximally effective paracrine actions at inflamed sites. An alternative to targeted delivery within peripheral lymphocytes is the possibility that freely circulating peptides, for example, ACTH and β-endorphin released from the anterior pituitary, might exert immunomodulatory effects. While this may be a plausible scenario in conditions of excessive ACTH secretion such as Cushing's syndrome, in normal circumstances concentrations are too low, and proteolytic degradation too rapid, for opioid peptides to exert physiologically relevant systemic effects.

Endomorphins and other opioid peptides are located in neuronal projections innervating respiratory tissues (Groneberg and Fischer, 2001) where they can exert bronchodilatory effects. Although not yet identified in immune cells in this system, circulating opioid-containing lymphocytes might also be sequestered in inflamed bronchial tissues in asthma. The role of opioids in mediating the Th1/Th2 cytokine switch in this predominantly Th2 disease is worthy of closer investigation.

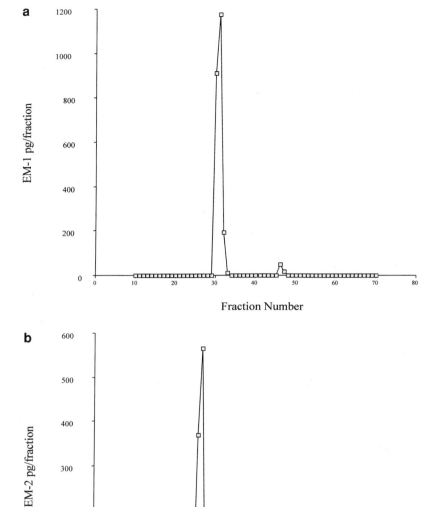

Fig. 1. Reversed-phase HPLC of (a) EM-1 and (b) EM-2 extracted from normal human peripheral blood lymphocytes. HPLC fractions were vacuum-dried and reconstituted in phosphate buffer for measurement of EM-1 and EM-2 by radioimmunoassays. Synthetic EM-1 and EM-2 eluted in fractions 31 to 32, and 28 to 29, respectively.

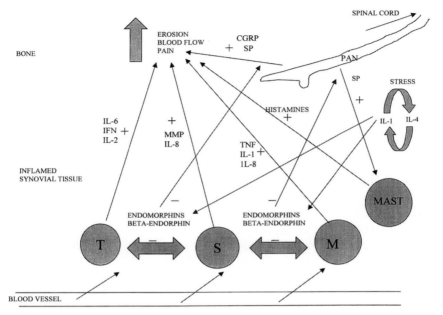

FIG. 2. During the onset of rheumatoid arthritis (RA), synovial tissue becomes inflamed with greatly increased blood flow. Increased expression of adhesion proteins on vascular endothelium leads to recruitment and infiltration of mature leukocytes which release a range of inflammatory compounds. Endomorphin (EM)-1 and EM-2, transported intact within peripheral blood lymphocytes (PBLs) (see Fig. 1), and β-endorphin are released from leukocytes and can act on T cells, synoviocytes (S), and macrophages (M) within synovial tissue to inhibit the release of pro-inflammatory mediators, e.g., IL-1, IL-2, IL-6, IL-8, TNF, IFN, and matrix metalloproteinases (MMP). This creates a dynamic homeostatic interaction of pro- and anti-inflammatory mediators within synovial tissues in RA. Opioid peptides also act through receptors on terminals of primary afferent nociceptive neurons (PAN) within synovial tissues and inhibit the release of algesic agents substance P (SP) and CGRP. Therefore immune opioids can act as antiinflammatory agents by directly inhibiting the release of Th1 cytokines, and they can attenuate neurogenic pain associated with RA by indirectly inhibiting SP-induced histamine secretion from mast cells. Stress, through release of catecholamines and glucocorticoids which alter the Th1/Th2 cytokine balance, can influence opioid secretion from immune cells and thereby modulate the onset and severity of inflammatory diseases. + = stimulatory; − = inhibitory.

V. Trophic Peptides

Many hypothalamic- and pituitary-derived regulatory peptides such as thyroid-stimulating hormone, somatostatin, prolactin, follicle-stimulating hormone, growth hormone (GH), and gonadotropin-releasing hormone are also expressed in immune tissues (Blalock, 1994; Clark, 1997). Receptors for GH and prolactin are present in immune tissues, and immunological changes may be induced by either pituitary- or immune-derived GH and prolactin, or both (Clark, 1997; Clevenger *et al.*, 1998). A role has been demonstrated for immune GH involvement in tonic inhibition of lymphocyte proliferation (Weigent *et al.*, 1991). The physiological relevance of GH in maintaining immunocompetence in humans has been questioned, since pituitary-GH-deficient children show no signs of impaired immune function, but it is possible that expression of immune-derived GH may be sufficient to compensate under such conditions.

It is unclear whether lymphocyte GH expression and secretion are regulated by growth hormone-releasing hormone (GHRH) and somatostatin as they are in the anterior pituitary. GHRH and somatostatin are present in immune tissues along with their receptors (Campbell and Scanes, 1995; ten Bokum *et al.*, 2000). Inhibition of somatostatin mRNA translation using antisense probes resulted in increased cell proliferation (Aguila *et al.*, 1996). Somatostatin peptide and receptor expression in synovial tissues are upregulated during inflammation (ten Bokum *et al.*, 1999a,b), probably in response to inflammatory agents TNF-α and IFN-γ (Elliott *et al.*, 1998), while SP downregulates somatostatin expression in splenocytes (Blum *et al.*, 1998). An anti-inflammatory action has been demonstrated for somatostatin (Sakane and Suzuki, 1998), possibly through inhibition of TNF-α, IFN-γ, and matrix metalloproteinase secretion from synovial cells.

The immunological effects of immune GH may be mediated through insulin-derived growth factors (IGF) which, along with their receptors, are widely distributed throughout immune tissues. It has been proposed that immune IGF-1 is a potent apoptotic agent which limits tissue damage induced by glucocorticoids released during stress (Clark, 1997). Immunoneutralization of IGF-1 prevents the proliferative effects of GH and prolactin on thymocyte stimulation, demonstrating a dependence of these actions on IGF-1, but GH stimulation of splenocytes does not require IGF-1 (Postel-Vinay *et al.*, 1997). Insulin and nerve growth factor also promote the normal development and function of immune cells (Savino *et al.*, 1995).

VI. Other Immunoneuropeptides

Calcitonin gene-related peptide (CGRP) and SP, released from peripheral afferent nociceptors, are important mediators of neurogenic inflammation. SP is also expressed in human mononuclear cells and circulating leukocytes (De Giorgio *et al.*, 1998), while CGRP has been reported in rat thymocytes (Wang *et al.*, 1999; Xing *et al.*, 2000). Expression of SP and CGRP is decreased in immune cells during the course of SLE (Bracci-Laudiero *et al.*, 1998). Although it is well established that regulation of SP release from peripheral neurons is inhibited by peptides such as β-endorphin (Levine *et al.*, 1993) and somatostatin (Karalis *et al.*, 1994), function and control of SP and CGRP within lymphocytes is poorly understood and the extent, if any, to which lymphocyte CGRP and SP contribute to neurogenic inflammation is unknown. A functional relationship has been established between over-expression of CGRP in pancreatic beta cells and a decreased immune response (Khachatryan *et al.*, 1997), which culminates in total resistance to diabetes mellitis in male mice and two-thirds reduced incidence in females. This transgenic model was one of the earliest to provide physiological evidence for a link between peripheral neuropeptide expression and a specific autoimmune disease. Immunosuppressive effects of CGRP can occur through inhibition of NF-κB complex aggregation in cytoplasm as observed in mouse thymocytes (Millet *et al.*, 2000), a process which promotes apoptotic cell death.

Structurally related vasoactive intestinal peptide (VIP) and pituitary adenylate cyclase activating polypeptide (PACAP) and their receptors are distributed throughout the immune system (Goetzl *et al.*, 1998; Gaytan *et al.*, 1994; Delgado and Ganea, 1999). Expression of VIP, like POMC and GH, is not constitutive and is dependent on thymocyte maturity (Delgado *et al.*, 1996). Evidence for anti-inflammatory effects of VIP and PACAP has been reported in animal models of lung (Said and Dickman, 2000) and joint (Delgado *et al.*, 2001) inflammation.

Related peptides vasopressin (AVP), oxytocin, and neurophysin are expressed in thymic epithelial cells where they have the potential to influence T-cell maturation (Martens *et al.*, 1996; Vanneste *et al.*, 1997). AVP has been proposed as a pro-inflammatory agent in RA (Chikanza and Grossman, 1998). AVP in the spleen is located predominantly in B lymphocytes (Jessop *et al.*, 1995b), and splenic contents of AVP increase around the time of inflammation in AA (Jessop *et al.*, 1995a). Levels of AVP in spleens and thymuses from female rats are two- to fivefold greater than in males (Chowdrey *et al.*, 1994), and AVP receptor density is significantly higher in peripheral blood lymphocytes from women (Elands *et al.*, 1990)—observations of sexual dimorphism which may be of relevance with regard to the considerably higher incidence of RA in women.

VII. Quantitative Analysis of Immunoneuropeptides by MACS

One area in the literature which has been beset with difficulties is the measurement of neuropeptides in immune tissues, especially in nonactivated leukocytes, since peptide contents are usually 100–1000 times lower than in the CNS. This has contributed to consistency and reproducibility

TABLE I

MEASUREMENT OF CRH BY RADIOIMMUNOASSAY IN MACS-PURIFIED
SUBPOPULATIONS OF NONACTIVATED RAT SPLENOCYTES[a]

Monoclonal antibody	Splenocyte subtype	Percentage purity	CRH (pg/million cells)
R 73	T cells	93	0.62
OX 33	B cells	80	0.62
ED 2	Macrophages	95	3.5

[a] Mouse anti-rat monoclonal antibodies were obtained from Serotec (U.K.) and magnetic microbeads labeled with anti-mouse IgG were obtained from Miltenyi (U.K.). Splenocyte subpopulation enrichment was determined by FACS analysis (Becton Dickinson U.K.).

problems within and between laboratories. These problems can be resolved by application of the technique of magnetic cell sorting and separation (MACS) which can be utilized to prepare T and B cells and macrophage subpopulations of high purity. Availability of a wide range of monoclonal antibodies to cell surface antigenic markers permits the application of the MACS technique to mouse, rat, or human leukocyte populations. Cell subtype purification by MACS combined with highly sensitive immunoassays can thus enhance the power and accuracy of neuropeptide measurement within immune cells. This combination of techniques can also determine neuropeptide colocalization within leukocyte subsets, an area of research which has provided a wealth of information about the synergy of neuropeptide interactions centrally and systemically, but about which little is known in the immune system.

MACS can also resolve any dispute over whether neuropeptides are present at inflamed sites in lymphocytes or in nerve terminals. Many peptides such as CRH, Ucn, and SP are probably synthesized in both locations but relative amounts are difficult to establish because of the small amounts of tissue involved. Harvesting of cells from inflamed tissues, lysing to eliminate synaptosomes, purification by MACS, and identification by FACS can provide purified lymphocyte subsets for analysis of mRNA by Northern blotting or PCR, or peptides by immunoassays. This technique has been utilized to confirm that most CRH in rat splenocytes is contained in macrophages (Table I).

VIII. Conclusions

It is now beyond dispute that neuropeptides possess potent immunomodulatory properties. Although glucocorticoids remain the most powerful of all anti-inflammatory agents, their clinical application in chronic inflammation and autoimmune diseases is limited by systemic side effects. The challenge now is to develop selective anti-inflammatory drugs with efficacy equal to or better than glucocorticoids. Immunoneuropeptides have exciting potential for development as a new generation of nonglucocorticoid anti-inflammatory drugs. One area of promise is the direct injection into inflamed tissue of small amounts of a long-acting compound, or local implants of slow-release polymer beads impregnated with an anti-inflammatory agent. This would be a useful strategy for the application of low dose/high potency opioid or CRH analogs to exert anti-inflammatory effects without systemic release and consequent nonspecific effects.

There is compelling evidence that expression of some immunoneuropeptides is responsive to glucocorticoids and catecholamines and that these neuropeptides can modulate the effects of stress on immune functions. Stress can influence the production of CRH and POMC peptides from leukocytes and also determine T-cell differentiation into distinctive Th1/Th2 cytokine secretory profiles. Therefore neuropeptides may be involved in mediating the effects of stress on production of specific cytokines with pro- or anti-inflammatory activity. It is now accepted that, while most stressors in common activate the HPA axis and the SNS, each type of stressor—acute, repeated or chronic, physical or psychological—is characterized by subtle and unique differences in the secretory profile of neuropeptides, neurotransmitters, and glucocorticoids which act as stimulatory or inhibitory agents (Jessop, 1999). It might be useful to view the immune system in this multifarious light whereby selective neuropeptides can mediate specific immune responses to different types of stress. Most autoimmune diseases are sensitive to stress, yet very little is known of the mechanisms whereby stress can exacerbate, or ameliorate, inflammation (Solomon, 1997). It is possible that neuropeptides in immune cells, like their counterparts in the HPA axis, may selectively determine stress-induced predisposition to, or onset and severity of, some inflammatory diseases through stimulation of a pro- or anti-inflammatory cytokine profile. Identification of these neuropeptides and their receptors, and their specific effects on cytokine secretion and actions, will represent an important step toward understanding how stress influences inflammation.

Acknowledgments

This work has been supported in part by the Wellcome Trust and the Neuroendocrinology Charitable Trust. We are grateful to Louise Richards for her skillful performance of the endomorphin HPLC studies and radioimmunoassays and to Kathryn Elsegood for MACS and FACS analysis.

References

Aguila, M. C., Rodriguez, A. M., Aguila-Mansilla, H. N., and Lee, W. T. (1996). Somatostatin an-
    tisense oligodeoxynucleotide-mediated stimulation of lymphocyte proliferation in culture.
    Endocrinology **137,** 1585–1590.
Aird, F., Clevenger, C. V., Prystowsky, M. B., and Redei, E. (1993). Corticotropin-releasing factor
    mRNA in rat thymus and spleen. *Proc. Natl. Acad. Sci. USA* **90,** 7104–7108.
Azuma, Y., Wang, P. L., Shinohara, M., and Ohura, K. (2000). Immunomodulation of the
    neutrophil respiratory burst by endomorphins 1 and 2. *Immunol. Lett.* **75,** 55–59.
Baigent, S. M. (2001). Peripheral corticotropin-releasing hormone and urocortin in the control
    of the immune response. *Peptides* **22,** 809–820.
Bhardwaj, R., Becher, E., Mahnke, K., Hartmeyer, M., Schwarz, T., Scholzen, T., and Luger, T. A.
    (1997). Evidence for the differential expression of the functional α-melanocyte-stimulating
    hormone receptor MC-1 on human monocytes. *J. Immunol.* **158,** 3378–3384.
Bhardwaj, R. S., Schwarz, A., Becher, E., Mahnke, K., Aragane, Y., Schwarz, T., and Luger, T. A.
    (1996). Pro-opiomelanocortin-derived peptides induce IL-10 production in human mono-
    cytes. *J. Immunol.* **156,** 2517–2521.
Blalock, J. E. (1994). The syntax of immune-neuroendocrine communication. *Immunol. Today*
    15, 504–511.
Blalock, J. E. (1999). Proopiomelanocortin and the immune-neuroendocrine connection. *Ann.
    N.Y. Acad. Sci.* **885,** 161–172.
Blum, A. M., Elliott, D. E., Metwali, A., Li, J., Qadir, K., and Weinstock, J. V. (1998). Substance
    P regulates somatostatin expression in inflammation. *J. Immunol.* **161,** 6316–6322.
Boehme, S. A., Gaur, A., Crowe, P. D., Liu, X. J., Wong, T., Pahuja, A., Ling, N., Vale, W.,
    De Souza, E. B., and Conlon, P. J. (1997). Immunosuppressive phenotype of corticotropin-
    releasing factor transgenic mice is reversed by adrenalectomy. *Cell Immunol.* **176,** 103–
    112.
Bracci-Laudiero, L., Aloe, L., Stenfors, C., Theodorsson, E., and Lundeberg, T. (1998). Devel-
    opment of systemic lupus erythematosus in mice is associated with alteration of neuropep-
    tide concentrations in inflamed kidneys and immunoregulatory organs. *Neurosci. Lett.* **248,**
    97–100.
Brouxhon, S. M., Prasad, A. V., Joseph, S. A., Felten, D. L., and Bellinger, D. (1998). Localization
    of corticotropin-releasing factor in primary and secondary lymphoid organs of the rat.
    Brain Behav. Immun. **12,** 107–122.
Buzzetti, R., McLoughlin, L., Lavender, P., Clark, A., and Rees, L. H. (1989). Expression of
    pro-opiomelanocortin gene and quantification of adrenocorticotropic hormonelike im-
    munoreactivity in human normal peripheral mononuclear cells and lymphoid and myeloid
    malignancies. *J. Clin. Invest.* **83,** 733–737.
Cabot, P. J., Carter, L., Gaiddon, C., Zhang, Q., Schafer, M., Loeffler, J. P., and Stein, C. (1997).
    Immune cell–derived β-endorphin. Production, release, and control of inflammatory pain
    in rats. *J. Clin. Invest.* **100,** 142–148.
Campbell, R. M., and Scanes, C. G. (1995). Endocrine hormones "moonlighting" as im-
    mune modulators: Roles for somatostatin and GH-releasing factor. *J. Endocrinol.* **147,** 383–
    396.
Carrigan, K. A., Nelson, C. J., and Lysle, D. T. (2000). Endomorphin-1 induces antinociception
    without immunomodulatory effects in the rat. *Psychopharmacology (Berlin)* **151,** 299–305.
Chancellor-Freeland, C., Zhu, G. F., Kage, R., Beller, D. I., Leeman, S. E., and Black, P. H.
    (1995). Substance P and stress-induced changes in macrophages. *Ann. N.Y. Acad. Sci.* **771,**
    472–484.

Chikanza, I. C., and Grossman, A. B. (1998). Hypothalamic-pituitary–mediated immunomodulation: Arginine vasopressin is a neuroendocrine immune mediator. *Br. J. Rheumatol.* **37,** 131–136.

Chowdrey, H. S., Lightman, S. L., Harbuz, M. S., Larsen, P. J., and Jessop, D. S. (1994). Contents of corticotropin-releasing hormone and arginine vasopressin immunoreactivity in the spleen and thymus during a chronic inflammatory stress. *J. Neuroimmunol.* **53,** 17–21.

Clarke, B. L., and Bost, K. L. (1989). Differential expression of functional adrenocorticotropic hormone receptors by subpopulations of lymphocytes. *J. Immunol.* **143,** 464–469.

Clark, R. (1997). The somatogenic hormones and insulin-like growth factor-1: Stimulators of lymphopoiesis and immune function. *Endocr. Rev.* **18,** 157–179.

Clevenger, C. V., Freier, D. O., and Kline, J. B. (1998). Prolactin receptor signal transduction in cells of the immune system. *J. Endocrinol.* **157,** 187–197.

Conti, F., Pittoni, V., Sacerdote, P., Priori, R., Meroni, P. L., and Valesini, G. (1998). Decreased immunoreactive β-endorphin in mononuclear leucocytes from patients with chronic fatigue syndrome. *Clin. Exp. Rheumatol.* **16,** 729–732.

Coventry, T. L., Jessop, D. S., Finn, D. P., Crabb, M. D., Kinoshita, H., and Harbuz, M. S. (2001). Endomorphins and activation of the hypothalamo-pituitary-adrenal axis. *J. Endocrinol.* **169,** 185–193.

Crofford, L. J., Sano, H., Karalis, K., Webster, E., Goldmuntz, E., Chrousos, G., and Wilder, R. (1992). Local secretion of corticotropin-releasing hormone in the joints of Lewis rats with inflammatory arthritis. *J. Clin. Invest.* **90,** 2555–2564.

Crofford, L. J., Sano, H., Karalis, K., Friedman, T. C., Epps, H. R., Remmers, E. F., Mathern, P., Chrousos, G. P., and Wilder, R. L. (1993). Corticotropin-releasing hormone in synovial fluids and tissues of patients with rheumatoid arthritis and osteoarthritis. *J. Immunol.* **151,** 1587–1596.

De Giorgio, R., Tazzari, P. L., Barbara, G., Stanghellini, V., and Corinaldesi, R. (1998). Detection of substance P immunoreactivity in human peripheral leukocytes. *J. Neuroimmunol.* **82,** 175–181.

Delgado, M., and Ganea, D. (1999). Vasoactive intestinal peptide and pituitary adenylate cyclase-activating polypeptide inhibit interleukin-12 transcription by regulating nuclear factor kappaB and Ets activation. *J. Biol. Chem.* **274,** 31930–31940.

Delgado, M., Martinez, C., Leceta, J., Garrido, E., and Gomariz, R. P. (1996). Differential VIP and VIP1 receptor gene expression in rat thymocyte subsets. *Peptides* **17,** 803–807.

Delgado, M., Abad, C., Martinez, C., Leceta, J., and Gomariz, R. P. (2001). Vasoactive intestinal peptide prevents experimental arthritis by downregulating both autoimmune and inflammatory components of the disease. *Nat. Med.* **7,** 563–568.

Elands, J., Van Woudenberg, A., Resink, A., and de Kloet, R. (1990). Vasopressin receptor capacity of human blood peripheral mononuclear cells is sex dependent. *Brain Behav. Immun.* **4,** 30–38.

Elenkov, I. J., and Chrousos, G. P. (1999). Stress, cytokine patterns and susceptibility to disease. *Baillieres Best Pract. Res. Clin. Endocrinol. Metab.* **13,** 583–595.

Elliott, D. E., Blum, A. M., Li, J., Metwali, A., and Weinstock, J. V. (1998). Preprosomatostatin messenger RNA is expressed by inflammatory cells and induced by inflammatory mediators and cytokines. *J. Immunol.* **160,** 3997–4003.

Felten, D. L., Felten, S. Y., Bellinger, S. L., Carlson, S. L., Ackerman, K. S., Madden, J. A., Olschowka, J. A., and Livnat, S. (1987). Noradrenergic sympathetic neural interactions with the immune system: Structure and function. *Immunol. Rev.* **100,** 225–260.

Fuchs, B. A., and Pruett, S. B. (1993). Morphine induces apoptosis in murine thymocytes *in vivo* but not *in vitro*: Involvement of both opiate and glucocorticoid receptors. *J. Pharmacol. Exp. Ther.* **266,** 417–423.

Fulford, A. J., Harbuz, M. S., and Jessop, D. S. (2000). Antisense inhibition of pro-opiomelanocortin and proenkephalin A messenger RNA translation alters rat immune cell function *in vitro*. *J. Neuroimmunol.* **106**, 6–13.

Gaillard, R. C., Daneva, T., Hadid, R., Muller, K., and Spinedi, E. (1998). The hypothalamo-pituitary-adrenal axis of athymic Swiss nude mice. The implications of T lymphocytes in the ACTH release from immune cells. *Ann. N. Y. Acad. Sci.* **840**, 480–490.

Galin, F. S., LeBoeuf, R. D., and Blalock, J. E. (1991). Corticotropin-releasing factor upregulates expression of two truncated pro-opiomelanocortin transcripts in murine lymphocytes. *J. Neuroimmunol.* **31**, 51–58.

Gaytan, F., Martinez-Fuentes, A. J., Garcia-Navarro, F., Vaudry, H., and Aguilar, E. (1994). Pituitary adenylate cyclase-activating peptide (PACAP) immunolocalization in lymphoid tissues of the rat. *Cell Tissue Res.* **276**, 223–227.

Getting, S. J., Gibbs, L., Clark, A. J., Flower, R. J., and Perretti, M. (1999). POMC gene-derived peptides activate melanocortin type 3 receptor on murine macrophages, suppress cytokine release, and inhibit neutrophil migration in acute experimental inflammation. *J. Immunol.* **162**, 7446–7453.

Getting, S. J., Allcock, G. H., Flower, R., and Perretti, M. (2001). Natural and synthetic agonists of the melanocortin receptor type 3 possess anti-inflammatory properties. *J. Leukoc. Biol.* **69**, 98–104.

Goetzl, E. J., Pankhaniya, R. R., Gaufo, G., Mu, Y., Xia, M., and Sreedharan, S. P. (1998). Selectivity of effects of vasoactive intestinal peptide on macrophages and lymphocytes in compartmental immune responses. *Ann. N. Y. Acad. Sci.* **840**, 540–550.

Groneberg, D. A., and Fischer, A. (2001). Endogenous opioids as mediators of asthma. *Pulm. Pharmacol. Ther.* **14**, 383–389.

Harbuz, M. S., Chowdrey, H. S., Lightman, S. L., Wei, E. T., and Jessop, D. S. (1996). An investigation into the effects of chronic infusion of corticotrophin-releasing factor on hind paw inflammation in adjuvant-induced arthritis. *Stress* **1**, 105–111.

Harbuz, M. S., Chover-Gonzalez, A., Gibert-Rahola, J., and Jessop, D. S. (2002). Protective effect of prior acute immune challenge, but not footshock, on inflammation in the rat. *Brain. Behav. Immun.* **16**, 439–449.

Hargreaves, K. M., Costello, A. H., and Joris, J. L. (1989). Release from inflamed tissue of a substance with properties similar to corticotropin-releasing factor. *Neuroendocrinology* **49**, 476–482.

Hashemi, F. B., Hughes, T. K., and Smith, E. M. (1998). Human immunodeficiency virus induction of corticotropin in lymphoid cells. *J. Clin. Endocrinol. Metab.* **83**, 4373–4381.

Hassan, A. H., Przewlocki, R., Herz, A., and Stein, C. (1992). Dynorphin, a preferential ligand for κ-opioid receptors, is present in nerve fibers and immune cells within inflamed tissue of the rat. *Neurosci. Lett.* **140**, 85–88.

Heijnen, C. J., Zijlstra, J., Kavelaars, A., Croiset, G., and Ballieux, R. E. (1987). Modulation of the immune response by POMC-derived peptides. I. Influence on proliferation of human lymphocytes. *Brain Behav. Immun.* **1**, 284–291.

Ientile, R., Ginoprelli, T., Cannavo, G., Picerno, I., and Piedimonte, G. (1997). Effect of β-endorphin on cell growth and cell death in human peripheral blood lymphocytes. *J. Neuroimmunol.* **80**, 87–92.

Jessop, D. S. (1998). Neuropeptides: Modulators of the immune system. *Curr. Opin. Endocrinol. Diabetes* **5**, 52–58.

Jessop, D. S. (1999). Central non-glucocorticoid inhibitors of the hypothalamo-pituitary-adrenal axis. *J. Endocrinol.* **160**, 169–180.

Jessop, D. S. (2001). Neuropeptides in the immune system: Functional roles in health and disease. *In* "Frontiers of Hormone Research" (R. Gaillard, ed.), Vol. 29, pp. 50–68. S. Karger AG, Basel.

Jessop, D. S., Lightman, S. L., and Chowdrey, H. S. (1994). Effects of a chronic inflammatory stress on levels of pro-opiomelanocortin-derived peptides in the rat spleen and thymus. *J. Neuroimmunol.* **49,** 197–203.

Jessop, D. S., Renshaw, D., Lightman, S. L., and Harbuz, M. S. (1995a). Changes in ACTH and β-endorphin immunoreactivity in immune tissues during a chronic inflammatory stress are not correlated with changes in corticotropin-releasing hormone and arginine vasopressin. *J. Neuroimmunol.* **60,** 29–35.

Jessop, D. S., Chowdrey, H. S., Lightman, S. L., and Larsen, P. J. (1995b). Vasopressin is located within lymphocytes in the rat spleen. *J. Neuroimmunol.* **56,** 219–223.

Jessop, D. S., Harbuz, M. S., Snelson, C. L., Dayan, C. M., and Lightman, S. L. (1997a). An antisense oligodeoxynucleotide complementary to corticotropin-releasing hormone mRNA inhibits rat splenocyte proliferation *in vitro. J. Neuroimmunol.* **75,** 135–140.

Jessop, D. S., Douthwaite, J. D., Conde, G. L., Lightman, S. L., Dayan, C. M., and Harbuz, M. S. (1997b). Effects of acute stress or centrally injected interleukin-1β on neuropeptide expression in the immune system. *Stress* **2,** 133–144.

Jessop, D. S., Major, G. N., Coventry, T. L., Kaye, S. J., Fulford, A. J., Harbuz, M. S., and de Bree, F. M. (2000). Opioid peptides endomorphin-1 and endomorphin-2 are present in mammalian immune tissues. *J. Neuroimmunol.* **106,** 53–59.

Jessop, D. S., Harbuz, M. S., and Lightman, S. L. (2001). CRH in chronic inflammatory stress. *Peptides* **22,** 803–807.

Jessop, D. S., Richards, L. J., and Harbuz, M. S. (2002). Opioid peptides endomorphin (EM)-1 and EM-2 in the immune system in a rodent model of inflammation. *Ann. N.Y. Acad. Sci.* **966,** 456–463.

Johnson, E. W., Hughes, T. K. Jr, and Smith, E. M. (2001). ACTH receptor distribution and modulation among murine mononuclear leukocyte populations. *J. Biol. Regul. Homeost. Agents* **15,** 156–162.

Jordan, B. A., and Devi, L. A. (1999). G-protein-coupled receptor heterodimerization modulates receptor function. *Nature (London)* **399,** 697–700.

Kageyama, K., Bradbury, M. J., Zhao, L., Blount, A. L., and Vale, W. W. (1999). Urocortin messenger ribonucleic acid: Tissue distribution in the rat and regulation in thymus by lipopolysaccharide and glucocorticoids. *Endocrinology* **140,** 5651–5658.

Kamphuis, S., Kavelaars, A., Brooimans, R., Kuis, W., Zegers, B. J. M., and Heijnen, C. J. (1997). T helper 2 cytokines induce preproenkephalin mRNA expression and proenkephalin A in human peripheral blood mononuclear cells. *J. Neuroimmunol.* **79,** 91–99.

Kamphuis, S., Eriksson, F., Kavelaars, A., Zijlstra, J., van de Pol, M., Kuis, W., and Heijnen, C. J. (1998). Role of endogenous pro-enkephalin A-derived peptides in human T cell proliferation and monocyte IL-6 production. *J. Neuroimmunol.* **84,** 53–60.

Karalis, K., Sano, H., Redwine, J., Listwak, J., Wilder, R., and Chrousos, G. (1991). Autocrine or paracrine inflammatory actions of corticotropin-releasing hormone *in vivo. Science* **254,** 421–423.

Karalis, K., Mastorakos, G., Chrousos, G. P., and Tolis, G. (1994). Somatostatin analogues suppress the inflammatory reaction *in vivo. J. Clin. Invest.* **93,** 2000–2006.

Karalis, K., Muglia, L. J., Bae, D., Hilderbrand, H., and Majzoud, J. A. (1997). CRH and the immune system. *J. Neuroimmunol.* **72,** 131–136.

Kavelaars, A., Ballieux, R. E., and Heijnen, C. J. (1989). The role of IL-1 in the corticotropin-releasing factor and arginine-vasopressin-induced secretion of immunoreactive β-endorphin by human peripheral blood mononuclear cells. *J. Immunol.* **142,** 2338–2342.

Kavelaars, A., Ballieux, R. E., and Heijnen, C. J. (1990). *In vitro* β-adrenergic stimulation of lymphocytes induces the release of immunoreactive β-endorphin. *Endocrinology* **126,** 3028–3032.

Khachatryan, A., Guerder, S., Palluault, F., Cote, G., Solimena, M., Valentijn, K., Millet, I., Flavell, R. A., and Vignery, S. (1997). Targeted expression of the neuropeptide calcitonin gene-related peptide to β cells prevents diabetes in NOD mice. *J. Immunol.* **158,** 1409–1416.

Khalil, Z., Sanderson, K., Modig, M., and Nyberg, F. (1999). Modulation of peripheral inflammation by locally administered endomorphin-1. *Inflamm. Res.* **48,** 550–556.

Kohno, M., Kawahito, Y., Tsubouchi, Y., Hashiramoto, A., Yamada, R., Inoue, K. I., Kusaka, Y., Kubo, T., Elenkov, I. J., Chrousos, G. P., Kondo, M., and Sano, H. (2001). Urocortin expression in synovium of patients with rheumatoid arthritis and osteoarthritis: Relation to inflammatory activity. *J. Clin. Endocrinol. Metab.* **86,** 4344–4352.

Kong, L. Y., McMillian, M. K., Hudson, P. M., Jin, L., and Hong, J. S. (1997). Inhibition of lipopolysaccharide-induced nitric oxide and cytokine production by ultralow concentrations of dynorphins in mixed glia cultures. *J. Pharmacol. Exp. Ther.* **280,** 61–66.

Kraus, J., Borner, C., Giannini, E., Hickfang, K., Braun, H., Mayer, P., Hoehe, M. R., Ambrosch, A., Konig, W., and Hollt, V. (2001). Regulation of μ-opioid receptor gene transcription by interleukin-4 and influence of an allelic variation within a STAT6 transcription factor binding site. *J. Biol. Chem.* **276,** 43901–43908.

Levine, J. D., Fields, H. L., and Basbaum, A. I. (1993). Peptides and the primary afferent nociceptor. *J. Neurosci.* **13,** 2273–2286.

Linner, K. M., Quist, H. E., and Sharp, B. M. (1995). Met-enkephalin-containing peptides encoded by proenkephalin A mRNA expressed in activated murine thymocytes inhibit thymocyte proliferation. *J. Immunol.* **154,** 5049–5060.

Lipton, J. M., and Catania, A. (1998). Mechanisms of anti-inflammatory action of the neuroimmunomodulatory peptide α-MSH. *Ann. N. Y. Acad. Sci.* **840,** 373–380.

Lolait, S., Clements, J. A., Markwick, A., Cheng, C., McNulty, M., Smith, A. I., and Funder, J. W. (1986). Pro-opiomelanocortin messenger ribonucleic acid and posttranslational processing of β-endorphin in spleen macrophages. *J. Clin. Invest.* **77,** 1776–1779.

Lyons, P. D., and Blalock, J. E. (1995). The kinetics of ACTH expression in rat leukocyte subpopulations. *J. Neuroimmunol.* **63,** 103–112.

Lyons, P. D., and Blalock, J. E. (1997). Pro-opiomelanocortin gene expression and protein processing in rat mononuclear leukocytes. *J. Neuroimmunol.* **78,** 47–56.

Machelska, H., Cabot, P. J., Mousa, S. A., Zhang, Q., and Stein, C. (1998). Pain control in inflammation governed by selectins. *Nat. Med.* **4,** 1425–1428.

Madden, J. J., Whaley, W. L., Ketelsen, D., and Donahoe, R. M. (2001). The morphine-binding site on human activated T-cells is not related to the μ-opioid receptor. *Drug Alcohol Depend.* **62,** 131–139.

Martens, H., Goxe, B., and Geenen, V. (1996). The thymic repertoire of neuroendocrine self-antigens: Physiological implications in T-cell life and death. *Immunol. Today* **17,** 312–317.

McCain, H. W., Lamster, I. B., Bozzone, J. M., and Grbic, J. T. (1982). β-Endorphin modulates human immune activity via non-opiate receptor mechanisms. *Life Sci.* **31,** 1619–1624.

McEvoy, A. N., Bresnihan, B., FitzGerald, O., and Murphy, E. P. (2001). Corticotropin-releasing hormone signaling in synovial tissue from patients with early inflammatory arthritis is mediated by the type 1 α-corticotropin-releasing hormone receptor. *Arthritis Rheum.* **44,** 1761–1767.

Millet, I., Phillips, R. J., Sherwin, R. S., Ghosh, S., Voll, R. E., Flavell, R. A., Vignery, A., and Rincon, M. (2000). Inhibition of NF-κB activity and enhancement of apoptosis by the neuropeptide calcitonin gene-related peptide. *J. Biol. Chem.* **275,** 15114–15121.

Mousa, S. A., Schafer, M., Mitchell, W. M., Hassan, A. H., and Stein, C. (1996). Local upregulation of corticotropin-releasing hormone and interleukin-1 receptors in rats with painful hindlimb inflammation. *Eur. J. Pharmacol.* **311,** 221–231.

Murphy, E. P., McEvoy, A., Conneely, O. M., Bresnihan, B., and FitzGerald, O. (2001). Involvement of the nuclear orphan receptor NURR1 in the regulation of corticotropin-releasing hormone expression and actions in human inflammatory arthritis. *Arthritis Rheum.* **44,** 782–793.

Ni, X., Lin, B. C., Song, C. Y., and Wang, C. H. (1999). Dynorphin A enhances mitogen-induced proliferative response and interleukin-2 production of rat splenocytes. *Neuropeptides* **33,** 137–143.

Nishioka, T., Kurokawa, H., Takao, T., Kumon, Y., Nishiya, K., and Hashimoto, K. (1996). Differential changes of corticotropin releasing hormone (CRH) concentrations in plasma and synovial fluids of patients with rheumatoid arthritis (RA). *Endocr. J.* **43,** 241–247.

Ohsawa, M., Mizoguchi, H., Narita, M., Nagase, H., Kampine, J. P., and Tseng, L. F. (2001). Differential antinociception induced by spinally administered endomorphin-1 and endomorphin-2 in the mouse. *J. Pharmacol. Exp. Ther.* **298,** 592–597.

Owen, D. L., Morley, J. S., Ensor, D. M., and Miles, J. B. (1998). The C-terminal tetrapeptide of β-endorphin (MPF) enhances lymphocyte proliferative responses. *Neuropeptides* **32,** 131–139.

Pampusch, M. S., Serie, J. R., Osinski, M. A., Seybold, V. S., Murtaugh, M. P., and Brown, D. R. (2000). Expression of nociceptin/OFQ receptor and prepro-nociceptin/OFQ in lymphoid tissues. *Peptides* **21,** 1865–1870.

Panerai, A. E., and Sacerdote, P. (1997). β-endorphin in the immune system: A role at last? *Immunol. Today* **18,** 317–319.

Panerai, A. E., Sacerdote, P., Bianchi, M., and Manfredi, B. (1997). Intermittent but not continuous inescapable footshock stress and intracerebroventricular interleukin-1 similarly affect immune responses and immunocyte β-endorphin concentrations in the rat. *Int. J. Clin. Pharmacol. Res.* **17,** 115–126.

Park, J. H., Lee, Y. J., Na, S. Y., and Kim, K. L. (2000). Genomic organization and tissue-specific expression of rat urocortin. *Neurosci. Lett.* **292,** 45–48.

Pasnik, J., Tchorzewski, H., Baj, Z., Luciak, M., and Tchorzewski, M. (1999). Priming effect of Met-enkephalin and β-endorphin on chemiluminescence, chemotaxis and CD11b molecule expression on human neutrophils *in vitro*. *Immunol. Lett.* **67,** 77–83.

Peluso, J., LaForge, K. S., Matthes, H. W., Kreek, M. J., Kieffer, B. L., and Gaveriaux-Ruff, C. (1998). Distribution of nociceptin/orphanin FQ receptor transcript in human central nervous system and immune cells. *J. Neuroimmunol.* **81,** 184–192.

Peluso, J., Gaveriaux-Ruff, C., Matthes, H. W., Filliol, D., and Kieffer, B. L. (2001). Orphanin FQ/nociceptin binds to functionally coupled ORL1 receptors on human immune cell lines and alters peripheral blood mononuclear cell proliferation. *Brain Res. Bull.* **54,** 655–660.

Poliak, S., Mor, F., Conlon, P., Wong, T., Ling, N., Rivier, J., Vale, W., and Steinman, L. (1997). Stress and autoimmunity: The neuropeptides corticotropin-releasing factor and urocortin suppress encephalomyelitis via effects on both the hypothalamic-pituitary-adrenal axis and the immune system. *J. Immunol.* **158,** 5751–5756.

Postel-Vinay, M. C., de Mello-Coelho, V., Gagnerault, M. C., and Dardenne, M. (1997). Growth hormone stimulates the proliferation of activated mouse T lymphocytes. *Endocrinology* **138,** 1816–1820.

Przewlocki, R., Hassan, A. H. S., Lason, W., Epplen, C., Herz, A., and Stein, C. (1992). Gene expression and localisation of opioid peptides in immune cells of inflamed tissue: Functional role in antinociception. *Neuroscience* **48,** 491–500.

Raap, T., Justen, H. P., Miller, L. E., Cutolo, M., Scholmerich, J., and Straub, R. H. (2000). Neurotransmitter modulation of interleukin 6 (IL-6) and IL-8 secretion of synovial fibroblasts in patients with rheumatoid arthritis compared to osteoarthritis. *J. Rheumatol.* **27,** 2558–2565.

Radulovic, M., Dautzenberg, F. M., Sydow, S., Radulovic, J., and Spiess, J. (1999). Corticotropin-releasing factor receptor 1 in mouse spleen: Expression after immune stimulation and identification of receptor-bearing cells. *J. Immunol.* **162**, 3013–3021.

Rittner, H. L., Brack, A., Machelska, H., Mousa, S. A., Bauer, M., Schafer, M., and Stein, C. (2001). Opioid peptide-expressing leukocytes: Identification, recruitment, and simultaneously increasing inhibition of inflammatory pain. *Anesthesiology* **95**, 500–508.

Roy, S., Ge, B. L., Ramakrishnan, S., Lee, N. M., and Loh, H. H. (1991). [3H]morphine binding is enhanced by IL-1-stimulated thymocyte proliferation. *FEBS Lett.* **287**, 93–96.

Sacerdote, P., di San Secondo, V. E., Sirchia, G., Manfredi, B., and Panerai, A. E. (1998). Endogenous opioids modulate allograft rejection time in mice: Possible relation with Th1/Th2 cytokines. *Clin. Exp. Immunol.* **113**, 465–469.

Sacerdote, P., Gaspani, L., and Panerai, A. E. (2000). The opioid antagonist naloxone induces a shift from type 2 to type 1 cytokine pattern in normal and skin-grafted mice. *Ann. N.Y. Acad. Sci.* **917**, 755–763.

Said, S. I., and Dickman, K. G. (2000). Pathways of inflammation and cell death in the lung: Modulation by vasoactive intestinal peptide. *Regul. Pept.* **93**, 21–29.

Sakane, T., and Suzuki, N. (1998). The role of somatostatin in the pathophysiology of rheumatoid arthritis. *Clin. Exp. Rheumatol.* **16**, 745–749.

Sakurada, S., Hayashi, T., Yuhki, M., Orito, T., Zadina, J. E., Kastin, A. J., Fujimura, T., Murayama, K., Sakurada, C., Sakurada, T., Narita, M., Suzuki, T., Tan-no, K., and Tseng, L. F. (2001). Differential antinociceptive effects induced by intrathecally administered endomorphin-1 and endomorphin-2 in the mouse. *Eur. J. Pharmacol.* **427**, 203–210.

Savino, W., de Mello-Coelho, V., and Dardenne, M. (1995). Control of the thymic microenvironment by growth hormone/insulin-like growth factor-1–mediated circuits. *Neuroimmunomodulation* **2**, 313–318.

Schafer, M., Mousa, S. A., and Stein, C. (1997). Corticotropin-releasing factor in antinociception and inflammation. *Eur. J. Pharmacol.* **323**, 1–10.

Serhan, C. N., Fierro, I. M., Chiang, N., and Pouliot, M. (2001). Nociceptin stimulates neutrophil chemotaxis and recruitment: Inhibition by aspirin-triggered-15-epi-lipoxin A4. *J. Immunol.* **166**, 3650–3654.

Shahabi, N. A., Linner, K. M., and Sharp, B. M. (1990). Murine splenocytes express a naloxone-insensitive binding site for β-endorphin. *Endocrinology* **126**, 1442–1448.

Sharp, B. M., Roy, S., and Bidlack, J. M. (1998). Evidence for opioid receptors on cells involved in host response and the immune system. *J. Neuroimmunol.* **83**, 45–56.

Skofitsch, G., Hamill, G. S., and Jacobowitz, D. M. (1984). Capsaicin depletes corticotropin-releasing factor-like immunoreactive neurons in the rat spinal cord and medulla oblongata. *Neuroendocrinology* **38**, 514–517.

Singh, V. (1989). Stimulatory effect of corticotropin-releasing neurohormone on human lymphocyte proliferation and interleukin-2 receptor expression. *J. Neuroimmunol.* **23**, 256–262.

Singh, V., and Leu, C. (1990). Enhancing effect of corticotropin-releasing neurohormone on the production of interleukin-1 and interleukin-2. *Neurosci. Lett.* **120**, 151–154.

Smith, E. M., Morrill, A. C., Meyer, W. J., and Blalock, J. E. (1986). Corticotropin releasing factor induction of leukocyte derived immunoreactive ACTH and endorphins. *Nature (London)* **321**, 881.

Smith, E. M., Brosnan, P., Meyer, W. J., and Blalock, J. E. (1987). An ACTH receptor on human mononuclear leukocytes. *New Engl. J. Med.* **317**, 1266–1269.

Solomon, G. F. (1997). Clinical and social implications of stress-induced neuroendocrine-immune interactions. *In* "Stress, Stress Hormones and the Immune System" (J. C. Buckingham, G. E. Gillies, and A-M. Cowell, eds.), pp. 225–239. John Wiley and Sons, Chichester.

Star, R. A., Rajora, N., Huang, J., Stock, R. C., Catania, A., and Lipton, J. M. (1995). Evidence of autocrine modulation of macrophage nitric oxide synthase by α-melanocyte-stimulating hormone. *Proc. Natl. Acad. Sci. USA* **92**, 8016–8020.

Stein, C., Hassan, A. H. S., Lehrberger, K., Giefing, J., and Yassouridis, A. (1993). Local analgesic effect of endogenous opioid peptides. *Lancet* **342**, 321–322.

Stenzel-Poore, M. P., Duncan, J. E., Rittenberg, M. B., Bakke, A. C., and Heinrichs, S. C. (1996). CRH overproduction in transgenic mice. *Ann. N.Y. Acad. Sci.* **780**, 36–48.

Stephanou, A., Jessop, D. S., Knight, R. A., and Lightman, S. L. (1990). Corticotropin-releasing factor-like immunoreactivity and mRNA in human leucocytes. *Brain Behav. Immun.* **4**, 67–73.

Straub, R. H., Dorner, M., Riedel, J., Kubitza, M., Van Rooijen, N., Lang, B., Scholmerich, J., and Falk, W. (1998). Tonic neurogenic inhibition of interleukin-6 secretion from murine spleen caused by opioidergic transmission. *Am. J. Physiol.* **274**, R997–R1003.

Takeba, Y., Suzuki, N., Kaneko, A., Asai, T., and Sakane, T. (2001). Endorphin and enkephalin ameliorate excessive synovial cell functions in patients with rheumatoid arthritis. *J. Rheumatol.* **28**, 2176–2183.

ten Bokum, A. M., Melief, M. J., Schonbrunn, A., van der Ham, F., Lindeman, J., Hofland, L. J., Lamberts, S. W., and van Hagen, P. M. (1999a). Immunohistochemical localization of somatostatin receptor sst2A in human rheumatoid synovium. *J. Rheumatol.* **26**, 532–535.

ten Bokum, A. M., Lichtenauer-Kaligis, E. G., Melief, M. J., van Koetsveld, P. M., Bruns, C., van Hagen, P. M., Hofland, L. J., Lamberts, S. W., and Hazenberg, M. P. (1999b). Somatostatin receptor subtype expression in cells of the rat immune system during adjuvant arthritis. *J. Endocrinol.* **161**, 167–175.

ten Bokum, A. M., Hofland, L. J., and van Hagen, P. M. (2000). Somatostatin and somatostatin receptors in the immune system: A review. *Eur. Cytokine Netw.* **11**, 161–176.

Udelsman, R., Harwood, J. P., Millan, M. A., Chrousos, G. P., Goldstein, D. S., Zimlichman, R., Catt, K. J., and Aguilera, G. (1986). Functional corticotropin releasing factor receptors in the primate peripheral sympathetic nervous system. *Nature (London)* **319**, 147–150.

Uzuki, M., Sasano, H., Muramatsu, Y., Totsune, K., Takahashi, K., Oki, Y., Iino, K., and Sawai, T. (2001). Urocortin in the synovial tissue of patients with rheumatoid arthritis. *Clin. Sci. (London).* **100**, 577–589.

Vanneste, Y., Thome, A. N., Vandermissen, E., Charlet, C., Franchimont, D., Martens, H., Lliaubet, A.-M., Schimpff, R.-M., Rostene, W., and Geenen, V. (1997). Identification of neurotensin-related peptides in human thymic epithelial cell membranes with major histocompatibility complex class 1 molecules. *J. Neuroimmunol.* **76**, 161–166.

Venihaki, M., Dikkes, P., Carrigan, A., and Karalis, K. P. (2001). Corticotropin-releasing hormone regulates IL-6 expression during inflammation. *J. Clin. Invest.* **108**, 1159–1166.

Wang, X., Xing, L., Xing, Y., Tang, Y., and Han, C. (1999). Identification and characterization of immunoreactive calcitonin gene-related peptide from lymphocytes of the rat. *J. Neuroimmunol.* **94**, 95–102.

Weigent, D. A., Blalock, J. E., and LeBoeuf, R. D. (1991). An antisense oligodeoxynucleotide to growth hormone messenger ribonucleic acid inhibits lymphocyte proliferation. *Endocrinology* **128**, 2053–2057.

Williamson, S. A., Knight, R. A., Lightman, S. L., and Hobbs, J. R. (1988). Effects of β-endorphin on specific immune responses in man. *Immunology* **65**, 47–51.

Xing, L., Guo, J., and Wang, X. (2000). Induction and expression of β-calcitonin gene-related peptide in rat T lymphocytes and its significance. *J. Immunol.* **165**, 4359–4366.

Yin, D., Mufson, A., Wang, R., and Shi, Y. (1999). Fas-mediated cell death promoted by opioids. *Nature (London)* **397**, 218.

BRAIN–IMMUNE INTERACTIONS IN SLEEP

Lisa Marshall and Jan Born

Department of Clinical Neuroendocrinology
Medical University of Lübeck, 23538 Lübeck, Germany

I. Sleep

Sleep research began in earnest in 1953 when Aserinsky and Kleitman discovered rapid eye movement (REM) sleep. The state of sleep is most frequently characterized by neurophysiological, vegetative, and endocrine changes. The functions of sleep appear to be manifold and related to the body as well as to the brain/mind. Whereas the former are most easily associated with cell growth, temperature reduction/energy restoration, and immune function, the latter involve memory formation, neuronal repair, and reorganization as well as cerebral metabolism in sleep (e.g., Drucker-Colin, 1995; Toth, 1995; Maquet, 2001). Moreover, memory formation is not only limited to cognitive processes, immunological memory also reportedly benefits from sleep (see Sections IV.A and IV.B.). Characteristic sleep-associated changes in neurophysiological (electroencephalographic), vegetative, and endocrine parameters with the main focus on human research are given in Sections I.A and I.B. For more on sleep regulatory mechanisms, their circadian and homeostatic components, and sleep functions refer to other

INTERNATIONAL REVIEW OF
NEUROBIOLOGY, VOL. 52

93

reviews (e.g., Horne, 1992; Borbely, 1998; Czeisler and Klerman, 1999; Peigneux, 2001).

A. ELECTROENCEPHALOGRAPHIC RECORDING IN SLEEP

In humans the standard procedure to assess sleep in the laboratory is the polysomnographical recording. This consists of the continuous measurement of electroencephalographic activity (EEG), electrooculographic activity (EOG), and electromyographic activity (EMG). Wakefulness and five stages of sleep, stages 1–4 and REM sleep, are discriminated, whereby stages 1 and 2 represent light Non-REM (NREM) sleep and stages 3 and 4 represent slow-wave sleep (SWS) or also deep NREM sleep. The criteria of Rechtschaffen and Kales (Rechtschaffen and Kales, 1968) are commonly used for classification. According to this scoring system, a subject is still awake when the alpha rhythm in the EEG (8–12 Hz) predominates and the EMG reveals enhanced activity. Stage 1 sleep is associated with the occurrence of low voltage, mixed frequency EEG with a prominence of 2–7 Hz and decreased EMG activity. Sleep stage 2 begins as soon as the first sleep spindles and/or K complexes appear in the EEG, which are both phasic events of about 0.5 s duration. Slow-wave sleep is characterized by large amplitude (>75 μV) slow waves in the delta frequency range (0.5–4 Hz), also termed EEG slow-wave activity (SWA). In sleep stage 3 such slow-wave activity must occur at least 20% of the time; when slow waves occur 50% or more of the time, stage 4 is present. EEG SWA is considered an indicator of sleep intensity. During REM sleep the EEG is characterized by low-voltage EEG with fast oscillations of mixed frequency, similar to the EEG during attention in awake subjects. However, the EMG is tonically suppressed and episodic REMs occur.

Hypnograms in humans are usually constructed by assigning each successive 30-s period of polysomnographic recordings a sleep score value, which thus reveals the temporal pattern (architecture) across the entire sleep period (Fig. 1). Undisturbed nocturnal sleep in humans typically reveals four to five NREM–REM sleep cycles of about 100 min duration. As Fig. 1 shows SWS is most prominent at the beginning of the night, during the first and second sleep cycles. REM sleep episodes, on the other hand, increase in length in the course of nocturnal sleep with marginal amounts during early sleep and very extended periods over 30 min toward the end of sleep.

Sleep architecture differs between species. Whereas in humans normal sleep occurs during the night, the rat spends about 70% of the light period and about 30% of the dark period asleep. As in humans, in rats SWS

A

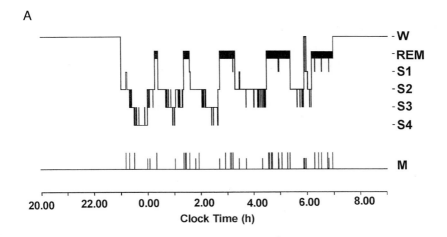

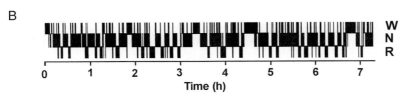

FIG. 1. Hypnograms of a human and rat subject. (A) Hypnogram with the sleep stages (W, wake; REM, rapid eye movement sleep; S1, S2, S3, S4, sleep stages 1–4; M, movements) of a healthy human subject. (B) Hypnogram with the sleep stages (W, wake; N, non rapid eye movement sleep; R, rapid eye movement sleep) of a rat in a baseline condition.

is high at the beginning of the major rest period and declines thereafter. A NREM–REM cycle is much shorter in rats than in humans, lasting typically 7–13 min. The most frequent sleep scoring method in rats is based on the ratio between EEG delta (0.75–4 Hz) and theta (6–9 Hz) bands, and EMG activity. NREM sleep is characterized by low EMG amplitude and EEG amplitude with high power density in the delta band, REM sleep is characterized by low EMG amplitude and low EEG amplitude with high values in the theta band, and waking is characterized by both high EMG and high EEG amplitudes (Trachsel *et al.*, 1991; Franken *et al.*, 1991). Rabbits, on a 12:12-h light:dark cycle, spend approximately only 40–50% of the light period asleep compared with about 25–35% of the dark period, whereby the periods of sleep are concentrated around the transitions between the dark and the light period. The 24-h sleep architecture of cats is even more variable, in general they sleep more during the light period (Opp *et al.*, 1992).

Neuronal substrates underlying sleep-associated brain rhythms are reported by Steriade (e.g., Steriade, 1993).

B. ENDOCRINE ACTIVITY IN SLEEP

Sleep is accompanied by distinct temporal patterns of endocrine activity that are probably especially relevant for mediating immunological effects. In particular, the interactions between sleep, immune activity, and several hormones of the hypothalamic-pituitary-adrenal axis—growth hormone-releasing hormone (GHRH), growth hormone (GH), adrenocorticotropin (ACTH), and cortisol—are well documented (Fig. 2). ACTH affects the immune system essentially through its stimulatory effect on the adreno-cortical release of glucocorticoids, which at high concentrations generally suppress immune activation. At the beginning of nocturnal sleep in humans, ACTH and cortisol release initially decreases to a minimum. Toward the second half of the night, around 180–200 min following sleep onset, episodic ACTH/cortisol secretion starts to increase reaching average concentrations that are five- to tenfold higher than during the early part of the night. The nocturnal pituitary–adrenal secretory pattern possesses circadian and sleep-associated components. Much evidence indicates that both components contribute to the strong suppression of the hormones at the beginning of the night (Born and Fehm, 1998). The episodic ACTH/cortisol release in later sleep occurs primarily during periods of NREM sleep; however, average ACTH/cortisol concentrations are highest in the REM sleep–dominated second half of the night. Exogenous cortisol administration suppresses REM sleep and, at low concentrations, increases time spent in SWS in humans (Born et al., 1989). Corticotropin-releasing hormone (CRH) induced non-significant effects in the same direction, a higher dose of CRH significantly reduced SWS, but enhanced sleep stages 1 and 2 (Tsuchiyama et al., 1995). In rabbits, cortisone administration also suppressed the amount of time spent in REM sleep but did not affect time spent in SWS. However, the amplitude of EEG SWA during SWS was reduced (Toth et al., 1992). In contrast to ACTH, GH usually peaks concurrent with the first extended epoch of nocturnal SWS at the beginning of normal nocturnal sleep. Exogenous systemic GHRH administration increases both GH concentration and time spent in SWS in humans (Steiger et al., 1992; Kerkhofs et al., 1993; Marshall et al., 1996) and in rats (Obál et al., 1996). A further sleep-related modulator of the immune system is melatonin, which is distinctly increased during the dark hours of nocturnal sleep and inhibited during daytime wakefulness (Maestroni, 2001). It has been suggested that the nocturnal increase in IL-2 concentrations could depend at least in part on the promoting actions of

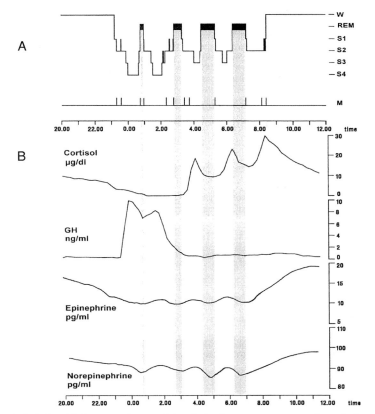

FIG. 2. Relationship between sleep and circulating hormone concentrations during normal nocturnal sleep in a healthy young man. (A) Hypnogram with the sleep stages (W, wake; REM, rapid eye movement sleep; S1, S2, S3, S4, sleep stages 1–4; M, movements). (B) Profiles of plasma concentrations of cortisol, growth hormone (GH), epinephrine, and norepinephrine. Minimum plasma concentrations of cortisol and maximum concentrations of GH occur during early sleep. Catecholamine concentrations are reduced during sleep, in particular during REM sleep. Gray vertical bars indicate periods of REM sleep.

melatonin (Lissoni *et al.,* 1998). In addition, light affects immunomodulatory neuropeptides such as vasointestinal peptide (VIP) and neuropeptide Y (NPY) as well as neurotransmitters such as norepinephrine, acetylcholine, and γ-aminobutyric acid (GABA; for reviews see Poon *et al.,* 1994; Plytycz and Seljelid, 1997; Roberts, 2000).

Cytokines are also known to stimulate endocrine systems, which in turn can affect sleep. For instance IL-1, IL-6, TNF-α, nerve growth factor (NGF), IFN-α, and in some cases IL-2 have been shown to stimulate the hypothalamic-pituitary-adrenocortical axis (Späth-Schwalbe *et al.*, 1989, 1998; reviewed in Turnbull and Rivier, 1999; Dunn, 2000). IL-1β is known for stimulating the somatotropic axis (Payne *et al.*, 1992). The stimulative actions of pro-inflammatory cytokines on GHRH have been considered essential for the somnogenic effects of these cytokines observed in rats and rabbits (Krueger *et al.*, 1999). Cytokines also influence neurotransmitter activity (De Simoni *et al.*, 1995); however, there is no clear picture on how neurotransmitters may mediate the effects of cytokines on sleep.

II. Microbial Products and Cytokines: A Brief Overview

Excessive sleepiness and fever are constitutional symptoms associated with systemic infection. Sleep patterns are modified by inoculation of animals with bacterial, viral, protozoan, and fungal organisms. The best-studied microorganisms associated with sleep enhancement are the bacteria. Muramyl peptides are fragments of peptidoglycan, a major component of the cell wall in gram-positive bacteria. They were first isolated from rabbit brains and human urine (Krueger *et al.*, 1982) and were revealed to exert somnogenic effects in rabbits, cats, rats, and monkeys (Wexler and Moore-Ede, 1984; Krueger and Majde, 1994). Muramyl peptides appear to induce sleep by binding to serotinergic receptors located peripherally on macrophages, or more likely on glial cells in the brain, which in turn releases somnogenic pro-inflammatory cytokines such as IL-1β (Karnovsky, 1986; Silverman *et al.*, 1989; Andren *et al.*, 1995; Pabst *et al.*, 1999).

Lipopolysaccharides (LPS), the main component of the outer cell wall of gram-negative bacteria together with its lipid moiety (lipid A) are responsible for the biological activity of endotoxin. On death, when the bacterial cell wall breaks down endotoxin is set free in large amounts. Both LPS and endotoxin are somnogenic in rats (Krueger *et al.*, 1986) and rabbits (Cady *et al.*, 1989) and, at only very low concentrations, also in humans (Mullington *et al.*, 2000). LPS injected into the general circulation may enter the brain by penetrating the circumventricular organs, which then allows the endotoxin to trigger locally the synthesis of its own receptor CD14 and stimulate the transcription of pro-inflammatory cytokines such as IL-1β and TNF-α (Rivest *et al.*, 2000).

Changes in sleep patterns also occur upon infection with viruses, such as influenza or HIV in humans. However, the viral factor responsible for triggering the acute phase response, or flulike symptoms, is not defined. One candidate viral factor is double-stranded RNA which is generated during viral infection (Kimura-Takeuchi *et al.*, 1992; Majde *et al.*, 1998). Fang *et al.* (1999)

demonstrated that intracerebroventricular injection of synthetic low molecular weight double-stranded RNAs increased NREM sleep and suppressed REM sleep, aside from inducing fever. The ability of double-stranded RNA to induce various cytokines (e.g., interferons) as well as other inflammatory mediators such as prostaglandins was reviewed by Krueger and Majde (1995). Infections with pathogenic fungi such as *Candida albicans* and protozoans such as *Trypanosoma brucei* (responsible for the sleeping sickness) can alter sleep patterns through immune stimulation of their hosts and subsequent cytokine induction (Krueger and Majde, 1994).

Cytokines are a heterogeneous group of polypeptide mediators involved in the regulation of the immune system. Immune cells are the main sources of cytokines; however, neurons and other nonimmune cells also produce cytokines and express cytokine receptors. Besides membrane-bound receptors, soluble receptors have been identified for many cytokines (e.g., sIL-1β, sTNF-αR, sIL-2R, sIL-4R, sIL-10R, sIFN-γR). Most soluble cytokine receptors *in vivo* inhibit the ability of cytokines to bind their membrane receptors and thus prevent them from generating a biological response. However, soluble IL-6 receptor (sIL-6R) is agonistic in that together with IL-6 it can activate target cells lacking the complete membrane-bound receptor form (trans-signaling) (Igaz *et al.*, 2000). Table I summarizes some diverse cytokines and their subdivision into families as suggested by Hopkins and Rothwell (1995), which appear to be also related to sleep.

There are several routes through which cytokines may influence the central nervous system sleep processes. When induced systemically, cytokines may communicate to the brain via four mechanisms: (1) carrier-mediated transport across the blood–brain barrier (BBB), (2) transport across the circumventricular organs which lack a BBB, (3) neuronal activation via primary sensory neurons with afferents to the brain such as the vagal afferents,

TABLE I
SELECTED CYTOKINE FAMILIES AND THEIR MAJOR FEATURES[a]

Selected cytokine families	Major features
Interleukins (e.g., IL-1β, IL-2)	Immunoregulatory
Tumor necrosis factors (e.g., TNF-α)	Immunoregulatory, tumor cytotoxicity
Interferons (e.g., IFN-α, -β, -γ)	Immunoregulatory
Colony-stimulating factors (e.g., G-CSF)	Colony cell formation
Growth factors (e.g., IGF-I)	Cell growth and differentiation
Neurotrophins (e.g., BDNF, NGF)	Growth and differentiation of neurons

[a] Modified from Hopkins and Rothwell (1995).

and/or (4) binding of cytokines to cerebral vascular endothelial-glial cells, thereby inducing secondary factors such as prostaglandins, nitric oxide, and other cytokines (Reyes and Coe, 1998; Kronfol and Remick, 2000; Dantzer *et al.*, 2001; Maier *et al.*, 2001; Dunn *et al.*, 2002). Alternatively, numerous cytokines have been shown to arise from cellular components of brain parenchyma (e.g., neurons, astrocytes, microglia, endothelia), the choroid plexus, and brain vasculature (Reichlin, 2001). The ability of cytokines (derived initially either from the periphery or from the brain) to stimulate endocrine systems and thereby modulate sleep was described previously in Section I.B.

A. Endogenous Sleep–Wake Fluctuations in Cytokine Patterns

A sleep–wake or diurnal variation is an indicator for a possible biological relevance of cytokines in sleep regulation. Unfortunately, the number of studies, especially on cytokine measurements within the brain are limited. Diurnal variations within cerebrospinal fluid have been reported for concentrations of IL-1 (Lue *et al.*, 1988), or within the brain for TNF-α mRNA (Bredow *et al.*, 1997; Floyd and Krueger, 1997) and brain-derived neurotrophic factor (BDNF) (Bova *et al.*, 1998; Berchtold *et al.*, 1999) as well as for the uptake of circulating IL-1α by the spinal cord and brain (Banks *et al.*, 1998). Enhanced expression of IL-1β mRNA in the hypothalamus, hippocampus, and cortex during the light period, the period when rats sleep the most, and decreased expression during the dark period was reported by Taishi *et al.* (Taishi *et al.*, 1997, 1998; Krueger *et al.*, 1998). Others failed to find a systematic 24-h variation in IL-1β mRNA in the rat brain, although levels were increased significantly in the hypothalamus and brain stem after sleep deprivation (Mackiewicz *et al.*, 1996).

Diurnal patterns of circulating cytokines have been most extensively investigated in humans (Sothern and Roitman-Johnson, 2001). Serum concentrations of most cytokines are, however, quite low and close to the detection limit of the assay. An exception is IL-6; however, studies reveal inconsistent results. Born *et al.* (1997) failed to reveal any sleep-related changes in plasma concentrations over a 48-h period, whereas Vgontzas and colleagues (1999) measured a biphasic circadian pattern. This is probably in part a methodological problem, since IL-6 measurements obtained by frequent blood sampling using an indwelling catheter over extended time periods have been shown to produce spurious results (Haack *et al.*, 2000). Serum levels of IL-2 have also been measured and revealed an early morning peak (Lissoni *et al.*, 1998), although this did not reach significance. The production of IL-2 measured after *in vitro* stimulation similarly increased distinctly in the early morning (Born *et al.*, 1997).

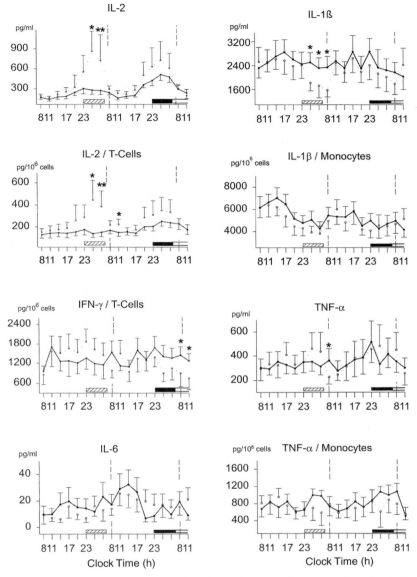

FIG. 3. Patterns of cytokine concentrations dependent upon sleep versus sleep deprivation. Mean (±SEM) production (IL-2, IFN-γ, IL-1β, TNF-α) or plasma concentration (IL-6) of cytokines are shown for a session consisting of two successive nights of normal nocturnal sleep (both horizontal bars); dotted lines, and a session with one night of sleep deprivation (hatched horizontal bar) followed by a night of recovery sleep (black horizontal bar); continuous line ($n = 10$). Cytokines IL-2, IFN-γ, IL-1β, and TNF-α are shown relative to the main cellular source of their production in peripheral blood. Asterisks indicate significant differences between the two sessions. Modified from Born *et al.* (1997). Copyright 1997. The American Association of Immunologists, Inc.

The production of IL-1β and TNF-α also revealed sleep-associated changes in concentration (Hohagen *et al.*, 1993; Uthgenannt *et al.*, 1995; Darko *et al.*, 1995; Born *et al.*, 1997). During the nocturnal sleep period IL-1β and IL-6 have been reported to increase (Moldofsky *et al.*, 1986; Gudewill *et al.*, 1992; Covelli *et al.*, 1992; Bauer *et al.*, 1994). However, when concentrations of IL-1β and TNF-α were put into relationship with the concentration of monocytes, the main cellular source of these cytokines in blood, which itself reveals a 24-h pattern, significant temporal cytokine patterns disappeared (Fig. 3). This suggests that respective 24-h cytokine variations do not necessarily reflect cytokine production or release per se, but rather the number of circulating cells producing these monokines.

Hence, when discussing circulating cytokine concentrations it is important to be aware of the concentrations of cytokine-producing cells, which are specific white blood cell subpopulations for most circulating cytokine families. In humans, across the 24-h day monocytes and lymphocytes, including the T helper (CD4$^+$), T cytotoxic (CD8$^+$), and B cell subsets, obtain maximal values during the night, whereby sleep deprivation shifted the nocturnal peak from the period before midnight to the early morning. Minimal lymphocyte and monocyte counts occurred within 1–5 h upon morning awakening. Natural killer cells obtained maximal counts during late afternoon. The decrease in natural killer cells around midnight was suppressed during sleep deprivation (Born *et al.*, 1997). Similar to humans, at the beginning of the rat's active period leukocyte and lymphocyte numbers were significantly reduced, the decrease in lymphocytes was mirrored by an increase in neutrophils, and peripheral blood leukocyte numbers were inversely related to plasma corticosterone levels (Dhabhar *et al.*, 1994). Collectively the data indicate a circadian increase in circulating monocytes and lymphocytes during the rest period that is enhanced by sleep. Also, the production of T-cell related cytokines appears to be increased by sleep, independent from migratory changes in T-cell distribution.

III. Effects of Microbial Products and Cytokines on Sleep

This section gives an overview of the effects on sleep of experimental administration of microbial products and cytokines in humans and laboratory animals. When possible, results are discussed comparatively. However, due to the strong interactions between immune, endocrine, circadian, and sleep–wake activity patterns and their species specificity, comparisons must always be treated with caution.

A. ENDOTOXIN

The effects of intravenously administered endotoxin on the sleep of healthy human subjects was studied extensively by the group of Pollmächer and colleagues (Pollmächer *et al.*, 1993, 1995). The administration of 0.4 ng/kg and 0.8 ng/kg *Salmonella abortus* endotoxin at 1900 h induced significant increases in circulating levels of TNF-α, IL-6, cortisol, and ACTH and in body temperature. The increase in TNF-α was the first of these responses commencing 1 h after endotoxin administration, followed by IL-6. After the increase in cytokine levels, heart rate and rectal temperature increased. The hormonal concentrations of cortisol, ACTH, and GH revealed bi- or multiphasic responses. Neither time spent in SWS nor EEG SWA were affected by endotoxin administration. However, time in sleep stage 2 increased and time spent awake decreased significantly over the entire night after the 0.4 ng/kg dosage. The first REM sleep period tended to occur later, and it was prolonged after endotoxin administration. All of these effects of endotoxin on sleep were most pronounced in the first half of the sleep period. In another study, the same research group compared the effects of the same endotoxin doses and placebo with an even lower dosage of 0.2 ng/kg administered at 2300 h, that is, directly before lights were turned off to begin the sleep period (Mullington *et al.*, 2000). The effects of the low 0.2 ng/kg dose on various cytokines, soluble cytokine receptors, cortisol, body temperature, and sleep results are shown in Fig. 4. Time spent in NREM sleep as well as EEG SWA was acutely enhanced only after this very subtle host defense activation, which was not accompanied by fever. On the other hand, all endotoxin concentrations decreased the time spent in REM sleep. REM sleep suppression was also the most consistent effect on sleep after daytime endotoxin administration (Korth *et al.*, 1996) and in depressed patients treated with endotoxin (Bauer *et al.*, 1995). REM sleep suppression is very likely a consequence of increased cortisol release after endotoxin administration (Born *et al.*, 1989).

The administration of LPS and lipid A enhanced similarly the duration of NREM sleep and SWS in rats and rabbits after either intravenous, intraperitoneal, or intraventricular administration, and LPS also suppressed REM sleep (Krueger *et al.*, 1986; Kapas *et al.*, 1998; Schiffelholz and Lancel, 2001). The effects of LPS on sleep have been shown to depend in part on the time of administration relative to the light or dark period (Kapas *et al.*, 1998; Mathias *et al.*, 2000) and age (Schiffelholz and Lancel, 2001). In the study by Mathias *et al.*, LPS typically increased the duration of NREM sleep while increasing the number of NREM sleep episodes and decreasing their duration, increased EEG SWA, and decreased REM sleep irrespective of time of LPS administration. Differences in response to LPS administration

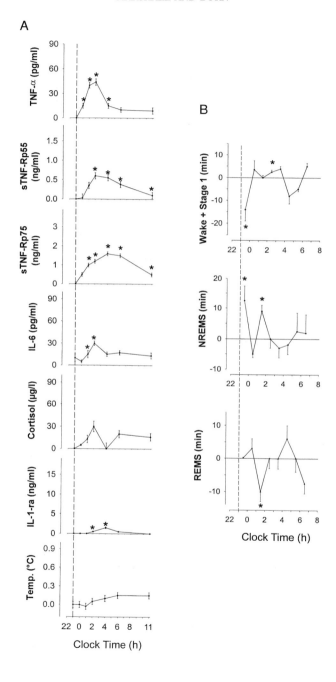

(i.e., at the beginning of the activity phase or the rest phase) were only reflected in the EEG frequencies >5 Hz during NREM sleep. In contrast, the response to LPS between young and middle-aged rats was only reflected in a decline in EEG SWA in middle-aged subjects (Schiffelholz and Lancel, 2001). Concerning possible mechanisms of LPS effects, it was shown that vagotomy can blunt the NREM sleep responses suggesting the involvement of subdiaphragmal vagi in mediating effects on sleep (Kapas *et al.*, 1998). On the other hand, TNF-α by binding to its p75 receptor does not appear to mediate the effects of endotoxin on either NREM or REM sleep (Lancel *et al.*, 1997). Together, the data reveal a somnogenic action of endotoxin depending strongly on dose. The mechanisms are unclear and may be linked to the cascade of the pro-inflammatory cytokine induced by the substance.

B. INTERLEUKINS, TUMOR NECROSIS FACTORS, INTERFERONS, COLONY-STIMULATING FACTORS, GROWTH FACTORS, AND NEUROTROPHINS

In humans the effects of cytokines on sleep were first investigated in patients receiving cytokines as treatment against cancer (TNF-α, IFN-α, IFN-γ) and in phase I clinical trials (IL-1β). Enhanced feelings of fatigue and sleepiness were observed, yet it was difficult to distinguish between effects resulting from the disease itself and those from the high doses used. Furthermore in these initial studies sleep was assessed subjectively and not through polysomnography. Polysomnographically monitored sleep has since been investigated in healthy humans after administration of IFN-α (Späth-Schwalbe *et al.*, 2000), IL-2 (Lange *et al.*, 2002a), IL-6 (Späth-Schwalbe *et al.*, 1998), and granulocyte colony-stimulating factor (G-CSF) (Schuld *et al.*, 1999). All cytokines were given subcutaneously to healthy male subjects prior to the experimental night, either at 1900 h (IFN-α, IL-2, IL-6) or at 2300 h (G-CSF). Dosages of IFN-α (1000 and 10,000 IU/kg body weight), IL-2 (1000 and 10,000 IU/kg), and IL-6 (0.5 μg/kg) were quite low, the intent being to

---

FIG. 4. Mean ($\pm$SEM) difference values between a very low dose of endotoxin and placebo. Immediately after endotoxin administration (0.2 ng/kg at 2300 h) lights were turned off and subjects allowed to sleep ($n = 7$). Time of endotoxin or placebo administration is indicated by the vertical dashed line. (A) Endotoxin-induced changes seen in cytokines and cytokine receptors, but not in cortisol or body temperature. TNF, tumor necrosis factor; sTNF-R, soluble TNF receptor; IL-1Ra, interleukin-1 receptor antagonist. (B) Endotoxin-induced changes seen in sleep disruption (top), non rapid eye movement sleep, NREMS, and rapid eye movement sleep, REMS. Asterisks indicate significant differences between endotoxin and placebo administration. Modified from Mullington *et al.* (2000), with permission.

induce plasma concentrations similar to those occurring during moderate bacterial and viral infections. Whereas none of these cytokines enhanced sleep acutely, fatigue was reported following IFN-α and IL-6. Following IL-6 the amount of polysomnographically recorded REM sleep revealed a highly significant reduction across the night, and REM sleep onset latency was prolonged. Most interesting was the reduced amount of SWS in the first half to be followed by a highly significant increase as compared to placebo in the second half of the night. Whether this increase reflected a rebound effect after acute SWS suppression or whether IL-6 induced a cascade of (biochemical) events resulting eventually in the increase in SWS can only be speculated on at present. Plasma concentrations of cortisol and ACTH were enhanced after IL-6 which probably explains the decrease in REM sleep (Born et al., 1989). The effects on sleep were paralleled by a moderate increase in body temperature (average 1°C). The effects of IL-6 correspond with results from a study on the effects of endotoxin on daytime sleep: NREM sleep within a 30-min nap 1-h following endotoxin injection was negatively correlated with the individual peak level of plasma IL-6 (Korth et al., 1996). Effects on sleep following IFN-α were similar to those for IL-6 (Späth-Schwalbe et al., 2000). The total amount of REM sleep across the night was reduced, REM sleep latency and light sleep stage 1 were increased. However, the strongest effect after the higher IFN-α dosage was a highly significant reduction in SWS in the first half of the night. IFN-α increased IL-6 plasma concentrations, ACTH, and cortisol in the first half of the night. Whereas the former may have mediated the decrement in SWS, the hormones of the hypothalamic-pituitary-adrenal system presumably accounted for the loss in REM sleep. IL-2 did not have any effect on sleep latencies, or on the amount of time spent in any sleep stage across the night or within either half of the night (Lange et al., 2002a).

The fact that a role in sleep regulation of TNF-α and IL-1β is well established in animals led Schuld and colleagues (1999) to investigate the effects of G-CSF on sleep. Subcutaneously administered G-CSF (300 μg) increases the circulating levels of both soluble TNF-α receptors, p55 and p75, and IL-1 receptor antagonist (IL-1ra) but, as the same group has revealed in a previous study (Pollmächer et al., 1996), does not affect plasma levels of cortisol and GH or temperature. G-CSF increased total time spent asleep, as well as REM sleep latency and amount of movement time. Differences in the amount of time spent in the different sleep stages and EEG power were only found after analyses over successive 2-h periods: within the first 2 h after lights off (2300–0100 h) the amount of SWS, in particular stage 3, and EEG SWA decreased by about 20%. This decrease occurred parallel to steep increases in the plasma levels of IL-1ra and both soluble TNF-α receptors. During the last 2 h of the night (0500–0700 h) sleep stage 3 and

EEG SWA tended to reveal an increase. Interestingly at this time the TNF-α plasma level was significantly increased. Overall, G-CSF, IFN-α, and IL-6 at least acutely suppress SWS and also REM sleep, although an enhancement may eventually occur at a later time period.

In animals the somnogenic effect of various cytokines have been investigated given either systemically (e.g., intravenously or intraperitoneally) or centrally (e.g., intracerebroventricularly). Table II gives an overview of administered substances, species, and effects. The pattern of effects of cytokine administration on sleep may, however, vary distinctly depending on concentration and time of administration in relation to sleep and/or the light–dark cycle. For instance, in the rat low doses of IL-1β (0.5 ng at lights out, 2.5 ng 5 h after lights on) given intracerebroventricularly increased both the duration of NREM sleep and EEG SWA. At higher doses ($\leq$10 ng and $\leq$25 ng, respectively) NREM sleep duration was promoted at night and EEG SWA during the day, whereas NREM sleep duration during the day and EEG slow wave amplitudes at night were suppressed. At the highest doses REM sleep was also decreased during both day and night (Opp *et al.*, 1991). A distinct example of the strong modulation of sleep responses by circadian processes in highly circadian species such as the rat is given by Opp and Imeri (1999). These authors summarize that during the light period of the light–dark cycle, the predominant effect of IL-1β is to increase EEG SWA without any great change in the duration of NREM sleep. In contrast, during the dark period NREM duration is generally enhanced with little effect on EEG SWA. Furthermore, intracerebroventricular and intravenous administration of IL-1β or TNF-α induced increases in EEG SWA; however, intraperitoneal injection was found to induce decreases in EEG SWA (Hansen and Krueger, 1997). These findings suggest that independent regulatory mechanisms exist for EEG SWA and NREM sleep duration.

Interestingly, four cytokines, IL-4, IL-10, IL-13, and TGF-β1 (cp. Table II) which inhibit pro-inflammatory cytokine production also inhibited spontaneous sleep. Furthermore, mice lacking a functional IL-10 gene spent a greater amount of time in SWS (Toth and Opp, 2001). These inhibitory actions are likely due to an inhibition in the production of endogenous sleep regulatory substances such as IL-1β, TNF-α, and nitric oxide (NO), and/or through antagonizing effects for instance via the production of IL-1Ra (Kubota *et al.*, 2000).

Studies on the ability to modulate sleep by cytokines have concentrated mainly on IL-1β and TNF-α. The ability to suppress sleep or sleep rebound following sleep deprivation by administering antibodies which inhibit the activity of IL-1β or TNF-α (Opp and Krueger, 1994; Takahashi *et al.*, 1995, 1997) led these authors to suggest that these cytokines are involved in physiological sleep regulation. Interestingly, in the case of IL-1β, central, but not

TABLE II

EFFECTS OF CYTOKINES ON SLEEP IN ANIMALS

Substance	Administration[a]	Effect[b]	Species	References
IL-1	Central, systemic	Enhanced duration in NREMS, enhanced EEG SWA amplitude, but also:	Rats, cats, rabbits, mice monkeys	Krueger et al. (1998), Krueger et al. (1999)
	Systemic	Enhanced duration in NREMS, no increase in EEG SWA		
IL-2	Central, systemic	Enhanced duration in NREMS, enhanced EEG SWA; suppressed REM sleep when administered at light onset	Rats, rabbits	Nistico and De Sarro (1991), Kubota et al. (2001a)
IL-4	Central	Suppressed NREMS, no change in EEG SWA, suppressed REM sleep at the highest dose (250 ng); no effect on sleep when administered at dark onset	Rabbits	Kushikata et al. (1998)
IL-6	Central	No effect on sleep	Rabbits	Opp et al. (1989)
IL-10	Central	Suppressed NREMS, no change in EEG SWA, suppressed REM sleep at the highest dose (250 ng); no effect on sleep when administered at dark onset	Rabbits	Kushikata et al. (1999b)
IL-13	Central	Suppressed NREMS, no change in EEG SWA, suppressed REMS when administered at dark onset	Rabbits	Kubota et al. (2000)
IL-15	Central	Enhanced duration in NREMS, enhanced EEG SWA, suppressed REM sleep; no effect on sleep when administered at dark onset	Rabbits	Kubota et al. (2001a)
IL-18	Central	Enhanced duration in NREMS, (rabbits, rats)	Rabbits, rats	Kubota et al. (2001b)
	Systemic	Enhanced duration in NREMS, decreased EEG SWA (rats), suppressed REM sleep		

	Administration[a]	Effect	Species	References
TNF-α	Central, systemic	Enhanced duration in NREMS, enhanced EEG SWA but also:	Rats, rabbits, mice, sheep	Krueger et al. (1998), Krueger et al. (1999), Dickstein et al., (1999)
	Systemic	Enhanced duration in NREMS, no increase in EEG SWA		
IFN-α	Central, systemic	Enhanced duration in NREMS, enhanced EEG SWA, slight reductions in REM sleep	Rabbits	Krueger et al. (1987), Kimura et al., (1994)
IFN-γ	Central	Enhanced duration in NREMS, enhanced EEG SWA, suppressed REM sleep at light onset	Rabbits	Kubota et al. (2001c)
IGF-I	Central	Enhanced duration in NREMS at the lowest dose (0.5 μg), acute NREMS suppression after the higher doses (≤5.0 μg), delayed increase in EEG SWA no effect on REMS	Rats	Obál et al. (1999)
TGF-β1	Central	Suppressed NREMS, when administered during the light period, no change in EEG SWA; no effect on sleep when administered at dark onset	Rabbits	Kubota et al. (2000)
NGF	Central	Enhanced duration in NREMS, decrease in EEG SWA, enhanced duration in REMS cats[c]: only enhanced duration in REMS	Rabbits, cats	Takahashi et al. (1999), Yamuy et al. (1995)
BDNF	Central	Enhanced duration in NREMS, no effect on EEG SWA; enhanced duration in REMS; rabbits: only enhanced duration in REMS	Rabbits, cats	Kushikata et al. (1999a)

[a] Administration is characterized as central (intracerebroventricular) or systemic (intravenous or intraperitoneal).
[b] NREMS, NREM sleep; REMS, REM sleep.
[c] Microinjection into the pons.

systemic, inhibition blocked the enhanced sleep responses in rabbits after sleep deprivation suggesting a primary central nervous system site of action for this cytokine (Opp and Krueger, 1994; Takahashi *et al.*, 1997).

Focusing on the somnogenic activity of IL-1β and TNF-α, Krueger and co-workers compiled a most integrative concept on the immunological regulation of sleep, essentially on the bases of their extensive animal studies. The concept depicted in Fig. 5 attempts to account not only for the effects of many of the cytokines depicted in Table II, but also adapts known feedback loops among the cytokines and activity of other sleep regulatory mediators. Two central pathways mediating the sleep-promoting effects of IL-1β and TNF-α involve NF-κB and the synthesis of NO and prostaglandin on the one hand, and the central nervous system induction of GHRH on the other hand. The promoting influence of GHRH on SWS and NREM sleep has been well documented in numerous animal and human studies (Krueger

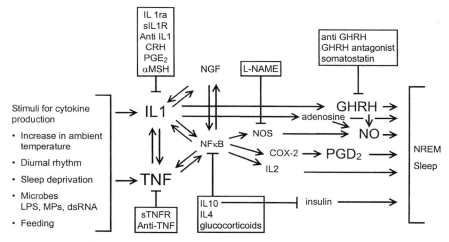

Fig. 5. Cascade of sleep regulatory events. Substances in boxes inhibit both NREMS and the production or action of substances in the somnogenic pathways. Inhibition of any one step does not result in complete sleep loss; animals likely compensate for the loss of any one step by relying on parallel somnogenic pathways. Such redundant pathways provide stability to the sleep regulatory system and alternative mechanisms by which a variety of sleep-promoting, or sleep inhibitory, stimuli may affect sleep. Our current knowledge of the biochemical events involved in sleep regulation is much more extensive than that shown here. For example, the involvement of prolactin and vasoactive intestinal peptide in REMS regulation are not illustrated. TNF, tumor necrosis factor; sTNFR, soluble TNF receptor; anti-TNF, anti-TNF antibody; NGF, nerve growth factor; NF-κB, nuclear factor-κB; IL-1Ra, IL-1 receptor antagonist; sIL-1R, soluble IL-1 receptor; anti-IL-1, antiarginine analog; NOS, nitric oxide synthase; GHRH, growth hormone-releasing hormone; →, stimulation; ⊣, inhibition. Modified from Krueger *et al.* (1999), with permission.

and Obál, 1993; Obál *et al.*, 1995, 1996; Zhang *et al.*, 1999). In addition, inhibition of GHRH using antibodies also attenuates IL-1β-induced increases in NREM sleep (Obál *et al.*, 1995). Also, the concept accounts for other findings; for instance, the enhancing effect of IFN-γ on NREM sleep appears to be mediated at least in part by TNF-α, since TNF-α receptor fragment attenuated the NREM sleep-promoting effect of IFN-γ (Kubota *et al.*, 2001c). However, although mechanisms through which sleep is modulated by these cytokines have been suggested (as shown in Fig. 5), the role of many other cytokines in physiological sleep regulation has yet to be proven. For instance anti-IL-18 antibody attenuated IL-18 muramyl dipeptide–induced sleep, but not spontaneous sleep (Kubota *et al.*, 2001b). Thus, from that study it can only be concluded that IL-18 is involved in sleep responses associated with microbial challenge.

IV. Effect of Sleep–Wake Activity on Cytokines and the Immune System

The influence of sleep–wake activity on the immune system is investigated by manipulating sleep—typically by sleep deprivation. The deprivation can involve the complete sleep period as in total sleep deprivation, partial sleep deprivation when sleep is prohibited during defined intervals (for instance a 2-h suppression of sleep in an 8-h sleep period), or the selective deprivation of a certain sleep stage (for instance REM sleep). A general problem of sleep deprivation studies is related to the nonspecific effects of stress mediated by increased hypothalamic-pituitary-adrenal axis and sympathetic nervous system activity. Whereas stress can be reduced by the voluntary participation of human subjects, particular older techniques used to prohibit sleep in animals frequently induced a state of enhanced arousal and stress (Suchecki *et al.*, 2002). An elegant method for investigating the effects of sleep activity on immune function is by measuring a long-term change in immunological function (e.g., to antibody production) subsequent to an acute suppression of sleep. Assessment in humans, in fact, indicates that plasma levels of cortisol and catecholamines during nocturnal sleep disruption, although enhanced in comparison with nocturnal sleep, by far do not reach the levels seen during daytime wakefulness or stress (e.g., Lange *et al.*, 2002b; Dodt *et al.*, 1994). Thus, generally ascribing differences in immunological parameters seen in humans between conditions of sleep and sleep deprivation to specific stress is probably not justified. However, in animals, under conditions of forced sleep deprivation, strong enhancements of the hypothalamic-pituitary-adrenocortical axis have been observed (Suchecki *et al.*, 2002). So the situation may differ from that in humans.

These considerations also underscore the strong circadian features of neu-roendocrine and immune activity in humans, which can normally be dis-sociated from effects of sleep. Accordingly, circadian rhythms interacting with the effects of sleep need to be taken into account. Circadian fluctua-tions in white blood cell subsets are of relevance when effects of sleep on cytokine concentrations are to be characterized, since the relationship be-tween number of cytokine-producing cells and cytokine concentration is important when measuring circulating cytokine levels (Section IV.A). On the other hand, apart from circulating blood cells, the brain itself also can serve as a secretory source of cytokines entering the blood (reviewed in Reichlin, 2001).

A. Effects of Sleep—Humans

To distinguish between circadian and sleep-related changes in white blood cell counts the sampling rate of blood should be frequent, at least every 4 h, and cover at least one 24-h period. Unfortunately, until recently these conditions have only rarely been met. Reviewing earlier studies on the effects of sleep deprivation on sleep Dinges et al. (1995) summarized that sleep deprivation is generally associated with an increase in signs of nonspecific immunity (e.g., phagocyte numbers, NK cell numbers, NK cell activity) with no effect on humoral immune functions (e.g., B-cell counts). In their own study, Dinges and co-workers (Dinges et al., 1994), based on a single measurement per day, revealed that numbers of circulating lympho-cytes, monocytes, granulocytes, and NK cells increased with partial sleep deprivation of 64-h, whereas T-cell numbers in blood decreased. However, in another study with samples collected every 3 h to enable a separation of circadian and sleep-related changes, gross changes in the total number of white blood cells were not observed (Born et al., 1997). The decrease in lymphocyte counts in peripheral blood observed during sleep in several human studies (Abo et al., 1981; Palm et al., 1996; Suzuki et al., 1997) as well as in animals (Haus and Smolensky, 1999) might be explained by a migration of these cells into extravascular lymphatic tissue. In fact an exper-iment in sheep reported a decreased flow rate and the release of cells into efferent lymphatic tissues during sleep (Dickstein et al., 2000). Compara-ble results revealing a decreased lymphocyte release into efferent lymphatic vessels during nocturnal sleep in humans have also been reported (Engeset et al., 1977). These last two results support a sleep-associated increase in the retention of lymphocytes in the lymph nodes. Taking the circadian rhythms in circulating blood cell counts into account, 24-h sleep deprivation was acutely associated with an enhanced number of monocytes, NK cells, and lymphocytes including B cells, T helper cells (CD4$^+$), T suppressor cells (CD8$^+$), and activated T cells (HLA-DR$^+$). In the afternoon and evening following the night of sleep deprivation the numbers of NK cells, lymphocytes, and T-cell subsets were, however, decreased. Figure 6 shows that sleep deprivation mainly induced a shift in the phase shifts of the circadian variations on circulating blood cell levels. Infrequent blood sampling would not only fail to identify this shift, but also, depending on the actual time of measurement, would reveal either higher, lower, or unchanged blood cell counts following sleep deprivation.

An obvious difficulty in assessing the effects of sleep on cytokine activity are the low levels of endogenous cytokines circulating in blood and their proximity to the lower detection limit of most assays. In fact, since most cytokines probably do not have a hormonelike function as bloodborne messengers, concentrations in serum of cytokines may not represent valid measures of the functional role of these substances. A measure more closely linked to the biological role, therefore, could be cytokine production, measured after *in vitro* stimulation of white blood cells. Such measures also revealed more consistent changes in the context of sleep studies. Lipopolysaccharides are common mitogen stimulants for cytokines released from monocytes and macrophages, such as IL-1β and TNF-α, and from B cells, whereas phytohemagglutinin and concanavalin A activate, in particular, the cytokine release from T lymphocytes. Most studies in humans have analyzed the influence of sleep on these cytokine systems comprising assessments of IL-1β, TNF-α, IL-2, and IL-6.

One night of sleep deprivation was associated with an enhanced production of the cytokines IL-1β and TNF-α after *in vitro* mitogen stimulation (Uthgenannt *et al.*, 1995; Dinges *et al.*, 1995; Born and Hansen, 1997). However, for cytokine production relative to the number of circulating monocytes, the main source of these cytokines in peripheral blood, effects were no longer significant (Born *et al.*, 1997). On the other hand, results of various studies do suggest interactions between TNF-α and IL-1β cytokine systems and human sleep regulation. For one, increased endogenous antagonists of IL-1β and TNF-α activity induced after G-CSF administration in humans have been found to suppress NREM sleep and EEG SWA and prolong REM latency (Schuld *et al.*, 1999). Since neither TNF-α, GH, nor cortisol increased during this time, and an earlier study found that G-CSF does not influence plasma levels of IL-1β, IL-6, or body temperature (Pollmächer *et al.*, 1996), the decrease in sleep intensity may be essentially attributed to the increased IL-1β and TNF-α antagonistic activity. In another study (Shearer *et al.*, 2001) in which the effects on the TNF-α cytokine system of extended sleep loss rather than an acute modulation of sleep were investigated, sleep loss was associated with enhanced endogenous antagonists of TNF-α activity.

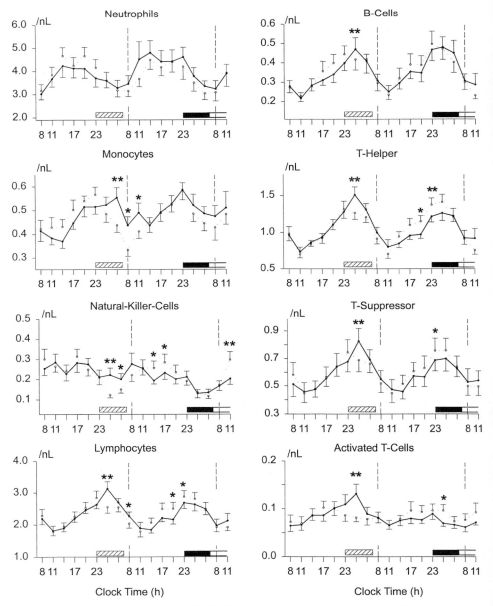

FIG. 6. Patterns of peripheral blood cell counts dependent on sleep versus sleep deprivation. Mean ($\pm$ SEM) numbers of neutrophils, monocytes, NK cells, lymphocytes, and major lymphocyte subsets—B cells, Th cells (CD4$^+$), T cytotoxic cells (CD8$^+$), and activated T cells

Comparing endocrine–immune changes in 4-day total versus partial sleep deprivation the authors concluded that the increase in plasma levels of the soluble TNF-α receptor I with total sleep deprivation reflected an elevation of the homeostatic drive for sleep. However, TNF-α levels did not differ between conditions of sleep deprivation. Whereas the latter two studies showed relations between sleep loss or suppression and cytokine systems, Marshall et al. (2001) investigated the relation between increased sleep and IL-1β production. In that study the enhancement of NREM sleep stage 4 during the first half of the night induced by a single GHRH bolus administration (50 μg) preceded a significant decrease in IL-1β production at 0300 h. This effect could not be attributed to changes in GH or cortisol. Although it would be premature to conclude a direct effect of sleep alone on IL-1β, the data do suggest that GHRH modulated immune functions through brain mechanisms also involved in the regulation of sleep. Altogether, results of the latter studies are consistent in that either an association between sleep and enhanced activity of the corresponding cytokine systems, or an association between sleep depression and enhanced activity of endogenous antagonistic of cytokines activity was found.

For IL-2 the most consistent finding with sleep deprivation is a decrease in IL-2 production (Uthgenannt et al., 1995; Dinges et al., 1995; Born et al., 1997). This reduction was evident for absolute cytokine production as well as for IL-2 production relative to the number of circulating T cells (Fig. 3; Born et al., 1997). Partial sleep deprivation from 2200 h to 0300 h also reduced IL-2 production (Irwin et al., 1996). Serum IL-2 concentrations, in contrast, were not altered by partial sleep deprivation in the second half of the night or at sleep onset (Irwin et al., 1999). Soluble IL-2 cytokine receptors which serve as endogenous IL-2 antagonists also remained unaffected by both 4-day partial and total sleep deprivation (Shearer et al., 2001), which agrees with the finding that fluctuations in soluble IL-2 receptors are mainly determined by a strong circadian rhythm (Lemmer et al., 1992).

Reports also vary for the effects of sleep on IL-6. In part, this may reflect methodological artifacts resulting from blood sampling with an indwelling catheter for an extended time period (Haack et al., 2002). In that study blood was drawn every 30 min for a 24-h period of sleep or sleep deprivation, either through an indwelling catheter or by a simple needle stick

---

(HLA-DR$^+$)—are shown for a session consisting of two successive nights of normal nocturnal sleep (both horizontal bars); dotted lines, and a session with one night of sleep deprivation (hatched horizontal bar) followed by a night of recovery sleep (black horizontal bar); continuous line ($n = 10$). Asterisks indicate significant differences between the two sessions. Modified from Born et al. (1997). Copyright 1997. The American Association of Immunologists, Inc.

A

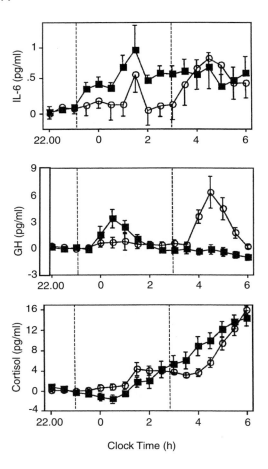

B

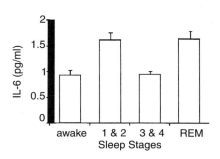

sample from the contralateral arm. Plasma IL-6 levels increased linearly with time for samples drawn from the indwelling catheter, but not for the stick samples. Furthermore, IL-6 levels were higher in samples rated as difficult to obtain, which occurred predominantly during sleep. Thus, it was concluded that local cytokine production at the indwelling catheter markedly influenced measurements. Comparing circadian and sleep–wake changes in IL-6 plasma concentration Born *et al.* (1997) did not find any sleep–wake or circadian changes in IL-6. In contrast, using a more sensitive assay and sampling blood every 30 min, IL-6 increased significantly as early as 30 min after sleep onset from a 2.5-h baseline level (Redwine *et al.*, 2000). Moreover, with partial sleep deprivation from 2300 to 0300 h IL-6 levels in that study did not deviate from baseline until after subjects were allowed to sleep (Fig. 7A). Furthermore, changes in IL-6 levels proved to be dependent on sleep stage (Fig. 7B). Circulating concentrations of IL-6 were higher during stages 1, 2, and REM sleep compared to average waking levels, whereas IL-6 concentration during SWS did not differ from waking. Using the same assay as Redwine *et al.* (2000), Vgontzas *et al.* (1999) reported an increase in IL-6 during the daytime after one night of total sleep deprivation. IL-6 levels across the 24-h postdeprivation level, however, did not differ from the control condition in which subjects had slept throughout the night. The finding by Shearer *et al.* (2001) that after the fourth day of total sleep deprivation IL-6 concentrations increased as compared to partial sleep deprivation over the same time period let the authors conclude that changes in IL-6 reflect elevations of the homeostatic drive for sleep. Many of the observed relations between increased IL-6 concentration and sleep or even sleep deprivation are, however, challenged by the previously mentioned finding on the effect of an indwelling catheter on local cytokine production (Haack *et al.*, 2002).

A beneficial effect of sleep on immune function was demonstrated by comparing the effect of acute sleep or sleep deprivation on the night following a vaccination with hepatitis A on the virus-specific antibody titer after 28 days. Subjects with regular sleep following vaccination, displayed a nearly twofold higher antibody titer four weeks later than subjects who stayed awake

---

FIG. 7. Effect of sleep and partial sleep deprivation on cytokine and hormone levels. (A) Mean (±SEM) changes from a pre-sleep (2200–2330 h) level of IL-6, growth hormone (GH), and cortisol plasma concentrations for normal nocturnal sleep (filled squares) and partial sleep deprivation (2200–0300 h) (open circles). The first vertical dashed line indicates the average time of sleep onset in the sleep session, and the second vertical dashed line indicates the end of sleep deprivation and sleep onset in the condition of partial sleep deprivation ($n = 31$). (B) Mean (±SEM) serum levels of IL-6 across the sleep stages: awake, sleep stages 1–4, and rapid eye movement sleep (REM). See text for comments on the results. Modified from Redwine *et al.* (2000), with permission of The Endocrine Society.

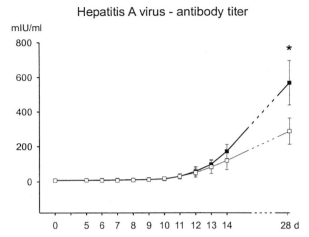

FIG. 8. Mean (±SEM) anti-hepatitis A virus titer before (day 0) and after (day 5–14, day 28) hepatitis A vaccination in subjects who had either regular nocturnal sleep after vaccination (thick line, filled square, $n = 9$) or were kept awake during that night (thin line, open square, $n = 10$). Asterisk indicates a significant difference between both conditions. Modified from Lange *et al.* (2002b).

on that one night (Fig. 8). Compared with sleep deprivation, sleep after vaccination also distinctly increased release of several immune-stimulating hormones including growth hormone, prolactin, and dopamine (Lange *et al.* 2002b). A positive relationship between sleep and infection was also reported for human immunodeficiency virus (HIV)–infected patients, as well as feline immunodeficiency virus–infected cats (Prospero-Garcia *et al.,* 1994; Gemma and Opp, 1999). Persons who were seropositive for HIV, but otherwise healthy demonstrated excessive stage 4 sleep, but that sleep deteriorates and becomes disrupted as the disease progresses (Norman *et al.,* 1990; Darko *et al.,* 1995). In conjunction with the increase in IL-2 observed rather consistently in humans, these data suggest that a supportive influence of sleep may in particular pertain to later stages of specific immune function as mediated by T cells.

B. EFFECTS OF SLEEP—ANIMALS

First systematic investigations on the effects of sleep deprivation reported on its lethal effects and on the deregulation of temperature homeostasis. Whereas a decrease in white and red blood cells was measured after a few days of sleep deprivation in an early study at the end of the nineteenth century by Manacaine in dogs (cited in Kleitman, 1963) negligible effects on immune parameters (e.g., *in vitro* lymphocyte proliferation and spleen cell counts) were found in rats at the end of a 32-day sleep deprivation period in a study by Rechtschaffen and colleagues (Benca *et al.*, 1989; Rechtschaffen *et al.*, 1989). An indication that sleep deprivation weakens the host defense mechanism is that rats which died after prolonged periods of sleep deprivation showed signs of infection and bacteremia (Everson, 1993).

A correlation between sleep enhancement, in particular in SWS, and survival rate was reported by Toth and colleagues (Toth *et al.*, 1993; Toth, 1995). Rabbits were inoculated with different bacterial antigens (*Escherichia coli, Staphylococcus aureus*) which enhanced sleep. Long periods of enhanced sleep were linked to a more favorable prognosis than short periods of enhanced sleep, and rabbits that eventually died had slept less than rabbits that survived the infection. However, the positive correlation between enhanced sleep and more efficient host defense could reflect a beneficial effect of sleep on immune defense functions as well as the consequence of a functionally strong immune system on sleep.

Similar to the study by Lange and colleagues (2002b) (Section IV.A) on the effects of sleep versus sleep deprivation on the response to immunization in humans the beneficial effect of sleep on such responses has also been observed in animals. However, the animal studies so far concentrated on the secondary response to antigen challenge. Brown *et al.* (1989) investigated mice orally immunized with the influenza virus and the response of these and unimmunized animals challenged with the influenza virus 1 week later. Within this week half of the animals were sleep deprived, in only a mildly stressful manner (i.e., by tapping gently on the cage as soon as the mice adopted a sleeping position). Serum antibody concentrations measured three days after the influenza challenge in the group allowed to sleep within the week were dramatically higher than in sleep-deprived immunized mice. The fact that antibody concentrations did not differ between both groups of unimmunized mice indicates that sleep within the 7-day period had not affected the strength of the influenza infection itself (e.g., viral replication), but instead the development of the host response system immediately after immunization. The experiment was, however, terminated too early, as the authors themselves state, to examine the primary response to antigen challenge, determining the generation of immunological long-term memories. Two further studies, however, attempting to demonstrate similar effects of sleep deprivation on secondary antibody responses to influenza infection in mice failed (Renegar *et al.*, 1998; Toth and Rehg, 1998). In fact, a beneficial effect of sleep deprivation on the mucosal anti-influenza defenses of immunized and unimmunized mice, due to an upregulation of antiviral cytokines IL-1β and IFN-α, was even suggested (Renegar *et al.*, 2000).

It has become recognized that neurotrophins are a component of the immunoregulatory system. Production and release of neurotrophins from immune cells including activated T lymphocytes, B lymphocytes, and monocytes has been demonstrated (Besser and Wank, 1999; Kerschensteiner *et al.*, 1999; Stocker *et al.*, 1999). Furthermore, stimulated human peripheral blood mononuclear cells express trkB and trkC, or functional NGF receptors, and receptor expression can be specifically influenced by interleukins (Brodie and Gelfand, 1992; Brodie, 1996; Besser and Wank, 1999). Moreover, immune cell functions such as B-cell proliferation (Melamed *et al.*, 1996; Torcia *et al.*, 1996), stimulation of phagocytosis, parasite killing, and IL-1 production by macrophages (Susaki *et al.*, 1996) are influenced by neurotrophins (Levi-Montalcini *et al.*, 1996). A major interest in neurotrophins of the brain arises from their increase in inflammatory infiltrates in the brain or cerebrospinal fluid of patients with acute disseminated encephalitis, multiple sclerosis, Alzheimer's disease, or after traumatic brain injury (Kossmann *et al.*, 1997; Kerschensteiner *et al.*, 1999; Otten *et al.*, 2000). Moreover, beneficial effects of neurotrophins have been reported in animal models of neurodegenerative disease (e.g., Mitsumoto *et al.*, 1994; Villoslada *et al.*, 2000). Also, a cross-talk between neurotrophins and other cytokines has been suggested, in particular the involvement of IL-6 has been hypothesized in physiological and pathophysiological processes in the CNS (Otten *et al.*, 2000). For instance, these authors suggest that IL-6 in conjunction with its soluble receptor modulate immune responses in the central nervous system by inducing a specific pattern of neurotrophins in astrocytes in defined brain regions. Aside from immunoregulatory functions the involvement of neurotrophins in synaptic plasticity and memory functions make the dependence of this cytokine family on sleep and sleep deprivation so very interesting. It should be noted here that the involvement in both sleep regulation and synaptic plasticity is a feature shared, for instance, with IL-1, TNF-α, nitric oxide, adenosine, prostaglandins, and nuclear factor-κB (summarized in Taishi *et al.*, 2001).

Taishi *et al.* (2001) revealed that brain-derived neurotrophic factor (BDNF) mRNA was increased by about 120% after 8 h of sleep deprivation maintained by gentle handling, in the rat cerebral cortex, whereas BDNF mRNA was indistinguishable from levels in control animals following 2 h of recovery sleep. However, in the hippocampus neither sleep deprivation nor recovery sleep, in that same study, influenced BDNF mRNA levels. Cirelli and Tononi (2000) similarily reported higher BDNF mRNA levels in the cerebral cortex after 8 h of sleep deprivation, yet not after 3 h of sleep deprivation as compared to sleep. Brain structure specific effects of two neurotrophins were observed as a result of 6 h of selective REM sleep deprivation (Sei *et al.*, 2000). In the cerebellum and brain stem BDNF protein levels were decreased, whereas in the hippocampus NGF protein was decreased. A most plausible explanation for this decreases in BDNF as compared to the increases after sleep deprivation observed in the former two studies is that it reflects structure-specific responses (i.e., cerbral cortex vs cerebellum and brain stem). In particular, the authors associated the decrease in NGF in the hippocampus with the lack of cholinergic input to the hippocampus which normally occurs during REM sleep (Marrosu *et al.*, 1995). Activity of the noradrenergic neurotransmitter system has proven essential for the increase in cortical BDNF expression observed during sleep deprivation and also during spontaneous periods of waking as compared to sleep (Cirelli and Tononi, 2000). Thus, sleep related neurotransmitter activity is involved and may even mediate sleep-dependent effects on neurotrophins in the brain. The interaction between sleep, neuronal afferent input, and the nature of ongoing synaptic reorganization was demonstrated in a study measuring NGF immunoreactivity in cortical pyramidal neurons of the rat cortex (Brandt *et al.*, 2001). The effect of unilateral trimming and 6 h of sleep deprivation depended on afferent input: on the side of the barrel cortex receiving input from the intact whiskers, NGF increased, whereas NGF immunoreactivity decreased to below the level observed in sleeping animals on the contralateral side which was now deprived of afferent input.

V. Summary

This chapter discusses various levels of interactions between the brain and the immune system in sleep. Sleep–wake behavior and the architecture of sleep are influenced by microbial products and cytokines. On the other hand, sleep processes, and perhaps also specific sleep states, appear to promote the production and/or release of certain cytokines. The effects of immune factors such as endotoxin and cytokines on sleep reveal species specificity and usually strong dependence on parameters such as substance concentration, time relative to administration or infection with microbial products, and phase relation to sleep and/or the light–dark cycle. For instance, endotoxin increased SWS and EEG SWA in humans only at very low concentrations, whereas higher concentrations increased sleep stage 2 only, but not SWS. In animals, increases in NREM sleep and SWA were more consistent over a wide range of endotoxin doses. Also, administration of pro-inflammatory cytokines such as IL-6 and IFN-α in humans acutely disturbed sleep while in rats such cytokines enhanced SWS and sleep. Overall, the findings in humans indicate that strong nonspecific immune responses are acutely linked to an arousing effect. Although subjects feel subjectively tired, their sleep flattens. However, some observations indicate a delayed enhancing effect on sleep which could be related to the induction of secondary, perhaps T-cell-related factors. This would also fit with results in animals in which the T-cell-derived cytokine IL-2 enhanced sleep while cytokines with immunosuppressive functions like IL-4 and L-10 suppressed sleep. The most straightforward similarity in the cascade of events inducing sleep in both animals and humans is the enhancing effect of GHRH on SWS, and possibly the involvement of the pro-inflammatory cytokine systems of IL-1β and TNF-α.

The precise mechanisms through which administered cytokines influence the central nervous system sleep processes are still unclear, although extensive research has identified the involvement of various molecular intermediates, neuropeptides, and neurotransmitters (cp. Fig. 5, Section III.B). Cytokines are not only released and found in peripheral blood mononuclear cells, but also in peripheral nerves and the brain (e.g., Hansen and Krueger, 1997; März *et al.*, 1998). Cytokines are thereby able to influence the central nervous system sleep processes through different routes. In addition, neuronal and glial sources have been reported for various cytokines as well as for their soluble receptors (e.g., Kubota *et al.*, 2001a). Links between the immune and endocrine systems represent a further important route through which cytokines influence sleep and, vice versa, sleep-associated processes, including variations in neurotransmitter and neuronal activity may influence cytokine levels.

The ability of sleep to enhance the release and/or production of certain cytokines was also discussed. Most consistent results were found for IL-2, which may indicate a sleep-associated increase in activity of the specific immune system. Furthermore, in humans the primary response to antigens following viral challenge is enhanced by sleep. In animals results are less consistent and have focused on the secondary response. The sleep-associated modulation in cytokine levels may be mediated by endocrine parameters. Patterns of endocrine activity during sleep are probably essential for the enhancement of IL-2 and T-cell diurnal functions seen in humans: Whereas prolactin and GH release stimulate Th1-derived cytokines such as IL-2, cortisol which is decreased during the beginning of nocturnal sleep inhibits Th1-derived cytokines. The immunological function of neurotrophins, in particular NGF and BDNF, has received great interest. Effects of sleep and sleep deprivation on this cytokine family are particularly relevant in view of the effects these endogenous neurotrophins can have not only on specific immune functions and the development of immunological memories, but also on synaptic reorganization and neuronal memory formation (e.g., Rachal *et al.*, 2001).

Acknowledgment

We gratefully thank Anja Otterbein for her skillful preparation of the figures.

References

Abo, T., Kawate, T., Itoh, K., and Kumagai, K. (1981). Studies on the bioperiodicity of the immune response. I. Circadian rhythms of human T, B, and K cell traffic in the peripheral blood. *J. Immunol.* **126**, 1360–1363.

Andren, J., Andrews, K., Brown, L., Chidgey, J., Geary, N., King, M. G., and Roberts, T. K. (1995). Muramyl peptides and the functions of sleep. *Behav. Brain Res.* **69**, 85–90.

Aserinsky, E., and Kleitman, W. (1953). Regularly occurring periods of eye motility and concomitant phenomena during sleep. *Science* **118**, 273–274.

Banks, W. A., Kastin, A. J., and Ehrensing, C. A. (1998). Diurnal uptake of circulating interleukin-1α by brain, spinal cord, testis and muscle. *Neuroimmunomodulation* **5**, 36–41.

Bauer, J., Hohagen, F., Ebert, T., Timmer, J., Ganter, U., Krieger, S., Lis, S., Postler, E., Voderholzer, U., and Berger, M. (1994). Interleukin-6 serum levels in healthy persons correspond to the sleep–wake cycle. *Clin. Invest.* **72**, 315.

Bauer, J., Hohagen, F., Gimmel, E., Bruns, F., Lis, S., Krieger, S., Ambach, W., Guthmann, A., Grunze, H., Fritsch-Montero, R., Weissbach, A., Ganter, V., Frommberger, U., Riemann, D., and Berger, M. (1995). Induction of cytokine synthesis and fever suppresses REM sleep and improves mood in patients with major depression. *Biol. Psychiatry* **38**, 611–621.

Benca, R. M., Kushida, C. A., Everson, C. A., Kalski, R., Bergmann, B. M., and Rechtschaffen, A. (1989). Sleep deprivation in the rat. VII. Immune function. *Sleep* **12**, 47–52.

Berchtold, N. C., Oliff, H. S., Isackson, P., and Cotman, C. W. (1999). Hippocampal BDNF mRNA shows a diurnal regulation, primarily in the exon III transcript. *Brain Res. Mol. Brain Res.* **71**, 11–22.

Besser, M., and Wank, R. (1999). Cutting edge: Clonally restricted production of the neurotrophins brain-derived neurotrophic factor and neurotrophin-3 mRNA by human immune cells and Th1/Th2-polarized expression of their receptors. *J. Immunol.* **162**, 6303–6306.

Borbely, A. A. (1998). Processes underlying sleep regulation. *Horm. Res.* **49**, 114–117.

Born, J., and Fehm, H. L. (1998). Hypothalamus-pituitary-adrenal activity during human sleep: A coordinating role for the limbic hippocampal system. *Exp. Clin. Endocrinol. Diabetes* **106**, 153–163.

Born, J., and Hansen, K. (1997). Dependence of human cytokine production and mononuclear cell subset counts on circadian rhythm and sleep. *In* "Immunological Alterations in Psychiatric Diseases." (A. E. Henneberg and W. P. Kaschka, eds.), pp. 18–31. Karger Basel.

Born, J., Spath-Schwalbe, E., Schwakenhofer, H., Kern, W., and Fehm, H. L. (1989). Influences of corticotropin-releasing hormone, adrenocorticotropin, and cortisol on sleep in normal man. *J. Clin. Endocrinol. Metab.* **68**, 904–911.

Born, J., Lange, T., Hansen, K., Molle, M., and Fehm, H. L. (1997). Effects of sleep and circadian rhythm on human circulating immune cells. *J. Immunol.* **158**, 4454–4464.

Bova, R., Micheli, M. R., Qualadrucci, P., and Zucconi, G. G. (1998). BDNF and trkB mRNAs oscillate in rat brain during the light–dark cycle. *Brain Res. Mol. Brain Res.* **57**, 321–324.

Brandt, J. A., Churchill, L., Guan, Z., Fang, J., Chen, L., and Krueger, J. M. (2001). Sleep deprivation but not a whisker trim increases nerve growth factor within barrel cortical neurons. *Brain Res.* **898**, 105–112.

Bredow, S., Guha-Thakurta, N., Taishi, P., Obál, F., Jr., and Krueger, J. M. (1997). Diurnal variations of tumor necrosis factor α mRNA and α- tubulin mRNA in rat brain. *Neuroimmunomodulation* **4**, 84–90.

Brodie, C. (1996). Differential effects of Th1 and Th2 derived cytokines on NGF synthesis by mouse astrocytes. *FEBS Lett.* **394**, 117–120.

Brodie, C., and Gelfand, E. W. (1992). Functional nerve growth factor receptors on human B lymphocytes. Interaction with IL-2. *J. Immunol.* **148**, 3492–3497.

Brown, R., Pang, G., Husband, A. J., and King, M. G. (1989). Suppression of immunity to influenza virus infection in the respiratory tract following sleep disturbance. *Reg. Immunol.* **2**, 321–325.

Cady, A. B., Kotani, S., Shiba, T., Kusumoto, S., and Krueger, J. M. (1989). Somnogenic activities of synthetic lipid A. *Infect. Immun.* **57**, 396–403.

Cirelli, C., and Tononi, G. (2000). Differential expression of plasticity-related genes in waking and sleep and their regulation by the noradrenergic system. *J. Neurosci.* **20**, 9187–9194.

Covelli, V., Massari, F., Fallacara, C., Munno, I., Jirillo, E., Savastano, S., Tommaselli, A. P., and Lombardi, G. (1992). Interleukin-1 β and β-endorphin circadian rhythms are inversely related in normal and stress-altered sleep. *Int. J. Neurosci.* **63**, 299–305.

Czeisler, C. A., and Klerman, E. B. (1999). Circadian and sleep-dependent regulation of hormone release in humans. *Recent Prog. Horm. Res.* **54**, 97–130.

Dantzer, R., Bluthé, R. M., Castanon, N., Chauvet, N., Capuron, L., Goodall, G., Kelley, K. W., Konsman, J. P., Layé, S., Parnet, P., and Pousset, F. (2001). Cytokine effects on behavior. *In* "Psychoneuroimmunology" (R. Ader, D. L. Felton, and N. Cohen, eds.), Vol. 1, pp. 703–727. Academic Press, London.

Darko, D. F., Miller, J. C., Gallen, C., White, J., Koziol, J., Brown, S. J., Hayduk, R., Atkinson, J. H., Assmus, J., and Munnell, D. T. (1995). Sleep electroencephalogram δ-frequency amplitude, night plasma levels of tumor necrosis factor α, and human immunodeficiency virus infection. *Proc. Natl. Acad. Sci. USA* **92**, 12080–12084.

De Simoni, M. G., Imeri, L., De Matteo, W., Perego, C., Simard, S., and Terazzino, S. (1995). Sleep regulation: Interactions among cytokines and classical neurotransmitters. *Adv. Neuroimmunol.* **5**, 189–200.

Dhabhar, F. S., Miller, A. H., Stein, M., McEwen, B. S., and Spencer, R. L. (1994). Diurnal and acute stress-induced changes in distribution of peripheral blood leukocyte subpopulations. *Brain Behav. Immun.* **8**, 66–79.

Dickstein, J. B., Moldofsky, H., Lue, F. A., and Hay, J. B. (1999). Intracerebroventricular injection of TNF-α promotes sleep and is recovered in cervical lymph. *Am. J. Physiol.* **276**, R1018–R1022.

Dickstein, J. B., Hay, J. B., Lue, F. A., and Moldofsky, H. (2000). The relationship of lymphocytes in blood and in lymph to sleep/wake states in sheep. *Sleep* **23**, 185–190.

Dinges, D. F., Douglas, S. D., Zaugg, L., Campbell, D. E., McMann, J. M., Whitehouse, W. G., Orne, E. C., Kapoor, S. C., Icaza, E., and Orne, M. T. (1994). Leukocytosis and natural killer cell function parallel neurobehavioral fatigue induced by 64 hours of sleep deprivation. *J. Clin. Invest.* **93**, 1930–1939.

Dinges, D. F., Douglas, S. D., Hamarman, S., Zaugg, L., and Kapoor, S. (1995). Sleep deprivation and human immune function. *Adv. Neuroimmunol.* **5**, 97–110.

Dodt, C., Theine, K. J., Uthgenannt, D., Born, J., and Fehm, H. L. (1994). Basal secretory activity of the hypothalamo-pituitary-adrenocortical axis is enhanced in healthy elderly. An assessment during undisturbed night-time sleep [see comments]. *Eur. J. Endocrinol.* **131**, 443–450.

Drucker-Colin, R. (1995). The function of sleep is to regulate brain excitability in order to satisfy the requirements imposed by waking. *Behav. Brain Res.* **69**, 117–124.

Dunn, A. J. (2000). Cytokine activation of the HPA axis. *Ann. N. Y. Acad. Sci.* **917**, 608–617.

Dunn, A. J. (2002). Mechanisms by which cytokines signal the brain. *In* "Neurobiology of the Immune System." (A. Clow, F. Hucklebridge, and P. Evans, eds.); this volume, ch. 3, pp. 43–66.

Engeset, A., Sokolowski, J., and Olszewski, W. L. (1977). Variation in output of leukocytes and erythrocytes in human peripheral lymph during rest and activity. *Lymphology* **10**, 198–203.

Everson, C. A. (1993). Sustained sleep deprivation impairs host defense. *Am. J. Physiol.* **265**, R1148–R1154.

Fang, J., Bredow, S., Taishi, P., Majde, J. A., and Krueger, J. M. (1999). Synthetic influenza viral double-stranded RNA induces an acute-phase response in rabbits. *J. Med. Virol.* **57**, 198–203.

Floyd, R. A., and Krueger, J. M. (1997). Diurnal variation of TNF α in the rat brain. *Neuroreport* **8**, 915–918.

Franken, P., Dijk, D. J., Tobler, I., and Borbely, A. A. (1991). Sleep deprivation in rats: Effects on EEG power spectra, vigilance states, and cortical temperature. *Am. J. Physiol.* **261**, R198–R208.

Gemma, C., and Opp, M. R. (1999). Human immunodeficiency virus glycoproteins 160 and 41 alter sleep and brain temperature of rats. *J. Neuroimmunol.* **97**, 94–101.

Gudewill, S., Pollmacher, T., Vedder, H., Schreiber, W., Fassbender, K., and Holsboer, F. (1992). Nocturnal plasma levels of cytokines in healthy men. *Eur. Arch. Psychiatry Clin. Neurosci.* **242**, 53–56.

Haack, M., Reichenberg, A., Kraus, T., Schuld, A., Yirmiya, R., and Pollmacher, T. (2000). Effects of an intravenous catheter on the local production of cytokines and soluble cytokine receptors in healthy men. *Cytokine* **12**, 694–698.

Haack, M., Kraus, T., Schuld, A., Dalal, M., Koethe, D., and Pollmächer, T. (2002). Diurnal variations of interleukin 6 plasma levels are confounded by blood drawing procedures. *Psychoneuroendocrinology.* DOI:10.1016/50306-4530(02)00006-9.

Hansen, M. K., and Krueger, J. M. (1997). Subdiaphragmatic vagotomy blocks the sleep- and fever-promoting effects of interleukin-1β. *Am. J. Physiol.* **273**, R1246–R1253.

Haus, E., and Smolensky, M. H. (1999). Biologic rhythms in the immune system. *Chronobiol. Int.* **16**, 581–622.

Hohagen, F., Timmer, J., Weyerbrock, A., Fritsch-Montero, R., Ganter, U., Krieger, S., Berger, M., and Bauer, J. (1993). Cytokine production during sleep and wakefulness and its relationship to cortisol in healthy humans. *Neuropsychobiology* **28**, 9–16.

Hopkins, S. J., and Rothwell, N. J. (1995). Cytokines and the nervous system. I. Expression and recognition. *Trends Neurosci.* **18**, 83–88.

Horne, J. (1992). Human slow wave sleep: A review and appraisal of recent findings, with implications for sleep functions, and psychiatric illness. *Experientia* **48**, 941–954.

Igaz, P., Horvath, A., Horvath, B., Szalai, C., Pallinger, E., Rajnavolgyi, E., Toth, S., Rose-John, S., and Falus, A. (2000). Soluble interleukin-6 receptor (sIL-6R) makes IL-6R negative T cell line respond to IL-6; it inhibits TNF production. *Immunol. Lett.* **71**, 143–148.

Irwin, M., McClintick, J., Costlow, C., Fortner, M., White, J., and Gillin, J. C. (1996). Partial night sleep deprivation reduces natural killer and cellular immune responses in humans. *FASEB J.* **10**, 643–653.

Irwin, M., Thompson, J., Miller, C., Gillin, J. C., and Ziegler, M. (1999). Effects of sleep and sleep deprivation on catecholamine and interleukin- 2 levels in humans: Clinical implications. *J. Clin. Endocrinol. Metab.* **84**, 1979–1985.

Kapas, L., Hansen, M. K., Chang, H. Y., and Krueger, J. M. (1998). Vagotomy attenuates but does not prevent the somnogenic and febrile effects of lipopolysaccharide in rats. *Am. J. Physiol.* **274**, R406–R411.

Karnovsky, M. L. (1986). Muramyl peptides in mammalian tissues and their effects at the cellular level. *Fed. Proc.* **45**, 2556–2560.

Kerkhofs, M., Van Cauter, E., Van Onderbergen, A., Caufriez, A., Thorner, M. O., and Copinschi, G. (1993). Sleep-promoting effects of growth hormone-releasing hormone in normal men. *Am. J. Physiol.* **264**, E594–E598.

Kerschensteiner, M., Gallmeier, E., Behrens, L., Leal, V. V., Misgeld, T., Klinkert, W. E., Kolbeck, R., Hoppe, E., Oropeza-Wekerle, R. L., Bartke, I., Stadelmann, C., Lassmann, H., Wekerle, H., and Hohlfeld, R. (1999). Activated human T cells, B cells, and monocytes produce brain-derived neurotrophic factor *in vitro* and in inflammatory brain lesions: A neuroprotective role of inflammation? *J. Exp. Med.* **189**, 865–870.

Kimura, M., Majde, J. A., Toth, L. A., Opp, M. R., and Krueger, J. M. (1994). Somnogenic effects of rabbit and recombinant human interferons in rabbits. *Am. J. Physiol.* **267**, R53–R61.

Kimura-Takeuchi, M., Majde, J. A., Toth, L. A., and Krueger, J. M. (1992). The role of double-stranded RNA in induction of the acute-phase response in an abortive influenza virus infection model. *J. Infect. Dis.* **166**, 1266–1275.

Kleitman, N. (1963). "Sleep and Wakefulness" University of Chicago Press, Chicago.

Korth, C., Mullington, J., Schreiber, W., and Pollmacher, T. (1996). Influence of endotoxin on daytime sleep in humans. *Infect. Immun.* **64**, 1110–1115.

Kossmann, T., Stahel, P. F., Lenzlinger, P. M., Redl, H., Dubs, R. W., Trentz, O., Schlag, G., and Morganti-Kossmann, M. C. (1997). Interleukin-8 released into the cerebrospinal fluid after brain injury is associated with blood–brain barrier dysfunction and nerve growth factor production. *J. Cereb. Blood Flow Metab.* **17**, 280–289.

Kronfol, Z., and Remick, D. G. (2000). Cytokines and the brain: Implications for clinical psychiatry. *Am. J. Psychiatry* **157**, 683–694.

Krueger, J. M., and Majde, J. A. (1994). Microbial products and cytokines in sleep and fever regulation. *Crit. Rev. Immunol.* **14**, 355–379.

Krueger, J. M., and Majde, J. A. (1995). Cytokines and sleep. *Int. Arch. Allergy Appl. Immunol.* **106**, 97–100.

Krueger, J. M., and Obál, F., Jr. (1993). Growth hormone-releasing hormone and interleukin-1 in sleep regulation. *FASEB J.* **7**, 645–652.

Krueger, J. M., Pappenheimer, J. R., and Karnovsky, M. L. (1982). Sleep-promoting effects of muramyl peptides. *Proc. Natl. Acad. Sci. USA* **79**, 6102–6106.

Krueger, J. M., Kubillus, S., Shoham, S., and Davenne, D. (1986). Enhancement of slow-wave sleep by endotoxin and lipid A. *Am. J. Physiol.* **251**, R591–R597.

Krueger, J. M., Dinarello, C. A., Shoham, S., Davenne, D., Walter, J., and Kubillus, S. (1987). Interferon α-2 enhances slow-wave sleep in rabbits. *Int. J. Immunopharmacol.* **9**, 23–30.

Krueger, J. M., Fang, J., Taishi, P., Chen, Z., Kushikata, T., and Gardi, J. (1998). Sleep. A physiologic role for IL-1 β and TNF-α. *Ann. N. Y. Acad. Sci.* **856**, 148–159.

Krueger, J. M., Obál, F., Jr., and Fang, J. (1999). Humoral regulation of physiological sleep: Cytokines and GHRH. *J. Sleep Res.* **8** Suppl. 1. 53–59.

Kubota, T., Fang, J., Kushikata, T., and Krueger, J. M. (2000). Interleukin-13 and transforming growth factor-β1 inhibit spontaneous sleep in rabbits. *Am. J. Physiol. Regul. Integr. Comp. Physiol.* **279**, R786–R792.

Kubota, T., Brown, R. A., Fang, J., and Krueger, J. M. (2001a). Interleukin-15 and interleukin-2 enhance non-REM sleep in rabbits. *Am. J. Physiol.* **281**, R1004–R1012.

Kubota, T., Fang, J., Brown, R. A., and Krueger, J. M. (2001b). Interleukin-18 promotes sleep in rabbits and rats. *Am. J. Physiol. Regul. Integr. Comp. Physiol.* **281**, R828–R838.

Kubota, T., Majde, J. A., Brown, R. A., and Krueger, J. M. (2001c). Tumor necrosis factor receptor fragment attenuates interferon-γ- induced non-REM sleep in rabbits. *J. Neuroimmunol.* **119**, 192–198.

Kushikata, T., Fang, J., Wang, Y., and Krueger, J. M. (1998). Interleukin-4 inhibits spontaneous sleep in rabbits. *Am. J. Physiol.* **275**, R1185–R1191.

Kushikata, T., Fang, J., and Krueger, J. M. (1999a). Brain-derived neurotrophic factor enhances spontaneous sleep in rats and rabbits. *Am. J. Physiol.* **276**, R1334–R1338.

Kushikata, T., Fang, J., and Krueger, J. M. (1999b). Interleukin-10 inhibits spontaneous sleep in rabbits. *J. Interferon Cytokine Res.* **19**, 1025–1030.

Lancel, M., Mathias, S., Schiffelholz, T., Behl, C., and Holsboer, F. (1997). Soluble tumor necrosis factor receptor (p75) does not attenuate the sleep changes induced by lipopolysaccharide in the rat during the dark period. *Brain Res.* **770**, 184–191.

Lange, T., Marshall, L., Späth-Schwalbe, E., Fehm, H. L., and Born, J. (2002a). Systemic immune parameters and sleep after ultra low dose administration of IL-2 in healthy men. *Brain Behav. Immun.* In press.

Lange, T., Perras, B., Fehm, H. L., and Born, J. (2002b). Sleep enhances the human antibody response to hepatitis A vaccination. *Psychosom. Med.* In press.

Lemmer, B., Schwulera, U., Thrun, A., and Lissner, R. (1992). Circadian rhythm of soluble interleukin-2 receptor in healthy individuals. *Eur. Cytokine Netw.* **3**, 335–336.

Levi-Montalcini, R., Skaper, S. D., Dal Toso, R., Petrelli, L., and Leon, A. (1996). Nerve growth factor: From neurotrophin to neurokine. *Trends Neurosci.* **19**, 514–520.

Lissoni, P., Rovelli, F., Brivio, F., Brivio, O., and Fumagalli, L. (1998). Circadian secretions of IL-2, IL-12, IL-6 and IL-10 in relation to the light/dark rhythm of the pineal hormone melatonin in healthy humans. *Nat. Immun.* **16**, 1–5.

Lue, F. A., Bail, M., Jephthah-Ochola, J., Carayanniotis, K., Gorczynski, R., and Moldofsky, H. (1988). Sleep and cerebrospinal fluid interleukin-1-like activity in the cat. *Int. J. Neurosci.* **42**, 179–183.

Mackiewicz, M., Sollars, P. J., Ogilvie, M. D., and Pack, A. I. (1996). Modulation of IL-1 β gene expression in the rat CNS during sleep deprivation. *Neuroreport* **7**, 529–533.

Maestroni, G. J. M. (2001). Melatonin and immune function. *In* "Psychoneuroimmunology" (R. Ader, D. L. Felten, and N. Cohen, eds.), Vol. 1, pp. 433–443. Academic Press, London.

Maier, S. F., Watkins, L. R., and Nance, D. M. (2001). Multiple routes of action of interleukin-1 on the nervous system. *In* "Psychoneuroimmunology" (R. Ader, D. L. Felten, and N. Cohen, eds.), 563–583. Academic Press, London.

Majde, J. A., Guha-Thakurta, N., Chen, Z., Bredow, S., and Krueger, J. M. (1998). Spontaneous release of stable viral double-stranded RNA into the extracellular medium by influenza virus-infected MDCK epithelial cells: Implications for the viral acute phase response. *Arch. Virol.* **143**, 2371–2380.

Maquet, P. (2001). The role of sleep in learning and memory. *Science* **294**, 1048–1052.

Marrosu, F., Portas, C., Mascia, M. S., Casu, M. A., Fa, M., Giagheddu, M., Imperato, A., and Gessa, G. L. (1995). Microdialysis measurement of cortical and hippocampal acetylcholine release during sleep–wake cycle in freely moving cats. *Brain Res.* **671**, 329–332.

Marshall, L., Molle, M., Boschen, G., Steiger, A., Fehm, H. L., and Born, J. (1996). Greater efficacy of episodic than continuous growth hormone-releasing hormone (GHRH) administration in promoting slow-wave sleep (SWS). *J Clin. Endocrinol. Metab.* **81**, 1009–1013.

Marshall, L., Perras, B., Fehm, H. L., and Born, J. (2001). Changes in immune cell counts and interleukin (IL)-1 production in humans after a somnogenically active growth hormone-releasing hormone (GHRH) administration. *Brain Behav. Immun.* **15**, 227–234.

März, P., Cheng, J. G., Gadient, R. A., Patterson, P. H., Stoyan, T., Otten, U., and Rose-John, S. (1998). Sympathetic neurons can produce and respond to interleukin 6. *Proc. Natl. Acad. Sci. USA* **95**, 3251–3256.

Mathias, S., Schiffelholz, T., Linthorst, A. C., Pollmacher, T., and Lancel, M. (2000). Diurnal variations in lipopolysaccharide-induced sleep, sickness behavior and changes in corticosterone levels in the rat. *Neuroendocrinology* **71**, 375–385.

Melamed, I., Kelleher, C. A., Franklin, R. A., Brodie, C., Hempstead, B., Kaplan, D., and Gelfand, E. W. (1996). Nerve growth factor signal transduction in human B lymphocytes is mediated by gp140trk. *Eur. J. Immunol.* **26**, 1985–1992.

Mitsumoto, H., Ikeda, K., Klinkosz, B., Cedarbaum, J. M., Wong, V., and Lindsay, R. M. (1994). Arrest of motor neuron disease in wobbler mice cotreated with CNTF and BDNF. *Science* **265**, 1107–1110.

Moldofsky, H., Lue, F. A., Eisen, J., Keystone, E., and Gorczynski, R. M. (1986). The relationship of interleukin-1 and immune functions to sleep in humans. *Psychosom. Med.* **48**, 309–318.

Mullington, J., Korth, C., Hermann, D. M., Orth, A., Galanos, C., Holsboer, F., and Pollmacher, T. (2000). Dose-dependent effects of endotoxin on human sleep. *Am. J. Physiol. Regul. Integr. Comp. Physiol.* **278**, R947–R955.

Nistico, G., and De Sarro, G. (1991). Behavioral and electrocortical spectrum power effects after microinfusion of lymphokines in several areas of the rat brain. *Ann. N. Y. Acad. Sci.* **621**, 119–134.

Norman, S. E., Chediak, A. D., Kiel, M., and Cohn, M. A. (1990). Sleep disturbances in HIV-infected homosexual men. *AIDS* **4**, 775–781.

Obál, Jr., F., Floyd, R., Kapas, L., Bodosi, B., and Krueger, J. M. (1996). Effects of systemic GHRH on sleep in intact and hypophysectomized rats. *Am. J. Physiol.* **270**, E230–E237.

Obál, Jr., F., Kapas, L., Gardi, J., Taishi, P., Bodosi, B., and Krueger, J. M. (1999). Insulin-like growth factor-1 (IGF-1)-induced inhibition of growth hormone secretion is associated with sleep suppression. *Brain Res.* **818**, 267–274.

Obál, F., Jr., Fang, J., Payne, L. C., and Krueger, J. M. (1995). Growth-hormone-releasing hormone mediates the sleep-promoting activity of interleukin-1 in rats. *Neuroendocrinology* **61**, 559–565.

Opp, M., Obál, F., Jr., Cady, A. B., Johannsen, L., and Krueger, J. M. (1989). Interleukin-6 is pyrogenic but not somnogenic. *Physiol. Behav.* **45**, 1069–1072.

Opp, M. R., and Imeri, L. (1999). Sleep as a behavioral model of neuro-immune interactions. *Acta Neurobiol. Exp.* (*Warsaw*) **59**, 45–53.

Opp, M. R., and Krueger, J. M. (1994). Anti-interleukin-1 β reduces sleep and sleep rebound after sleep deprivation in rats. *Am. J. Physiol.* **266**, R688–R695.

Opp, M. R., Obál, F., Jr., and Krueger, J. M. (1991). Interleukin 1 alters rat sleep: Temporal and dose-related effects. *Am. J. Physiol.* **260**, R52–R58.

Opp, M. R., Kapas, L., and Toth, L. A. (1992). Cytokine involvement in the regulation of sleep. *Proc. Soc. Exp. Biol. Med.* **201**, 16–27.

Otten, U., März, P., Heese, K., Hock, C., Kunz, D., and Rose-John, S. (2000). Cytokines and neurotrophins interact in normal and diseased states. *Ann. N.Y. Acad. Sci.* **917**, 322–330.

Pabst, M. J., Beranova-Giorgianni, S., and Krueger, J. M. (1999). Effects of muramyl peptides on macrophages, monokines, and sleep. *Neuroimmunomodulation* **6**, 261–283.

Palm, S., Postler, E., Hinrichsen, H., Maier, H., Zabel, P., and Kirch, W. (1996). Twenty-four-hour analysis of lymphocyte subpopulations and cytokines in healthy subjects. *Chronobiol. Int.* **13**, 423–434.

Payne, L. C., Obál, F., Jr., Opp, M. R., and Krueger, J. M. (1992). Stimulation and inhibition of growth hormone secretion by interleukin-1 β: The involvement of growth hormone-releasing hormone. *Neuroendocrinology* **56**, 118–123.

Peigneux, P. (2001). Sleeping brain, learning brain. The role of sleep and memory systems. *Neuroreport* **12**, A111–A124.

Plytycz, B., and Seljelid, R. (1997). Rhythms of immunity. *Arch. Immunol. Ther. Exp.* (*Warsaw*) **45,** 157–162.

Pollmächer, T., Schreiber, W., Gudewill, S., Vedder, H., Fassbender, K., Wiedemann, K., Trachsel, L., Galanos, C., and Holsboer, F. (1993). Influence of endotoxin on nocturnal sleep in humans. *Am. J. Physiol.* **264,** R1077–R1083.

Pollmächer, T., Mullington, J., Korth, C., and Hinze-Selch, D. (1995). Influence of host defense activation on sleep in humans. *Adv. Neuroimmunol.* **5,** 155–169.

Pollmächer, T., Korth, C., Mullington, J., Schreiber, W., Sauer, J., Vedder, H., Galanos, C., and Holsboer, F. (1996). Effects of granulocyte colony-stimulating factor on plasma cytokine and cytokine receptor levels and on the *in vivo* host response to endotoxin in healthy men. *Blood* **87,** 900–905.

Poon, A. M., Liu, Z. M., Pang, C. S., Brown, G. M., and Pang, S. F. (1994). Evidence for a direct action of melatonin on the immune system. *Biol. Signals* **3,** 107–117.

Prospero-Garcia, O., Herold, N., Phillips, T. R., Elder, J. H., Bloom, F. E., and Henriksen, S. J. (1994). Sleep patterns are disturbed in cats infected with feline immunodeficiency virus. *Proc. Natl. Acad. Sci. USA* **91,** 12947–12951.

Rachal, Pugh C., Fleshner, M., Watkins, L. R., Maier, S. F., and Rudy, J. W. (2001). The immune system and memory consolidation: A role for the cytokine IL-1β. *Neurosci. Biobehav. Rev.* **25,** 29–41.

Rechtschaffen, A., and Kales, A. (1968). A manual of standardized terminology, techniques and scoring system for sleep stages of human subjects. NIH Publ. 204, U.S. Government Printing Office, Washington, D.C.

Rechtschaffen, A., Bergmann, B. M., Everson, C. A., Kushida, C. A., and Gilliland, M. A. (1989). Sleep deprivation in the rat. X. Integration and discussion of the findings. *Sleep* **12,** 68–87.

Redwine, L., Hauger, R. L., Gillin, J. C., and Irwin, M. (2000). Effects of sleep and sleep deprivation on interleukin-6, growth hormone, cortisol, and melatonin levels in humans. *J. Clin. Endocrinol. Metab.* **85,** 3597–3603.

Reichlin, S. (2001). Secretion of immunomodulatory mediators from the brain: An alternative pathway of neuroimmunomodulation. *In* "Psychoneuroimmunology" (R. Ader, D. L. Felten, and N. Cohen, eds.), Vol. 1, pp. 499–516. Academic Press, London.

Renegar, K. B., Floyd, R. A., and Krueger, J. M. (1998). Effects of short-term sleep deprivation on murine immunity to influenza virus in young adult and senescent mice. *Sleep* **21,** 241–248.

Renegar, K. B., Crouse, D., Floyd, R. A., and Krueger, J. (2000). Progression of influenza viral infection through the murine respiratory tract: The protective role of sleep deprivation. *Sleep* **23,** 859–863.

Reyes, T. M., and Coe, C. L. (1998). The proinflammatory cytokine network: Interactions in the CNS and blood of rhesus monkeys. *Am. J. Physiol.* **274,** R139–R144.

Rivest, S., Lacroix, S., Vallieres, L., Nadeau, S., Zhang, J., and Laflamme, N. (2000). How the blood talks to the brain parenchyma and the paraventricular nucleus of the hypothalamus during systemic inflammatory and infectious stimuli. *Proc. Soc. Exp. Biol. Med.* **223,** 22–38.

Roberts, J. E. (2000). Light and immunomodulation. *Ann. N. Y. Acad. Sci.* **917,** 435–445.

Schiffelholz, T., and Lancel, M. (2001). Sleep changes induced by lipopolysaccharide in the rat are influenced by age. *Am. J. Physiol. Regul. Integr. Comp. Physiol.* **280,** R398–R403.

Schuld, A., Mullington, J., Hermann, D., Hinze-Selch, D., Fenzel, T., Holsboer, F., and Pollmacher, T. (1999). Effects of granulocyte colony-stimulating factor on night sleep in humans. *Am. J. Physiol.* **276,** R1149–R1155.

Sei, H., Saitoh, D., Yamamoto, K., Morita, K., and Morita, Y. (2000). Differential effect of short-term REM sleep deprivation on NGF and BDNF protein levels in the rat brain. *Brain Res.* **877,** 387–390.

Shearer, W. T., Reuben, J. M., Mullington, J. M., Price, N. J., Lee, B. N., Smith, E. O., Szuba, M. P., Van Dongen, H. P., and Dinges, D. F. (2001). Soluble TNF-α receptor 1 and IL-6 plasma levels in humans subjected to the sleep deprivation model of spaceflight. *J Allergy Clin. Immunol.* **107,** 165–170.

Silverman, D. H., Imam, K., and Karnovsky, M. L. (1989). Muramyl peptide/serotonin receptors in brain-derived preparations. *Pept. Res.* **2,** 338–344.

Sothern, R. B., and Roitman-Johnson, B. (2001). Biological rhythms and immune function. *In* "Psychoneuroimmunology" (R. Ader, D. L. Felten, and N. Cohen, eds.), Vol. 1, pp. 445–479. Academic Press, London.

Späth-Schwalbe, E., Porzsolt, F., Digel, W., Born, J., Kloss, B., and Fehm, H. L. (1989). Elevated plasma cortisol levels during interferon-γ treatment. *Immunopharmacology* **17,** 141–145.

Späth-Schwalbe, E., Hansen, K., Schmidt, F., Schrezenmeier, H., Marshall, L., Burger, K., Fehm, H. L., and Born, J. (1998). Acute effects of recombinant human interleukin-6 on endocrine and central nervous sleep functions in healthy men. *J. Clin. Endocrinol. Metab.* **83,** 1573–1579.

Späth-Schwalbe, E., Lange, T., Perras, B., Fehm, H. L., and Born, J. (2000). Interferon-α acutely impairs sleep in healthy humans. *Cytokine* **12,** 518–521.

Steiger, A., Guldner, J., Hemmeter, U., Rothe, B., Wiedemann, K., and Holsboer, F. (1992). Effects of growth hormone-releasing hormone and somatostatin on sleep EEG and nocturnal hormone secretion in male controls. *Neuroendocrinology* **56,** 566–573.

Steriade, M. (1993). Cellular substrates of brain rhythms. *In* "Electroencephalography: Basic Principles, Clinical Applications, and Related Fields." (E. Niedermeyer, and F. Lopes da Silva, eds.), pp. 27–62. Lippincott, Williams & Wilkins, Baltimore.

Stocker, M., Hellwig, M., and Kerschensteiner, D. (1999). Subunit assembly and domain analysis of electrically silent K^+ channel α-subunits of the rat Kv9 subfamily. *J. Neurochem.* **72,** 1725–1734.

Suchecki, D., Tiba, P. A., and Tufik, S. (2002). Paradoxical sleep deprivation induces facilitation of the hypothalamic-pituitary-adrenal axis in rats. *Neurosci Lett.* **320,** 45–48.

Susaki, Y., Shimizu, S., Katakura, K., Watanabe, N., Kawamoto, K., Matsumoto, M., Tsudzuki, M., Furusaka, T., Kitamura, Y., and Matsuda, H. (1996). Functional properties of murine macrophages promoted by nerve growth factor. *Blood* **88,** 4630–4637.

Suzuki, S., Toyabe, S., Moroda, T., Tada, T., Tsukahara, A., Iiai, T., Minagawa, M., Maruyama, S., Hatakeyama, K., Endoh, K., and Abo, T. (1997). Circadian rhythm of leucocytes and lymphocytes subsets and its possible correlation with the function of the autonomic nervous system. *Clin. Exp. Immunol.* **110,** 500–508.

Taishi, P., Bredow, S., Guha-Thakurta, N., Obál, F., Jr., and Krueger, J. M. (1997). Diurnal variations of interleukin-1 β mRNA and β-actin mRNA in rat brain. *J. Neuroimmunol.* **75,** 69–74.

Taishi, P., Chen, Z., Obál, F., Jr., Hansen, M. K., Zhang, J., Fang, J., and Krueger, J. M. (1998). Sleep-associated changes in interleukin-1β mRNA in the brain. *J. Interferon Cytokine Res.* **18,** 793–798.

Taishi, P., Sanchez, C., Wang, Y., Fang, J., Harding, J. W., and Krueger, J. M. (2001). Conditions that affect sleep alter the expression of molecules associated with synaptic plasticity. *Am. J. Physiol. Regul. Integr. Comp. Physiol.* **281,** R839–R845.

Takahashi, S., Kapas, L., Fang, J., and Krueger, J. M. (1995). An anti-tumor necrosis factor antibody suppresses sleep in rats and rabbits. *Brain Res.* **690,** 241–244.

Takahashi, S., Fang, J., Kapas, L., Wang, Y., and Krueger, J. M. (1997). Inhibition of brain interleukin-1 attenuates sleep rebound after sleep deprivation in rabbits. *Am. J. Physiol.* **273,** R677–R682.

Takahashi, T., Yamashita, H., Nakamura, S., Ishiguro, H., Nagatsu, T., and Kawakami, H. (1999). Effects of nerve growth factor and nicotine on the expression of nicotinic acetylcholine receptor subunits in PC12 cells. *Neurosci. Res.* **35,** 175–181.

Torcia, M., Bracci-Laudiero, L., Lucibello, M., Nencioni, L., Labardi, D., Rubartelli, A., Cozzolino, F., Aloe, L., and Garaci, E. (1996). Nerve growth factor is an autocrine survival factor for memory B lymphocytes. *Cell* **85,** 345–356.

Toth, L. A. (1995). Sleep, sleep deprivation and infectious disease: Studies in animals. *Adv. Neuroimmunol.* **5,** 79–92.

Toth, L. A., and Opp, M. R. (2001). Cytokine- and microbially induced sleep responses of interleukin-10 deficient mice. *Am. J. Physiol. Regul. Integr. Comp. Physiol.* **280,** R1806–R1814.

Toth, L. A., and Rehg, J. E. (1998). Effects of sleep deprivation and other stressors on the immune and inflammatory responses of influenza-infected mice. *Life Sci.* **63,** 701–709.

Toth, L. A., Gardiner, T. W., and Krueger, J. M. (1992). Modulation of sleep by cortisone in normal and bacterially infected rabbits. *Am. J. Physiol.* **263,** R1339–R1346.

Toth, L. A., Tolley, E. A., and Krueger, J. M. (1993). Sleep as a prognostic indicator during infectious disease in rabbits. *Proc. Soc. Exp. Biol. Med.* **203,** 179–192.

Trachsel, L., Tobler, I., Achermann, P., and Borbely, A. A. (1991). Sleep continuity and the REM–nonREM cycle in the rat under baseline conditions and after sleep deprivation. *Physiol. Behav.* **49,** 575–580.

Tsuchiyama, Y., Uchimura, N., Sakamoto, T., Maeda, H., and Kotorii, T. (1995). Effects of hCRH on sleep and body temperature rhythms. *Psychiatry Clin. Neurosci.* **49,** 299–304.

Turnbull, A. V., and Rivier, C. L. (1999). Regulation of the hypothalamic-pituitary-adrenal axis by cytokines: Actions and mechanisms of action. *Physiol. Rev.* **79,** 1–71.

Uthgenannt, D., Schoolmann, D., Pietrowsky, R., Fehm, H. L., and Born, J. (1995). Effects of sleep on the production of cytokines in humans. *Psychosom. Med.* **57,** 97–104.

Vgontzas, A. N., Papanicolaou, D. A., Bixler, E. O., Lotsikas, A., Zachman, K., Kales, A., Prolo, P., Wong, M. L., Licinio, J., Gold, P. W., Hermida, R. C., Mastorakos, G., and Chrousos, G. P. (1999). Circadian interleukin-6 secretion and quantity and depth of sleep. *J. Clin. Endocrinol. Metab.* **84,** 2603–2607.

Villoslada, P., Hauser, S. L., Bartke, I., Unger, J., Heald, N., Rosenberg, D., Cheung, S. W., Mobley, W. C., Fisher, S., and Genain, C. P. (2000). Human nerve growth factor protects common marmosets against autoimmune encephalomyelitis by switching the balance of T helper cell type 1 and 2 cytokines within the central nervous system. *J. Exp. Med.* **191,** 1799–1806.

Wexler, D. B., and Moore-Ede, M. C. (1984). Effects of a muramyl dipeptide on the temperature and sleep–wake cycles of the squirrel monkey. *Am. J. Physiol.* **247,** R672–R680.

Yamuy, J., Morales, F. R., and Chase, M. H. (1995). Induction of rapid eye movement sleep by the microinjection of nerve growth factor into the pontine reticular formation of the cat. *Neuroscience* **66,** 9–13.

Zhang, J., Obál, F., Jr., Zheng, T., Fang, J., Taishi, P., and Krueger, J. M. (1999). Intrapreoptic microinjection of GHRH or its antagonist alters sleep in rats. *J. Neurosci.* **19,** 2187–2194.

NEUROENDOCRINOLOGY OF AUTOIMMUNITY

Michael Harbuz

University Research Center for Neuroendocrinology
University of Bristol
Bristol BS2 8HW, United Kingdom

I. Introduction

Autoimmune diseases such as rheumatoid arthritis (RA) and multiple sclerosis (MS) have a major impact on the quality of life of individuals with these diseases and are a burden on the health services treating them. Despite much effort the origins of these diseases remain obscure. The factors underlying the fundamental question of susceptibility are poorly understood as are those responsible for the differences in severity between individuals. Of considerable interest has been the suggestion that susceptibility may be related to impaired responsiveness of the hypothalamic-pituitary-adrenal (HPA) axis. This hypothesis suggested that an inability to mount an appropriate cortisol response with which to downregulate the immune system allowed the immune system to rampage unchecked and attack the self. This hypothesis links regulation of the release from the adrenal gland of the potent immunosuppressive glucocorticoids to the disease process itself. That glucocorticoids are fundamental in regulating the severity of the disease process is without question. Blocking endogenous glucocorticoid production using metyrapone results in a flare in disease activity in patients with RA. Furthermore, surgical removal of the adrenal gland in patients with Cushing's disease has resulted in the onset of autoimmune diseases.

Although the hypothesis proposing a link between a hyporesponsive HPA axis and susceptibility to disease is intriguing, various strands of evidence have suggested that this may be too simplistic a notion. In patients with RA, for example, alterations in HPA axis activity have not been consistently observed. However, there are both subtle changes associated with the development of autoimmune disease and evidence to link behavioral responses and neurotransmitter systems to differences in susceptibility. This chapter will review the importance of the HPA axis in determining susceptibility and in regulating the severity of inflammatory processes in autoimmune disease.

II. The Hypothalamic-Pituitary-Adrenal (HPA) Axis

In 1950, Philip S. Hench received the Nobel Prize for his pioneering work demonstrating the anti-inflammatory effects of glucocorticoids in patients with RA (Hench, 1949; Hench *et al.*, 1949). Sadly, the initial optimism of this "magic bullet" cure for RA was misplaced, as the unfortunate side effects of long-term, high-dose glucocorticoid use became apparent. Despite the problems associated with their long-term use, glucocorticoids remain an important weapon in the clinical armory for treating inflammatory episodes in a variety of conditions. Attention has focused on the importance of endogenous glucocorticoids, and hence the HPA axis, in regulating the inflammatory process. The endpoint of the activation of the HPA axis is the release of glucocorticoids (cortisol in man, corticosterone in rodents), from the adrenal cortex into the general circulation (Fig. 1) (Buckingham *et al.*, 1997). This release is under the control of ACTH secreted from the anterior pituitary gland. ACTH is generated from the parent peptide pro-opiomelanocortin (POMC), hence activation of this pathway can be monitored by determining POMC mRNA expression. ACTH release in turn is regulated by the corticotropin-releasing factors— corticotropin-releasing factor$_{1-41}$ (CRF) and arginine vasopressin (AVP). These corticotropin-releasing factors are synthesized in the parvocellular cells of the paraventricular nucleus (PVN) and under normal conditions approximately 50% of these CRF neurons also contain AVP; the remainder contain CRF but no AVP (Whitnall *et al.*, 1985; Whitnall *et al.*, 1987).

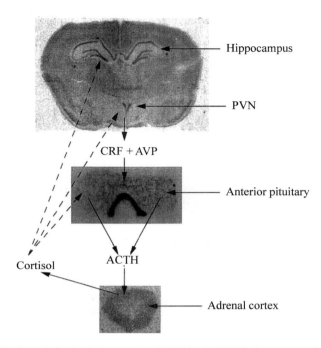

FIG. 1. The hypothalamic-pituitary-adrenal (HPA) axis. PVN is the paraventricular nucleus located within the hypothalamus. It is within the parvocellular cells of this nucleus that the major corticotropin-releasing factors, CRF and AVP, are synthesized. ACTH is adrenocorticotropin, the principal stress hormone released from the corticotrophs of the anterior pituitary. The solid lines represent stimulatory feedforward pathways, and the dashed lines represent inhibitory feedback pathways.

The axons of these neurons terminate in the external zone of the median eminence where the neurosecretory peptides are released into the hypophysial portal blood to be carried to the corticotrophs of the anterior pituitary. CRF and AVP exert a synergistic action on the release of ACTH. Currently, CRF is the only releasing factor known to stimulate synthesis of POMC mRNA. The PVN acts as a coordinating center, receiving signals from many brain areas including the hippocampus and the amygdala, together with a variety of brain stem nuclei including the raphe nucleus, nucleus tractus solitarius, and the locus coeruleus. Not only does the PVN coordinate signals to initiate activation of the HPA axis, but also neurons arising in the PVN signal to many brain areas including the brain stem where these can influence activation of the sympathetic nervous system (SNS), itself an important contributor to the stress response. The innervation of the immune tissues by the SNS and the role of neurotransmitters, especially nor-adrenaline and serotonin, are increasingly being investigated as important regulators of immune activity (Mossner and Lesch, 1998; Baerwald *et al.*, 2000; Kohm and Sanders, 2000; Stefulj *et al.*, 2000).

A. NEGATIVE FEEDBACK

Activation of the HPA axis in response to a challenge is important for long-term survival. Of equal importance is the ability to terminate the response once the threat has passed. The major mechanism involved is negative feedback where glucocorticoid released from the adrenal cortex is able to act at the pituitary, at the PVN, and at higher centers to inhibit the synthesis of precursor mRNAs such as POMC, CRF, and AVP and also to inhibit the release of the mature peptide products. This activity is medi-ated via two corticosteroid receptors—type I (mineralocorticoid) and type II (glucocorticoid) receptors. The type I receptor has a high affinity for gluco-corticoids and is present in the hippocampus. The type II receptor is found in the pituitary, PVN, and hippocampus and has a relatively low affinity for glucocorticoids. The presence of both type I and type II receptors in the hippocampus makes this area an important target for negative feedback in addition to the direct actions at the pituitary and hypothalamic levels. The hippocampus has an inhibitory input to the PVN, as lesions between these areas result in an upregulation of HPA axis activity.

B. DIURNAL RHYTHM

In normal humans under baseline conditions the release of ACTH and cortisol follows a diurnal rhythm with peak release occurring in the early hours of the morning, at or around the time of awakening, with levels falling through the day. Studies have demonstrated that there is a further peak in cortisol release that occurs within 30 min of waking (Edwards *et al.*, 2001a,b). Under resting conditions the type I, high-affinity receptors are fully occu-pied in the hippocampus. At the peak of cortisol secretion (i.e., in the early morning) and in response to stress the type II receptors are also occupied. This increase in cortisol release is believed to be involved in activating metabolic systems providing energy substrates prior to increased activity after waking. The diurnal rhythm of cortisol is, however, likely to have far wider reaching consequences. As will be discussed later, many disease condi-tions are associated with a flattened cortisol rhythm and this has implications for the regulation of disease activity.

C. ACUTE STRESS

Stress results in activation of the HPA axis that is reflected by increased CRF and AVP release into the hypophysial portal blood, by increased release of ACTH and cortisol into the blood, by increased POMC mRNA in the anterior pituitary, and by increased CRF and AVP mRNAs in the PVN (Buckingham *et al.*, 1997). Stress can be predominantly psychological (e.g., exams, interviews, public speaking); physical (e.g., broken limbs or painful stimuli); or immunological (e.g., following infection). The proportion of CRF-positive/AVP-negative to CRF-positive/AVP-positive cells can be altered in response to stress or other neuroendocrine manipulations suggesting the possibility of a level of control at this point (Whitnall, 1989). In addition to these two releasing factors, other peptides such as enkephalins, oxytocin, etc., are colocalized in these cells. These can be differentially activated depending on the type of stimulus and its duration and frequency. The net effect of this release will determine the signal received by the corticotrophs of the anterior pituitary. The potential therefore exists for exquisite control of ACTH release through the regulation of the synthesis and release of these factors into the hypophysial portal blood in response to different signals. Appropriate activation of the HPA axis to a stimulus is crucial. However, termination of the stress response once the challenge has passed is considered important in the regulation of the HPA axis. In addition to glucocorticoid negative feedback, a number of neurotransmitters have been identified as exerting negative effects on the HPA axis. For a review of central nonglucocorticoid inhibitors of the HPA axis, see Jessop (1999).

D. LONG-TERM EFFECTS OF A SINGLE STRESSOR

For a long time it has been assumed that the HPA axis represented a homeostatic mechanism; that following activation of the HPA axis in response to stress, and subsequent termination of the stressor, homeostatic mechanisms would return hormonal levels back to baseline within an hour or so and that mRNA levels would return to baseline within a few hours. It is generally accepted that the HPA axis would not be permanently altered by a single acute stress. Studies by a number of groups have, however, questioned this assumption and instead have suggested that following exposure to stress there are long-term alterations to the HPA axis. For example, it has been shown that exposure to the same stressor days or even weeks after initial exposure can alter responsiveness to a variety of stressors including interleukin-1, footshock, social defeat, immobilization, etc. (Schmidt *et al.*, 1995; Tilders *et al.*, 1999; Marti *et al.*, 2001). These reports highlight the important contribution of the stress history of individuals in human and animal stress studies and of the absolute need for contemporary controls in all such experiments.

E. Chronic Stress

When considering rodent chronic stress studies it is important to define the type of chronic stress. There are few suitable models of chronic stress and most of these utilize an acute stress repeated daily for a number of days. In addition to the problems mentioned earlier of the long-term implications of a single acute stress, repeated stress may produce habituation or adaptation to the stressor. Repeated footshock, restraint, and ethanol administration all elevate plasma corticosterone for up to 7 days but levels return toward baseline despite repeated exposure. This habituation is stressor-specific, as a novel stress will elicit a normal or even an exaggerated stress response. This is likely to be due to stimulation of different afferent pathways resulting in an altered activation profile in the PVN and alteration of factors released into the hypophysial portal blood. Although hormone levels may be returned to normal, POMC mRNA levels in the anterior pituitary and adrenal weight are increased reflecting chronic stimulation.

One major alteration observed in repeated stress studies is the increased importance of AVP. Chronically restrained rats do not increase plasma ACTH in response to exogenous CRF despite being able to respond to an acute stress (Hashimoto *et al.*, 1988). However, these animals do respond to AVP, suggesting that AVP is necessary for sustaining the stress response when the pituitary is refractory to CRF stimulation (Scaccianoce *et al.*, 1991). Further evidence is provided by reports of an increase in AVP stores and of increased AVP colocalization within CRF-positive terminals in the median eminence together with loss of anterior pituitary CRF receptors following chronic re-straint stress (de Goeij *et al.*, 1991). Together these data suggest AVP may become more important as a regulator of the HPA axis following repeated stress as CRF is downregulated.

III. The HPA Axis in Autoimmune Disease

A. A Crucial Role for the HPA Axis

A functioning HPA axis is crucial for regulating autoimmune disease and inflammatory conditions. In the absence of a fully functioning HPA axis the prognosis is not good. In RA, treatment with metyrapone, which prevents synthesis of endogenous glucocorticoids by the adrenal cortex, results in an exacerbation of disease activity (Panayi, 1992). Removal of the adrenal gland, as a treatment for Cushing's syndrome, has resulted in the development of RA and autoimmune thyroid disease (Yakushiji *et al.*, 1995; Takasu *et al.*, 1990). Similarly, in adjuvant-induced arthritis (AA) and experimental allergic encephalomyelitis (EAE—the model of choice for multiple sclerosis research), surgical removal of the adrenal glands results in an earlier onset of disease, a rapid increase in the severity of the disease (irrespective of the timing of the adrenalectomy), and a fatal outcome (Mason *et al.*, 1990; Harbuz *et al.*, 1993a). These effects can be prevented if the animals are treated with steroids. Further evidence is provided by observations that the severity of streptococcal cell wall–induced arthritis in the Lewis rat can be reduced by treatment with corticosterone at physiological doses and that the resistant Fischer strain can be made susceptible if treated with a corticosteroid receptor antagonist (Sternberg *et al.*, 1989a).

B. DIURNAL RHYTHM

As previously noted, a feature of many diseases is the loss of the normal diurnal rhythm, and this is often associated with increased disease severity in the early hours of the morning. Individuals with RA are likely to experience an increase in joint stiffness in the early hours of the morning. It has been suggested that the variation in disease activity seen throughout the day may be related to cortisol effects on the immune system. Not only are diurnal variations noted for ACTH and cortisol, but these have also been reported for pro-inflammatory cytokines such as IL-1, IL-6, IL-12, interferon-γ and TNF-α in normal subjects, with peak production in the early morning. Levels of these cytokines are also increased in the circulation of patients with RA although the profiles exhibit a time shift. Thus, in patients with RA, peak ACTH and cortisol concentrations tend to occur earlier in the night whereas IL-6 concentrations peak later than in controls (Masi and Chrousos, 1997). Not only is the IL-6 peak later in RA but it is also much greater than that seen in controls. This exaggerated IL-6 response has been suggested to be responsible for the increased symptoms such as joint stiffness present in the early morning in RA. This increase in IL-6 has been targeted by giving low-dose steroids (up to 5-mg prednisolone in women and up to 7.5 mg in men) at 2 a.m. with positive benefits (Arvidson *et al.*, 1997). We have conducted a preliminary study using lower doses of prednisolone (1 mg and 5 mg) delivered at 2 a.m. and found similar positive effects on morning stiffness (M. Harbuz, unpublished observations). Although there is some controversy over the use of low-dose steroids in RA (Masi and Chrousos, 1997), they have been shown to be beneficial in substantially reducing joint destruction in patients with early, active RA (Kirwan, 1995). There is therefore a real and exciting prospect that such studies may elucidate mechanisms allowing for selective timing of drug delivery using lower doses of steroids that would thus minimize the adverse effects of prolonged steroid use.

A loss in diurnal cortisol secretion has also been reported in RA with the effects dependent on the severity of the disease. With medium severity of RA there is a shift in the rhythm with peak concentrations occurring earlier in the night. With high disease activity the rhythm is blunted. It is therefore likely that modification of treatment regimes and dosage reduction may be more beneficial in the early stages of disease when the delivery of steroids can augment the normal diurnal rhythm. Once the disease is established and the diurnal rhythm lost, such augmentation may be less effective. Other components of the immune system also follow marked diurnal variation in normal individuals (e.g., natural killer cell activity), and this rhythmicity is also lost in patients with RA. Such changes in immune parameters are likely to contribute to the exacerbation in disease activity.

Alterations in HPA axis activity are found in many diseases. Not only in autoimmune diseases, but also in psychiatric and other immune-mediated conditions. One study has highlighted the importance of appropriate regulation of the HPA axis with respect to cancer survival. For individuals able to maintain a normal cortisol diurnal rhythm, survival time was increased. Where the rhythm was blunted, survival time was decreased. Where a flattened profile was recorded, the patients were at increased risk of dying within two years (Sephton et al., 2000). These alterations in long-term survival were correlated to the changes in cortisol and therefore HPA axis regulation. It is intriguing to speculate whether treatments targeting the HPA axis and reimposing a normal cortisol rhythm, may not only improve quality of life of the individuals but also increase their ability to fight cancer. This has been suggested by a number of studies but needs to be rigorously investigated (see Chapter 14 and 15, this volume).

Rats with AA also exhibit a loss of circadian rhythm (Persellin et al., 1972; Sarlis et al., 1992). The normal rhythm is blunted with a shallow peak and trough occurring 6 h earlier than in nonarthritic animals, similar to that reported in humans. Studies have demonstrated that there is a profound alteration in the profile of corticosterone release in rats with AA (Windle et al., 2001). Under normal conditions, naive rats release corticosterone in a pulsatile fashion with approximately 13 pulses occurring within 24 h. A greater number of pulses occurs in the early evening (peak of the diurnal rhythm in rodents), and these are of greater magnitude contributing to the peak in corticosterone release at this time. In AA, there is a doubling in the number of pulses over 24 h and although the individual peaks in release of corticosterone are not greater than those seen in controls, the greater number contributes to the reported increase in plasma corticosterone associated with AA. Whether similar mechanisms underlie the reported elevations in plasma corticosterone associated with peak clinical symptoms in EAE and other autoimmune disease models remains to be determined.

It is apparent that there is a link between alterations in the regulation of the HPA axis and disease severity over a wide range of disorders. It is therefore pertinent to consider how the axis is regulated and how this alters in respect to the disease process at a central level. Such considerations are fraught with difficulties in humans and so a review of findings in preclinical models with suitable corroborative examples from human studies with be provided.

C. HYPOTHALAMIC REGULATION AND INFLAMMATION

As previously noted, the development of AA results in profound changes in the release of ACTH and corticosterone into the general circulation. In addition, there is also a significant increase in POMC mRNA in the anterior pituitary (Neidhart and Fluckiger, 1992; Harbuz *et al.*, 1992; Stephanou *et al.*, 1992). CRF remains the only factor demonstrated to increase POMC mRNA. It was therefore a considerable surprise to determine a significant decrease in CRF mRNA associated with the onset of inflammation in AA and further decreases in CRF mRNA correlated with increased disease activity (Harbuz *et al.*, 1992). In addition to the decrease in CRF mRNA there was also a significant decrease in CRF peptide release into the hypophysial portal blood. This paradoxical decrease in CRF activity has been reported in a number of strains of rat: the Piebald-Viral-Glaxo (PVG) (Harbuz *et al.*, 1992), the Lewis (Brady *et al.*, 1994), and the Wistar (Chover-Gonzalez *et al.*, 2000), suggesting it is a common mechanism in response to the development of inflammation in rodents. This decrease is first apparent at Day 11 after adjuvant injection, when the first indications of inflammation are apparent in the model, and reaches a nadir at Day 21, the peak in inflammatory activity.

In contrast to CRF, AVP concentrations in the portal blood and AVP mRNA in the parvocellular cells of the PVN are both increased suggesting that in the presence of permissive levels of CRF, AVP is able to take over as the major stimulator of the HPA axis in the AA rat (Harbuz *et al.*, 1992; Chowdrey *et al.*, 1995a). These data support those from other repeated stress studies that have proposed an increased role for AVP in chronic stress situations as noted earlier. These data also support the notion that despite the colocalization of AVP within CRF neurons and indeed the colocalization of CRF and AVP within the same secretory vesicles of these neurons they appear to be under independent control and regulation. The role of AVP as a mediator in RA, MS, and other autoimmune diseases has been reviewed (Michelson and Gold, 1998; Chikanza and Grossman, 1998; Chikanza et al., 2000).

This alteration in hypothalamic regulation with a change from CRF control of the HPA axis is not just a feature of AA, but rather appears to be common to a wide range of immune-mediated disorders that crosses species barriers. The adoptive transfer model of EAE involves injecting activated splenocytes into naive recipients which develop peak symptoms after 6–7 days and show full recovery after 11 days. Peak clinical symptoms are associated with elevated plasma corticosterone and elevated POMC mRNA in the anterior pituitary. At the level of the PVN there is a paradoxical decrease in CRF activity (Harbuz et al., 1993b). These alterations in HPA axis activity return to normal following recovery at Day 11. Increased HPA axis activity has been reported in patients with MS under basal conditions and in dynamic testing (Michelson et al., 1994; Reder et al., 1994; Grasser et al., 1996; Wei and Lightman, 1997; Fassbender et al., 1998; Then Bergh et al., 2001). Postmortem analysis of MS-affected brains has revealed an increase in the number of CRF-containing neurons in the PVN. This increase was specifically located in the CRF-positive/AVP-positive neurons of the PVN (Erkut et al., 1995; Purba et al., 1995) suggesting an increased role for AVP in MS. Further support for an increased role for AVP is provided by the observation that MS patients have an increased relative HPA axis response to AVP compared with controls (Michelson et al., 1994).

Eosinophilia-myalgia syndrome (EMS) reached epidemic proportions on the West Coast of the United States in 1989. EMS is an immune-mediated disease characterized by eosinophilia, myalgias, edema, fasciitis, and neuropathies. The source of the epidemic was eventually traced to a rogue batch of impure L-tryptophan (L-Trp) derived from a genetically modified bacteria. This L-Trp was being used as a diet supplement and found its way into a number of health food preparations. To confirm the agent responsible, Lewis rats were treated with comparable doses (w/w) of impure L-Trp to those ingested by humans. These animals also developed fasciitis and perimyositis similar to the pathological features of human EMS. The appearance of symptoms of EMS in the rat model was associated with a decrease in CRF mRNA in the PVN (Brady et al., 1994; Crofford et al., 1990).

These observations of altered hypothalamic regulation are not solely confined to rat models. The MRL strain of mouse has been used as a model of the autoimmune disease systemic lupus erythematosus. As the animals age, they develop symptoms of lupus. As the symptoms develop, CRF mRNA levels in the PVN fall (Shanks et al., 1999). Leishmaniasis is a parasitic liver infection caused by the protozoan *Leishmania donovani* which is associated with T-cell activation. This parasite is prevalent in the tropics and in some Mediterranean areas. Certain mouse strains are susceptible to infection with the parasite, and following the inoculation of a susceptible strain there is a decrease in CRF mRNA associated with infection (Harbuz *et al.*, 1995).

These data suggest that an alteration in the hypothalamic control of the HPA axis may be part of the adaptive response to chronic immune-mediated disease/inflammatory stressors. This change can be seen in a number of different disease models in different rodent species and may also occur in humans. From the available evidence it would appear that AVP takes over as the major stimulator of the HPA axis while CRF provides a secondary permissive role in chronic inflammatory stress.

D. IMPLICATIONS FOR THE STRESS RESPONSE

These alterations in hypothalamic regulation, resulting in a decrease in CRF activity, have a profound influence on the ability of AA rats to respond to acute stress. This is, perhaps, not surprising as CRF is considered to be the major stimulator of the acute stress response. We have noted an inability of rats with AA to respond to the predominantly psychological stressors of restraint and noise or to the physical stress of ip hypertonic saline injection (Harbuz *et al.*, 1993a; Aguilera *et al.*, 1997; Windle *et al.*, 2001). This inability to respond to acute stress appears to be intimately related to the alteration in pulsatile corticosterone release patterns in AA rats (Windle *et al.*, 2001). A stress response can only be generated either in the rising phase of a corticosterone peak or in the plateau phase immediately prior to the initiation of the rise. In the AA rat, the increased number of pulses associated with inflammation results in the rats spending a considerably greater time in the falling (inhibitory) phase of the release cycle (Fig. 2). It appears therefore that the AA rat is not entirely refractory and unable to respond to acute stress, but rather the increased time spent in the inhibitory phase makes it unlikely that the animal will be able to respond to a stressor.

An inability to respond to an acute challenge has been reported in humans with RA. Chikanza and colleagues (1992) reported no increase in plasma cortisol in RA patients following joint replacement surgery. They also found that these patients responded normally to a CRF stimulation test indicating that the pituitary and adrenal responses were intact and further supporting the case for a hypothalamic defect associated with inflammatory disease. However, another study was unable to find any such hyporesponsiveness (Eijsbouts *et al.*, 1998) suggesting that further investigation is required to tease apart these discrepancies in humans with RA.

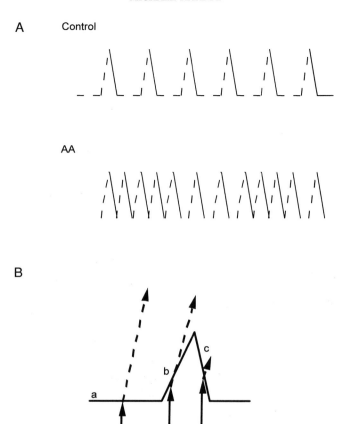

FIG. 2. (A) Control rats release corticosterone in approximately 13 ± 1 pulses in 24 h. During the plateau or interpulse phase or the rising phase (dashed lines) animals are able to respond to stress and increase corticosterone release. During the falling phase (solid lines) no response to stress is generated. Following development of inflammation in rats with adjuvant-induced arthritis (AA) the number of pulses increases to 22 ± 1 in 24 h. There is a corresponding decrease in the interpulse phase, and the rising phase is also reduced. (B) Stress is represented by the solid arrows associated with (a) the interpulse phase, (b) the rising phase, and (c) the falling phase. During (a) and (b) a stress response will be generated as represented by the dashed arrows. During (c) there is no significant increase in plasma corticosterone release due to inhibition of release. See Windle *et al.*, 2001 for details.

These observations are of particular interest as an inability to mount an HPA axis response to either acute or chronic immune stimulation is likely to be life-threatening. However, no alteration in susceptibility to infection has been noted in animal disease models or in patients with autoimmune disease. We have demonstrated an intact HPA axis response at all levels of the axis in response to acute injection with endotoxin (Harbuz et al., 1999). These data show that although the ability to respond to acute stress is impaired in rats with AA, the ability to respond to acute immune stimulation is intact reflecting the potentially life-threatening nature of the latter challenge. These data suggest activation of alternative pathway(s) to these stimuli and provide a further example of differential control of the response to acute stress and acute immune stimulation.

E. DOES A DEFECT IN THE HPA AXIS PREDISPOSE TO INFLAMMATORY DISEASE?

A compelling hypothesis first postulated in the late 1980s suggested that susceptibility to autoimmune disease resides in an inability to mount an appropriate response to stress. The hypothesis developed from studies using the rat model of streptococcal cell wall–induced arthritis and EAE (Sternberg et al., 1989a,b; Mason, 1991). The Lewis strain, which is susceptible to both of these disease models has a hyporesponse to a variety of acute stressors. In contrast, the Fischer and PVG strains of rat are resistant in these models. The major factor influencing this effect was determined to reside at the level of the hypothalamus and involved a defect in the responsiveness of CRF (Sternberg et al., 1989a). This defect resulted in low basal corticosterone levels, a blunted circadian rhythm, and an inability to respond to stress. The Lewis rat also has smaller adrenal glands and a larger thymus than the other strains reflecting lower corticosteroid production. These differences correlate well with the increased susceptibility to a wide range of T-cell-mediated autoimmune-like diseases seen in the Lewis rat.

However, this hypothesis has been questioned. First, despite having a hyperresponse to stress that is believed to protect the PVG strain of rat from EAE, this strain is susceptible to AA suggesting that the hypothesis does not hold true for all disease models (Harbuz et al., 1993a). Second, a number of groups have failed to replicate the difference in stress response between the Lewis and the Fischer strains (Rivest and Rivier, 1994; Dhabhar et al., 1993; Grota et al., 1997) suggesting that this difference may only occur in the immature females used in the original studies. Third, comparing susceptibility and resistance to EAE in a number of rat strains failed to identify HPA axis characteristics as a predictor of disease susceptibility (Stefferl et al., 1999). Fourth, the use of different inbred strains has an inherent disadvantage in that by definition they are genetically different. As a consequence they are likely to have a number of significant variations in a variety of neuroendocrine and immune parameters that may contribute to alterations in susceptibility. To rigorously test the hypothesis that a difference in HPA axis responsivity can alter disease susceptibility or severity, it is pertinent to do so in a single population of the same strain. The use of an outbred strain guarantees genotypic heterogeneity and is likely to be closer to the human condition than comparison of inbred strains.

Dividing a single population into subpopulations can be done using behavioral tests such as the open-field test. This is an established measure of anxiety and can easily be used to separate animals with high and low anxiety from within a population. Subsequent analysis of stress-responsiveness revealed clear differences between these groups. However, despite the differences in plasma corticosterone response to stress no difference in susceptibility or severity of AA were determined (Chover-Gonzalez et al., 1998). The learned helplessness (LH) paradigm has been used as an animal model of depression. Following exposure to uncontrollable footshock, subsequent exposure to escapable footshock reveals a group of rats that make no attempt to escape the shock. These are termed learned helpless (LH+) (i.e., exhibiting depression-like symptoms). In contrast, another subpopulation very rapidly learn to escape and receive few footshocks (LH−). There is a striking difference in corticosterone responses comparing these groups with a further significant increase in the LH− rats compared to the increase in the LH+ rats. According to the hypothesis the LH− rats should be protected by the relatively larger increase in plasma corticosterone. In fact, the LH− rats develop inflammation sooner and with increased severity when compared to the LH+ animals suggesting HPA axis responsiveness is not a good predictor of disease activity (Chover-Gonzalez et al., 2000). In another behavioral study it was demonstrated that the latency of an animal to attack an intruder correlated significantly with the disease score in EAE. Animals that did not attack an intruder were more resistant to the disease (Kavelaars et al., 1999). Together, these data suggest a link between behavioral characteristics of an individual and subsequent disease severity.

While there is evidence to support a hyporesponsive HPA axis both in animal models and in autoimmune disease in humans once the disease is established, it is difficult in humans to determine if this hyporesponsiveness was present prior to the onset of diseases. However, impaired HPA axis function has been noted in a number of immune-mediated diseases in humans. African sleeping sickness is a potentially lethal parasitic disease in humans caused by the protozoan *Trypanosoma brucei*. A variety of symptoms occur and in the later stages the parasite may invade the CNS and this can produce a broad spectrum of neurologic and psychiatric symptoms. Left untreated the disease is eventually fatal. Individuals with sleeping sickness are unable to mount a cortisol response to ACTH or CRF injection (Reincke et al., 1994). However, following antiparasitic treatment and recovery, ACTH and cortisol responses to CRF are normal suggesting that the defective HPA axis responsiveness was in response to the disease and not inherent to the individual prior to infection. Overall these data from animal and human studies suggest that HPA axis responsivity is unlikely to be a good predictor of susceptibility to autoimmune disease.

F. Is Susceptibility Predetermined?

Much of the literature has been concerned with whether a strain of rat is resistant or susceptible to autoimmune disease models. This may not be entirely appropriate as studies have demonstrated changes in susceptibility dependent on early life events, challenges in adulthood, and housing conditions. Housing and the associated microbiological status of rats have been highlighted as important factors in determining resistance or susceptibility to experimental arthritis. Germ-free Fischer rats are highly susceptible to experimental arthritis whereas conventionally housed rats are resistant (van den Broek *et al.*, 1992; Gripenberg-Lerche and Toivanen 1993; van de Langerijt *et al.*, 1993). These observations lend further support to the notion that there is no inherent resistance of the Fischer strain compared to the Lewis strain, suggesting instead that the environment plays an important role in determining these observations. In one study we investigated whether there was a difference in susceptibility in rats transferred from a clean breeding facility (fitted with HEPA filters but not a germ-free SPF unit) to a conventional stock room. These animals were resistant to AA for up to 2 days after transfer. However, 8 days after transfer and up to 27 days after transfer they were susceptible (Fig. 3). These effects are unlikely to be related to the stress-related changes 48 h after moving from the colony rooms to the experimental rooms. These data support previous studies concerning the importance of bacterial flora in determining susceptibility to, and the ability to reactivate, experimental arthritis (van de Langerijt *et al.*, 1993; Lichtman *et al.*, 1995). Indeed, these data extend these observations to suggest that changes in bacterial flora in the environment may subtly alter susceptibility and resistance to AA depending on the length of exposure. This subtle change suggests this need not be an all-or-none phenomenon whereby pathogen-free rats are highly susceptible and conventionally housed rats are relatively resistant. The timescale of this alteration in susceptibility suggests these changes may involve the immune system. The likely mediator appears to involve the colonization of the individual by novel bacteria.

These data raise the question of whether acute immune challenges might influence susceptibility to models of autoimmune disease. In this regard it has been shown that injection of LPS in neonates confers resistance to AA on these animals when adjuvant is injected in adulthood (Shanks *et al.*, 2000). These data are in keeping with the well-established effects of neonatal treatments affecting HPA axis and behavioral activities in adulthood (Francis

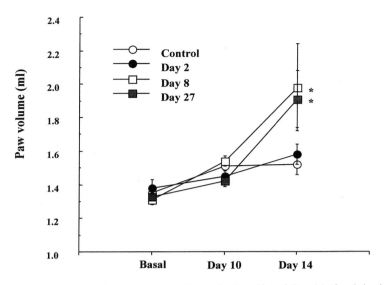

FIG. 3. Paw volume (ml) of rats on Day 0 (basal), Day 10 and Day 14 after injection of adjuvant. Control rats received vehicle alone. Adjuvant was injected either 2 days, 8 days, or 27 days after transfer of the animals from a breeding facility to the stockroom. Data represent mean $\pm$ SEM for $n = 5$ animals. $^*p < 0.05$ compared to control rats at Day 14. Unpublished observations courtesy of A. J. Chover-Gonzalez, Department of Neurosciences, University of Cadiz, Spain.

et al., 1999). Rather more surprising is the observation that exposure of adults to LPS can confer resistance to AA when the adjuvant is injected weeks later (Harbuz *et al.*, 2002a). This is not an adaptive stress-related effect as exposure of rats to footshock has no effect on the development of AA in this paradigm. Taken together, these data suggest that questions of susceptibility and resistance are relative and depend on a variety of factors involving behavioral responses, exposure to pathogens in early or later life, and the stress history of the individual. The cause of autoimmune diseases remains unknown, but, bacteria and viruses continue to be implicated although there is little consensus on which may be involved or the mechanisms through which they may operate.

G. A HANDLE ON THE HPA AXIS?

Despite the evidence highlighting the importance of the HPA axis in autoimmune disease the precise relationship remains obscure. A review of the literature regarding the HPA axis in RA concluded that HPA axis activity is not significantly different compared with normal subjects (Harbuz and Jessop, 1999). This was qualified by the observation that any inherent abnormality might lie in the inability of the axis to respond to the onset of inflammation and that therefore challenging the system might highlight these differences. Certainly there are differences in diurnal variation and these are likely to impact on disease as discussed earlier. Teasing out these differences and the underlying mechanisms may yield valuable clues for the containment of disease activity. Changes in HPA axis responsiveness have been investigated by injection of the hypothalamic releasing factors CRF and AVP and have suggested that in MS AVP may play a more significant role than CRF in maintaining the activity of the HPA axis (Michelson *et al.*, 1994). Wei and Lightman (1997) found a difference in response to ovine CRF depending on the disease status of the individual with a lower cortisol response in patients with secondary progressive MS but a normal response in those with the less severe primary progressive MS. Cortisol secretion in patients with RA was also impaired in response to challenge with releasing hormones (Gudbjornsson *et al.*, 1996). In contrast patients with newly diagnosed, untreated RA showed no difference in plasma ACTH and cortisol to a similar challenge (Templ *et al.*, 1996). These authors note, however, that given the presence of inflammation the response might be considered inappropriately low.

The dexamethasone-CRF test has been used extensively to assess HPA axis dysfunction in a variety of psychiatric disorders (Holsboer *et al.*, 1985; Deuschle *et al.*, 1998; Hundt *et al.*, 2001; Zobel *et al.*, 2001). Individuals with depression or alcoholism appear to have an altered feedback regulatory system such that dexamethasone suppression of endogenous HPA axis activity is incomplete. When challenged with CRF, these patients are able to escape the dexamethasone suppression and increases in cortisol are observed. It has been suggested that the underlying mechanism involves a shift in the dose–response curve toward lower glucocorticoid receptor sensitivity in depressed patients reflecting feedback resistance at the level of the PVN and the pituitary (Modell *et al.*, 1997). The involvement of corticosteroid receptors is supported by evidence of decreased type I and type II receptor mRNAs in suicide victims with a history of depression (Lopez *et al.*, 1998). Following successful treatment dexamethasone suppression prevents the response to CRF. These data provide further evidence that alterations in HPA axis activity are not necessarily inherent to these individuals but are instead a consequence of the disorder itself.

In an attempt to tease out the subtleties underlying HPA regulation in autoimmune disease, investigations have used the dexamethasone-CRF test. Early studies had demonstrated abnormalities in dexamethasone suppression with approximately 50% of MS patients not suppressing cortisol

(Reder *et al.*, 1987). This is similar to patients with major depression and in excess of that found in the normal control population. There was no apparent difference in depression scores between the suppressors and the nonsuppressors suggesting this was not a major factor. Combining the dexamethasone suppression with a CRF challenge revealed a heterogeneous response with 15 out of 19 patients responding to CRF with increased cortisol (i.e., escaping the dexamethasone suppression) (Grasser *et al.*, 1996). Six of these mounted an exaggerated response. Four of the subjects did not respond. Once again these patterns were not obviously related to psychopathological features. Subsequent studies have confirmed that MS patients are able to escape from dexamethasone suppression following challenge with CRF suggesting a hyperactivity of the HPA axis that is significantly correlated to disease activity (Fassbender *et al.*, 1998; Kumpfel *et al.*, 1999; Then Bergh *et al.*, 2001). However, although one study has related this escape to higher depression and anxiety scores in MS patients compared to controls (Fassbinder *et al.*, 1998), others have not found any evidence to relate the cortisol response to CRF with increased incidence of depression (Kumpfel *et al.*, 1999; Then Bergh *et al.*, 2001). The cortisol response to the dexamethasone-CRF test has been investigated in patients with RA (Harbuz *et al.*, 2002b). Three out of seven patients escaped the dexamethasone suppression while four remained suppressed (Fig. 4). No correlation between disease activity or severity and nonsuppression was noted in this study suggesting the existence of subpopulations of patients with RA. The impact of these differences on disease progression have yet to be elucidated.

The observation that the alterations in HPA axis activity, including the escape from cortisol suppression in the dexamethasone-CRF test, seen in patients with depression can be reversed by antidepressants led one group to investigate the use of an antidepressant in patients with MS (Then Bergh *et al.*, 2001). This study noted that treatment with glucocorticoids alone did not alter the escape from dexamethasone-CRF when this was tested. However, if the treatment was combined with the antidepressant moclobemide then the cortisol response to the test was normalized. Due to the small sample size and the use of patients with relapsing–remitting MS only, (the mildest form of MS), no significant difference in clinical outcome was observed, although it was noted that this was not the purpose of the study. It is of interest that the cortisol response in the dexamethasone-CRF test has been suggested to predict relapse in patients with remitted depression (Zobel *et al.*, 2001). Whether such a mechanism may be operating in patients with autoimmune disease or whether the escape can be predictive of alterations in severity of inflammation requires further investigation.

Together these data suggest an alteration in the responsiveness of the HPA axis associated with autoimmune disease. However, it appears that the

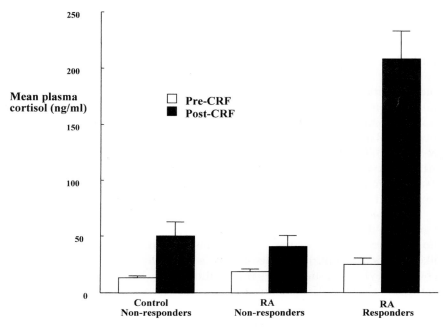

FIG. 4. Mean plasma cortisol in patients with active rheumatoid arthritis admitted to hospital for bed rest and age-matched control subjects. All subjects were given 1.5 mg dexamethasone orally at 2300 h on Day 1. On Day 2 at 1300 h an iv catheter was fitted and baseline blood samples were taken at 1400 h and 1430 h. The mean of these two samples comprise the Pre-CRF group. At 1500 h, $100\mu g$ CRF was infused and samples collected every 15 min for 2 h. The mean of these values comprise the Post-CRF group. The values represent mean + SEM for 6 controls, 4 RA nonresponders, and 3 RA responders.

mechanism involved is not simply a predisposition to autoimmune disease secondary to a defect in the HPA axis but a more complex interaction between the HPA axis and the disease process associated with the development of the disease.

IV. Neurotransmitters in Disease

The observation that monoamine oxidase antidepressants may be beneficial in MS is intriguing. MS is seen as a neurological condition whereas most other autoimmune diseases are considered to predominantly involve mechanisms in the periphery. Increasing evidence suggests that neurotransmitters may be important in a variety of inflammatory conditions. Substance

P antagonists have been reported to have protective effects in AA (Chowdrey *et al.*, 1995a), probably through an action at the level of the CRF neurons in the PVN (Jessop *et al.*, 2000). The role of serotonin and substance P in the HPA axis response to stress and inflammation has been reviewed (Harbuz and Jessop, 2001). The role of neurotransmitters in modulating, for example, arthritic disease, has been investigated principally at the level of the joint that is densely innervated by fibers containing catecholamines, serotonin, substance P, and opioids (Levine *et al.*, 1993). So, for example, at the peripheral level adrenal medullectomy decreases the severity of AA (Levine *et al.*, 1988; Coderre *et al.*, 1990), which can be reversed by treatment with adrenaline or β_2-agonists (Coderre *et al.*, 1990). The role of a number of catecholamine receptor agonists and antagonists have been investigated in this regard (Levine *et al.*, 1988; Coderre *et al.*, 1991).

The role of central neurotransmitters has been less extensively investigated to date. However, the increased importance of the SNS innervation of the immune tissues and the central control of this activation is likely to focus attention on this area. It is generally accepted that the catecholamines exert a stimulatory role at the hypothalamic level on HPA axis activation to acute stress. There have, however, been relatively few studies on the role of catecholamines in chronic inflammatory stress/autoimmune disease models. Endogenous catecholamine depletion following neonatal treatment with 6-OHDA has been shown to increase the severity of experimental autoimmune myasthenia gravis and EAE (Agius *et al.*, 1987; Chelmicka-Schoor *et al.*, 1988, 1992). This treatment results in whole body depletion of catecholamines producing major effects on peripheral sympathetic catecholaminergic systems in addition to central lesions. Irrespective of whether this action is exerted at the peripheral or the central level, or involves a combination of both, these data demonstrate a protective effect of catecholamines. A similar protective effect was observed following direct catecholamine depletion in the PVN by targeted application of 6-OHDA in AA. This resulted in an increase in the severity of the hind paw inflammation supporting a protective role for central catecholamines in this disease model (Harbuz *et al.*, 1994). In a monoarthritic rat model, tyrosine hydroxylase mRNA levels were increased in the pontine noradrenergic cell groups which project to the spinal cord suggesting a possible involvement of spinal cord projections (Cho *et al.*, 1995). Noradrenaline concentrations are decreased in the lumbar and sacral portions of the spinal cord in chronic relapsing EAE and are decreased in the cranio-thoracal spinal cord following relapses (Krenger *et al.*, 1986). It has also been demonstrated that agonists directed against the β-adrenergic receptor alleviate EAE (Weigmann *et al.*, 1995). In contrast to these studies demonstrating an increase in severity following depletion of catecholamines, other groups have reported that both central and peripheral depletion of catecholamines can reduce the severity of EAE (Leonard *et al.,* 1991). These latter data are not easily resolved and suggest the need for further investigation. The mechanism by which these actions are mediated is unknown but may involve alterations in the HPA axis and/or SNS activity. One study has demonstrated that immune depression induced by protein calorie malnutrition can be suppressed following central catecholaminergic lesions suggesting that central catecholaminergic hyperactivity is one of the mechanisms involved in immunodepression. These data provide evidence relating central catecholaminergic activity to an improvement of lymphocyte proliferation and IL-1 production (Schlesinger *et al.,* 1995). While noradrenaline has received the most attention, the depleting agents used would have effects on other catecholamines (i.e., dopamine and adrenaline). Dopamine may also be important as pergolide (a dopamine receptor agonist), has anti-inflammatory properties in both an acute model of carrageenan-induced paw inflammation and in an arthritis model (Bendele *et al.,* 1991). These effects are partially inhibited by a centrally acting dopamine antagonist but not peripherally acting antagonists suggesting the involvement of centrally acting dopamine receptors.

Serotonin has a pro-inflammatory role. Direct injection of serotonin into the hind paw of the rat produces a dose-dependent increase in paw edema (Sufka *et al.,* 1992). An increase in serotonin in plasma, spinal cord, and some midbrain areas has been noted in AA although the location of these central changes was not specified (Weil-Fugazza *et al.,* 1979; Garzon *et al.,* 1990; Marlier *et al.,* 1991; Pertsch *et al.,* 1993; Sofia and Vassar., 1974; Godefroy *et al.,* 1987). Depletion of endogenous serotonin using *p*-chlorophenylalanine (PCPA; a reversible inhibitor of the rate-limiting enzyme of serotonin synthesis), at the time of development of inflammation, resulted in a reduction of hind paw inflammation (Harbuz *et al.,* 1996). In addition, the decrease in CRF mRNA was reversed in the serotonin-depleted animals. These data suggest that serotonin might be implicated in the reduced expression of CRF mRNA in AA. The role of central serotonin was addressed by depleting serotonin using 5'7'-dihydroxytryptamine, a toxin of serotonin neurons, injected into the lateral ventricle. These central lesions also significantly reduced inflammation, suggesting that central serotonin may be important in modulating inflammatory disease activity (Harbuz *et al.,* 1998). Despite measuring concentrations of serotonin and the major metabolite 5HIAA in a number of brain areas, no difference in either of these parameters of serotonin activity were determined comparing control and AA brains. The location of the changes in central serotonin remain to be determined.

Of particular interest is the possibility that serotonin antagonists may be of clinical benefit in reducing the severity of acute inflammation. Alterations in central 5HT2C receptor mRNA have been noted in the CA1–CA3 hippocampal regions in a monoarthritic rat model suggesting that 5HT2C receptors may be involved in mediating central serotoninergic changes in arthritis (Holmes *et al.*, 1995). However, a 5HT2 antagonist did not alter inflammation in the AA model (unpublished observations, M. Harbuz), while other groups have demonstrated an increase in severity of streptococcal cell wall–induced arthritis in the Lewis rat treated with a 5HT2 antagonist (Sternberg *et al.*, 1989b). The large number of serotonin receptor subtypes identified to date and the lack of specificity of the drugs available make further determination of the role of serotonin complicated.

If serotonin is indeed pro-inflammatory then it is likely that selective serotonin reuptake inhibitors (SSRIs), through their action in increasing available serotonin at the site of release, would increase inflammation. This has been shown in the rat model of AA (Harbuz *et al.*, 1998) and also reported in patients with arthritis receiving SSRIs as treatment for depression (Hood *et al.*, 2001). Indeed, in one person the onset of RA occurred following SSRI treatment for depression. These data suggest that SSRIs may be contraindicated in patients with underlying inflammation.

V. Summary

The HPA axis is fundamental for long-term survival and protection from the ravages of autoimmune disease. Continuing investigations suggest that the hypothesis linking susceptibility to autoimmune disease and a hyporesponsive HPA axis is somewhat simplistic. Instead, data from a number of different human diseases and from preclinical studies in a variety of models have suggested a more complicated picture. Alterations in the diurnal rhythms of ACTH, cortisol, and immune parameters appear to be linked to severity of disease. The use of low doses of steroids timed to target disrupted diurnal immune system changes in patients with RA may reduce the unfortunate side effects of long-term steroid use. Studies in cancer patients have related alterations in diurnal cortisol to survival. Whether differences in individual cortisol profiles are predictive of a deterioration in symptoms of autoimmune disease remains to be established. Responsiveness of the HPA axis to subtle challenges such as the dexamethasone suppression test and the related dexamethasone-CRF test suggest that there are different subpopulations of patients with RA and MS and these may have confounded earlier, apparently contradictory, studies. These different responses may be related to the severity of the disease. That these HPA axis differences can be altered beneficially through the use of antidepressants, as has been shown in MS, may impact on future health care strategies. However, reports of negative developments in arthritis associated with SSRI use suggest that the SSRIs may be unsuitable under some circumstances. The link of behavioral differences to alterations in neurotransmitter changes associated with disease is intriguing and opens new avenues of research. These future studies will require input from neuroscientists, neuroendocrinologists, psychologists, and immunologists working with the clinical specialities already involved in treating patients with autoimmune disease. These multidisciplinary studies reflecting the increased importance of hormonal and neurotransmitter involvement with the immune system hold great promise for the future.

References

Agius, M. A., Checinski, M. E., Richman, D. P., and Chelmicka-Schorr, E. (1987). Sympathectomy enhances the severity of experimental autoimmune myasthenia gravis (EAMG). *J. Neuroimmunol.* **16**, 11–12.

Aguilera, G., Jessop, D. S., Harbuz, M. S., Kiss, A., and and Lightman, S. L. (1997). Biphasic regulation of hypothalamic-pituitary corticotropin releasing hormone receptors during development of adjuvant-induced arthritis in the rat. *J. Endocrinol.* **153**, 185–191.

Arvidson, N. G., Gudbjornsson, B., Larsson, A., and Hallgren, R. (1997). The timing of glucocorticoid administration in rheumatoid arthritis. *Ann. Rheum. Dis.* **56**, 27–31.

Baerwald, C. G., Burmester, G. R., and Krause, A. (2000). Interactions of autonomic nervous, neuroendocrine, and immune systems in rheumatoid arthritis. *Rheum. Dis. Clin. North Am.* **26**, 841–857.

Bendele, A. M., Spaethe, S. M., Benslay, D. N., and Bryant, H. U. (1991). Anti-inflammatory activity of pergolide, a dopamine receptor agonist. *J. Pharmacol. Ther.* **259**, 169–175.

Brady, L. S., Page, S. W., Thomas, F. S., Rader, J. L., Lynn, A. B., Misiewicz-Poltorak, B., Zelazowski, E., Crofford, L. J., Zelazowski, P., Smith, C., Raybourne, R. B., Love, L. A., Gold, P. W., and Sternberg, E. M. (1994). 1'1-Ethylidenebis[L-tryptophan], a contaminant implicated in L-tryptophan eosinophilia myalgia syndrome, suppresses mRNA expression of hypothalamic corticotropin-releasing hormone in Lewis (LEW/N) rat brain. *Neuroimmunomodulation* **1**, 59–65.

Buckingham, J. C., Cowell, A-M., Gillies, G., Herbison, A. E., and Steel, J. H. (1997). The neuroendocrine system: Anatomy, physiology and responses to stress. *In* "Stress, Stress Hormones and the Immune System" (J. C. Buckingham, A.-M. Cowell, G. Gillies, eds.), pp. 9–47, John Wiley and Sons, Chichester.

Chelmicka-Schoor, E., Checinski, M. E., and Arnason, B. G. W. (1988). Chemical sympathectomy augments the severity of experimental allergic encephalomyelitis in Lewis rats. *J. Neuroimmunol.* **17**, 347–350.

Chelmicka-Schoor, E., Kwasniewski, M. N., and Wollmann, R. L. (1992). Sympathectomy augments adoptively transferred experimental allergic encephalomyelitis. *J. Neuroimmunol.* **37**, 99–103.

Chikanza, I. C., and Grossman, A. S. (1998). Hypothalamic-pituitary-mediated immunomodulation: Arginine vasopressin is a neuroendocrine immune mediator. *Br. J. Rheumatol.* **37**, 131–136.

Chikanza, I. C., Petrou, P., Kingsley, G., Chrousos, G., and Panayi, G. S. (1992). Defective hypothalamic response to immune and inflammatory stimuli in patients with rheumatoid arthritis. *Arthritis Rheum.* **35,** 1281–1288.

Chikanza, I. C, Petrou, P, and Chrousos, G. (2000). Perturbations of arginine vasopressin secretion during inflammatory stress. Pathophysiologic implications. *Ann. N.Y. Acad. Sci.* **917,** 825–834.

Cho, H.-J., Lee, H.-S., Bae, M.-A., and Joo, K. (1995). Chronic arthritis increases tyrosine hydroxylase mRNA levels in the pontine noradrenergic cell groups. *Brain Res.* **695,** 96–99.

Chover-Gonzalez, A. J., Tejedor-Real, P., Harbuz, M. S., Gibert-Rahola, J., Larsen, P. J., and Jessop, D. S. (1998). A differential response to stress is not a prediction of susceptibility or severity in adjuvant-induced arthritis. *Stress* **2,** 221–226.

Chover-Gonzalez, A. J., Jessop, D. S., Tejedor-Real, P., Gibert-Rahola, J., and Harbuz, M. S. (2000). Onset and severity of inflammation in rats exposed to the learned helplessness paradigm. *Rheumatology* (*Oxford*) **39,** 764–771.

Chowdrey, H. S., Larsen, P. J., Harbuz, M. S., Lightman, S. L., and Jessop, D. S. (1995a). Endogenous substance P inhibits the expression of corticotropin-releasing hormone during a chronic inflammatory stress. *Life Sci.* **57,** 2021–2029.

Chowdrey, H. S., Larsen, P. J., Harbuz, M. S., Jessop, D. S., Aguilera, G., Eckland, D. J. A., and Lightman, S. L. (1995b). Evidence for arginine vasopressin as the primary activator of the HPA axis during adjuvant-induced arthritis. *Br. J. Pharmacol.* **116,** 2417–2424.

Coderre, T. J., Basbaum, A. I., Dallman, M. F., Helms, C., and Levine, J. D. (1990). Epinephrine exacerbates arthritis by an action at presynaptic β_2-adrenoceptors. *Neuroscience* **34,** 521–523.

Coderre, T. J., Chan, A. K., Helms, C., Basbaum, A. I., and Levine, J. D. (1991). Increasing sympathetic nerve terminal-dependent plasma extravasation correlates with decreased arthritic joint injury in rats. *Neuroscience* **40,** 185–189.

Crofford, L. J., Rader, J. I., Dalakas, M. C., Hill, R. H., Page, S. W., Needham, L. L., Brady, L. S., Heyes, M. P., Wilder, R. L., Gold, P. W., Illa, I., Smith, C., and Sternberg, E. M. (1990). L-Tryptophan implicated in human eosinophilia-myalgia syndrome causes fasciitis and perimyositis in the Lewis rat. *J. Clin. Invest.* **86,** 1757–1763.

de Goeij, D. C., Kvetnansky, R., Whitnall, M. H., Jezova, D., Berkenbosch, F., and Tilders, F. J. (1991). Repeated stress-induced activation of corticotropin-releasing factor neurons enhances vasopressin stores and colocalization with corticotropin-releasing factor in the median eminence of rats. *Neuroendocrinology* **53,** 150–159.

Deuschle, M., Schweiger, U., Gotthardt, U., Weber, B., Korner, A., Schmider, J., Standhardt, H., Lammers, C. H., Krumm, B., and Heuser, I. (1998). The combined dexamethasone/corticotropin-releasing hormone stimulation test is more closely associated with features of diurnal activity of the hypothalamo-pituitary-adrenocortical system than the dexamethasone suppression test. *Biol. Psychiatry.* **43,** 762–766.

Dhabhar, F. S., McEwen, B. S., and Spencer, R. L. (1993). Stress response, adrenal steroid receptor levels and corticosteroid-binding globulin levels—A comparison between Sprague-Dawley, Fischer 344 and Lewis rats. *Brain Res.* **16,** 89–98.

Edwards, S., Evans, P., Hucklebridge, F., and Clow, A. (2001a). Association between time of awakening and diurnal cortisol secretory activity. *Psychoneuroendocrinology* **26,** 613–622.

Edwards, S., Clow A., Evans P., and Hucklebridge, F. (2001b). Exploration of the awakening cortisol response in relation to diurnal cortisol secretory activity. *Life Sci.* **68,** 2093–2103.

Eijsbouts, A., van den Hoogen, F., Laan, R, de Waal Malefijt, M., Hermus, A., Sweep, C., de Rooij, D. J., and van de Putte, L. (1998). Similar response of adrenocorticotrophic hormone, cortisol and prolactin to surgery in rheumatoid arthritis and osteoarthritis. *Br. J. Rheumatol.* **37,** 1138–1139.

Erkut, Z. A., Hofman, M. A., Ravid, R., and Swaab, D. F. (1995). Increased activity of hypothalamic corticotropin-releasing hormone neurons in multiple sclerosis. *J. Neuroimmunol.* **62,** 27–33.

Fassbender, K., Schmidt, R., Mossner, R., Kischka, U., Kuhnen, J., Schwartz, A., and Hennerici, M. (1998). Mood disorders and dysfunction of the hypothalamic-pituitary-adrenal axis in multiple sclerosis: Association with cerebral inflammation. *Arch. Neurol.* **55,** 66–72.

Francis, D. D., Champagne, F. A., Liu, D., and Meaney, M. J. (1999). Maternal care, gene expression, and the development of individual differences in stress reactivity. *Ann. N.Y. Acad. Sci.* **896,** 66–84.

Garzon, J., Lerida, M., and Sanchez-Blazquez, P. (1990). Effect of intrathecal injection of pertussis toxin on substance P, norepinephrine and serotonin contents in various neural structures of arthritic rats. *Life Sci.* **47,** 1915–1923.

Godefroy, F., Weil-Fugazza, J., and Besson, J.-M. (1987). Complex temporal changes in 5–hydroxytryptamine synthesis in the central nervous system induced by experimental polyarthritis in the rat. *Pain* **28,** 223–238.

Grasser, A., Moller, A., Backmund, H., Yassouridis, A., and Holsboer, F. (1996). Heterogeneity of hypothalamic-pituitary-adrenal system response to a combined dexamethasone-CRH test in multiple sclerosis. *Exp. Clin. Endocrinol. Diabetes* **104,** 31–37.

Gripenberg-Lerche, C., and Toivanen, P. (1993). Yersinia associated arthritis in SHR rats: Effect of the microbial status of the host. *Ann. Rheum. Dis.* **52,** 223–228.

Grota, L. J., Bienen, T., and Felten, D. L. (1997). Corticosterone responses of adult Lewis and Fischer rats. *J. Neuroimmunol.* **74,** 95–101.

Gudbjornsson, B., Skogseid, B., Oberg, K., Wide, L., and Hallgren, R. (1996). Intact adrenocorticotropic hormone-secretion but impaired cortisol response in patients with active rheumatoid arthritis-effect of glucocorticoids. *J. Rheumatol.* **23,** 596–602.

Harbuz, M. S., and Jessop, D. S. (1999). Is there a defect in cortisol production in rheumatoid arthritis? *Rheumatology (Oxford)* **38,** 298–302.

Harbuz, M. S., and Jessop, D. S. (2001). Stress and inflammatory disease: Widening roles for serotonin and substance P. *Stress* **4,** 57–70.

Harbuz, M. S., Rees, R. G., Eckland, D., Jessop, D. S., Brewerton, D., and Lightman, S. L. (1992). Paradoxical responses of hypothalamic CRF mRNA and CRF-41 peptide and adenohypophyseal POMC mRNA during chronic inflammatory stress. *Endocrinology* **130,** 1394–1400.

Harbuz, M. S., Rees, R. G., and Lightman, S. L. (1993a). Hypothalamo-pituitary responses to acute stress and changes in circulating glucocorticoids during chronic adjuvant-induced arthritis in the rat. *Am. J. Physiol.* **264,** R179–R185.

Harbuz, M. S., Leonard, J. P., Lightman, S. L., and Cuzner, M. L. (1993b). Changes in hypothalamic corticotrophin-releasing factor and anterior pituitary pro-opiomelanocortin mRNA during the course of experimental allergic encephalomyelitis. *J. Neuroimmunol.* **45,** 127–132.

Harbuz, M. S., Chover-Gonzalez, A. J., Biswas, S., Lightman, S. L., and Chowdrey, H. S. (1994). Role of central catecholamines in the modulation of corticotrophin-releasing factor mRNA during adjuvant-induced arthritis in the rat. *Br. J. Rheumatol.* **33,** 205–209.

Harbuz, M. S., Jessop, D. S., Chowdrey, H. S., Blackwell, J. M., Larsen, P. J., and Lightman, S. L. (1995). Evidence for altered control of hypothalamic CRF in immune-mediated diseases. *Ann. N.Y. Acad. Sci.* **771,** 449–458.

Harbuz, M. S., Perveen-Gill, Z., Lalies, M. D., Jessop, D. S., Lightman, S. L., and Chowdrey, H. S. (1996). The role of endogenous serotonin in adjuvant-induced arthritis in the rat. *Br. J. Rheumatol.* **35,** 112–116.

Harbuz, M. S., Marti, O., Lightman, S. L., and Jessop, D. S. (1998). Alteration of central sero-tonin modifies onset and severity of adjuvant-induced arthritis in the rat. *Br. J. Rheumatol.* **37,** 1077–1083.

Harbuz, M. S., Rooney, C., Jones, M., and Ingram, C. D. (1999). Hypothalamo-pituitary-adrenal axis responses to lipopolysaccharide in male and female rats with adjuvant-induced arthritis. *Brain Behav. Immun.* **13,** 335–347.

Harbuz, M. S., Chover-Gonzalez, A. J., Gibert-Rahola, J., and Jessop, D. S. (2002a). Protective effect of prior acute immune challenge, but not stress, on inflammation in the rat. *Brain Behav. and Immun.* **16,** 439–449.

Harbuz, M. S., Korendowych, E., Jessop, D. S., Crown, A., Lightman, S. L., and Kirwan, J. (2002b). HPA axis dysregulation in patients with rheumatoid arthritis following the dexamethasone-CRF test. Submitted for publication.

Hashimoto, K., Suemaru, S., Takao, T., Sugarwara, M., Makino, S., and Ota, S. (1988). Corticotropin-releasing hormone and pituitary-adrenocortical responses in chronically stressed rats. *Regul. Pep.* **23,** 117–126.

Hench, P. S. (1949). The potential reversibility of rheumatoid arthritis. *Proc. Mayo Clinic* **24,** 167–180.

Hench, P. S., Kendall, E. C., Slocumb, C. H., and Polley, H. F. (1949). The effect of a hormone of the adrenal cortex (17-hydroxy-11-dehydrocorticosterone: compound E) and of pitui-tary adrenocorticotropic hormone on rheumatoid arthritis. *Proc. Mayo Clinic* **24,** 181–197.

Holmes, M. C., French, K. L., and Seckl, J. R. (1995). Modulation of serotonin and corticosteroid receptor gene expression in the rat hippocampus with circadian rhythm and stress. *Mol. Brain Res.* **28,** 186–192.

Holsboer, F., Gerken, A., Stalla, G. K., and Muller, O. A. (1985). ACTH, cortisol, and cortico-sterone output after ovine corticotropin-releasing factor challenge during depression and after recovery. *Biol. Psychiatry.* **20,** 276–286.

Hood, S. D., Argyropoulos, S. V., and Nutt, D. J. (2001). Arthritis and serotonergic antidepres-sants. *J. Clin. Psychopharmacol.* **21,** 458–461.

Hundt, W., Zimmermann, U., Pottig, M., Spring, K., and Holsboer, F. (2001). The combined dexamethasone-suppression/CRH-stimulation test in alcoholics during and after acute withdrawal. *Alcohol Clin. Exp. Res.* **25,** 687–691.

Jessop, D. S. (1999). Central non-glucocorticoid inhibitors of the hypothalamo-pituitary-adrenal axis. *J Endocrinol.* **160,** 169–180.

Jessop, D. S., Renshaw, D., Larsen, P. J., Chowdrey, H. S., and Harbuz, M. S. (2000). Substance P is involved in terminating the hypothalamo-pituitary-adrenal axis response to acute stress through centrally located neurokinin-1 receptors. *Stress* **3,** 209–220.

Kavelaars, A., Heijnen, C. J., Tennekes, R., Bruggink, J. E., and Koolhaas, J. M. (1999). In-dividual behavioral characteristics of wild-type rats predict susceptibility to experimental autoimmune encephalomyelitis. *Brain Behav. Immun.* **13,** 279–286.

Kirwan, J. R. (1995). The effect of glucocorticoids on joint destruction in rheumatoid arthritis. The Arthritis and Rheumatism Council Low-Dose Glucocorticoid Study Group. *New Engl. J. Med.* **333,** 142–146.

Kohm, A. P., and Sanders, V. M. (2000). Norepinephrine: A messenger from the brain to the immune system. *Immunol. Today* **21,** 539–542.

Krenger, W., Honegger, C. G., Feurer, C., and Cammiuli, S. (1986). Changes of neurotransmit-ter systems in chronic relapsing experimental allergic encephalomyelitis in rat brain and spinal cord. *J. Neurochem.* **47,** 1247–1254.

Kumpfel, T., Then Bergh, F., Friess, E., Uhr, M., Yassouridis, A., Trenkwalder, C., and Holsboer, F. (1999). Dehydroepiandrosterone response to the adrenocorticotropin test and the combined dexamethasone and corticotropin-releasing hormone test in patients with mul-tiple sclerosis. *Neuroendocrinology* **70,** 431–438.

Leonard, J. P., MacKenzie, F. J., Patel, H. A., and Cuzner, M. L. (1991). Hypothalamic nora-drenergic pathways influence neuroendocrine and clinical status in experimental allergic encephalomyelitis. *Brain Behav. Immun.* **5**, 328–338.

Levine, J. D., Coderre, T. J., Helms, C., and Basbaum, A. I. (1988). β_2-Adrenergic mechanisms in experimental arthritis. *Proc. Natl. Acad. Sci. USA* **85**, 4553–4556.

Levine, J. D., Fields, H. L., and Basbaum, A. I. (1993). Peptides and the primary afferent nociceptor. *J. Neurosci.* **13**, 2273–2286.

Lichtman, S. N., Wang, J., Sartor, R. B., Zhang, C., Bender, D., Dalldorf, F. G., and Schwab, J. H. (1995). Reactivation of arthritis induced by small bowel bacterial overgrowth in rats: Role of cytokines, bacteria, and bacterial polymers. *Infect. Immun.* **63**, 2295–2301.

Lopez, J. F., Chalmers, D. T., Little, K. Y., and Watson, S. J. (1998). A. E. Bennett Research Award. Regulation of serotonin1A, glucocorticoid, and mineralocorticoid receptor in rat and human hippocampus: Implications for the neurobiology of depression. *Biol. Psychiatry.* **43**, 547–573.

Marlier, L., Poulat, P., and Rajaofetra, N. (1991). Modifications of serotonin-, substance P- and calcitonin gene-related peptide-like immunoreactivities in the dorsal horn of the spinal cord of arthritic rats: A quantitative immunocytochemical study. *Exp. Brain Res.* **85**, 482–490.

Marti, O., Garcia, A., Valles, A., Harbuz, M. S., and Armario, A. (2001). Evidence that a single exposure to aversive stimuli triggers long-lasting effects in the hypothalamus-pituitary-adrenal axis that consolidate with time. *Eur. J. Neurosci.* **13**, 129–136.

Masi, A. T., and Chrousos, G. P. (1997). Dilemmas of low dosage glucocorticoid treatment in rheumatoid arthritis: Considerations of timing. *Ann. Rheum. Dis.* **56**, 1–4.

Mason, D. (1991). Genetic variation in the stress response; susceptibility to experimental aller-gic encephalomyelitis and implications for human inflammatory disease. *Immunol. Today* **12**, 57–60.

Mason, D., MacPhee, I., and Antoni, F. (1990). The role of the neuroendocrine system in de-termining genetic susceptibility to experimental encephalomyelitis in the rat. *Immunology* **70**, 1–5.

Michelson, D., Stone, L., Galliven, E., Magiakou, M. A., Chrousos, G. P., Sternberg, E. M., and Gold, P. W. (1994). Multiple sclerosis is associated with alterations in hypothalamic-pituitary-adrenal axis function. *J. Clin. Endocrinol. Metab.* **79**, 848–853.

Michelson, D., and Gold, P. W. (1998). Pathophysiologic and somatic investigations of hypothalamo-pituitary-adrenal axis activation in patients with depression. *Ann. N.Y. Acad. Sci.* **840**, 717–722.

Modell, S., Yassouridis, A., Huber, J., and Holsboer, F. (1997). Corticosteroid receptor function is decreased in depressed patients. *Neuroendocrinology* **65**, 216–222.

Mossner, R., and Lesch, K.-L. (1998). Role of serotonin in the immune system and in neuroim-mune interactions. *Brain Behav. Immun.* **12**, 249–271.

Neidhart, M., and Fluckiger, E. W. (1992). Hyperprolactinaemia in hypophysectomized or intact male rats and the development of adjuvant arthritis. *Immunology* **77**, 449–455.

Panayi, G. S. (1992). Neuroendocrine modulation of disease expression in rheumatoid arthritis. *Eular Congress Reports* **2**, 2–12.

Persellin, R. H., Kittinger, G. W., and Kendall, J. W. (1972). Adrenal response to experimental arthritis in the rat. *Am. J. Physiol.* **222**, 1545–1549.

Pertsch, M., Krause, E., and Hirschelmann, R. (1993). A comparison of serotonin (5-HT) blood levels and activity of 5-HT2 antagonists in adjuvant arthritic Lewis and Wistar rats. *Agents Actions* **38**, C98–C101.

Purba, J. S., Raadsheer, F. C., Hofman, M. A., Ravid, R., Polman, C. H., Kamphorst, W., and Swaab, D. F. (1995). Increased number of corticotropin-releasing hormone expressing neurons in the hypothalamic paraventricular nucleus of patients with multiple sclerosis. *Neuroendocrinology* **62**, 62–70.

Reder, A. T., Lowy, M. T., Meltzer, H. Y., and Antel, J. P. (1987). Dexamethasone suppression test abnormalities in multiple sclerosis: relation to ACTH therapy. *Neurology* **37**, 849–853.

Reder, A. T., Mackowiec, R. L., and Lowy, M. T. (1994). Adrenal size is increased in multiple sclerosis. *Arch. Neurol.* **51**, 151–154.

Reincke, M., Heppner, C., Petske, F., Allolio, B., Arlt, W., Mbulamberi, D., Siekmann, L., Vollmer, D., Winkelmann, W., and Chrousos, G. P. (1994). Impairment of adrenocortical function associated with increased plasma tumour necrosis factor-α and interleukin-6 concentrations in African trypanosomiasis. *Neuroimmunomodulation* **1**, 14–22.

Rivest, S., and Rivier, C. (1994). Stress and interleukin-1 β-induced activation of c-fos, NGFI-B and CRF gene expression in the hypothalamic PVN: Comparison between Sprague-Dawley, Fisher-344 and Lewis rats. *J. Neuroendocrinol.* **6**, 101–117.

Sarlis, N. J., Chowdrey, H. S., Stephanou, A., and Lightman, S. L. (1992). Chronic activation of the hypothalamo-pituitary-adrenal axis and loss of circadian rhythm during adjuvant-induced arthritis in the rat. *Endocrinology* **130**, 1775–1779.

Scaccianoce, S., Muscolo, L. A. A., Cigliana, G., Navarra, D., Nicolai, R., and Angelucci, L. (1991). Evidence for a specific role of vasopressin in sustaining pituitary-adrenocortical stress response in the rat. *Endocrinology* **128**, 3138–3143.

Schlesinger, L., Arevalo, M., Simon, V., Lopez, M., Munoz, C., Hernandez, A., Carreno, P., Belmar, J., White, A., and Haffnercavaillon, N. (1995). Immune depression induced by protein-calorie malnutrition can be suppressed by lesioning central noradrenaline systems. *J. Neuroimmunol.* **57**, 1–7.

Schmidt, E. D., Janszen, A. W., Wouterlood, F. G., and Tilders, F. J. (1995). Interleukin-1-induced long-lasting changes in hypothalamic corticotropin-releasing hormone (CRH)—Neurons and hyperresponsiveness of the hypothalamus-pituitary-adrenal axis. *J. Neurosci.* **15**, 7417–7426.

Sephton, S. E., Sapolsky, R. M., Kraemer, H. C., and Spiegel, D. (2000). Diurnal cortisol rhythm as a predictor of breast cancer survival. *J. Natl. Cancer Inst.* **92**, 994–1000.

Shanks, N., Moore, P. M., Perks, P., and Lightman, S. L. (1999). Alterations in hypothalamic-pituitary-adrenal function correlated with the onset of murine SLE in MRL +/+ and lpr/lpr mice. *Brain Behav. Immun.* **13**, 348–360.

Shanks, N., Windle, R. J., Perks, P. A., Harbuz, M. S., Jessop, D. S., Ingram, C. D., and Lightman, S. L. (2000). Early-life exposure to endotoxin alters hypothalamic-pituitary-adrenal function and predisposition to inflammation. *Proc. Natl. Acad. Sci. USA* **97**, 5645–5650.

Sofia, R. D., and Vassar, H. B. (1974). Changes in serotonin (5HT) concentrations in brain tissue of rats with adjuvant-induced polyarthritis. *Arch. Int. Pharmacodyn.* **211**, 74–79.

Stefferl, A., Linington, C., Holsboer, F., and Reul, J. M. (1999). Susceptibility and resistance to experimental allergic encephalomyelitis: Relationship with hypothalamic-pituitary-adrenocortical axis responsiveness in the rat. *Endocrinology* **140**, 4932–4938.

Stefulj, J., Jernej, B., Cicin-Sain, L., Rinner, I., and Schauenstein, K. (2000). mRNA expression of serotonin receptors in cells of the immune tissues of the rat. *Brain Behav. Immun.* **14**, 219–224.

Stephanou, A., Sarlis, N. J., Knight, R. A., Lightman, S. L., and Chowdrey, H. S. (1992). Glucocorticoid mediated responses of plasma ACTH and anterior pituitary pro-opiomelanocortin, growth hormone and prolactin mRNAs during adjuvant-induced arthritis in the rat. *J. Mol. Endocrinol.* **9**, 273–281.

Sternberg, E. M., Young, W. S., Bernardini, R., Calogero, A. E., Chrousos, G. P., Gold, P. W., and Wilder, R. L. (1989a). A central nervous system defect in biosynthesis of corticotropin-releasing hormone is associated with the susceptibility to streptococcal cell wall-induced arthritis in Lewis rats. *Proc. Natl. Acad. Sci. USA* **86,** 4771–4775.

Sternberg, E. M., Hill, J. M., Chrousos, G. P., Kamilaris, T., Listwak, S. J., Gold, P. W., and Wilder, R. L. (1989b). Inflammatory mediator-induced hypothalamic-pituitary-adrenal axis activation is defective in streptococcal cell wall arthritis susceptible Lewis rats. *Proc. Natl. Acad. Sci. USA* **86,** 2374–2378.

Sufka, K. J., Schomburg, F. M., and Giordano, J. (1992). Receptor mediation of 5-HT induced inflammation and nociception in rats. *Pharmacol. Biochem. Behav.* **41,** 53–56.

Takasu, N., Komiya, I., Nagasawa, Y., Asawa, T., and Yamada, T. (1990). Exacerbation of autoimmune thyroid dysfunction after unilateral adrenalectomy in patients with Cushing's syndrome due to an adrenocortical adenoma. *New Engl. J. Med.* **322,** 1708–1712.

Templ, E., Koeller, M., Riedl, M., Wagner, O., Graninger, W., and Luger, A. N. A. (1996). Anterior pituitary function in patients with newly-diagnosed rheumatoid arthritis. *Br. J. Rheumatol.* **35,** 350–356.

Then Bergh, F., Kumpfel, T., Grasser, A., Rupprecht, R., Holsboer, F., and Trenkwalder, C. (2001). Combined treatment with corticosteroids and moclobemide favors normalization of hypothalamo-pituitary-adrenal axis dysregulation in relapsing-remitting multiple sclerosis: A randomized, double blind trial. *J. Clin. Endocrinol. Metab.* **86,** 1610–1615.

Tilders, F. J., Schmidt, E. D., Hoogendijk, W. J., and Swaab, D. F. (1999). Delayed effects of stress and immune activation. *Baillieres Best Pract. Res. Clin. Endocrinol. Metab.* **13,** 523–540.

van de Langerijt, A. G., van Lent, P. L., Hermus, A. R., Sweep, C. G., Cools, A. R., and van den Berg, W. B. (1993). Susceptibility to adjuvant arthritis: Relative importance of adrenal activity and bacterial flora. *Clin. Exp. Immunol.* **94,** 150–155.

van den Broek, M. F., van Bruggen, M. C., Koopman, J. P., Hazenberg, M. P., and van den Berg, W. B. (1992). Gut flora induces and maintains resistance against streptococcal cell wall–induced arthritis in F344 rats. *Clin. Exp. Immunol.* **88,** 313–317.

Wei, T., and Lightman, S. L. (1997). The neuroendocrine axis in patients with multiple sclerosis. *Brain* **120,** 1067–1076.

Weigmann, K., Muthyala, S., Kim, D. H., Arnason, B. G. W., and Chelmicka-Schorr E. (1995). β-Adrenergic agonists suppress chronic/relapsing experimental allergic encephalomyelitis (CREAE) in Lewis rats. *J. Neuroimmunol.* **56,** 201–206.

Weil-Fugazza, J., Godefroy, F., and Besson, J. M. (1979). Changes in brain and spinal tryptophan and 5–hydroxyindoleacetic acid levels following acute morphine administration in normal and arthritic rats. *Brain Res.* **175,** 291–301.

Whitnall, M. H. (1989). Stress selectively activates the vasopressin-containing subset of corticotropin-releasing hormone neurons. *Neuroendocrinology* **50,** 702–707.

Whitnall, M. H., Mezey, E., and Gainer, H. (1985). Co-localization of corticotropin-releasing factor and vasopressin in median eminence neurosecretory vesicles. *Nature (London)* **317,** 248–250.

Whitnall, M. H., Smith, D., and Gainer, H. (1987). Vasopressin coexists in half of the corticotropin-releasing factor axons present in the external zone of the median eminence in normal rats. *Neuroendocrinology* **45,** 420–424.

Windle, R. J., Wood, S. A., Kershaw, Y. M., Lightman, S. L., Ingram, C. D., and Harbuz, M. S. (2001). Increased corticosterone pulse frequency during adjuvant-induced arthritis and its relationship to alterations in stress responsiveness. *J. Neuroendocrinol.* **13,** 905–911.

Yakushiji, F., Kita, M., Hiroi, N., Ueshiba, H., Monma, I., and Miyachi, Y. (1995). Exacerbation of rheumatoid arthritis after removal of adrenal adenoma in Cushing's syndrome. *Endocr. J.* **42,** 219–223.

Zobel, A. W., Nickel, T., Sonntag, A., Uhr, M., Holsboer, F., and Ising, M. (2001). Cortisol response in the combined dexamethasone/CRH test as predictor of relapse in patients with remitted depression: A prospective study. *J. Psychiatr. Res.* **35,** 83–94.

SYSTEMIC STRESS-INDUCED Th2 SHIFT AND ITS CLINICAL IMPLICATIONS

Ilia J. Elenkov

Division of Rheumatology, Immunology, and Allergy
Georgetown University Medical Center,
Washington, D.C. 20007

I. Introduction

During an immune response the brain and the immune system "talk to each other" in a process which is essential for maintaining *homeostasis*. Two major pathway systems are involved in this *cross-talk*: the hypothalamic-pituitary-adrenal (HPA) axis and the systemic/adrenomedullary sympathetic nervous system (SNS) (Besedovsky *et al.*, 1983; Chrousos, 1995; Elenkov *et al.*, 2000). The HPA axis and the SNS represent the peripheral limbs of the stress system. Activation of the stress system occurs within the central nervous system (CNS) in response to distinct bloodborne, neurosensory, and limbic signals. The central components of the stress system are the corticotropin-releasing factor (CRF) and locus ceruleus-norepinephrine (LC-NE)/autonomic (sympathetic) neurons of the hypothalamus and brain stem, which respectively regulate the peripheral activities of the HPA axis and the SNS (Chrousos, 1995; Elenkov *et al.*, 2000). The stress-induced release of hypothalamic CRF leads ultimately to systemic secretion of

163

glucocorticoids and catecholamines (CAs), mainly epinephrine and nor-epinephrine (NE), which in turn influence immune responses. An immune challenge that threatens the stability of the internal milieu can be regarded as a stressor. Thus, cell products from an activated immune system, predominately the cytokines tumor necrosis factor (TNF)-α, interleukin (IL)-1, and IL-6 stimulate CRF secretion and, hence, activate both the HPA axis and the SNS (Besedovsky *et al.*, 1986; Kovacs and Elenkov, 1995; Chrousos, 1995; Elenkov *et al.*, 2000).

Several studies during the 1970s and the 1980s revealed that stress hormones inhibit lymphocyte proliferation and cytotoxicity and the secretion of certain cytokines, such as IL-2 and interferon (IFN)-γ (Boumpas *et al.*, 1993). These early observations, in the context of the broad clinical use of glucocorticoids as potent anti-inflammatory and immunosuppressive agents, initially led to the conclusion that stress was, in general, immunosuppressive. However, there has been convincing evidence that glucocorticoids and CAs, at levels that can be achieved during stress, selectively suppress cellular but potentiate humoral immunity. This chapter focuses on this new concept that helps explain some well-known, but often contradictory, effects of stress on the immune system and on the onset and course of certain infectious, autoimmune/inflammatory, allergic, and neoplastic diseases.

II. Th1/Th2 Paradigm: Role of Type 1 and Type 2 Cytokines

Immune responses are regulated by antigen-presenting cells (APC), such as monocytes/macrophages, dendritic cells, and other phagocytic cells, that are components of *innate immunity,* and by the T *helper* (Th) lymphocyte subclasses Th1 and Th2 that are components of *acquired (adaptive) immunity.* Th1 cells primarily secrete IFN-γ, IL-2, and TNF-β, which promote cellular immunity, whereas Th2 cells secrete a different set of cytokines, primarily IL-4, IL-10, and IL-13 which promote humoral immunity (Mosmann and Sad, 1996; Fearon and Locksley, 1996; Trinchieri, 1995) (Fig. 1).

Naive CD4[+] (antigen-inexperienced) Th0 cells are clearly bipotential and serve as precursors of Th1 and Th2 cells. Among the factors currently known to influence the differentiation of these cells toward Th1 or Th2, cytokines produced by cells of the innate immune system are the most important. Thus, IL-12, produced by activated monocytes/macrophages or other APCs, is a major inducer of Th1 differentiation and hence cellular immunity. This cytokine acts in concert with NK-derived IFN-γ to further promote Th1 responses (Trinchieri, 1995). APC-derived IL-12 and TNF-α, in concert with natural killer (NK) cell– and Th1-derived IFN-γ, stimulate the functional activity of T cytotoxic cells (Tc), NK cells, and activated macrophages, which

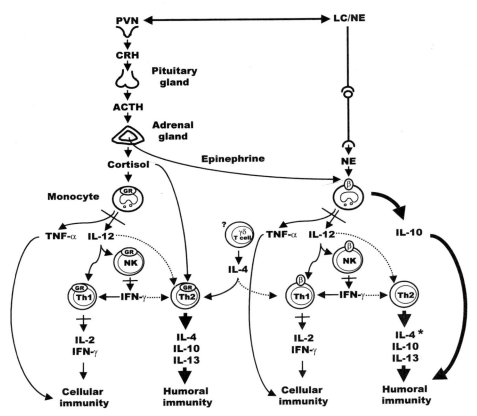

FIG. 1. Systemic effects of stress hormones on Th1/Th2 balance. Hypothalamic CRF stimulates the secretion of pituitary ACTH, which in turn triggers the secretion of cortisol from the adrenal cortex. Upon CRF and LC/NE activation, NE is released from the postganglionic sympathetic nerve terminals in blood vessels and lymphoid organs, and epinephrine is secreted from the adrenal medulla. Th1 cells primarily secrete IFN-γ and IL-2, which promote cellular immunity, whereas Th2 cells secrete primarily IL-4, IL-10, and IL-13, which promote humoral immunity. IL-12, a 75–kDa heterodimeric cytokine produced mostly by monocytes/macrophages is a central inducer of Th1 responses and cell-mediated immunity by favoring Th1 cell proliferation and differentiation, and by suppressing Th2 responses. The cellular source of IL-4 that stimulates the Th2 differentiation is not well defined. Note that glucocorticoids do not affect the production of IL-10 by monocytes/macrophages but upregulate the production of IL-10, IL-4, and IL-13 by Th2 cells. Catecholamines upregulate IL-10 production by monocytes/macrophages; however, they do not affect directly Th2 cells (Th2 cells have very few or no β-adrenergic receptors). *Indirectly, however, catecholamines may potentiate the cytokine production by Th2 cells, since they remove the inhibitory restraints on these cells exerted mainly by IL-12 and IFN-γ. Solid lines represent stimulation, while dashed lines represent inhibition. ACTH, Adrenocorticotropic hormone; CRH, corticotropin-releasing hormone; GR, glucocorticoid receptor; IL, interleukin; IFN, interferon; LC, locus ceruleus; NE, norepinephrine; NK, natural killer cell; PVN, paraventricular nucleus; Th, T helper lymphocyte; TNF, tumor necrosis factor.

are the major components of cellular immunity. The type 1 cytokines IL-12, TNF-α, and IFN-γ also stimulate the synthesis of nitric oxide (NO) and other inflammatory mediators that drive chronic delayed-type inflammatory responses. Because of these crucial and synergistic roles in inflammation IL-12, TNF-α, and IFN-γ are considered the major pro-inflammatory cytokines (Mosmann and Sad, 1996; Fearon and Locksley, 1996; Trinchieri, 1995).

Th1 and Th2 responses are mutually inhibitory. Thus, IL-12 and IFN-γ inhibit Th2 cell activities, while IL-4 and IL-10 inhibit Th1 responses. IL-4 and IL-10 promote humoral immunity by stimulating the growth and activation of mast cells and eosinophils, the differentiation of B cells into antibody-secreting B cells, and B-cell immunoglobulin switching to IgE. Importantly, these cytokines also inhibit macrophage activation, T-cell proliferation, and the production of pro-inflammatory cytokines (Mosmann and Sad, 1996; Fearon and Locksley, 1996; Trinchieri, 1995). Therefore, the Th2 (type 2) cytokines IL-4 and IL-10 are the major anti-inflammatory cytokines.

III. Stress Hormones and Th1/Th2 Balance

A. Systemic Effects

Previous studies have shown that glucocorticoids suppress the production of TNF-α, IFN-γ, and IL-2 *in vitro* and *in vivo* in animals and humans (Beutler *et al.*, 1986; Boumpas *et al.*, 1993). Evidence indicates that glucocorticoids also act through their classic cytoplasmic/nuclear receptors on APCs to suppress the production of the main inducer of Th1 responses, IL-12, *in vitro* and *ex vivo* (Elenkov *et al.*, 1995; Blotta *et al.*, 1997). Since IL-12 is extremely potent in enhancing IFN-γ and inhibiting IL-4 synthesis by T cells, the inhibition of IL-12 production by APCs may represent a major mechanism by which glucocorticoids affect the Th1/Th2 balance. Thus, glucocorticoid-treated monocytes/macrophages produce significantly less IL-12, leading to their decreased capacity to induce IFN-γ production by antigen-primed CD4$^+$ T cells. This is also associated with an increased production of IL-4 by T cells, probably resulting from disinhibition from the suppressive effects of IL-12 on Th2 activity (DeKruyff *et al.*, 1998). Furthermore, glucocorticoids potently downregulate the expression of IL-12 receptors on T and NK cells. This explains why human peripheral blood mononuclear cells (PBMCs) stimulated with immobilized anti-CD3 lose their ability to produce IFN-γ in the presence of glucocorticoids (Wu *et al.*, 1998). Thus, although glucocorticoids may have a direct suppressive effect on Th1 cells, the overall inhibition of IFN-γ production by these cells appears to result mainly from the inhibition of IL-12 production by APCs and from the loss of IL-12 responsiveness of NK and Th1 cells (Fig. 1).

It is particularly noteworthy that glucocorticoids have no effect on the production of the potent anti-inflammatory cytokine IL-10 by monocytes (Elenkov *et al.*, 1995; van der Poll *et al.*, 1996); yet, lymphocyte-derived IL-10 production appears to be upregulated by glucocorticoids (Fig. 1). Thus, rat CD4$^+$ T cells pretreated with dexamethasone exhibit increased levels of mRNA for IL-10 (Ramierz *et al.*, 1996). Similarly, during experimental endotoxemia or cardiopulmonary bypass, or in multiple sclerosis patients having an acute relapse, treatment with glucocorticoids is associated with increased plasma IL-10 secretion (van der Poll *et al.*, 1996; Tabardel *et al.*, 1996; Gayo *et al.*, 1998). This could be the result of a direct stimulatory effect of glucocorticoids on T-cell IL-10 production and/or a block on the restraining inputs of IL-12 and IFN-γ on monocyte/lymphocyte IL-10 production.

Catecholamines drive a Th2 shift at the level of both APCs and Th1 cells (Fig. 1). Norepinephrine and epinephrine potently inhibit or enhance the production of IL-12 and IL-10, respectively, in human whole blood cultures stimulated with bacterial lipopolysaccharide (LPS) *ex vivo* (Elenkov *et al.*, 1996). These effects are mediated by stimulation of β-adrenoreceptors (ARs) since they are completely prevented by propranolol, a β-AR antagonist. In addition, the nonselective β- and selective β_2-AR agonists inhibit the production of IL-12 *in vitro* and *in vivo* (Panina-Bordignon *et al.*, 1997; Hasko *et al.*, 1998). In conjunction with their ability to suppress IL-12 production, β_2-AR agonists also inhibit the development of Th1-type cells, while promoting Th2 cell differentiation (Panina-Bordignon *et al.*, 1997).

β_2-ARs are expressed on Th1 cells but not on Th2 cells (Sanders *et al.*, 1997). This might provide an additional mechanistic basis for the differential effect of CAs on Th1/Th2 functions. In fact, in both murine and human systems, β_2-AR agonists inhibit IFN-γ production by Th1 cells but do not affect IL-4 production by Th2 cells (Sanders *et al.*, 1997; Borger *et al.*, 1998). Importantly, the differential effect of CAs on type1/type2 cytokine production also operates in *in vivo* conditions. Thus, increasing sympathetic outflow in mice by selective α_2-AR antagonists or application of β-AR agonists results in inhibition of LPS-induced TNF-α and IL-12 production (Haskó *et al.*, 1994; Elenkov *et al.*, 1995). In humans, the administration of the β_2-AR agonist salbutamol results in inhibition of IL-12 production *ex vivo* (Panina-Bordignon *et al.*, 1997). Also, acute brain trauma that is followed by massive release of CAs triggers secretion of substantial amounts of systemic IL-10 (Woiciechowsky *et al.*, 1998).

CAs exert tonic inhibition on the production of pro-inflammatory cytokines *in vivo*. Application of propranolol, a β-AR antagonist which blocks their inhibitory effect on cytokine-producing cells, results in a substantial increase of LPS-induced secretion of TNF-α and IL-12 in mice (Elenkov *et al.*, 1995; Haskó *et al.*, 1998). Thus, systemically, both glucocorticoids and CAs, through inhibition and stimulation of type 1 and type 2 cytokine secretion, respectively, cause selective suppression of cellular immunity and a shift toward Th2-mediated humoral immunity. This is further substantiated by studies showing that stress hormones inhibit effector function of cellular immunity components (i.e., the activity of NK, Tc, and activated macrophages). For example, CAs are potent inhibitors of NK cell activity, both directly, acting on β_2-ARs expressed on these cells, or indirectly, through suppression of the production of IL-12 and INF-γ, which are essential for NK cell activity (Hellstrand and Hermodsson, 1989; Elenkov *et al.*, 1995). It appears that NK cells are the ones most "sensitive" to the suppressive effect of stress; indeed, NK cell activity has been used as an index of stress-induced immunosuppression in many studies (Irwin, 1994; Elenkov *et al.*, 2000).

B. LOCAL EFFECTS

The just-mentioned general conclusion on the effects of stress hormones on Th1/Th2 balance may not pertain to certain conditions or local responses in specific compartments of the body. Thus, the synthesis of transforming growth factor (TGF)-β, another type 2 cytokine with potent antiinflammatory activities, is differentially regulated by glucocorticoids: it is enhanced in human T cells but suppressed in glial cells (Batuman *et al.*, 1995). In addition, NE, via stimulation of α_2-ARs can augment LPS-stimulated production of TNF-α from mouse peritoneal macrophages (Spengler *et al.*, 1990), while hemorrhage, a condition associated with elevations of systemic CA concentrations, increases the expression of TNF-α and IL-1 by lung mononuclear cells via stimulation of α-ARs (Le Tulzo *et al.*, 1997). Because the response to β-AR agonist stimulation wanes during maturation of human monocytes into macrophages (Baker and Fuller, 1995), it is possible that in certain compartments of the body, the α-AR-mediated effect of CAs becomes transiently dominant. Interestingly, *in vitro*, long-term incubation with low concentrations of the synthetic glucocorticoid dexamethasone can indeed activate alveolar macrophages, leading to increased LPS-induced IL-1β production (Broug-Holub and Kraal, 1996. Exposure of rats to mild inescapable electrical footshock stress also results in increased IL-1β and TNF-α production by alveolar macrophages (Broug-Holub *et al.*, 1998). The upregulation of pro-inflammatory cytokine production, in *in vivo* conditions, appears to be dependent on intact sympathetic innervation and β-ARs. However, the effect is most likely indirect, since, *in vitro*, a direct modulatory effect of CAs on LPS-induced IL-1β by alveolar macrophages was not demonstrated. It can be envisaged that the stress-induced changes in alveolar macrophage activity result from β-AR-mediated alveolar type II epithelial cell activation, leading to release of surfactant and/or other factors (Broug-Holub *et al.*, 1998).

Through the just-mentioned mechanisms, CAs may actually boost local cellular immune responses in a transitory fashion. This is further substantiated by the finding that CAs potentiate the production of IL-8 by monocytes and epithelial cells of the lung (Linden, 1996; Kavelaars *et al.*, 1997), thus probably promoting recruitment of polymorphonuclear leukocytes in local inflammation. Again, this effect appears to be an indirect one. Evidence indicates that epinephrine promotes IL-8 production by human leukocytes via an effect on platelets. Thus, IL-8 levels in samples containing platelets and stimulated with LPS and epinephrine were significantly higher than control samples containing no platelets (Engstad *et al.*, 1999). In fact, as shown by Kaplanski *et al.* (1993), activated platelets are able to induce endothelial secretion of IL-8.

Anatomically, a close spatial relationship between sympathetic and peptidergic nerve fibers on one hand, and macrophages and mast cells on the other, is frequently observed (cf. Elenkov *et al.*, 2000). Neuro–macrophage and neuro–mast cell connections not only are restricted to the preformed lymphoid organs and tissues but also are regularly encountered in virtually all somatic and visceral tissues. Substance P (SP) and peripheral CRF (see also next section), which are released from sensory peptidergic neurons, are two of the most potent mast cell secretagogues (Foreman, 1987; Church *et al.*, 1989; Theoharides *et al.*, 1995, 1998). Furthermore, evidence indicates that SP upregulates both TNF-α and IL-12 production by human and murine monocytes and macrophages (Lotz *et al.*, 1988; Kincy-Cain and Bost, 1997; Ho *et al.*, 1998). This adds further complexity to the local immunomodulatory effects of stress hormones, in conjunction with other neurotransmitters and/or mediators (for more details see (Elenkov *et al.*, 2000; Straub and Cutolo, 2001). Thus, in summary, although stress hormones suppress Th1 responses and pro-inflammatory cytokine secretion and boost Th2 responses systemically, they may affect certain local responses differently. Further studies are needed to address this question.

C. CRF–MAST CELL–HISTAMINE AXIS

CRF is also secreted peripherally at inflammatory sites (*peripheral* or *immune* CRF) and influences the immune system directly, through local modulatory actions (Karalis *et al.*, 1991). This adds another complexity to the stress–immune system interactions. Immunoreactive CRF is identified locally in experimental carrageenan-induced subcutaneous aseptic inflammation, streptococcal cell wall– and adjuvant-induced arthritis, and retinol-binding protein (RBP)–induced uveitis, and in human tissues from patients with various autoimmune/inflammatory diseases, including rheumatoid arthritis, autoimmune thyroid disease, and ulcerative colitis. The demonstration of CRF-like immunoreactivity in the dorsal horn of the spinal cord, dorsal root ganglia, and sympathetic ganglia supports the hypothesis that most of the immune CRF in early inflammation is of peripheral nerve rather than immune cell origin (cf. Elenkov *et al.*, 1999).

Peripheral CRF has pro-inflammatory, vascular permeability-enhancing, and vasodilatory actions. Systemic administration of specific CRF antiserum reduces the inflammatory exudate volume and cell number in carrageenan-induced inflammation and RBP-induced uveitis, and inhibits stress-induced intracranial mast cell degranulation (Karalis *et al.*, 1991; Theoharides *et al.*, 1995). In addition, CRF administration to humans or nonhuman primates causes major peripheral vasodilation manifested as flushing and increased blood flow and hypotension (Udelsman *et al.*, 1986). An intradermal CRF injection induces a marked increase of vascular permeability and mast cell degranulation (Theoharides *et al.*, 1998). Importantly, this effect is mediated through CRF type 1 receptors and is stronger than the effect of an equimolar concentration of C48/80, a potent mast cell secretagogue (Theoharides *et al.*, 1998). Therefore, it appears that the mast cell is a major target of immune CRF. This concept has an anatomic prerequisite: in blood vessels, periarterial sympathetic plexuses are closely associated with mast cells lining the perivascular regions, and plexuses of nerve fibers (noradrenergic and peptidergic) within lymphoid parenchyma are also closely associated with clusters of mast cells.

Interestingly, evidence suggests that urocortin, a newly discovered member of the CRF family, which binds to the same receptors as CRF, is produced by human lymphocytes and Jurkat T lymphoma cells (Bamberger *et al.*, 1998; Bamberger and Bamberger, 2000). High expression of urocortin immunoreactivity was demonstrated in the synovial lining cell layer, subsynovial stromal cells, blood vessels endothelial cells and mononuclear inflammatory cells from joints of rheumatoid arthritis (RA) patients. In addition, urocortin also stimulated IL-1β and IL-6 secretion by PBMC *in vitro* (Kohno *et al.*, 2001). Thus, this peptide may also participate in the peripheral CRF receptor–mediated inflammatory response.

Histamine, a major product of mast cell degranulation, is a well recognized mediator of acute inflammation and allergic reactions. These actions are mainly mediated by activation of H1 histamine receptors and include vasodilation, increased permeability of the vessel wall, edema, and, in the lungs, bronchoconstriction. Thus, it is conceivable that CRF activates mast cells via a CRF receptor type 1–dependent mechanism leading to release of histamine and other contents of the mast cell granules that subsequently cause vasodilation, increased vascular permeability, and other manifestations of inflammation.

The past 10–15 years have provided strong evidence that histamine may have important immunoregulatory functions via H2 receptors expressed on immune cells. We have found that histamine, via stimulation of H2 receptors on peripheral monocytes and subsequent elevation of cAMP, inhibits the secretion of human IL-12 and stimulates the production of IL-10 (Elenkov *et al.*, 1998). Our data are consistent with previous studies showing that histamine, via H2 receptors, also inhibits TNF-α production from monocytes and IFN-γ production by Th1-like cells but has no effect on IL-4 production from Th2 clones (Lagier *et al.*, 1997). Histamine then, similarly to CAs, appears to drive a Th2 shift at the level of both APCs and Th1 cells. Thus, the activation of CRF–mast cell–histamine axis through stimulation of H1 receptors may induce acute inflammation and allergic reactions, whereas through activation of H2 receptors it may induce suppression of Th1 responses and a Th2 shift.

IV. Clinical Implications

A. Intracellular Infections

A major factor governing the outcome of infectious diseases is the selection of Th1- versus Th2-predominant adaptive responses during and after the initial invasion of the host by the pathogen. A stress-induced Th2 shift may, therefore, have a profound effect on the susceptibility of the host to infections and/or it may influence the course of infections—particularly the intracellular ones, the defense against which is primarily through cellular immunity mechanisms (Table I).

Cellular immunity, particularly IL-12 and IL-12-dependent IFN-γ secretion in humans, seems essential in the control of mycobacterial infections (Altare *et al.*, 1998). In the 1950s, Thomas Holmes (cf. Lerner, 1996) reported that individuals who had experienced stressful life events were more likely to develop tuberculosis and less likely to recover from it. Although it is still a matter of some speculation, stress hormone–induced inhibition of IL-12 and IFN-γ production and the consequent suppression of cellular immunity, might explain the pathophysiological mechanisms of these observations (Elenkov *et al.*, 1996).

Helicobacter pylori infection is the most common cause of chronic gastritis, which in some cases progresses to peptic ulcer disease. The role of stress in promoting peptic ulcers has been recognized for many years (Levenstein

TABLE I
Systemic Stress-Induced Th2 Shift and Its Clinical Implications[a]

Activity of the stress system	Th1/Th2 balance	Clinical implications
Hyperactive stress system	*Th2 shift*	
Acute/chronic stress	Short/long lasting Th2 shift	Increased susceptibility to intracellular infections, flares of SLE, allergic reactions, and tumor growth
Excessive exercise	Besides stress hormones, other mediators such as PgE2, adenosine, and histamine may contribute to the Th2 shift	Postexercise infection
Major injury		Infectious complications
Pregnancy (3rd trimester)	Additive effects of estradiol, progesterone, and 1,25-dihydroxyvitamin D3	Suppression of RA and MS disease activity
Fischer rat	Th2 dominance	Resistance to experimentally induced Th1-related autoimmune disease models
Hypoactive stress system	*Th1 shift*	
Postpartum period	Rebound Th1 shift	Exacerbation/onset of
Post–chronic stress	?Rebound Th1 shift	RA and MS
Post–glucocorticoid therapy	?Rebound Th1 shift	
Lewis rat	Th1 dominance	Increased susceptibility to experimentally induced Th1-related autoimmune disease models
Rheumatoid arthritis	?Facilitates/sustains the Th1 shift	Exacerbation of RA

[a] Th, T Helper; RA, rheumatoid arthritis; MS, multiple sclerosis; SLE, systemic lupus erythematosus.

et al., 1999). Thus, increased systemic stress hormone levels, in concert with an increased local concentration of histamine, induced by inflammatory or stress-related mediators, may skew the local responses toward Th2 and thus might allow the onset or progression of a *H. pylori* infection.

HIV+ patients have IL-12 deficiency, while disease progression has been correlated with a Th2 shift. The innervation (primarily sympathetic/noradrenergic) of lymphoid tissue may be particularly relevant to HIV infection, since lymphoid organs represent the primary site of HIV pathogenesis. In fact, it has been shown that NE, the major sympathetic neurotransmitter released locally in lymphoid organs (Elenkov and Vizi, 1991; Vizi *et al.*, 1995), is able to directly accelerate HIV-1 replication by up to 11-fold in acutely infected human PBMCs (Cole *et al.*, 1998). The effect of NE on viral replication is transduced via the β-AR–adenylyl cyclase–cAMP–PKA signaling cascade

(Cole *et al.*, 1998). In another study, Haraguchi *et al.* found that the induction of intracellular cAMP by a synthetic, immunosuppressive, retroviral envelope peptide caused a shift in the cytokine balance and led to suppression of cell-mediated immunity by inhibiting IL-12 and stimulating IL-10 production (Haraguchi *et al.*, 1995).

Progression of HIV infection is also characterized by increased cortisol secretion in both the early and the late stages of the disease. Increased glucocorticoid production, triggered by the chronic infection, was proposed to contribute to HIV progression (Clerici *et al.*, 1994). Kino *et al.* (1999) found that one of the HIV-1 accessory proteins, Vpr, acts as a potent coactivator of the host glucocorticoid receptor rendering lymphoid cells hyperresponsive to glucocorticoids. Thus, on one hand, stress hormones suppress cellular immunity and directly accelerate HIV replication, while, on the other hand, retroviruses may suppress cell-mediated immunity using the same pathways by which stress hormones, including CAs and glucocorticoids, alter the Th1/Th2 balance.

In one study, an association was demonstrated between stress and susceptibility to the common cold among 394 persons who had been intentionally exposed to five different upper respiratory viruses. Psychological stress was found to be associated in a dose-dependent manner with an increased risk of acute infectious respiratory illness, and this risk was attributed to increased rates of infection rather than to an increased frequency of symptoms after infection (Cohen *et al.*, 1991). Thus, stress hormones, through their selective inhibition of cellular immunity, might play important roles in the increased risk to an individual of acute respiratory infections caused by common cold viruses.

B. Major Injury

Major injury (serious traumatic injury and major burns) or major surgical procedures often lead to severe immunosuppression, which contributes to infectious complications and, in some cases, to sepsis, the most common cause of late death after trauma. A strong stimulation of the SNS and the HPA axis correlates with the severity of both cerebral and extracerebral injury and an unfavorable prognosis (cf. Woiciechowsky *et al.*, 1998). In patients with traumatic major injury, and in animal models of burn injury, the suppressed cellular immunity is associated with diminished production of IFN-γ and IL-12 and increased production of IL-10 (i.e., a Th2 shift) (O'Sullivan *et al.*, 1995). One study indicated that systemic release of IL-10 triggered by SNS activation might be a key mechanism of immunosuppression after injury. Thus, high levels of systemic IL-10 documented in patients with "sympathetic storm" due to acute accidental or iatrogenic brain trauma were associated with a high incidence of infection (Woiciechowsky *et al.,* 1998). In a rat model, the increase of IL-10 was prevented by β-AR blockade (Woiciechowsky *et al.,* 1998). Therefore, stress hormones and histamine secretion triggered by major injury, via an induction of a Th2 shift, may contribute to the severe immunosuppression observed in these conditions.

C. AUTOIMMUNITY

Several autoimmune diseases are characterized by common alterations to the Th1 versus Th2 and the IL-12/TNF-α versus IL-10 balance. In rheumatoid arthritis (RA), multiple sclerosis (MS), type 1 diabetes mellitus, autoimmune thyroid disease (ATD), and Crohn's disease (CD) the balance is skewed toward Th1 and an excess of IL-12 and TNF-α production, whereas Th2 activity and the production of IL-10 are deficient. This appears to be a critical factor that determines the proliferation and differentiation of Th1-related autoreactive cellular immune responses in these disorders (Segal *et al.,* 1998). On the other hand, systemic lupus erythematosus (SLE) is associated with a Th2 shift and an excessive production of IL-10, while production of IL-12 and TNF-α appears to be deficient. Taking into consideration the Th2-driving effects of stress hormones systemically, one could postulate that a *hypoactive* stress system may facilitate or sustain the Th1 shift in MS or RA, and, vice versa, stress system *hyperactivity* may intensify the Th2 shift and induce or facilitate flares of SLE (Table I). Animal studies and certain clinical observations support this hypothesis.

1. *Stress System Activity in RA and MS*

Studies suggest that suboptimal production of cortisol is involved in the onset and/or progression of RA (cf. Wilder, 1995; Wilder and Elenkov, 1999; Straub and Cutolo, 2001). Most patients with RA have relatively "inappropriately normal" plasma cortisol levels in the setting of severe, chronic inflammation, characterized by increased production of TNF-α, IL-1, and IL-6. Since these cytokines are powerful stimulants to the HPA axis and cortisol production, we would have expected significantly elevated plasma cortisol levels in RA patients. The available data suggest that the HPA axis response is blunted in these patients. Whether this abnormality is primary or secondary has not been established (cf. Wilder and Elenkov, 1999).

Several lines of evidence indicate that the sympathetic–immune interface is defective in MS and its experimental model, the experimental allergic encephalomyelitis (EAE). Thus, sympathetic skin responses are decreased and lymphocyte β-ARs are increased in progressive MS (Karaszewski *et al.*, 1990). The density of β-ARs on CD8$^+$ T cells is increased between two- to threefold, compared with age-matched controls (Arnason *et al.*, 1988a; Karaszewski *et al.*, 1993). Similarly, in the preclinical stage of EAE the NE content in spleen is reduced, accompanied by an increase of splenocyte β-AR density (Mackenzie *et al.*, 1989). A defective or hypoactive SNS is most likely a "causative" factor for the upregulation of β-ARs observed in MS (Arnason *et al.*, 1988b). Furthermore, isoproterenol and terbutaline, β-AR and β_2-AR agonists, respectively, were reported to suppress chronic/ relapsing EAE in Lewis rats (Chelmicka-Schorr *et al.*, 1989; Wiegmann *et al.*, 1995). The latter observation might have resulted from the previously discussed effects of CAs and β-AR agonists on the production of type 1 cytokines.

Data from several studies suggest a "protective" role of the SNS in RA or its experimental models in animals. Thus, in the arthritis-prone Lewis rats, sympathectomy with 6-OHDA enhanced the severity of adjuvant-induced arthritis (Felten *et al.*, 1992; Lorton *et al.*, 1996). In this animal model of arthritis, selective sympathetic denervation of the reactive secondary lymphoid organs (the popliteal and inguinal lymph nodes) was achieved with local injection into the fat pads surrounding these lymph nodes (Felten *et al.*, 1992; Lorton *et al.*, 1996). This denervation resulted in earlier onset and enhanced severity of inflammation and bone erosions compared with nondenervated rats. The "protective" role of SNS is further substantiated by the study of Malfait *et al.* (1999) demonstrating that the β_2-AR agonist salbutamol is a potent suppressor of established collagen-induced arthritis in mice. This drug had a profound protective effect as assessed by clinical score, paw thickness, and joint histology. Additionally, in *in vitro* experiments salbutamol reduced IL-12 and TNF-α release by peritoneal macrophages and blocked mast cell degranulation in joint tissues.

Studies in humans also suggest a defective SNS in RA. In patients with RA, diminished autonomic responses were observed after cognitive discrimination and the Stroop Color-Word Interference Tests (Geenen *et al.*, 1996). Miller *et al.* (2000) demonstrated that patients with long-term RA had a highly significant reduction of sympathetic nerve fibers in synovial tissues, which was dependent on the degree of inflammation. Thus, the reduction of sympathetic nerve fibers in the chronic disease may lead to uncoupling of the local inflammation from the anti-inflammatory input of SNS. Interestingly, in RA synovial tissues it appears there is a preponderance of about 10:1 for primary sensory, substance P–positive fibers as compared with sympathetic fibers (Miller *et al.*, 2000). Since substance P is a powerful pro-inflammatory agent, via release of histamine, TNF-α, and IL-12, such a preponderance may lead to an unfavorable pro-inflammatory state, supporting the disease process of RA. In addition, Lombardi *et al.* (1999) demonstrated a decrease in G-protein-coupled receptor kinase (GRK) activity in lymphocytes of RA patients, particularly GRK2 and GRK6 subtypes. The GRKs are responsible for the rapid loss of receptor responsiveness despite continuous presence of the agonist, a process known as homologous desensitization. The decrease in GRK2 activity in RA appears to be mediated by cytokines such as IL-6 and IFN-γ. Local pro-inflammatory cytokines or a hypoactive SNS may, therefore, mediate the changes in coupling of β-ARs to G-proteins as observed in RA patients.

2. *Lewis/Fischer Paradigm*

Among inbred rats, Lewis (LEW) rats are highly susceptible to type II collagen-induced arthritis (CIA), adjuvant-induced arthritis (AIA), EAE, and experimental autoimmune uveitis (EAU), whereas Fischer (F344) rats are highly resistant to these diseases (Wilder, 1995; Joe *et al.*, 1999). These experimentally induced diseases are mediated by Th1-dominant immune responses. In ocular tissues of EAU, LEW express type 1/pro-inflammatory cytokines (IL-12p40, IFN-γ, and TNF-α), coincident with the peak of the response, whereas F344 express high basal IL-10 levels of mRNA in the eyes (Sun *et al.*, 2000). Sakamoto *et al.* (2001) have also demonstrated that lymph node cells from LEW rats express high levels of IL-12 p40 and there is upregulation of the expression of IL-12 receptors β_1 and β_2. These data suggest that LEW mounts a more polarized Th1 response that makes them more susceptible to Th1-mediated diseases, whereas F344 overproduces IL-10, which may contribute to a higher resistance to induction of these diseases. LEW rats have globally blunted stress system responses and fail, in response to a wide variety of stressors, to activate the hypothalamic CRF neuron appropriately (Wilder, 1995). Since F344 are known to have hyperresponsive stress responses and high corticosteroid production, but LEW rats have blunted stress responses with subnormal corticosteroid production, it is postulated that corticosteroids contribute to the differences in the development of pathogenic T cells in these two strains (Wilder, 1995).

3. *Pregnancy/Postpartum and Autoimmune Disease Activity*

Some autoimmune diseases like RA and MS often remit during pregnancy, particularly during the third trimester, but have an exacerbation or their initial onset during the postpartum period (Buyon, 1998; Buyon *et al.*, 1996; Confavreux *et al.*, 1998; Wilder, 1995). The risk of developing new onset RA during pregnancy, compared to nonpregnancy, is decreased by about 70%. In contrast, the risk of developing RA is markedly increased in the postpartum period, particularly in the first 3 months (odds ratio of 5.6 overall and 10.8 after first pregnancy). In women with multiple sclerosis, the rate of relapses declines during pregnancy, especially in the third trimester, increases during the first 3 months of the postpartum, and then returns to the prepregnancy rate (Confavreux *et al.*, 1998). In apparent contrast, autoimmune diseases that present with symptoms associated predominantly with antibody-mediated damage, such as systemic lupus erythematosus (SLE) and specifically the immune complex–mediated glomerulonephritis, tend to develop or flare during pregnancy (Buyon, 1998; Khamashta *et al.*, 1997; Petri, 1997; Petri *et al.*, 1991).

A decrease in the production of IL-2 and IFN-γ by antigen- and mitogen-stimulated peripheral blood mononuclear cells, accompanied by an increase in the production of IL-4 and IL-10, is observed in normal pregnancy. The lowest quantities of IL-2 and IFN-γ and the highest quantities of IL-4 and IL-10 are present in the third trimester of pregnancy (Marzi *et al.*, 1996). Placental tissues from mothers at term express high levels of IL-10 (Cadet *et al.*, 1995); although IL-10 is present in the amniotic fluid of the majority of pregnancies, higher concentrations are found at term compared with the second trimester (Greig *et al.*, 1995). We have found that during the third trimester of pregnancy, *ex vivo* monocytic IL-12 production was about three-fold and TNF-α production approximately 40% lower than postpartum values (Elenkov *et al.*, 2001). These studies suggest that type 1/pro-inflammatory cytokine production and cellular immunity are suppressed, and there is a Th2 shift during normal pregnancy, particularly in the third trimester (see Table I).

The third trimester of pregnancy and the early postpartum are also associated with abrupt changes of several hormones. Thus, during the third trimester of pregnancy, urinary cortisol and NE excretion and serum levels of 1,25-dihydroxyvitamin D3 are about two- to threefold higher than postpartum values (Elenkov *et al.*, 2001). This is accompanied by the well-known marked elevations of estradiol and progesterone serum concentrations. The data reviewed here are consistent with the view that the increased levels of cortisol, NE, 1,25-dihydroxyvitamin D3, estrogens, and progesterone in the third trimester of pregnancy might orchestrate the improvement in autoimmune diseases, such as RA and MS, via suppression of type 1/pro-inflammatory (IL-12, IFN-γ and TNF-α) and potentiation of type 2/anti-inflammatory (IL-4 and IL-10) cytokine production. Conversely, this particular type of hormonal control of pro-/anti-inflammatory cytokine balance might contribute to the flare ups of SLE observed during pregnancy. Postpartum, the hormonal state abruptly shifts. The deficit in hormones that inhibit Th1-type cytokines and cell-mediated immunity might permit autoimmune diseases such as RA and MS to first develop or established diseases to flare up (Wilder, 1995; Elenkov *et al.*, 1997; Elenkov *et al.*, 2001).

D. ALLERGY/ATOPY

Allergic reactions of type 1 hypersensitivity (atopy), such as asthma, eczema, hay fever, urticaria and food allergy, are characterized by dominant Th2 responses, overproduction of histamine, and a shift to IgE production. The effects of stress on atopic reactions are complex, at multiple levels, and can be in either direction. Stress hormones acting at the level of APCs and lymphocytes may induce a Th2 shift, and, thus, facilitate or sustain atopic reactions; however, this can be antagonized by their effects on the mast cell. Glucocorticoids and CAs (through β_2-ARs) suppress the release of histamine by mast cells, thus abolishing its pro-inflammatory, allergic, and bronchoconstrictor effects. Thus, reduced levels of epinephrine and cortisol in the very early morning could contribute to nocturnal wheezing and have been linked to high circulating histamine levels in asthmatics (Barnes et al., 1980). This might also explain the beneficial effect of glucocorticoids and β_2-AR agonists in asthma. Infusion of high doses of epinephrine, however, causes a rise in circulating histamine levels that may be due to an α-adrenergic-mediated increase in mediator release (cf. Barnes et al., 1980). Thus, severe acute stress associated with high epinephrine concentrations and/or high local secretion of CRF could lead to mast cell degranulation. As a result, a substantial amount of histamine could be released, which consequently would not antagonize, but rather would amplify, the Th2 shift through H2 receptors, while in parallel, by acting on H1 receptors it could initiate a new episode or exacerbate a chronic allergic condition.

Glucocorticoids alone or in combination with β_2-AR agonists are broadly used in the treatment of atopic reactions, and particularly asthma. In vivo, ex vivo, and in vitro exposure to glucocorticoids and β_2-AR agonists results in a reduction of IL-12 production, which persists for at least several days (Elenkov et al., 1996; DeKruyff et al., 1998; Panina-Bordignon et al., 1997). Thus, glucocorticoid and/or β_2-AR-agonist therapy is likely to reduce the capacity of APC to produce IL-12, to suppress greatly the synthesis of type 2 cytokine in activated but not resting T cells, and to abolish eosinophilia (DeKruyff et al., 1998). If, however, resting (cytokine-uncommitted) T cells are subsequently activated by APCs preexposed to glucocorticoids and/or β_2-AR agonists, enhanced IL-4 production could be induced, with limited IFN-γ synthesis, (DeKruyff et al., 1998). Thus, although in the short term the effect of glucocorticoids and β_2-AR agonists might be beneficial, their long-term effects might be to sustain the increased vulnerability of the patient to the allergic condition. This is substantiated by the observations that both glucocorticoids and β_2-AR agonists potentiate IgE production in vitro and in vivo (Zieg et al., 1994; Coqueret et al., 1995).

E. TUMOR GROWTH

The amount of IL-12 available at the tumor site appears to be critical for tumor regression (cf. Colombo *et al.*, 1996). Thus, low levels of IL-12 have been associated with tumor growth, as opposed to tumor regression observed with administration of IL-12 delivered in situ or systemically. On the other hand, local overproduction of IL-10 and TGF-β, by inhibiting the production of IL-12 and TNF-α, and the cytotoxicity of NK and Tc cells, seems to play an inappropriate immunosuppressive role, allowing increased malignant tumor growth, as seen for example in melanoma (Chouaib *et al.*, 1997). These and others studies suggest that Th1 function is locally down-regulated during tumor growth.

Several lines of evidence suggest that stress can increase the susceptibility to tumors, tumor growth, and metastases. In animals, β-ARs stimulation suppresses NK cell activity and compromises resistance to tumor metastases (Shakhar and Ben-Eliyahu, 1998); stress decreases the potential of spleen cells to turn into antitumor Tc against syngeneic B16 melanoma, and it significantly suppresses the ability of tumor-specific CD4$^+$ cells to produce IFN-γ and IL-2 (Li *et al.*, 1997). In humans, the augmentation of the rate of tumor progression and cancer-related death has been associated with stress (cf. Li *et al.*, 1997), while treatment with cimetidine, an H2 histamine antagonist, correlated with increased survival in patients with gastric and colorectal cancer ((Matsumoto, 1995). In fact, high concentrations of histamine have been measured within colorectal and breast cancer tissues and large numbers of mast cells have been identified within certain tumor tissues (cf. Elenkov *et al.*, 1998). These data suggest that stress hormone/histamine-induced suppression of cellular immunity may contribute to increased growth of certain tumors.

V. Conclusions

Evidence accumulated over the past decade strongly suggests that stress hormones selectively suppress Th1 responses and cause a Th2 shift rather than generalized immunosuppression. The stress hormone–induced Th2 shift may have both beneficial and detrimental consequences. Although interest in the Th2 response was initially directed at its protective role in helminthic infections and its pathogenic role in allergy, this response may have important regulatory functions in countering the tissue-damaging effects of macrophages and Th1 cells (Fearon and Locksley, 1996). Thus, an excessive immune response, through activation of the stress system, may trigger a mechanism that inhibits Th1 but potentiates Th2 responses. This important feedback mechanism may protect the organism from "overshooting" by type 1/pro-inflammatory cytokines and other products of activated macrophages with tissue damaging potential.

The substantial Th2-driving force, however, of endogenous glucocorticoids and CAs can be amplified to a great extent during certain conditions such as severe acute or chronic stress, excessive exercise, or pregnancy (see Table I). For example, in major injury, a condition followed by substantial activation of the stress system and a "sympathetic storm", the effects of stress hormones may contribute to serious infectious complications. On the other hand, in the early postpartum period, a hypoactive stress system may determine a rebound increase of pro-inflammatory/type 1 cytokine production that may be the driving force for the exacerbations and/or the onset of RA and MS during this period. An abnormal stress system activity, therefore, in either direction, might contribute to the pathophysiology of common human diseases, where a selection of Th1 (type 1) versus Th2 (type 2) responses plays a significant role. These include several intracellular infections, major injury and its complications, allergic (atopic) reactions, autoimmune/inflammatory diseases, and tumor growth. Clearly these hypotheses require further investigation, but the answers should provide critical insights into mechanisms underlying a variety of common human diseases.

References

Altare, F., Durandy, A., Lammas, D., Emile, J. F., Lamhamedi, S., Le Deist, F., Drysdale, P., Jouanguy, E., Doffinger, R., Bernaudin, F., Jeppsson, O., Gollob, J. A., Meinl, E., Segal, A. W., Fischer, A., Kumararatne, D., and Casanova, J. L. (1998). Impairment of mycobacterial immunity in human interleukin-12 receptor deficiency. *Science* **280**, 1432–1435.

Arnason, B. G., Brown, M., Maselli, R., Karaszewski, J., and Reder, A. (1988a). Blood lymphocyte β-adrenergic receptors in multiple sclerosis. *Ann. N.Y. Acad. Sci.* **540**, 585–588.

Arnason, B. G., Noronha, A. B., and Reder, A. T. (1988b). Immunoregulation in rapidly progressive multiple sclerosis. *Ann. N.Y. Acad. Sci.* **540**, 4–12.

Baker, A. J., and Fuller, R. W. (1995). Loss of response to β-adrenoceptor agonists during the maturation of human monocytes to macrophages *in vitro. J. Leukoc. Biol.* **57**, 395–400.

Bamberger, C. M., and Bamberger, A. M. (2000). The peripheral CRF/urocortin system. *Ann. N.Y. Acad. Sci.* **917**, 290–296.

Bamberger, C. M., Wald, M., Bamberger, A. M., Ergun, S., Beil, F. U., and Schulte, H. M. (1998). Human lymphocytes produce urocortin, but not corticotropin-releasing hormone. *J. Clin. Endocrinol. Metab.* **83**, 708–711.

Barnes, P., FitzGerald, G., Brown, M., and Dollery, C. (1980). Nocturnal asthma and changes in circulating epinephrine, histamine, and cortisol. *New Engl. J. Med.* **303**, 263–267.

Batuman, O. A., Ferrero, A., Cupp, C., Jimenez, S. A., and Khalili, K. (1995). Differential regulation of transforming growth factor β-1 gene expression by glucocorticoids in human T and glial cells. *J. Immunol.* **155**, 4397–4405.

Besedovsky, H., Del Rey, A., Sorkin, E., and Dinarello, C. A. (1986). Immunoregulatory feedback between interleukin-1 and glucocorticoid hormones. *Science* **233**, 652–654.

Besedovsky, H. O., del Rey, A. E., and Sorkin, E. (1983). What do the immune system and the brain know about each other? *Immunol. Today* **4**, 342–346.

Beutler, B., Krochin, N., Milsark, I. W., Luedke, C., and Cerami, A. (1986). Control of cachectin (tumor necrosis factor) synthesis: Mechanisms of endotoxin resistance. *Science* **232**, 977–980.

Blotta, M. H., DeKruyff, R. H., and Umetsu, D. T. (1997). Corticosteroids inhibit IL-12 production in human monocytes and enhance their capacity to induce IL-4 synthesis in CD4$^+$ lymphocytes. *J. Immunol.* **158**, 5589–5595.

Borger, P., Hoekstra, Y., Esselink, M. T., Postma, D. S., Zaagsma, J., Vellenga, E., and Kauffman, H. F. (1998). β-Adrenoceptor-mediated inhibition of IFN-γ, IL-3, and GM-CSF mRNA accumulation in activated human T lymphocytes is solely mediated by the β$_2$-Adrenoceptor subtype. *Am. J. Respir. Cell Mol. Biol.* **19**, 400–407.

Boumpas, D. T., Chrousos, G. P., Wilder, R. L., Cupps, T. R., and Balow, J. E. (1993). Glucocorticoid therapy for immune-mediated diseases: Basic and clinical correlates. *Ann. Intern. Med.* **119**, 1198–1208.

Broug-Holub, E., and Kraal, G. (1996). Dose- and time-dependent activation of rat alveolar macrophages by glucocorticoids. *Clin. Exp. Immunol.* **104**, 332–336.

Broug-Holub, E., Persoons, J. H., Schornagel, K., Mastbergen, S. C., and Kraal, G. (1998). Effects of stress on alveolar macrophages: A role for the sympathetic nervous system. *Am. J. Respir. Cell Mol. Biol.* **19**, 842–848.

Buyon, J. P. (1998). The effects of pregnancy on autoimmune diseases. *J. Leukoc. Biol.* **63**, 281–287.

Buyon, J. P., Nelson, J. L., and Lockshin, M. D. (1996). The effects of pregnancy on autoimmune diseases. *Clin. Immunol. Immunopathol.* **78**, 99–104.

Cadet, P., Rady, P. L., Tyring, S. K., Yandell, R. B., and Hughes, T. K. (1995). Interleukin-10 messenger ribonucleic acid in human placenta: Implications of a role for interleukin-10 in fetal allograft protection. *Am. J. Obstet. Gynecol.* **173**, 25–29.

Chelmicka-Schorr, E., Kwasniewski, M. N., Thomas, B. E., and Arnason, B. G. (1989). The β-adrenergic agonist isoproterenol suppresses experimental allergic encephalomyelitis in Lewis rats. *J. Neuroimmunol.* **25**, 203–207.

Chouaib, S., Asselin-Paturel, C., Mami-Chouaib, F., Caignard, A., and Blay, J. Y. (1997). The host–tumor immune conflict: From immunosuppression to resistance and destruction. *Immunol. Today* **18**, 493–497.

Chrousos, G. P. (1995). The hypothalamic-pituitary-adrenal axis and immune-mediated inflammation [see comments]. *New Engl. J. Med.* **332**, 1351–1362.

Church, M. K., Lowman, M. A., Robinson, C., Holgate, S. T., and Benyon, R. C. (1989). Interaction of neuropeptides with human mast cells. *Int. Arch. Allergy Appl. Immunol.* **88**, 70–78.

Clerici, M., Bevilacqua, M., Vago, T., Villa, M. L., Shearer, G. M., and Norbiato, G. (1994). An immunoendocrinological hypothesis of HIV infection [see comments]. *Lancet* **343**, 1552–1553.

Cohen, S., Tyrrell, D. A., and Smith, A. P. (1991). Psychological stress and susceptibility to the common cold [see comments]. *New Engl. J. Med.* **325**, 606–612.

Cole, S. W., Korin, Y. D., Fahey, J. L., and Zack, J. A. (1998). Norepinephrine accelerates HIV replication via protein kinase A–dependent effects on cytokine production. *J. Immunol.* **161**, 610–616.

Colombo, M. P., Vagliani, M., Spreafico, F., Parenza, M., Chiodoni, C., Melani, C., and Stoppacciaro, A. (1996). Amount of interleukin 12 available at the tumor site is critical for tumor regression. *Cancer Res.* **56,** 2531–2534.

Confavreux, C., Hutchinson, M., Hours, M. M., Cortinovis-Tourniaire, P., and Moreau, T. (1998). Rate of pregnancy-related relapse in multiple sclerosis. Pregnancy in Multiple Sclerosis Group. *New Engl. J. Med.* **339,** 285–291.

Coqueret, O., Dugas, B., Mencia-Huerta, J. M., and Braquet, P. (1995). Regulation of IgE production from human mononuclear cells by β_2- adrenoceptor agonists [see comments]. *Clin. Exp. Allergy* **25,** 304–311.

DeKruyff, R. H., Fang, Y., and Umetsu, D. T. (1998). Corticosteroids enhance the capacity of macrophages to induce Th2 cytokine synthesis in $CD4^+$ lymphocytes by inhibiting IL-12 production. *J. Immunol.* **160,** 2231–2237.

Elenkov, I. J., and Vizi, E. S. (1991). Presynaptic modulation of release of noradrenaline from the sympathetic nerve terminals in the rat spleen. *Neuropharmacology* **30,** 1319–1324.

Elenkov, I. J., Hasko, G., Kovacs, K. J., and Vizi, E. S. (1995). Modulation of lipopolysaccharide-induced tumor necrosis factor-α production by selective α- and β-adrenergic drugs in mice. *J. Neuroimmunol.* **61,** 123–131.

Elenkov, I. J., Papanicolaou, D. A., Wilder, R. L., and Chrousos, G. P. (1996). Modulatory effects of glucocorticoids and catecholamines on human interleukin-12 and interleukin-10 production: Clinical implications. *Proc. Assoc. Am. Physicians.* **108,** 374–381.

Elenkov, I. J., Hoffman, J., and Wilder, R. L. (1997). Does differential neuroendocrine control of cytokine production govern the expression of autoimmune diseases in pregnancy and the postpartum period? *Mol. Med. Today* **3,** 379–383.

Elenkov, I. J., Webster, E., Papanicolaou, D. A., Fleisher, T. A., Chrousos, G. P., and Wilder, R. L. (1998). Histamine potently suppresses human IL-12 and stimulates IL-10 production via H2 receptors. *J. Immunol.* **161,** 2586–2593.

Elenkov, I. J., Webster, E. L., Torpy, D. J., and Chrousos, G. P. (1999). Stress, corticotropin-releasing hormone, glucocorticoids, and the immune/inflammatory response: Acute and chronic effects. *Ann. N.Y. Acad. Sci.* **876,** 1–11.

Elenkov, I. J., Wilder, R. L., Chrousos, G. P., and Vizi, E. S. (2000). The sympathetic nerve—An integrative interface between two supersystems: The brain and the immune system. *Pharmacol. Rev.* **52,** 595–638.

Elenkov, I. J., Wilder, R. L., Bakalov, V. K., Link, A. A., Dimitrov, M. A., Fisher, S., Crane, M., Kanik, K. S., and Chrousos, G. P. (2001). IL-12, TNF-α, and hormonal changes during late pregnancy and early postpartum: Implications for autoimmune disease activity during these times. *J. Clin. Endocrinol. Metab.* **86,** 4933–4938.

Engstad, C. S., Lund, T., and Osterud, B. (1999). Epinephrine promotes IL-8 production in human leukocytes via an effect on platelets. *Thromb. Haemost.* **81,** 139–145.

Fearon, D. T., and Locksley, R. M. (1996). The instructive role of innate immunity in the acquired immune response. *Science* **272,** 50–53.

Felten, D. L., Felten, S. Y., Bellinger, D. L., and Lorton, D. (1992). Noradrenergic and peptider-gic innervation of secondary lymphoid organs: Role in experimental rheumatoid arthritis. *Eur. J. Clin. Invest.* **22** Suppl 1, 37–41.

Foreman, J. C. (1987). Substance P and calcitonin gene-related peptide: Effects on mast cells and in human skin. *Int. Arch. Allergy Appl. Immunol.* **82,** 366–371.

Gayo, A., Mozo, L., Suarez, A., Tunon, A., Lahoz, C., and Gutierrez, C. (1998). Glucocorticoids increase IL-10 expression in multiple sclerosis patients with acute relapse. *J. Neuroimmunol.* **85,** 122–130.

Geenen, R., Godaert, G. L., Jacobs, J. W., Peters, M. L., and Bijlsma, J. W. (1996). Diminished autonomic nervous system responsiveness in rheumatoid arthritis of recent onset. *J. Rheumatol.* **23,** 258–264.

Greig, P. C., Herbert, W. N., Robinette, B. L., and Teot, L. A. (1995). Amniotic fluid interleukin-10 concentrations increase through pregnancy and are elevated in patients with preterm labor associated with intrauterine infection. *Am. J. Obstet. Gynecol.* **173,** 1223–1227.

Haraguchi, S., Good, R. A., James-Yarish, M., Cianciolo, G. J., and Day, N. K. (1995). Induction of intracellular cAMP by a synthetic retroviral envelope peptide: A possible mechanism of immunopathogenesis in retroviral infections. *Proc. Natl. Acad. Sci. USA* **92,** 5568–5571.

Haskó, G., Elenkov, I. J., Kvetan, V., and Vizi, E. S. (1995). Differential effect of selective block of α_2-adrenoreceptors on plasma levels of tumour necrosis factor-α, interleukin-6 and corticosterone induced by bacterial lipopolysaccharide. *J. Endocr.* **144,** 457–462.

Haskó, G., Szabo, C., Nemeth, Z. H., Salzman, A. L., and Vizi, E. S. (1998). Stimulation of β-adrenoceptors inhibits endotoxin-induced IL-12 production in normal and IL-10 deficient mice. *J. Neuroimmunol.* **88,** 57–61.

Hellstrand, K., and Hermodsson, S. (1989). An immunopharmacological analysis of adrenaline-induced suppression of human natural killer cell cytotoxicity. *Int. Arch. Allergy Appl. Immunol.* **89,** 334–341.

Ho, W. Z., Stavropoulos, G., Lai, J. P., Hu, B. F., Magafa, V., Anagnostides, S., and Douglas, S. D. (1998). Substance P C-terminal octapeptide analogues augment tumor necrosis factor-α release by human blood monocytes and macrophages. *J. Neuroimmunol.* **82,** 126–132.

Irwin, M. (1994). Stress-induced immune suppression: Role of brain corticotropin releasing hormone and autonomic nervous system mechanisms. *Adv. Neuroimmunol.* **4,** 29–47.

Joe, B., Griffiths, M. M., Remmers, E. F., and Wilder, R. L. (1999). Animal models of rheumatoid arthritis and related inflammation. *Curr. Rheumatol. Rep.* **1,** 139–148.

Kaplanski, G., Porat, R., Aiura, K., Erban, J. K., Gelfand, J. A., and Dinarello, C. A. (1993). Activated platelets induce endothelial secretion of interleukin-8 *in vitro* via an interleukin-1-mediated event. *Blood* **81,** 2492–2495.

Karalis, K., Sano, H., Redwine, J., Listwak, S., Wilder, R. L., and Chrousos, G. P. (1991). Autocrine or paracrine inflammatory actions of corticotropin-releasing hormone *in vivo*. *Science* **254,** 421–423.

Karaszewski, J. W., Reder, A. T., Maselli, R., Brown, M., and Arnason, B. G. (1990). Sympathetic skin responses are decreased and lymphocyte β-adrenergic receptors are increased in progressive multiple sclerosis. *Ann. Neurol.* **27,** 366–372.

Karaszewski, J. W., Reder, A. T., Anlar, B., and Arnason, G. W. (1993). Increased high affinity β-adrenergic receptor densities and cyclic AMP responses of CD8 cells in multiple sclerosis. *J. Neuroimmunol.* **43,** 1–7.

Kavelaars, A., van, d.P., Zijlstra, J., and Heijnen, C. J. (1997). β_2-adrenergic activation enhances interleukin-8 production by human monocytes. *J. Neuroimmunol.* **77,** 211–216.

Khamashta, M. A., Ruiz-Irastorza, G., and Hughes, G. R. (1997). Systemic lupus erythematosus flares during pregnancy. *Rheum. Dis. Clin. North Am.* **23,** 15–30.

Kincy-Cain, T., and Bost, K. L. (1997). Substance P-induced IL-12 production by murine macrophages. *J. Immunol.* **158,** 2334–2339.

Kino, T., Gragerov, A., Kopp, J. B., Stauber, R. H., Pavlakis, G. N., and Chrousos, G. P. (1999). The HIV-1 virion-associated protein vpr is a coactivator of the human glucocorticoid receptor. *J. Exp. Med.* **189,** 51–62.

Kohno, M., Kawahito, Y., Tsubouchi, Y., Hashiramoto, A., Yamada, R., Inoue, K. I., Kusaka, Y., Kubo, T., Elenkov, I. J., Chrousos, G. P., Kondo, M., and Sano, H. (2001). Urocortin expression in synovium of patients with rheumatoid arthritis and osteoarthritis: Relation to inflammatory activity. *J. Clin. Endocrinol. Metab.* **86,** 4344–4352.

Kovacs, K. J., and Elenkov, I. J. (1995). Differential dependence of ACTH secretion induced by various cytokines on the integrity of the paraventricular nucleus. *J. Neuroendocrinol.* **7,** 15–23.

Lagier, B., Lebel, B., Bousquet, J., and Pene, J. (1997). Different modulation by histamine of IL-4 and interferon-γ (IFN-γ) release according to the phenotype of human Th0, Th1 and Th2 clones. *Clin. Exp. Immunol.* **108,** 545–551.

Lerner, B. H. (1996). Can stress cause disease? Revisiting the tuberculosis research of Thomas Holmes, 1949–1961. *Ann. Intern. Med.* **124,** 673–680.

Le Tulzo, Y., Shenkar, R., Kaneko, D., Moine, P., Fantuzzi, G., Dinarello, C. A., and Abraham, E. (1997). Hemorrhage increases cytokine expression in lung mononuclear cells in mice: Involvement of catecholamines in nuclear factor-κB regulation and cytokine expression. *J. Clin. Invest.* **99,** 1516–1524.

Levenstein, S., Ackerman, S., Kiecolt-Glaser, J. K., and Dubois, A. (1999). Stress and peptic ulcer disease. *JAMA* **281,** 10–11.

Li, T., Harada, M., Tamada, K., Abe, K., and Nomoto, K. (1997). Repeated restraint stress impairs the antitumor T cell response through its suppressive effect on Th1-type CD4$^+$ T cells. *Anticancer Res.* **17,** 4259–4268.

Linden, A. (1996). Increased interleukin-8 release by β-adrenoceptor activation in human transformed bronchial epithelial cells. *Br. J. Pharmacol.* **119,** 402–406.

Lombardi, M. S., Kavelaars, A., Schedlowski, M., Bijlsma, J. W., Okihara, K. L., Van de, P. M., Ochsmann, S., Pawlak, C., Schmidt, R. E., and Heijnen, C. J. (1999). Decreased expression and activity of G-protein-coupled receptor kinases in peripheral blood mononuclear cells of patients with rheumatoid arthritis. *FASEB J.* **13,** 715–725.

Lorton, D., Bellinger, D., Duclos, M., Felten, S. Y., and Felten, D. L. (1996). Application of 6-hydroxydopamine into the fatpads surrounding the draining lymph nodes exacerbates adjuvant-induced arthritis. *J. Neuroimmunol.* **64,** 103–113.

Lotz, M., Vaughan, J. H., and Carson, D. A. (1988). Effect of neuropeptides on production of inflammatory cytokines by human monocytes. *Science* **241,** 1218–1221.

Mackenzie, F. J., Leonard, J. P., and Cuzner, M. L. (1989). Changes in lymphocyte β-adrenergic receptor density and noradrenaline content of the spleen are early indicators of immune reactivity in acute experimental allergic encephalomyelitis in the Lewis rat. *J. Neuroimmunol.* **23,** 93–100.

Malfait, A. M., Malik, A. S., Marinova-Mutafchieva, L., Butler, D. M., Maini, R. N., and Feldmann, M. (1999). The β2-adrenergic agonist salbutamol is a potent suppressor of established collagen-induced arthritis: Mechanisms of action. *J. Immunol.* **162,** 6278–6283.

Marzi, M., Vigano, A., Trabattoni, D., Villa, M. L., Salvaggio, A., Clerici, E., and Clerici, M. (1996). Characterization of type 1 and type 2 cytokine production profile in physiologic and pathologic human pregnancy. *Clin. Exp. Immunol.* **106,** 127–133.

Matsumoto, S. (1995). Cimetidine and survival with colorectal cancer. *Lancet* **346,** 115.

Miller, L. E., Justen, H. P., Scholmerich, J., and Straub, R. H. (2000). The loss of sympathetic nerve fibers in the synovial tissue of patients with rheumatoid arthritis is accompanied by increased norepinephrine release from synovial macrophages. *FASEB J.* **14,** 2097–2107.

Mosmann, T. R., and Sad, S. (1996). The expanding universe of T-cell subsets: Th1, Th2 and more. *Immunol. Today* **17,** 138–146.

O'Sullivan, S. T., Lederer, J. A., Horgan, A. F., Chin, D. H., Mannick, J. A., and Rodrick, M. L. (1995). Major injury leads to predominance of the T helper-2 lymphocyte phenotype and diminished interleukin-12 production associated with decreased resistance to infection [see comments]. *Ann. Surg.* **222,** 482–490.

Panina-Bordignon, P., Mazzeo, D., Lucia, P. D., D'Ambrosio, D., Lang, R., Fabbri, L., Self, C., and Sinigaglia, F. (1997). β2-Agonists prevent Th1 development by selective inhibition of interleukin 12. *J. Clin. Invest.* **100,** 1513–1519.

Petri, M. (1997). Hopkins Lupus Pregnancy Center: 1987 to 1996. *Rheum. Dis. Clin. North Am.* **23,** 1–13.

Petri, M., Howard, D., and Repke, J. (1991). Frequency of lupus flare in pregnancy. The Hopkins Lupus Pregnancy Center experience. *Arthritis Rheum.* **34,** 1538–1545.

Ramierz, F., Fowell, D. J., Puklavec, M., Simmonds, S., and Mason, D. (1996). Glucocorticoids promote a Th2 cytokine response by CD4$^+$ T cells *in vitro. J. Immunol.* **156,** 2406–2412.

Sakamoto, S., Fukushima, A., Ozaki, A., Ueno, H., Kamakura, M., and Taniguchi, T. (2001). Mechanism for maintenance of dominant T helper 1 immune responses in Lewis rats. *Microbiol. Immunol.* **45,** 373–381.

Sanders, V. M., Baker, R. A., Ramer-Quinn, D. S., Kasprowicz, D. J., Fuchs, B. A., and Street, N. E. (1997). Differential expression of the β_2-adrenergic receptor by Th1 and Th2 clones: Implications for cytokine production and B cell help. *J. Immunol.* **158,** 4200–4210.

Segal, B. M., Dwyer, B. K., and Shevach, E. M. (1998). An interleukin (IL)-10/IL-12 immunoregulatory circuit controls susceptibility to autoimmune disease. *J. Exp. Med.* **187,** 537–546.

Shakhar, G., and Ben-Eliyahu, S. (1998). *In vivo* β-adrenergic stimulation suppresses natural killer activity and compromises resistance to tumor metastasis in rats. *J. Immunol.* **160,** 3251–3258.

Spengler, R. N., Allen, R. M., Remick, D. G., Strieter, R. M., and Kunkel, S. L. (1990). Stimulation of α-adrenergic receptor augments the production of macrophage-derived tumor necrosis factor. *J. Immunol.* **145,** 1430–1434.

Straub, R. H., and Cutolo, M. (2001). Involvement of the hypothalamic-pituitary-adrenal/gonadal axis and the peripheral nervous system in rheumatoid arthritis: Viewpoint based on a systemic pathogenetic role. *Arthritis Rheum.* **44,** 493–507.

Sun, B., Sun, S. H., Chan, C. C., and Caspi, R. R. (2000). Evaluation of *in vivo* cytokine expression in EAU-susceptible and resistant rats: A role for IL-10 in resistance? *Exp. Eye Res.* **70,** 493–502.

Tabardel, Y., Duchateau, J., Schmartz, D., Marecaux, G., Shahla, M., Barvais, L., LeClerc, J. L., and Vincent, J. L. (1996). Corticosteroids increase blood interleukin-10 levels during cardiopulmonary bypass in men. *Surgery* **119,** 76–80.

Theoharides, T. C., Spanos, C., Pang, X., Alferes, L., Ligris, K., Letourneau, R., Rozniecki, J. J., Webster, E., and Chrousos, G. P. (1995). Stress-induced intracranial mast cell degranulation: A corticotropin-releasing hormone–mediated effect. *Endocrinology* **136,** 5745–5750.

Theoharides, T. C., Singh, L. K., Boucher, W., Pang, X., Letourneau, R., Webster, E., and Chrousos, G. (1998). Corticotropin-releasing hormone induces skin mast cell degranulation and increased vascular permeability, a possible explanation for its proinflammatory effects. *Endocrinology* **139,** 403–413.

Trinchieri, G. (1995). Interleukin-12: A proinflammatory cytokine with immunoregulatory functions that bridge innate resistance and antigen-specific adaptive immunity. *Annu. Rev. Immunol.* **13,** 251–276.

Udelsman, R., Gallucci, W. T., Bacher, J., Loriaux, D. L., and Chrousos, G. P. (1986). Hemodynamic effects of corticotropin releasing hormone in the anesthetized cynomolgus monkey. *Peptides* **7,** 465–471.

van der Poll, T., Barber, A. E., Coyle, S. M., and Lowry, S. F. (1996). Hypercortisolemia increases plasma interleukin-10 concentrations during human endotoxemia—A clinical research center study. *J. Clin. Endocrinol. Metab.* **81,** 3604–3606.

Vizi, E. S., Orso, E., Osipenko, O. N., Hasko, G., and Elenkov, I. J. (1995). Neurochemical, electrophysiological and immunocytochemical evidence for a noradrenergic link between the sympathetic nervous system and thymocytes. *Neuroscience* **68,** 1263–1276.

Wiegmann, K., Muthyala, S., Kim, D. H., Arnason, B. G., and Chelmicka-Schorr, E. (1995). β-Adrenergic agonists suppress chronic/relapsing experimental allergic encephalomyelitis (CREAE) in Lewis rats. *J. Neuroimmunol.* **56,** 201–206.

Wilder, R. L. (1995). Neuroendocrine–immune system interactions and autoimmunity. *Annu. Rev. Immunol.* **13,** 307–338.

Wilder, R. L., and Elenkov, I. J. (1999). Hormonal regulation of tumor necrosis factor-α, interleukin-12 and interleukin-10 production by activated macrophages. A disease-modifying mechanism in rheumatoid arthritis and systemic lupus erythematosus? *Ann. N.Y. Acad. Sci.* **876,** 14–31.

Woiciechowsky, C., Asadullah, K., Nestler, D., Eberhardt, B., Platzer, C., Schoning, B., Glockner, F., Lanksch, W. R., Volk, H. D., and Docke, W. D. (1998). Sympathetic activation triggers systemic interleukin-10 release in immunodepression induced by brain injury. *Nat. Med.* **4,** 808–813.

Wu, C. Y., Wang, K., McDyer, J. F., and Seder, R. A. (1998). Prostaglandin E2 and dexamethasone inhibit IL-12 receptor expression and IL-12 responsiveness. *J. Immunol.* **161,** 2723–2730.

Zieg, G., Lack, G., Harbeck, R. J., Gelfand, E. W., and Leung, D. Y. (1994). *In vivo* effects of glucocorticoids on IgE production. *J. Allergy Clin. Immunol.* **94,** 222–230.

NEURAL CONTROL OF SALIVARY S-IgA SECRETION

Gordon B. Proctor and Guy H. Carpenter

Salivary Research Group
Guy's, King's and St. Thomas' School of Dentistry
King's College London
The Rayne Institute
London SE5 9NU, United Kingdom

I. Introduction

The concept of a common mucosal immune system is now established, and the mouth with its associated exocrine (salivary) glands is a part of this system (Mestecky, 1993). Immunoglobulin A (IgA) is the most abundant specific defense factor in saliva and most salivary IgA is in the form of secretory IgA (S-IgA) which is secreted by three pairs of major salivary glands (the parotids, submandibulars, and sublinguals) and numerous minor salivary glands. It is apparent from a variety of studies that the amounts of S-IgA secreted onto mucosal surfaces can be regulated (Norderhaug *et al.*, 1999). A clear example is salivary S-IgA secretion and its regulation by autonomic nerves. In this chapter we review the evidence of neural regulation of salivary S-IgA secretion and the mechanisms that might be operating in such regulation.

187

A. Mucosal Defense in the Mouth

Saliva forms a film over teeth and mucosal surfaces in the mouth (Dawes, 1996). The film of saliva performs a number of functions, including lubrication, prevention of soft tissue desiccation, prevention of hard tissue decalcification, and maintenance of an ecological balance (Bowen, 1996). Most of these functions are dependent on salivary proteins. Microbial colonization of oral surfaces is influenced by the interactions of salivary proteins with bacteria, fungi, and viruses. Some salivary proteins are overtly antimicrobial, for example, peroxidase in association with thiocyanate and hydrogen peroxide (Tenovuo, 1989). Other salivary proteins can aggregate or trap microorganisms in saliva or reduce microbial adherence to teeth and the oral mucosa. The submandibular mucins MUC7 and MUC5B are good examples of innate defense factors that can aggregate microorganisms and modify microbial adherence (Wu *et al.*, 1994). Binding of these salivary glycoproteins by microorganisms is frequently mediated by a microbial adhesin that recognizes a specific sugar sequence (Murray *et al.*, 1992). *In vitro* studies have demonstrated that, through specific antigen recognition, S-IgA also aggregates and prevents adherence of oral microorganisms (Marcotte and Lavoie, 1998). Since S-IgA has been shown to be present in the acquired protein pellicle present on teeth and in saliva coating mucosal surfaces it is likely to play a similar role *in vivo*. The role played by S-IgA in oral defense is likely to overlap with innate defense factors, and together these create an environment favoring the presence of the indigenous microbiota and preventing colonization by pathogens (Cole, 1985; Marcotte and Lavoie, 1998).

B. Control of Salivary Function by Autonomic Nerves

Salivary gland secretion of fluid and proteins is controlled by autonomic nerves and once these nerves have been sectioned secretion ceases almost entirely. A minority of salivary glands are also capable of secreting small amounts of saliva in the absence of impulses from nerves (Emmelin, 1981). Parasympathetic and sympathetic nerves form the efferent arms of salivary reflexes elicited mainly by gustatory and mechanical stimuli in the mouth. Eating elicits a large increase in salivary flow over and above a resting flow that is present throughout the day and night (Hector and Linden, 1999). Parasympathetic nerve–mediated stimuli play the principle role, through the release of the neurotransmitter acetylcholine, in activating salivary acinar cell fluid secretion. Sympathetic innervation varies between different salivary glands in different species. In the parotid and submandibular glands of the rat, the most extensively studied experimental models, the sympathetic nerves play an important role in evoking protein secretion. Neuropeptide cotransmitters present in autonomic nerves also modulate protein secretion by salivary glands (Ekström, 1999). Similar mechanisms to those just outlined also appear to operate in humans. A variety of autonomic receptors exist on salivary epithelial cells and the intracellular mechanisms which couple receptor occupation to secretion of fluid or proteins appear to be similar to those described for other systems. Thus, β-adrenoceptor activation is linked to increased intracellular cyclic adenosine monophosphate (cAMP) leading principally to stored protein secretion. Muscarinic cholinergic receptor activation is linked to formation of inositol triphosphate (IP_3) and diacylglycerol (DAG) and subsequent rises in the intracellular calcium that open membrane ion channels leading to fluid secretion (Baum and Wellner, 1999).

II. Nerve-Mediated Increases in Salivary Secretion of S-IgA

Unlike other major protein components of saliva, IgA is not the product of salivary epithelial cells but is produced by plasma cells within glands. It is generally accepted that the main mechanism by which IgA is secreted onto mucosal surfaces across glandular and nonsquamous epithelial cells is via the epithelial polymeric immunoglobulin receptor (pIgR). Paracellular movement of IgA is unlikely to make more than a minor contribution except during inflammation or when the epithelial cell layer is disrupted (Brandtzaeg *et al.*, 1999a). Figure 1 shows the proposed IgA secretory pathway from plasma cell to saliva.

The influences of autonomic nerves on salivary secretion can be conveniently studied by electrically stimulating the parasympathetic and sympathetic nerve supplies. Such studies have been performed in a number of different species but the rat parotid and submandibular glands have been most frequently studied (e.g., Anderson *et al.*, 1995; Ekström, 1999). Carpenter *et al.* (1998) used a nerve stimulation protocol that allowed calculation of the unstimulated rate of IgA secretion from rat submandibular gland. When the parasympathetic and sympathetic nerve supplies were stimulated, IgA output into saliva was increased above the calculated unstimulated rate (Fig. 2). Similar results were subsequently obtained in the rat parotid gland (Proctor *et al.*, 2000a).

In the previously discussed studies, IgA secretion was compared with markers of secretion from submandibular and parotid acinar cells—the

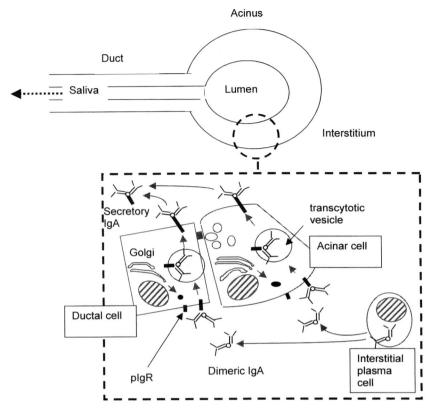

FIG. 1. Proposed pathway of S-IgA secretion at the salivary gland epithelium. Production of J chain–containing polymeric IgA by plasma cells in glandular interstitium is followed by binding to the polymeric immunoglobulin receptor (pIgR) on the basolateral membrane of epithelial cells. Bound IgA is then endocytosed and transported in a membrane bound compartment to the apical membrane, a process referred to as receptor-mediated transcytosis.

main protein-secreting cells in salivary glands. Such comparisons revealed that IgA secretion was not increased to the same extent as secretory peroxidase, a useful enzymatic marker of submandibular acinar cell protein secretion (Anderson *et al.*, 1995; Proctor and Chan, 1994), nor amylase, a stored secretory product of parotid acinar cells. These results suggest that IgA is not stored and secreted as are other acinar cell secretory proteins. Figure 2 shows that sympathetic nerve stimulation increased IgA secretion from the submandibular gland to a greater extent than the parotid. The reason for this difference is unclear but may be due to granular duct cells that are present in the rat submandibular gland but absent from the parotid

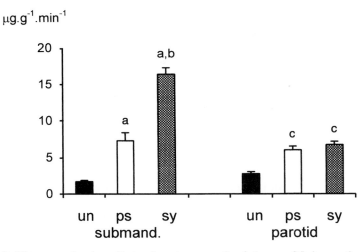

FIG. 2. Histogram showing effects of acute nerve stimulation on S-IgA secretion in rat submandibular and parotid glands. S-IgA secretion is increased by parasympathetic (ps) and sympathetic (sy) stimulation compared to the unstimulated (un) rate. (a) Statistically significantly different ($p < 0.05$) compared to the unstimulated rate in the submandibular gland. (b) Statistically significantly different ($p < 0.05$) from parasympathetically stimulated rate in the submandibular gland. (c) Statistically significantly different ($p < 0.05$) compared to the unstimulated rate in the parotid gland.

gland. Granular duct cells can contribute substantial amounts of protein to saliva, particularly in response to high frequencies of impulses in the sympathetic nerve supply (Anderson *et al.*, 1995). Further experiments examined submandibular secretion of S-IgA in response to different frequencies of impulses in the sympathetic nerve supply on a background of parasympathetic nerve impulses (Carpenter *et al.*, 2000). Maximum secretion of IgA was obtained at sympathetic nerve impulse frequencies that evoked maximal activation of granular duct cell secretion. The experiments noted earlier again indicated that the mechanism of IgA secretion differs from acinar cell stored protein secretion. Electrophoresis and Western blotting confirmed that IgA was present in the form of S-IgA in these experiments and therefore that pIgR-mediated transport was responsible for increased IgA secretion (see Fig. 9 as example of Western blot of salivary S-IgA). Graded increases in frequency of either parasympathetic or sympathetic nerve stimulation produced increased acinar cell protein secretion while S-IgA secretion remained relatively unchanged over a range of nerve stimulation frequencies (Carpenter *et al.*, 2000). The use of dual nerve stimulation, that is, sympathetic nerve stimulation superimposed on a background of parasympathetic nerve stimulation, may provide a more approximate model of events in conscious animals under reflex conditions (Emmelin, 1987). While such dual stimulation in anesthetized rats produced a synergistic protein secretory response from acinar and ductal cells, S-IgA secretion showed no evidence of synergism. The lack of synergism is likely to reflect differences in the intracellular mechanisms coupling cell stimulation with protein and IgA secretion.

III. Autonomimetic Stimulation of pIgR-Mediated IgA Transcytosis

In vivo nerve stimulation of salivary glands has shown that the pIgR-mediated transcytosis of IgA has two rates—an unstimulated rate and a faster stimulated rate. The question addressed in this section is how pIgR transport of IgA can be upregulated in an almost immediate way that does not require *de novo* synthesis of proteins. Figure 3 shows different points in the pIgR transcytosis pathway at which upregulation might be achieved and this section will try to address each of these points.

Salivary glands are heavily dependent on nerve stimulation for secretion of both protein and fluid and also for the general well-being of the cells (see Section V). However, in other systems where IgA trancytosis and secretion

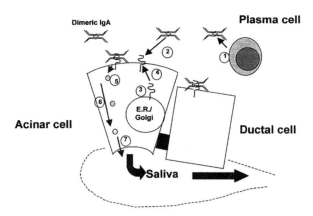

FIG. 3. Potential regulation of salivary S-IgA secretion by autonomic nerve–mediated stimuli may occur at different steps in the IgA secretory pathway: (1) Secretion of IgA by plasma cells within the glandular interstition; (2) binding of dimeric IgA to pIgR; (3) production and processing of pIgR in the endoplasmic reticulum and Golgi apparatus; (4) delivery of pIgR to basolateral membrane; (5) endocytosis of pIgR inside the cell; (6) transcytosis of pIgR-containing vesicles; and (7) delivery of vesicle to the apical membrane and the cleavage of pIgR to release secretory IgA into saliva.

is dependent on pIgR, other factors predominate; for example, sex hormones affect steady-state levels of pIgR in rodent uteri (Kaushic *et al.*, 1995), and in the kidney, water loading appears to affect pIgR levels (Rice *et al.*, 1998). Secretion from the rat lacrimal gland is dependent upon a background of androgen stimulation which has previously been found to upregulate pIgR expression in lacrimal cells (Sullivan *et al.*, 1984, 1990; Kelleher *et al.*, 1991). Superimposed on tissue specific regulation are the influences of cytokines such as interferon-γ, interleukin (IL)-4 and IL-6 (Piskurich *et al.*, 1993; Loman, 1997; Nilsen *et al.*, 1999; Norderhaug *et al.*, 1999) which, *in vitro* at least, increased pIgR levels. Most of these factors affected steady-state synthesis of pIgR, requiring *de novo* synthesis of pIgR, a process taking hours rather than minutes. However, do increasing amounts of pIgR increase the amount of IgA transcytosed? In one study pIgR was genetically overexpressed in mouse mammary epithelial cells with steady-state levels increased 60 to 270 times greater than normal and yet S-IgA secretion into milk was only increased one- to twofold (de Groot *et al.*, 2000). To date, there appear to be only two published studies of fast stimulated (i.e., less than 5 min) increases in pIgR-mediated IgA secretion: the rat salivary glands (Carpenter *et al.*, 1998) and the porcine ileum (Schmidt *et al.*, 1999). Increases in S-IgA release into both saliva and intestinal secretions were evoked by nerve stimulation. We have therefore used the rat submandibular gland to study the process of upregulation of pIgR-mediated IgA transport in more detail.

A. STUDIES ON SALIVARY CELLS *in vitro*

Digested salivary glands have been used many times before for studies of protein secretion as they allow secretory mechanisms to be studied in more detail than can easily be achieved *in vivo*. Primary salivary cell preparations of salivary glands are most suitable for such studies since cells are responsive to a complete range of autonomic agonists and contain measurable amounts of pIgR, although only short-term studies may be attempted as these cells do not survive for long *in vitro*. In comparison, most immortalized salivary cell lines are neither polarized nor express pIgR in sufficient quantities to allow studies of IgA transcytosis. Most cell lines used for IgA transcytosis studies have been transfected with a pIgR gene to increase pIgR expression levels (Cardone *et al.*, 1996; Hirt *et al.*, 1993; Su and Stanley, 1998), raising questions about functional expression of the pIgR molecule.

To prepare cells for *in vitro* studies submandibular glands were excised from the rat and digested with collagenase P (Boehringer Mannheim Biochemicals, Indianapolis, Ind.) for 1 h in 100% oxygen, producing a mixture of intact acinar and ductal cell units but also containing other cells

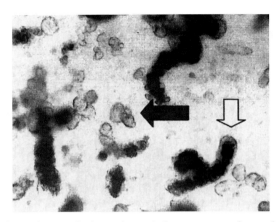

FIG. 4. The photomicrograph shows the typical appearance of rat submandibular cells prepared by enzymatic digestion and used for examining the effects of autonomimetics on S-IgA secretion *in vitro*. Cells are mostly present in intact units of acini (black arrow) and ducts (white arrow), the most prominent ductal component being granular ducts.

(see Fig. 4) such as plasma cells. Given that salivary cells do not synthesize IgA, an initial question for study was whether parenchymal cells still contain IgA following such preparation.

Assays of IgA in cells and the digestion medium before and after cell preparation revealed that only 5–10% of total glandular IgA is present within final cell preparations. The amount of IgA released during digestion can be estimated as 2–5% of total glandular IgA. It can therefore be concluded that most IgA present in the rat submandibular gland is in the extracellular spaces (but within the glandular capsule) near the site of production by plasma cells but outside of acinar and ductal cells. The extracellular pool of IgA is quite large (~0.3 mg/gland) and presumably sufficient for normal reflex secretion in conscious rats, although it can be diminished following extensive nerve stimulation when under anesthesia (Carpenter *et al.*, 1998). Thus under normal conditions plasma cells maintain a level of IgA within the glands sufficient for normal secretion, the excess presumably draining via the lymphatics. It is unlikely that short-term increases in salivary S-IgA secretion are due to increased secretion from the plasma cells since there already exists an excess of IgA.

Since we had shown IgA to be present in the salivary cells after digestion from the gland, we stimulated primary cell preparations with autonomimetics and revealed an immediate secretion of IgA (all extracellular IgA having been washed away) with cholinergic, adrenergic, and peptidergic

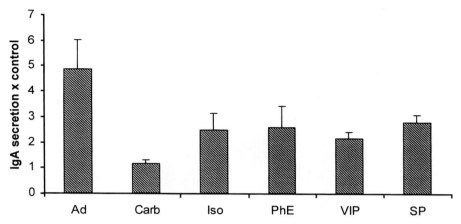

FIG. 5. S-IgA secretion by *in vitro* preparations of rat submandibular cells in response to carbachol (Carb), adrenergic (Ad, adrenaline; Iso, isoprenaline; PhE, phenylephrine), and peptidergic (VIP- vasoactive intestinal peptide; SP, substance P) autonomimetics at concentrations that gave maximal responses. S-IgA secretion is given as a proportion of that secreted from control cells in the absence of an agonist. Each agonist significantly ($p < 0.05$) increased S-IgA secretion.

stimulation (see Fig. 5). As these cell preparations are unfractionated it was unclear whether parenchymal and/or plasma cells secrete IgA in response to the agonists. To answer this question submandibular gland plasma cells were isolated on a reverse Percoll gradient following a more extensive digestion than described earlier (Mega *et al.*, 1995), and then stimulated with the same agonists for 30 min. Over this relatively short period of time there was no agonist-induced increase in IgA secretion. Thus despite being present within the unfractionated primary cell preparations plasma cells do not appear to increase IgA secretion in response to autonomimetics. We can conclude that the immediate secretion of IgA evoked with cholinergic, adrenergic, and peptidergic stimulation (see Fig. 5) was due to parenchymal secretion of IgA taken into the cells while still *in vivo*.

Although immunocytochemistry of human parotid glands confirms that both acinar and ductal cells express pIgR and contain IgA (for a review see Brandtzaeg *et al.*, 1999a) we wanted to confirm whether both cell types secrete S-IgA in response to autonomimetics and if so to which type of nerve signal. To answer this question rat submandibular glands were digested with collagenase P (as earlier) and acinar and ductal units separated using a Percoll density gradient (Amsallem *et al.*, 1996). Morphologically this procedure produced a good (but not complete) separation of acinar and ductal

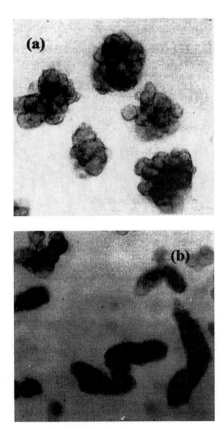

FIG. 6. The photomicrograph shows the typical appearance of rat submandibular cells prepared by enzymatic digestion and then subjected to Percoll density gradient centrifugation to separate (a) acinar and (b) ductal cells. Separated acinar and ductal cells were then used to examine S-IgA secretion in response to autonomimetics.

units (see Fig. 6). Stimulation with adrenaline (an α- and β-adrenoceptor agonist) caused a similar secretion of S-IgA from acinar and ductal cell preparations suggesting that both cell types are capable of secreting S-IgA. Furthermore, the use of α- and β-adrenoceptor specific blockers with adrenaline indicates both acinar and ductal units can secrete S-IgA by stimulation of either adrenergic receptor.

To confirm that S-IgA secretion was via transcytosis during stimulation with autonomimetics, confocal microscopy of live cells prepared as described earlier has been performed. To date, these studies have shown that FITC- or TRITC-labeled polymeric IgA is not transported paracellularly by either acinar or ductal cells (G. B. Proctor *et al.*, 2002, submitted). These results are supported by earlier confocal microscopy studies which showed that large macromolecules such as fluorescently labeled dextrans, do not move by a paracellular route and only small fluorescent dyes such as Lucifer Yellow show a paracellular movement with autonomimetic stimulation of rat parotid cells (Segawa and Yamashina, 1999).

B. *In vivo* Studies

The use of specific agonists and inhibitors with cell preparations of rat submandibular glands *in vitro* showed that the secretion of S-IgA from parenchymal cells is mainly in response to adrenergic stimuli. However, these cell preparations only give information about the release of S-IgA from epithelial cells and not the earlier process of IgA uptake. By combining agonists with an *in vivo* collection of saliva from rat submandibular glands we could examine the effect of specific α- and β-adrenoceptor and cholinergic stimuli on the whole process of S-IgA transport into saliva. To do this, the parasympathetic nerve supply to submandibular glands of anesthetized rats was electrically stimulated at three different frequencies and S-IgA secretion rates compared with those obtained during intravenous injection of methacholine (cholinergic stimulus), isoprenaline (β-adrenergic), or phenylephrine (α-adrenergic), each at three increasing doses. The adrenergic agonists increased the rate of S-IgA secretion which was sustained above an unstimulated rate for all doses of agonist. However, methacholine did not sustain increased S-IgA secretion at the highest doses (see Fig. 7). S-IgA secretion was relatively constant for each agonist across a range of doses while other markers of secretion including flow rate, acinar peroxidase, and ductal cell proteinase all increased with increasing dose.

The use of *in vivo* autonomimetics also highlighted the differences between parasympathetic and cholinergic stimulation on IgA secretion, implying a role for neuropeptides that can be co-released from parasympathetic nerves upon stimulation (Ekström, 1987). Assessing the influence of neuropeptides on the secretion of IgA into saliva is rather difficult as peptides in isolation often do not evoke sufficient saliva to enable analyses. An alternative strategy therefore was to give study subjects neuropeptide on top of parasympathetic stimulation as previously reported (Anderson *et al.*, 1996). Although parasympathetic stimulation and substance P (SP) increased acinar cell fluid secretion and parasympathetic stimulation and vasoactive intestinal peptide (VIP) increased ductal cell proteinase secretion neither of the peptides further increased S-IgA output (G. H. Carpenter *et al.*, 2002, unpublished). Unlike the stimulus-dependent increase in stored protein

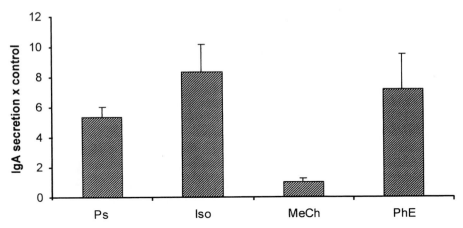

FIG. 7. S-IgA secretion by submandibular gland of the anesthetized rat in response to intravenous infusion of isoprenaline (Iso; β-adrenergic), phenylephrine (PhE; α-adrenergic), and methacholine (MeCh; cholinergic). Comparison is made with the effects of electrical stimulation of the parasympathetic nerve (Ps). With the exception of MeCh, each stimulus significantly ($p < 0.05$) increased S-IgA secretion above the calculated unstimulated rate.

(peroxidase and proteinase) secretion, S-IgA appears to be secreted either at an unstimulated rate or at a higher stimulated rate independently of dose, suggesting a simpler regulatory system compared to stored proteins. Thus, it may be that the addition of a neuropeptide stimulus to parasympathetic stimulation could not further increase S-IgA secretion. Studies with preparations of digested salivary glands *in vitro* (see Section III.A) confirmed that the peptides alone can increase secretion of S-IgA (vasoactive intestinal peptide and substance P increased IgA secretion by three and four times the control value, respectively). Overall, therefore, from these data we may conclude that S-IgA secretion is increased in response to adrenergic, cholinergic, and peptidergic stimuli, which apparently do not modulate each other's effects.

C. Intracellular Coupling of Salivary Cell Stimulation to IgA Secretion

The data described earlier (Sections III.A and III.B) suggest that S-IgA secretion can be stimulated via α- and β-adrenoceptors, peptidergic receptors, and, to a lesser extent, muscarinic cholinergic receptors. The mechanisms of intracellular signaling within salivary cells which couple autonomic receptor occupation to salivary secretion have been well studied (see Baum and Wellner, 1999). Stimulation through β-adrenoceptors raises intracellular levels of cAMP leading to protein storage granule exocytosis. Stimulation of either muscarinic cholinergic or α-adrenergic receptors initiates an elevation in intracellular calcium. How such a rise in intracellular calcium causes increased S-IgA secretion is uncertain; it may be by interacting with pIgR directly, or via calmodulin to phosphorylate pIgR, as found in MDCK cells (Cardone *et al.*, 1996). Likewise, elevations in cAMP may increase S-IgA secretion by activating the relevant protein kinase. Overall it appears that S-IgA secretion from salivary cells is mediated by both calcium and cAMP signal transduction pathways from either adrenergic or muscarinic receptor occupation.

These studies indicate that parenchymal cells are responsible for the increased S-IgA secretion into saliva with nerve stimulation. On a cellular level this may be achieved by increased availability of pIgR at the basolateral surface or an increased rate of transcytosis and secretion. Following studies of a different *in vitro* model, it was also concluded that basolateral expression of the pIgR molecule was the regulating factor of IgA transcytosis (Hirt *et al.*, 1993).

IV. Studies of Reflex IgA Secretion

Studies utilizing electrical nerve stimulation or autonomimetics have revealed the nature of mechanisms controlling salivary S-IgA secretion, but is S-IgA secretion increased by reflex stimulation in conscious animals? In humans, reflex stimulation of salivary secretion evokes an increased secretion of S-IgA (Brandtzaeg, 1971; Mandel and Khurana, 1969; Proctor and Carpenter, 2001). Results from each of these studies indicate that S-IgA concentration decreases during stimulated salivary secretion although S-IgA output increases. Proctor and Carpenter (2001) collected parotid saliva under resting conditions followed by a number of samples collected during a 10-min period of chewing and compared the pattern of S-IgA secretion with that of the stored acinar cell protein amylase and the ductal cell protein tissue kallikrein. As can be seen from Fig. 8 the patterns of S-IgA and amylase secretion were very similar whereas tissue kallikrein differed.

Brandtzaeg (1971) found previously that the patterns of S-IgA and amylase secretion differed when parotid secretion was stimulated with citric acid. Amylase output increased 16-fold whereas IgA output increased only 2.5-fold. Subsequently the effects of chewing and citric acid on S-IgA secretion

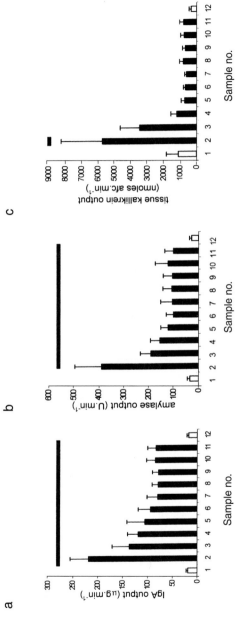

FIG. 8. Secretion of (a) IgA, (b) amylase, and (c) tissue kallikrein from the human parotid gland at rest and in response to chewing on inert plastic tubing. One 10 min period of rest (open bar) was followed by 10 periods of chewing (filled bars) for 1 min and a final 10 min period of rest (open bar). IgA was assayed by ELISA, amylase and tissue kallikrein were assayed by enzymatic activity, and secretion was expressed as outputs per minute, that is, the product of concentration or activity and salivary flow rate. Solid horizontal bars indicate values that are statistically significantly different ($p < 0.05$) from the initial unstimulated rate.

TABLE I

	Rest	Chewing	2% Citric Acid
Flow rate (ml min^{-1})	7.26 ± 2.48	91.72 ± 11.13^a	98.04 ± 14.02^a
IgA (μg min^{-1})	4.9 ± 1.31	96.77 ± 11.29^a	101.37 ± 15.29^a
Amylase (Units min^{-1})	3.48 ± 1.34	53.18 ± 9.31^a	$318.9 \pm 109^{a,b}$

[a] Statistically significantly different ($p < 0.05$) from mean value at rest.
[b] Statistically significantly different ($p < 0.05$) from mean value with chewing.

were compared and the results are shown in Table I (G. B. Proctor *et al.*, 2002, unpublished observations). As found previously, chewing increased S-IgA and amylase secretion to similar extents compared to resting (unstimulated) secretion. However, citric acid increased stored protein secretion to a much greater extent than S-IgA. The secretory mechanism leading to increased S-IgA secretion was examined in both published studies. Western blotting of rest- and chew-stimulated saliva samples demonstrated that IgA was secreted in the form of S-IgA, that is, complete with secretory component (Fig. 9). Interestingly, scanning of Western blots and quantification of band intensities revealed that the proportion of free secretory component remained unchanged with stimulation (Fig. 9). Thus it might be that the epithelial transcytosis mechanism can always cope with the glandular polymeric IgA available for secretion (Proctor and Carpenter, 2001). Brandtzaeg (1971) assayed free secretory component by single radial immunodiffusion and found an excess in both stimulated and resting saliva samples. In experiments on rat liver Giffroy *et al.* (1998) found that increased availability of polymeric IgA increased transcytosis across hepatocytes with corresponding decreases in the ratio of free to bound secretory component. Therefore, it can be tentatively concluded that reflexly stimulated increases in salivary S-IgA secretion are not driven by increased availability of polymeric IgA since a relatively constant proportion of pIgR is transcytosed without IgA attached.

Since nerve stimulation studies have revealed that both parasympathetic and sympathetic nerve–mediated stimuli can potentially increase the transcytosis of IgA into saliva it is of interest to examine the role played by these nerves under reflex conditions in conscious rats. Such a study of the influence of the sympathetic nerve under reflex conditions was undertaken by Matsuo *et al.* (2000). The concentrations of S-IgA in salivas evoked by grooming, feeding, rejection behavior, and heat were determined before and after acute sectioning of the sympathetic chain caudal to the superior

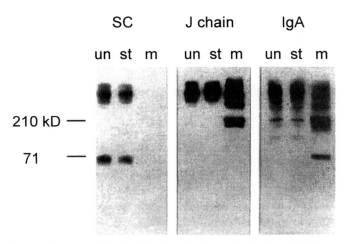

FIG. 9. Western blot of unstimulated (un) and citric acid stimulated (st) human parotid salivary proteins separated on a gradient SDS gel and probed with antibodies, showing immunoglobulin A, secretory component (SC), and J chain. A sample of J chain–containing myeloma IgA(m) has been electrophoresed for comparison and the positions of molecular weight standard proteins are shown. High relative molecular weight (Mr) protein in saliva samples contains IgA, J chain, and SC, while the myeloma, which has not been transcytosed, is SC negative. Free SC is seen as a more mobile band (at Mr of 71 kD) in both salivas.

cervical ganglion creating a unilateral sympathetic decentralization (sx). S-IgA secretion on reflex stimulation remained unchanged following decentralization except for the grooming stimulus that showed a decreased output. These results contrasted with total protein secretion and acinar cell peroxidase that showed a decrease in both concentration and output following sx. The latter indicated that in the rat submandibular gland, as has previously been found in the rat and rabbit parotid glands (Speirs and Hodgson, 1974; Gjorstrup, 1980), sympathetic nerve–mediated stimuli make an important contribution to reflexly induced secretion of salivary proteins. When the results of Matsuo et al. (2000) are considered with those obtained in nerve stimulation studies in anesthetized animals it can be concluded that high-frequency sympathetic nerve–mediated stimuli are unlikely to be contributing significantly to S-IgA secretion under reflex conditions. In the nerve stimulation studies, higher S-IgA secretion was evoked by high-frequency (50 Hz) electrical stimulation and it would seem that such high frequencies were not evoked by the reflex stimuli. It may therefore be that reflexly induced increases in S-IgA secretion into saliva are regulated mainly by parasympathetic nerve–mediated impulses. In humans, sweet or salty, tastes evoke more protein-rich salivas than sour taste, possibly due to differing proportions of impulses mediated by parasympathetic and sympathetic nerves. The results with chewing and citric acid stimulation in humans (see earlier) suggest that unlike stored protein secretion, S-IgA secretion is not stimulated to different extents by different afferent stimuli.

A. EFFECTS OF STRESS ON REFLEX IgA SECRETION

There is considerable interest in stress-related immune impairment and the function of the hypothalamic pituitary-adrenocortical (HPA) axis. A number of indices of immune competence have been utilized in the field of psychoneuroimmunology, including assays of immunoglobulins as indices of the humoral immune response (Vedhara et al., 1999). Salivary S-IgA has been widely assayed since whole mouth saliva is easy to collect and S-IgA might provide an index of the mucosal immune response relevant to the increased incidence of upper respiratory tract infections that have been associated with stress (e.g., Drummond and HewsonBower, 1997). Individuals with increased S-IgA levels in upper respiratory tract secretions, including saliva, appear to develop fewer symptoms of infection during viral infection (Rossen et al., 1970; McClelland et al., 1980). In general, saliva samples collected in such studies are not overtly stimulated but resting samples collected using absorbant material placed in the mouth (e.g., Deinzer et al., 2000). Kugler et al. (1992) studied resting salivary secretion of S-IgA in 85 subjects and found a negative correlation between S-IgA concentration and salivary flow rate.

There is a well-recognized drying of the mouth during times of acute stress and it might be expected that saliva sampled at such times would have higher concentrations of S-IgA. Some studies have tried to account for changes in salivary flow by expressing secretion as output (i.e., concentration × flow rate). It would appear that there are effects of stress on S-IgA secretion which are not dependent on changed flow rate, although the sampling techniques used in many studies makes estimating salivary flow difficult (Deinzer et al., 2000). Previous studies suggest that salivary S-IgA levels can be raised or lowered by stress and the differences between studies may depend on the nature and definition of stress. Results suggest that the levels of other stored salivary proteins are affected by stress (J. Bosch et al., 2002, personal communication). It may therefore be that both salivary fluid and protein secretion are subject to central inhibition from other centers in the brain. Certainly central connections between the salivary centers and other centers in the rat brain have been demonstrated (Matsuo, 1999).

B. Effects of Exercise on Reflex S-IgA Secretion

A study of elite cross-country skiers found that their salivary IgA levels were lower than those of recreational athletes (Tomasi *et al.*, 1982). From a number of subsequent studies it has been concluded that exhaustive exercise can reduce S-IgA levels, although there is usually recovery within 24 h (Gleeson and Pyne, 2000). Occasional studies have examined other aspects of mucosal immunity, thus it has been found that breast milk S-IgA levels were decreased in one study (Gregory *et al.*, 1997). It has been suggested, though not proven, that reduced levels of IgA in mucosal secretions, particularly in salivary S-IgA1 levels, are associated with upper respiratory tract infections in athletes undertaking large volumes and intensities of exercise (Gleeson and Pyne, 2000). The mechanism underlying the effects of exercise on salivary S-IgA levels remains unknown. In general, studies in this area have expressed S-IgA levels as concentrations, which are decreased. Few studies have addressed whether salivary flow rate is affected by exercise and how secretion rates of S-IgA are affected, although results of one study suggest that secretion rates are reduced (Gleeson *et al.*, 2000).

V. Long-Term Influences of Nerves on S-IgA Secretion

The studies described earlier established that nerves can acutely regulate S-IgA transcytosis by salivary epithelial cells. What are the long-term effects of nerves on S-IgA secretion? Autonomic nerves have previously been shown to exert long-term influences on salivary glands. Removal of the parasympathetic nerve supply in particular causes atrophy while secretion of fluid and the synthesis and secretion of major salivary proteins are altered by denervation (Proctor, 1999).

The effects of 1 week of sympathetic decentralization on S-IgA secretion were examined (Proctor *et al.*, 2000b). The right cervical sympathetic chain was sectioned, leaving the superior cervical ganglion intact and thus creating sx without destruction of postganglionic nerves and the induction of supersensitivity. Saliva was collected 1 week later from decentralized and contralateral control glands and as in the studies previously described the protocol enabled both unstimulated and nerve stimulated secretion of S-IgA to be estimated. The results are summarized in Figure 10a. Sympathetic decentralization had little effect on the unstimulated rate of S-IgA secretion but completely abolished the increase evoked by stimulation of the intact parasympathetic nerve. In contrast the secretion of stored secretory proteins increased in parasympathetically evoked saliva from decentralized

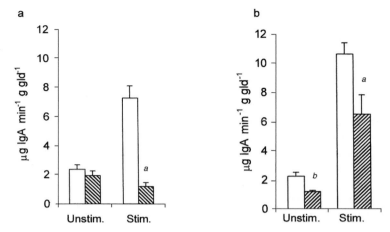

FIG. 10. Secretion of IgA into submandibular saliva from denervated (hatched bars) compared to control contralateral (open bars) glands of anesthetized rats. Saliva was collected from denervated glands during electrical stimulation of the remaining intact nerve supply following 7 days of either (a) sympathectomy or (b) parasympathectomy. The stimulation protocol allowed calculation of the unstimulated rate of IgA secretion for comparison with the outputs in response to nerve stimulation. The unstimulated rate of IgA secretion was not significantly changed by sympathectomy whereas the parasympathetically stimulated rate was greatly reduced ($p < 0.05$) compared to the contralateral control gland.[a] Both the unstimulated[b] and sympathetically stimulated[a] rates of IgA secretion were significantly reduced ($p < 0.05$) by parasympathectomy.

compared to intact glands. Assay of free SC in the same samples showed that it too was reduced by decentralization and therefore that the effects of decentralization were mediated, at least in part, through a reduced rate of IgA transcytosis in epithelial cells. Certainly decentralization did not lead to a reduction in IgA production by glandular plasma cells since the levels of IgA in homogenates of decentralized and intact glands were similar (Proctor *et al.*, 2000b). The effects of parasympathectomy (px), achieved by sectioning branches of the chorda lingual nerve which innervate the submandibular gland, have also been investigated. Stimulation of the intact sympathetic nerve 1 week following px revealed that both the calculated unstimulated and the sympathetically stimulated rates of S-IgA secretion were decreased by approximately 30% (Figure 10b). Again this effect contrasted with that observed for stored secretory proteins, which showed little change in secretion rate following px. The effects of px might be explained by a decrease in glandular levels of pIgR and preliminary evidence suggests that this is so (G. H. Carpenter *et al.*, 2002, unpublished observations). There appeared to be no decrease in plasma cell production of IgA since homogenates of parasympathectomized and intact glands contained similar levels of IgA.

VI. Effects of Nerves on Plasma Cell Production of IgA

Salivary gland epithelial cells are subject to acute regulation by autonomic nerves that have a dominant role in controlling secretion. The investigations described earlier have established that this role extends to a control of IgA secretion by salivary glands, most likely through regulation of epithelial cell transcytosis. Do nerves also influence IgA production by plasma cells within salivary glands? Following extensive nerve stimulation of IgA secretion from the rat salivary glands, glandular levels of IgA were assayed in homogenates of both stimulated and contralateral control glands and compared with the total IgA secreted into saliva. In both glands IgA levels were reduced following stimulation compared to control glands suggesting that during prolonged stimulation, the rate of IgA production by plasma cells does not match the rate of IgA transcytosis and secretion into saliva. In the submandibular gland there was evidence of IgA production within the gland since IgA levels did not decrease to the same extent as would be indicated by the amounts secreted. Immunocytochemistry of the rat and human salivary glands suggests that there is a reservoir of extracellular IgA within the gland (G. B. Proctor *et al.*, 2002, unpublished observations; Brandtzaeg, 1974). Acute isolation of epithelial cell units as described previously (Section III.A) supports the concept of such an extracellular reservoir since only a minor proportion of total glandular IgA levels were found within cells. However, this data does not indicate whether nerves upregulate IgA production by plasma cells. Following 1 week of px or sx as described earlier there was no reduction in the IgA content of submandibular glands. In fact, determination of the intracellular and extracellular levels of IgA in parasympathectomized glands suggests that there has been no reduction in IgA production.

Neuropeptides can increase IgA production by B cells isolated from different sources (see Pascual *et al.*, 1994). Substance P may act directly on B and/or T lymphocytes, both of which have been shown to express substance P receptors (Stanisz *et al.*, 1987). Substance P may act as a co-stimulatory signal for B cells activated by antigen or T-cell-derived (Th2 subset) cytokines such as interleukin 5 (IL-5) and IL-6 that have roles in humoral immunity. It may be that at mucosal sites receiving an autonomic innervation, nerves can upregulate IgA production through release of neuropeptides. Salivary glands are effector sites of the mucosal immune system that receive a rich autonomic innervation, and substance P and other neuropeptides,

such as vasoactive intestinal peptide, have been localized in the parasympathetic nerves supplying salivary glands in a number of species (Ekström,
1999). Peptide release from parasympathetic nerves might upregulate IgA
production by plasma cells in salivary glands. Cytokines, such as Il-6 and
transforming growth factor–β that are also produced by intestinal epithelial cells, have been demonstrated to regulate mucosal B-cell IgA secretion
(Goodrich and McGee, 1998). If the regulation of salivary epithelial cells
by autonomic nerves extends to cytokine production then it may be that
autonomic nerves can influence salivary gland plasma cell IgA secretion
indirectly through effects on epithelial cell cytokine production.

There have been few studies of the effects of neurotransmitters on non-
transformed plasma cells since the latter tend to be widely distributed in
small numbers at mucosal effector sites, which makes isolation of significant
numbers difficult. Brink *et al.* (1994) overcame this problem by studying the
Harderian gland of the chicken, an accessory lacrimal gland that is rich
in plasma cells and innervated by cholinergic parasympathetic nerves. In
contrast to previous studies of an IgA-producing myeloma (Tartakoff and
Vassalli, 1977), the Harderian gland–derived plasma cells responded to carbachol with increased IgG secretion within 20 min of stimulation. IgG was
assayed rather than IgA in order to be able to discount the influence of glandular epithelial cells and transcytosis on IgA secretion. Muscarinic cholinergic receptors were immunocytochemically localized on the cells and carbachol, although at high doses (100 μM) raised levels of intracellular calcium
and induced IgG secretion. Neuropeptides, the cholinergic agonist pilo-
carpine, and electrical stimulation of the extrinsic innervation of intestinal
segments have each been found to increase luminal IgA secretion (Schmidt
et al., 1999; Wilson *et al.,* 1982). The mechanism operating is not certain but
may represent a direct effect on plasma cells of the lamina propria.

The mechanisms regulating plasma cell migration to and retention at
mucosal sites are beginning to be unraveled (Brandtzaeg *et al.,* 1999b). It
seems likely that mucosal epithelial cells have a role to play, probably through
the local release of cytokines, in retaining plasma cells at mucosal sites.
Following systemic chemical sympathectomy using 6-hydroxydopamine, retention and accumulation of ovalbumin-primed B cells at mucosal sites was
reduced suggesting a change in postextravasation signals at effector sites
(González-Ariki and Husband, 1998). Autonomic nerves may well influence, either directly or indirectly through epithelial cells, the retention
of plasma cells within salivary glands. Following 1 week of either sx or
px there was no obvious change in plasma cell numbers, as indicated by
immunocytochemistry or IgA levels in submandibular glands. However, it
may well be that periods of denervation lasting longer than 1 week are required before an effect on plasma cell migration is apparent. As mentioned
previously, surgical denervation did not appear to alter total IgA-producing

plasma cells in the rat submandibular gland. However, a further study might examine the effect of denervation on new recruitment of B cells by salivary glands.

VII. Conclusion

S-IgA secretion into saliva of humans is rapidly (within 15 s) increased by autonomic nerve–mediated reflex stimuli through increased glandular epithelial cell transcytosis of IgA by pIgR. Investigations in rat salivary glands have revealed that such autonomic regulation is mediated through both sympathetic and parasympathetic nerves and that a number of epithelial cell autonomic receptors are involved, although cholinergic stimuli appear to have a lesser effect than adrenergic and peptidergic stimuli. The autonomic regulation of S-IgA secretion is different from the major stored secretory proteins of salivary glands and we can tentatively conclude from *in vivo* studies that stimulated secretion is on or off and does not show dose dependency or additive effects with combinations of agonists. Autonomic nerves have longer term effects on S-IgA secretion through effects on pIgR-mediated transcytosis as demonstrated by denervation experiments. Stress or extensive, intense physical training have been linked with increased susceptibility to upper respiratory tract infections and it has been suggested that decreased S-IgA secretion may play a role. The studies described earlier establish that S-IgA secretion is under the influence of the autonomic nervous system but further studies are required to determine whether altered autonomic activity can change susceptibility to mucosal infection.

Acknowledgments

The authors are grateful to colleagues who appear as coauthors on cited references, to the Wellcome Trust for support through a project grant, and to GlaxoSmithKline for support of studies of human salivary secretion.

References

Amsallem, H., Metioui, M., van den Abeele, A., Elyamani, A., Moran, A., and Dehae, J. P. (1996). Presence of metabotropic and an ionotrophic purinergic receptor on rat submandibular ductal cells. *Cell. Physiol.* **40,** C1546–C1555.
Anderson, L. C., Garrett, J. R., Zhang, X., Proctor, G. B., and Shori, D. K. (1995). Differential secretion of proteins by rat submandibular acini and granular ducts on graded autonomic nerve stimulations. *J. Physiol.* **485.2,** 503–511.

Anderson, L. C., Garrett, J. R., Zhang, X., Proctor, G. B., and Shori, D. K. (1996). Protein secretion from rat submandibular acini and granular ducts: Effects of exogenous VIP and substance P during parasympathetic nerve stimulation. *Comp. Biochem. Physiol.* **119A,** 327–331.

Baum, J. B., and Wellner, R. B. (1999). Receptors in salivary glands. *In* "Neural Mechanisms of Salivary Glands" (J. R. Garrett, J. Ekström, and L. C. Anderson, eds.). Frontiers in Oral Biology, pp. 44–58. Karger, Basel.

Bowen, W. H. (1996). Salivary influences on the oral microflora. *In* "Saliva and Oral Health" (W. M. Edgar and D. M. O'Mullane, eds.), pp. 95–103. British Dental Association, London.

Brandtzaeg, P. (1971). Human secretory immunoglobulins, VII: Concentrations of parotid IgA and other secretory proteins in relation to the rate of flow and duration of secretory stimulus. *Arch. Oral. Biol.* **16,** 1295–1310.

Brandtzaeg, P. (1974). Mucosal and glandular distribution of immunoglobulin components. *Immunology* **26,** 1101–1113.

Brandtzaeg, P., Farstad, I. N., Johansen, F.-E., Morton, H. C., Norderhaug, I. N., and Yamanaka, T. (1999a). The B-cell system of human mucosae and exocrine glands. *Immunol. Rev.* **171,** 45–87.

Brandtzaeg, P., Baekkevold, E. S., Farstad, I. N., Jahnsen, F. L., Johansen, F.-E., Nilsen, E. M., and Yamanaka, T. (1999b). Regional specialization in the mucosal immune system: What happens in the microcompartments? *Immunol. Today* **20,** 141–151.

Brink, P. R., Walcott, B., Roemer, E., Grine, E., Pastor, M., Christ, G. J., and Cameron, R. H. (1994). Cholinergic modulation of immunoglobulin secretion from avian plasma cells: The role of calcium. *J. Neuroimmunol.* **51,** 113–121.

Cardone, M. H., Smith, B. L., Mennitt, P. A., Mochly-Rosen, D., Silver, R. B., and Mostov, K. E. (1996). Signal transduction by the polymeric immunoglobulin receptor suggests a role in regulation of receptor transcytosis. *J. Cell. Biol.* **133,** 997–1005.

Carpenter, G. H., Garrett, J. R., Hartley, R. H., and Proctor, G. B. (1998). The influence of nerves on the secretion of immunoglobulin A into submandibular saliva in rats. *J. Physiol.* **512.2,** 567–573.

Carpenter, G. H., Proctor, G. B., Anderson, L. C., Zhang, X. S., and Garrett, J. R. (2000). Immunoglobulin A secretion into saliva during dual sympathetic and parasympathetic nerve stimulation of rat submandibular glands. *Exp. Physiol.* **85.3,** 281–286.

Cole, M. F. (1985). Influence of secretory immunoglobulin A on ecology of oral bacteria. *In* "Molecular Basis of Oral Microbial Adhesion" (S. E. Mergenhagen and B. Rosan, eds.), pp. 125–130. Am. Soc. Microbiologists, Washington, D.C.

Dawes, C. (1996). Clearance of substances from the oral cavity—Implications for oral health. *In* "Saliva and Oral Health" (W. M. Edgar and D. M. O'Mullane, eds.), pp. 67–79. British Dental Association, London.

De Groot, N., van Kuik-Romeiln, P., Lee, S. H., and de Boer, H. A. (2000). Increased immunoglobulin A levels in milk by overexpressing the murine polymeric immunoglobulin receptor gene in the mammary gland epithelial cells of transgenic mice. *Immunology* **101,** 218–224.

Deinzer, R., Kleineidam, C., Stiller-Winkler, R., Edel, H., and Bachg, D. (2000). Prolonged reduction of salivary immunoglobulin A (S-IgA) after a major academic exam. *Int. J. Psychophysiol.* **37,** 219–232.

Drummond, P. D., and HewsonBower, B. (1997). Increased psychosocial stress and decreased mucosal immunity in children with recurrent upper respiratory tract infections. *J. Psychosom. Res.* **43,** 271–278.

Ekström, J. (1999). Role of nonadrenergic, noncholinergic autonomic transmitters in salivary glandular activities *in vivo*. *In* "Neural Mechanisms of Salivary Gland Secretion"

(J. R. Garrett, J. Ekström, and L. C. Anderson, eds.), Frontiers in Oral Biology, Volume 11, Chapter 6, pp. 196–218. Karger, Basel.

Emmelin, N. (1981). Salivary glands: Secretory mechanisms. *In* "Scientific Foundations of Gastroenterology" (W. Sircus and A. N. Smith, eds.), pp. 219–225. Heineman, London.

Emmelin, N. (1987). Nerve interactions in salivary glands. *J. Dent. Res.* **66,** 509–517.

Giffroy, D., Langendries, A., Maurice, M., Daniel, F., Lardeux, B., Courtoy, P. J., and Vaerman, J. P. (1998). *In vivo* stimulation of polymeric Ig receptor transcytosis by circulating polymeric IgA in rat liver. *Int. Immunol.* **10,** 347–354.

Gjorstrup, P. (1980). Taste and chewing as stimuli for the secretion of amylase from the parotid gland of the rabbit. *Acta Physiol. Scand.* **110,** 295–301.

Gleeson, M., and Pyne, D. B. (2000). Exercise effects on mucosal immunity. *Immunol. and Cell Biol.* **78,** 536–544.

Gleeson, M., Ginn, E., and Francis, J. L. (2000). Salivary immunoglobulin monitoring in an elite kayaker. *Clin. J. Sport Med.* **10,** 206–208.

González-Ariki, S., and Husband, A. J. (1998). The role of sympathetic innervation of the gut in regulating mucosal immune responses. *Brain Behav. Immun.* **12,** 53–63.

Goodrich, M., and McGee, D. W. (1998). Regulation of mucosal B cell immunoglobulin secretion by intestinal epithelial cell-derived cytokines. *Cytokine* **10,** 948–955.

Gregory, R. L., Wallace, J. P., Gfell, L. E., Marks, J., and King, B. A. (1997). Effect of exercise on milk immunoglobulin A. *Med. Sci. Sport Exer.* **29,** 1596–1601.

Hector, M. P., and Linden, R. W. A. (1999). Reflexes of salivary secretion. *In* "Neural Mechanisms of Salivary Gland Secretion" (J. R. Garrett, J. Ekström, and L. C. Anderson, eds.). Frontiers in Oral Biology, Volume 11, Chapter 12, pp. 196–218. Karger, Basel.

Hirt, R. P., Hughes, G. J., Frutiger, S., Michett, P., Perregaux, C., Poulain-Goaefroy, O., Jeangvenat, N., Neutra, M. R., Kraehenbuhl, J. P. (1993). Transcytosis of the polymeric Ig receptor requires phosphorylation of serine 664 in the absence but not the presence of dimeric IgA. *Cell* **74,** 245–255.

Kaushic, C., Richardson, J. M., and Wira, C. R. (1995). Regulation of polymeric immunoglobulin A receptor messenger ribonucleic acid expression in rodent uteri: Effect of sex hormones. *Endocrinology* **136,** 2836–2844.

Kelleher, R. S., Hann, L. E., Edwards, J. A., and Sullivan, D. A. (1991). Endocrine, neural, and immune control of secretory component output by lacrimal gland acinar cells. *J. Immunol.* **146,** 3405–3412.

Kugler, J., Hess, M., and Haake, D. (1992). Secretion of salivary immunoglobulin A in relation to age, saliva flow, mood states, secretion of albumin, cortisol, and catecholamines in saliva. *J. Clin. Immunol.* **12,** 45–49.

Loman, S., Radl, J., and Jansen, H. M. *et al.* (1997). Vectorial transcytosis of dimeric IgA by the Calu-3 human epithelial cell line: Upregulation by IFN-γ. *Am. J. Physiol.* **16,** L951–L958.

Mandel, I. D., and Khurana, H. S. (1969). The relation of human salivary γ A globulin and albumin to flow rate. *Arch. Oral Biol.* **14,** 1433–1435.

Marcotte, H., and Lavoie, M. C. (1998). Oral microbial ecology and the role of salivary immunoglobulin A. *Microbiol. Mol. Biol.* **62,** 71–105.

Matsuo, R. (1999). Central connections for salivary innervations and efferent impulse formation. *In* "Neural Mechanisms of Salivary Gland Secretion" (J. R. Garrett, J. Ekström, and L. C. Anderson, eds.). Frontiers in Oral Biology, Volume 11, Chapter 2, pp. 26–43. Karger, Basel.

Matsuo, R., Garrett, J. R., Proctor, G. B., and Carpenter, G. H. (2000). Reflex secretion of proteins into submandibular saliva in conscious rats, before and after preganglionic sympathectomy. *J. Physiol.* **527,** 175–184.

McClelland, D. C., Davison, R. J., Floor, E., and Saron, C. (1980). Stressed power motivation, sympathetic activation, immune function and illness. *J. Human. Stress.* **6**, 11–19.

Mega, J., McGhee, J. R., and Kiyono, H. (1995). Characterisation of cytokine producing T cells, TCR expression, and IgA plasma cells in salivary gland associated tissues. *In* "Advances in Mucosal Immunology" (Mestecky, J., Russell, M. W., Jackson, S., Michalek, S. M., Tlaskolová-Hogenorá, H., and Sterzl, J. eds.). Plenum Press, New York.

Mestecky, J. (1993). Saliva as a manifestation of the common mucosal immune system. *Ann. N.Y. Acad. Sci.* **694**, 184–194.

Murray, P. A., Prakobphol, A., Lee, T., Hoover, C. I., and Fisher, S. J. (1992). Adherence of oral streptococci to salivary glycoproteins. *Infect. Immun.* **60**, 31–38.

Nilsen, E. M., Johansen, F. E., and Kvale, D. (1999). Different regulatory pathways employed in cytokine enhanced expression of secretory component and epithelial, HLA class I genes. *Eur. J. Immunol.* **29**, 168–179.

Norderhaug, I. N., Johansen, F.-E., Schjerven, H., and Brandtaeg, P. (1999). Regulation of the formation and external transport of secretory immunoglobulins. *Crit. Rev. Immunol.* **19**, 481–508.

Pascual, D. W., Stanisz, A. M., and Bost, K. L. (1994). Functional aspects of the peptidergic circuit in mucosal immunity. *In* "Handbook of Mucosal Immunology" (P. L. Ogra, Mestecky, J., Lamm, M. F., Strober, W., McGhee, J. R., Bienenstock, J., eds.), pp. 203–216. Academic Press, San Diego.

Piskurich, J. F., France, J. A., Tamer, C. M., Willmer, C. A., Kaetzel, C. S., and Kaetzel, D. M. (1993). Interferon-γ induces polymeric immunoglobulin receptor messenger RNA in human intestinal epithelial cells by a protein synthesis dependent mechanism. *Mol. Immunol.* **30**, 413–421.

Proctor, G. B. (1999). Effects of autonomic denervations on protein secretion and synthesis by salivary glands. *In* "Neural Mechanisms of Salivary Gland Secretion" (J. R. Garrett, J. Ekström, and L. C. Anderson, eds.), Frontiers in Oral Biology, Volume 11, Chapter 8, pp. 150–165. Karger, Basel.

Proctor, G. B., and Carpenter, G. H. (2001). Chewing stimulates secretion of human salivary secretory immunoglobulin A. *J. Dent. Res.* **80**, 909–913.

Proctor, G. B., and Chan, K.-M. (1994). A fluorometric assay of peroxidase activity utilizing 2′,7′dichlorofluorescin with thiocyanate: Application to the study of salivary secretion. *J. Biochem. Biophys. Methods.* **28**, 329–336.

Proctor, G. B., Carpenter, G. H., Anderson, L. C., and Garrett, J. R. (2000a). Nerve-evoked secretion of immunoglobulin A in relation to other secretory proteins of rat parotid glands. *Exp. Physiol.* **85**, 511–518.

Proctor, G. B., Carpenter, G. H., and Garrett, J. R. (2000b). Sympathetic decentralization abolishes increased secretion of immunoglobulin A evoked by parasympathetic stimulation of rat submandibular glands. *J. Neuroimmunol.* **109**, 147–154.

Rice, J. C., Spence, J. S., Megyesi, J., Safirstein, R. L., and Goldblum, R. M. (1998). Regulation of the polymeric immunoglobulin receptor by water intake and vasopressin in the rat kidney. *Am. J. Physiol.* **274**, F966–F977.

Rossen, R. D., Butler, W. T., and Waldman, R. H. (1970). The proteins in nasal secretion, II. A longitudinal study of IgA and neutralizing antibodies in nasal washings of men infected with influenza virus. *JAMA* **211**, 1157–1160.

Schmidt, P. T., Eriksen, L., Loftager, M., Rasmussen, T. N., and Holst, J. J. (1999). Fast acting nervous regulation of immunoglobulin A secretion from isolated perfused porcine ileum. *Gut* **45**, 679–685.

Segawa, A., and Yamashina, S. (1999). The dynamics of exocytosis of preformed secretory granules from acini in rat salivary glands. *In* "Glandular Mechanisms of Salivary Gland

Secretion" (J. R. Garrett, J. Ekström, and L. C. Anderson, eds.), Frontiers in Oral Biology, Volume 10, Chapter 6, pp. 89–100. Karger, Basel.

Speirs, R. L., and Hodgson, C. (1974). Control of amylase secretion in the rat parotid gland during feeding. *Arch. Oral Biol.* **21,** 539–544.

Stanisz, A. M., Scicchitano, R., Dazin, P., Bienenstock, J., and Payan, D. G. (1987). Distribution of substance P receptors on murine spleen and Peyer's patch T and B cells. *J. Immunol.* **139,** 749–754.

Su, T., and Stanley, K. K. (1998). Opposite sorting and transcytosis of the polymeric immunoglobulin receptor in transfected endothelial and epithelial cells. *J. Cell Sci.* **111,** 1197–1206.

Sullivan, D. A., Bloch, K. J., and Allansmith, M. R. (1984). Hormonal influence on the secretory immune system of the eye: Androgen regulation of secretory component levels in rat tears. *J. Immunol.* **132,** 1130.

Sullivan, D. A., Kelleher, R. S., and Vaerman, J. P. (1990). Androgen regulation of secretory component synthesis by lacrimal gland acinar cells *in vitro. J. Immunol.* **145,** 4238–4244.

Tartakoff, A. M., and Vassalli, P. (1977). Plasma cell immunoglobulin secretion arrest is accompanied by alteration of the Golgi apparatus. *J. Exp. Med.* **146,** 1332–1345.

Tenovuo, J. O. (1989). Non-immunoglobulin defense factors in human saliva. *In* "Human Saliva: Clinical Chemistry and Microbiology" (J. O. Tenovuo, ed.), Volume 2, pp. 55–91. CRC Press, Boca Raton.

Tomasi, T. B., Trudeau, F. B., Czerwinski, D., and Erredge, S. (1982). Immune parameters in athletes before and after strenuous exercise. *J. Clin. Immunol.* **2,** 173–178.

Vedhara, K., Fox, J. D., and Wang, E. C. Y. (1999). The measurement of stress-related immune dysfunction in psychoneuroimmunology. *Neurosci. Biobehav. Rev.* **23,** 699–715.

Wilson, I. D., Soltis, R. D., and Olson, R. E. (1982). Cholinergic stimulation of immunoglobulin A secretion in rat intestine. *Gastroenterology* **83,** 881–888.

Wu, A. M., Csako, G., and Herp, A. (1994). Structure, biosynthesis and function of salivary mucins. *Mol. Cell. Biochem.* **137,** 39–55.

STRESS AND SECRETORY IMMUNITY

Jos A. Bosch

Department of Oral Biology
College of Dentistry, The Ohio State University
Columbus, Ohio 43218
and
Academic Centre for Dentistry Amsterdam Section of Oral Biochemistry
Vrije Universiteit
1081 BT Amsterdam, The Netherlands

Christopher Ring

School of Sport and Exercise Sciences
University of Birmingham
Edgbaston, Birmingham B15 2TT, United Kingdom

Eco J. C. de Geus

Department of Biological Psychology
Vrije Universiteit
1081 BT Amsterdam, The Netherlands

Enno C. I. Veerman and Arie V. Nieuw Amerongen

Academic Centre for Dentistry Amsterdam, Section of Oral Biochemistry
Vrije Universiteit
1081 BT Amsterdam, The Netherlands

I. Introduction

Human and animal studies have provided convincing evidence that psychosocial stress is associated with increased susceptibility to infectious diseases (Cohen and Herbert, 1996; Bonneau *et al.*, 2001). It has been estimated that up to 95% of all infections are initiated at mucosal surfaces. These mucosal surfaces are protected by the secretions of various exocrine glands, including the salivary glands (Nieuw Amerongen *et al.*, 1995; Rudney, 1995; Schenkels *et al.*, 1995; Nieuw Amerongen *et al.*, 1998) and lacrimal glands (Schenkels *et al.*, 1995; McClellan, 1997), as well as those of the respiratory (Kaliner, 1991; Rogers, 1994; Nieuw Amerongen *et al.*, 1998; Finkbeiner, 1999; Travis *et al.*, 2001) and gastrointestinal tracts (Bevins, 1994; Schenkels *et al.*, 1995; Nieuw Amerongen *et al.*, 1998; Hecht, 1999; Baron *et al.*, 2000). The antimicrobial proteins (e.g., immunoglobulins, mucins, cystatins, lysozyme, lactoferrin) secreted by these glands constitute a first line of defense that prevents infection and disease by interfering with microbial entry and multiplication. The secretion of these protective proteins in under strong neurohormonal control (Garrett, 1987; Rogers, 1994; Berczi *et al.*, 1998; Finkbeiner, 1999; Garrett, 1999b; Shimura, 2000). This presents a potential psychoneuroimmunological (PNI) pathway linking psychosocial stress and infectious disease.

This review will focus on the relationship between stress and salivary secretory immunity in humans. There are compelling reasons for studying salivary immunity. First, the salivary glands, being strategically located at the portal of entry to the respiratory and gastrointestinal tracts, provide a first line of defense against microorganisms that may develop their pathogenic potential elsewhere in the body. Second, salivary secretory immunity plays a crucial role in the maintenance of oral health. This is particularly relevant from a PNI perspective, as several pathologies of the oral cavity have been associated with stress, including periodontal disease (Rugh *et al.*, 1984; da Silva *et al.*, 1995; Genco *et al.*, 1998; Breivik and Thrane, 2001), acute necrotizing gingivitis (Rugh *et al.*, 1984; da Silva *et al.*, 1995), dental caries (Sutton, 1966; Rugh *et al.*, 1984; Beck *et al.*, 1987; Vanderas *et al.*, 1995, 2000), herpes labialis (Rugh *et al.*, 1984; Cohen and Herbert, 1996; Logan *et al.*, 1998), recurrent aphthous ulcerations (Rugh *et al.*, 1984; McCartan *et al.*, 1996), and impaired oral wound healing (Marucha *et al.*, 1998). The relevance of these observations may not be limited to dental science, as impaired oral health is a risk factor for infective endocarditis, atherosclerosis, coronary thrombosis, stroke, and respiratory infection (Herzberg, 1996; Page, 1998; Scannapieco, 1999; Valtonen, 1999; Li *et al.*, 2000; Wu *et al.*, 2000).

II. Saliva as a Model for Mucosal Secretory Defense

A third reason for studying salivary immunity is that it may serve as a representative model for studying both innate and adaptive secretory immune defenses. The salivary glands and other exocrine glands share similarities in morphology and architecture; neuronal regulation (Rogers, 1994; Berczi *et al.*, 1998; Finkbeiner, 1999; Garrett, 1999b; Shimura, 2000); the major functions of their secretions, such as host defense and lubrication (Kaliner, 1991; Levine, 1993; Hecht, 1999; Travis *et al.*, 2001); and the composition of these secretions (Schenkels *et al.*, 1995). While the concentrations may vary from one anatomical location to another, most salivary proteins can also be found in other mucosal fluids where they serve similar functions. Moreover, the salivary glands form a common effector site for IgA-secreting plasma cells: IgA-secreting immunocytes that are activated at the various mucosal induction sites where the antigen presentation takes place, all migrate to the salivary glands (Mestecky, 1993). Thus, analysis of immune responses in saliva may provide a general picture of the entire secretory immune system.

The most common route for microbial acquisition and infection is through colonization of the mucosa. Gibbons (Gibbons, 1989) has argued that the mouth, which is readily accessible and harbors an extremely rich microflora, provides a model to monitor basic colonization processes such as microbial adherence, microbial growth, and microbial interactions. Given that saliva is a representative mucosal secretion and that mucosal fluids are important determinants of mucosal microbial colonization, saliva–microbe interactions may act as a model for studying basic microbial colonization processes.

III. An Outline of Secretory Immunity

A. SECRETORY IMMUNOGLOBULIN A

The two main secretory proteins that represent adaptive immunity are secretory immunoglobulins A (S-IgA) and M (S-IgM) (Lamm, 1997; Brandtzaeg *et al.*, 1999; Norderhaug *et al.*, 1999). S-IgA, which is by far the most abundant immunoglobulin, can bind selectively to antigens, such as microbes and toxins. This binding prevents such antigens from attaching to or penetrating the mucosal surface, for example, by blocking interactions between microbial ligands and their receptors on epithelial cells. The antimicrobial effects of S-IgA may be potentiated by interaction with innate antimicrobial secretory proteins such as lactoperoxidase and MUC7 (Tenovuo *et al.*, 1982; Biesbrock *et al.*, 1991).

The B cells responsible for local IgA production originate from mucosa-associated lymphoid tissue (MALT). After antigen exposure in the MALT, the B cells migrate to the exocrine glands, including the salivary glands, where they reside in the interstitium as IgA-producing plasma cells (Brandtzaeg, 1998). The presence of plasma cells that originate from many different induction sites in the salivary glands ensures that the buccal membrane is furnished with a wide spectrum of secretory antibodies (Mestecky, 1993). This is particularly relevant considering that the oral cavity forms a portal of entry to the respiratory and gastrointestinal tracts.

Most IgA-secreting plasma cells in the salivary glands produce a polymeric form of IgA. This polymeric IgA (pIgA) is transported and secreted into saliva as S-IgA. The formation and secretion of S-IgA is a two-step process (Norderhaug *et al.*, 1999). First, the pIgA locks on to a receptor molecule, referred to as the polymeric immunoglobulin receptor (pIgR), present on the basolateral surface of a secretory glandular cell. The pIgR, with or without an immunoglobulin attached to it, is translocated from the basolateral to the apical surface of the glandular cell. Here the receptor, with or without an immunoglobulin attached to it, is cleaved off and secreted into saliva as secretory IgA (S-IgA) or secretory component (SC), respectively. Accordingly, both IgA secretion by the plasma cells and the availability of pIgR for translocation by the glandular cells are rate-determining steps in the secretion of S-IgA. As well as transporting IgA into the external secretions, SC also protects the immunoglobulin against proteolytic breakdown in this hostile milieu exterieur.

Human IgA occurs in two isotypic forms, IgA1 and IgA2 (Brandtzaeg *et al.*, 1999). Several structural and functional differences between the two types have been detected. The most notable difference is a longer hinge region in the IgA1 antibody molecule (Kilian *et al.*, 1996). Some bacteria, including the viridans group streptococci that are abundant in saliva, have proteinases that split human IgA1 in the hinge region (Kilian *et al.*, 1996). These IgA1 proteinases allow bacteria to colonize mucosal surfaces in spite of adherence-inhibiting S-IgA1 antibodies. This may explain why the less vulnerable IgA2 subclass is more abundant in mucosal secretions: while the IgA1 subclass makes up some 90% of total IgA in serum, it comprises 60% in saliva, and only 35% in the lower gastrointestinal tract (Brandtzaeg *et al.*, 1999).

B. INNATE SECRETORY IMMUNITY

S-IgA is only one of many protective proteins secreted in the fluids covering the mucosae. Most proteins secreted by the mucosal exocrine glands belong to the innate immune system. The secretion of these innate immune factors (e.g., mucins, cystatins, lysozyme, lactoferrin) is predominantly under neurohormonal control (Garrett, 1987; Rogers, 1994; Berczi *et al.*, 1998; Finkbeiner, 1999; Garrett, 1999b; Shimura, 2000), and therefore, antimicrobial secretory proteins are candidates for PNI investigation.

1. *The Mucins*

The mucins form a group of 14 very large glycoproteins that consist largely of carbohydrate. They are present in nearly all exocrine mucosal fluids and give these fluids their slimy characteristic (Schenkels *et al.*, 1995; Nieuw Amerongen *et al.*, 1998). Most mucins form an adhesive gel-like film on the mucosal epithelium that protects the epithelium against noxious substances, prevents dehydration, and traps microorganisms (Nieuw Amerongen *et al.*, 1995). Mucins, such as MUC7, which lack gel-forming properties perform their protective functions in solution where they bind to bacterial and viral adhesion molecules (Nieuw Amerongen *et al.*, 1995). This binding inactivates these microorganisms and prevents their attachment to the mucosal surfaces (Ligtenberg *et al.*, 1992; Liu *et al.*, 2000).

2. *The Cystatins*

The cystatins, present in every cell and body fluid, inhibit cysteine proteinases, a class of enzymes involved in virus replication and tissue invasion by bacteria (Henskens *et al.*, 1996). Some cystatins, particularly those in mucosal secretions, also have bactericidal (killing) and bacteriostatic (growth inhibiting) effects (Blankenvoorde *et al.*, 1998).

3. *Lactoferrin*

Lactoferrin is a bacteriostatic protein that is secreted by virtually all exocrine glands and by neutrophils. It is found in particularly high concentrations in lacrimal fluid and bronchial and nasal mucus (Schenkels *et al.*, 1995; Vorland, 1999). The bacteriostatic properties of lactoferrin derive from its capacity to bind iron which deprives microorganisms of this essential substrate. Lactoferrin also has other protective functions (e.g., bactericidal, fungicidal, anti-inflammatory) (Vorland, 1999).

4. *Lysozyme*

Lysozyme (muramidase) is known for its capacity to lyse certain bacteria by cleaving the polysaccharide component of their cell walls. This is, however, just one of its antimicrobial activities, some of which depend on the presence of other components, such as the complement system, antibodies, and peroxidase, for full effectiveness. Its common occurrence in secretions is very similar to that of lactoferrin (Schenkels *et al.*, 1995).

5. *α-Amylase*

The observation that α-amylase is found in many exocrine fluids besides saliva suggests that it may have functions other than degrading starch (Schenkels *et al.*, 1995). It affects the adherence and growth of various streptococcal bacteria and may play a role in regulating the normal microflora (Scannapieco *et al.*, 1993).

IV. The Neurobiological Basis for Stress-Induced Changes in Salivary Secretory Immunity

Saliva, also denoted as "whole saliva," is the product of three pairs of major glands (the parotid, submandibular, and sublingual glands) and hundreds of minor glands in the lip, cheek, tongue, and palate. Each of these glands, as well as the various cell types within these glands, synthesize and secrete a characteristic set of proteins. Before secretion, most salivary proteins (with the secretory immunoglobulins as a notable exception, see earlier) follow the common secretory pathway: synthesis on the ribosomes on the endoplasmic reticulum, further assembly and modification in the Golgi apparatus, and subsequent storage in membrane secretory granules (Proctor, 1998). Exocytosis of these granules occurs upon neuronal stimulation—β-adrenergic stimulation in particular (Carpenter *et al.*, 1998; Garrett, 1998; Proctor, 1998; Segawa and Yamashina, 1998). Since the whole process of protein biosynthesis and cellular transport of secretory proteins takes at least 20–40 min, it can be inferred that the immediate effects (i.e., within minutes) of a stressor or other neuroendocrine stimulus on protein secretion represent the release of preformed protein, whereas later effects (i.e., over hours or even days) may involve, in addition, protein synthesis and posttranslational modification (Kousvelari *et al.,* 1988).

Salivary gland function is largely under autonomic control; the parasympathetic nerves mainly govern salivary fluid secretion, whereas the sympathetic nerves regulate protein secretion (Garrett, 1987). However, there are nuances to this general rule. First, the parasympathetic nerves also affect salivary protein secretion. Protein secretion of some glands, such as the sublingual and some of the minor glands, may even be entirely under parasympathetic control. Second, sympathetic activation also causes some stimulation of salivary flow rate (Garrett, 1987, 1999b). This implies that sympathetic activation during stress is not the cause of the well-known inhibition of salivary flow (Garrett, 1987, 1999b). We may add that rather than acting antagonistically, the two branches of the autonomic nervous system exert relatively independent effects in which the activity of one branch may synergistically augment the other (Emmelin, 1987).

A. Autonomic and Central Regulation of the Salivary Glands

Parasympathetic control of the submandibular and sublingual glands originates in the inferior salivatory nucleus, located in the pons. The axons of these neurons follow the facial nerve, where they directly, or indirectly via the submandibular ganglion, synapse in the glands. Parasympathetic control of the parotid gland is governed by the superior salivatory nucleus, situated in the caudal medulla. The efferent nerves enter the periphery via the glossopharyngeal nerve, and terminate on the parotid gland via the otic ganglion. While the precise location of the sympathetic salivary center has yet to be identified, it is known that the sympathetic preganglionic neurons in the upper thoracic nerves connect with the salivary glands (Brown, 1970; Matsuo, 1999).

The hypothesis that psychological processes modulate salivary gland function assumes that the salivatory brain stem centers receive inputs from higher neural structures. Two lines of evidence support this assumption: (1) studies that have stimulated specific brain areas and recorded changes in saliva (Wang, 1962; Brown, 1970), (2) and histological studies that have identified central connections by tracing the retrograde axonal transport of horseradish peroxidase (Matsuo, 1999). These studies show that the primary parasympathetic salivary centers receive direct inputs from the forebrain (e.g., paraventricular and lateral hypothalamus, central nucleus of the amygdala, bed nucleus of the stria terminalis) (Matsuo, 1999) and the lower brain centers (e.g., nucleus solitarius) (Wang, 1962; Matsuo, 1999). Besides governing salivary functions related to drinking and feeding behavior, these centers are also concerned with parasympathetic cardiac and visceral control, and they play a pivotal role in orchestrating neuroendocrine stress responses (Gray, 1987; Matsuo, 1999).

Less is known about central sympathetic salivary regulation, although the paraventricular and lateral hypothalamus, central gray matter, and rostral ventrolateral medulla have been implicated (Brown, 1970; Matsuo, 1999). The central gray matter is also an important relay station for coordinating fight–flight responses, whereas the rostral ventrolateral medulla is known for its role in the integration of cardiovascular and respiratory reflexes (Matsuo, 1999).

In sum, functional and histological studies have shown that the primary salivary centers in the brain stem receive inhibitory and excitatory inputs from neural structures in the forebrain and brain stem. As well as governing typical salivary functions, these structures are also involved in generating bodily changes associated with stress. It is therefore reasonable to assume that salivary changes during stress are an integral part of a centrally coordinated stress response that encompasses many other bodily functions (e.g., systemic immunity, cardiovascular activity, visceral functions).

B. LOCAL REGULATION

The convergence of synaptic inputs from many brain loci appears to permit different patterns of salivary responses. For example, in humans the salivary glands are capable of producing differentiated protein responses to different stressors (Bosch *et al.*, 2002). Matsuo (Matsuo *et al.*, 2000) showed that different patterns of protein secretion could be achieved within one gland with only a functional parasympathetic branch. This versatile regulation is made possible by the local presence of multiple autonomic messenger substances and autonomic receptors. These different autonomic transmitters are released at different rates and with different patterns of neuronal activation: acetylcholine and noradrenaline are released with every nerve impulse, whereas neuropeptides are released at higher frequencies of nerve stimulation (Ekstrom *et al.*, 1998; Ekstrom, 1999). Thus different neuronal stimulation patterns may cause a differential glandular protein release (Ekstrom, 1999; Garrett, 1999a,b).

Moreover, among the glands and cell types within glands there are marked differences in the density and patterning of receptors that are responsive to the messenger substances released by the autonomic nerves.

The autonomic receptors in the salivary glands can be divided into two main groups: the classic autonomic receptor types, which respond to either noradrenaline or acetylcholine; and the nonadrenergic–noncholinergic (NANC) receptors that respond to other autonomic messenger substances, such as peptides, nitric oxide, and purines. Differential activation of these receptor types can cause additive, synergistic, or antagonistic intracellular responses, ultimately resulting in a protein release that is capable of being differentially regulated both between and within glands.

C. NEUROENDOCRINE REGULATION OF S-IgA

Studies with rats have shown that stimulation of either autonomic branch innervating the salivary glands gives a rapid (within minutes) increase in the secretion of IgA into saliva (El-Mofty and Schneyer, 1978; Carpenter et al., 1998, 2000; Matsuo et al., 2000; Proctor et al., 2000a,b). Similar rapid increases in S-IgA secretion are observed in the gastrointestinal tract after stimulation of the local autonomic nerves (Schmidt et al., 1999). These autonomic effects may involve both IgA release by the plasma cells and epithelial translocation, as both processes can be enhanced by autonomic transmitter substances (Brink et al., 1994; Wilson et al., 1982; Freier et al., 1989; Kelleher et al., 1991; McGee et al., 1995; Schmidt et al., 1999). However, studies in rats show that autonomic stimulation of the salivary glands primarily affects the IgA translocation process, and has little, if any, effect on IgA release by B lymphocytes (Carpenter et al., 1998)

The exact transmitter substances that mediate the autonomic effects on S-IgA output are not yet known. Pharmacological studies suggest that parasympathetic stimulation of S-IgA secretion may involve the classic transmitter acetylcholine and various neuropeptides, whereas the same studies also suggest that the effects of sympathetic stimulation at least involve peptidergic transmitters (Wilson et al., 1982; Freier et al., 1989; Kelleher et al., 1991; McGee et al., 1995; Schmidt et al., 1999). It is still unclear whether the classic sympathetic transmitter noradrenaline is a major factor in the sympathetic stimulation of salivary S-IgA secretion. For example, the enhanced S-IgA output during moderate physical exercise, which is a strong sympathetic stimulus, is not attenuated by either α- or β-adrenergic blockade in humans (Winzer et al., 1999; Ring et al., 2000). Moreover, sympathetic S-IgA secretion by the submandibular salivary gland in rats is enhanced by high-frequency (>20 Hz) nerve stimulation but not by lower stimulation frequencies (Carpenter et al., 2000). Such high-frequency stimulation is associated with the release of neuropeptides (Ekstrom et al., 1998), whereas noradrenaline is released at all frequencies (see also Chapter 8 in this volume).

Whereas autonomic stimulation may rapidly affect S-IgA levels, slower effects have been reported for various hormones (e.g., steroids) and cytokines (Sullivan et al., 1983; Kelleher et al., 1991; Gleeson et al., 1995; Norderhaug et al., 1999), many of which are influenced by stress. In particular, the release of glucocorticoid hormones (cortisol in humans, cortisone in rodents) is considered central to the stress response. In rats, the synthetic glucocorticoid dexamethasone reduces both total salivary S-IgA and antigen-specific S-IgA levels, whereas dexamethasone increases serum IgA levels (Wira et al., 1990). These effects can be observed within 24–48 h. Paradoxically, the expression of SC, the molecule that transports IgA into the secretions, is also increased (Wira and Rossoll, 1991).

Most of the studies described earlier have been conducted in rats, a species in which several aspects of the S-IgA system are considerably different from that of humans. For example, in rats pIgR–IgA binding enhances S-IgA translocation through activation of a signal-transduction pathway, whereas this is not the case in humans (Norderhaug et al., 1999). As a result, in humans the enhancement of salivary S-IgA output is determined solely by an enhancement of S-IgA translocation, and cannot be caused by an increased IgA release from the plasma cells (since this does not affect translocation in humans). Moreover, in rats, polymeric IgA is cleared from serum through a specific, SC-dependent hepatic pump. However, this transporter mechanism, although present, exerts little activity in humans. As glucocorticoid treatment affects the activity of this hepatic pump (Wira and Colby, 1985; Wira et al., 1990), it is unclear to what extent the findings on IgA and dexamethasone generalize to humans. Thus, as far as the S-IgA system is concerned, animal findings should be interpreted with some caution in relation to human processes.

D. SUMMARY

The available data provide a sound biological basis for expecting a modulatory effect of stress on salivary secretory immunity. Salivary glands, as with other mucosal glands (Rogers, 1994; Finkbeiner, 1999; Shimura, 2000), are largely under autonomic nervous system control. The preganglionic autonomic centers in the brain stem that regulate salivary gland activity receive direct inhibitory and excitatory inputs from neural structures in the forebrain that are part of recognized "stress circuits" and centers for homeostatic regulation. The salivary glands form a highly sophisticated endpoint in the CNS control of local immune defenses, capable of responding instantly and with a high level of specificity to potential sources of harm (e.g., stress, inflammation). This remarkable ability, together with their strategic location at the portal of entry to the respiratory and gastrointestinal tract, make these glands ideally suited to provide the host with a first line of defense. However, the intricate complexity of multiple interacting factors (e.g., various central inputs, autonomic interactions, local variation in messengers and receptors) makes the effects of stress-related autonomic regulation difficult to predict *a priori*. Experimental research may therefore provide a useful vehicle for exploring this intricate system.

V. Stress and Secretory IgA

The impact of psychosocial factors on salivary S-IgA has been investigated extensively, and includes the effects of mood, relaxation, crying, humor, and stress. The literature on stress and S-IgA appears to be inconsistent (Stone *et al.*, 1987b; Kugler, 1991; Rood *et al.*, 1993; Gleeson *et al.*, 1995; Valdimarsdottir and Stone, 1997); some studies report that S-IgA decreases, whereas other studies, even when studying the same type of stressor (e.g., academic examinations; see later), report that S-IgA increases. In other areas of research within PNI it has repeatedly been noted that opposite results are obtained on the same parameters for acute and chronic stressors (e.g., Kusnecov *et al.*, 2001; Dhabhar and McEwen, 2001). Therefore, we adopted as a working hypothesis that the apparent inconsistencies in the literature may be resolved by discerning acute from chronic forms of stress.

An updated review of the literature was performed by interrogating PubMed (National Library of Medicine) and Web of Science (Institute for Scientific Information) databases, using the search terms "immunoglobulin A," "iga," "s-iga," or "siga." These search terms were combined with the terms "stress*,"[1] "mood," "psycho*," "emotion*," "anxi*," "depress*." In addition, the references in the articles that were generated by using the aforementioned search strategy were checked. Conference abstracts, dissertations, and book chapters were not considered here. As this review focused on the relation between stress and S-IgA, we did not include reports on positive moods, relaxation, and humor. Neither did we include literature on personality variables, with the exception of trait anxiety.

---

[1] The symbol * is a wild card, which is used to search databases for any word containing a particular string of letters. For example, the string "stress*" would, besides "stress," also yield documents containing the words "stressful," "stressed," etc.

TABLE I

ACADEMIC STRESS (SALIVA SAMPLES TAKEN DURING EXAM PERIOD)

N	Methodology and Design					Outcome[a]		References
	Stress measurement	Baseline measurement	Control for circadian effects?	Verification of stress response	Control group?	S-IgA concentration	S-IgA output	
64	During exam periods (3 ×)	Periods with low academic stress at start and end of academic year	Yes	Questionnaire	No	→	→	Jemmott et al., 1983
32	First day of exams	1 month before exams	Yes	Questionnaire	No	↔	—	Kiecolt-Glaser et al., 1984
15	Day of third exam	5 days before exams and 2 weeks after exams	Yes	Questionnaire	No	→	→	Jemmott and Magloire, 1988
44	During exam periods (2 ×)	After summer and midterm break	Yes	Questionnaire	No	→	↔	Mouton et al., 1989
57	During exam period	4 weeks before exams	Yes	Questionnaire	No	→	—	Li et al., 1997
38	Daily sampling, immediately after awakening, during two consecutive exam weeks	4 weeks before exams	No	Questionnaire	No	↓ During exam weeks, up to 6 days after last exam	↓ During exam weeks, up to 6 days after last exam	Deinzer and Schuller, 1998
27 experimental 27 control	Daily sampling, immediately after awakening, during exam week and following 2 weeks	4 weeks before exams	No	Questionnaire	Yes	↔During exam week ↓ 2 weeks postexam	↔ During exam week ↔ Postexam	Deinzer et al., 2000

[a] ↓, Decreases; ↔, no change; —, not determined.

A. The Academic Puzzle: Academic Stress or Examination Stress?

A majority of studies concerned with the association between stress and S-IgA have examined the impact of academic stressors. Their results paint a confusing picture; aside from occasional null findings, approximately one-half reported increased S-IgA and the other half reported decreased S-IgA. A number of reasons may account for such inconsistencies. For example, there are substantial differences among academic systems. In some institutions where testing is spread over the academic year, examinations are routine events, whereas others have final examination periods lasting between one week and one month when all courses are examined. Further, academic stressors can combine acute (e.g., just before an examination) and chronic (e.g., during the review period) characteristics.

The current review of the literature takes account of the distinction between short-term and long-term features of academic stressors. Accordingly, Tables I and II categorize academic stress studies based on whether, as stated in their methods, saliva samples were collected *minutes before, during,* or *after* a single examination (referred to as acute *examination* stress) or sometime during the extended examination period (referred to as chronic *academic* stress). Sorting the studies according to this single criterion reveals a remarkably consistent picture: all acute examination stress studies were associated with *increased* S-IgA (Table II), whereas most chronic academic stress studies were associated with *decreased* S-IgA (Table I).

To further explore the proposition that apparently conflicting S-IgA findings can be reconciled by simply distinguishing between acute and chronic stress exposure, the results yielded by other stress paradigms are presented in Tables III, IV, and V. Because timing is a critical factor, special emphasis is given to temporal features, such as when baseline and stress samples were collected and the duration of the stressor. Salivary S-IgA exhibits a diurnal rhythm, with a peak in the morning hours (Hucklebridge *et al.*, 1998; Park and Tokura, 1999), and therefore, the extent to which studies controlled for time-of-day was noted (e.g., whether the time of saliva sampling was held constant). Also, independent verification of a stress response (e.g., by self-report or physiological activity) was determined for each study. Finally, because changes in S-IgA concentration may be secondary to changes in salivary flow rate, both the results on S-IgA concentration (μg/ml) and S-IgA output (μg/min) are listed[2] (not all studies, however, reported both

---

[2] Other methodological issues specific to salivary research, including the appropriate methods for collection and handling of saliva, are discussed elsewhere (Navazesh and Christensen, 1982; Stone *et al.*, 1987b; Jemmott and McClelland, 1989; Navazesh, 1993; Rudney, 1995; Veerman *et al.*, 1996; Brandtzaeg, 1998).

TABLE II

EXAMINATION STRESS (SAMPLES TAKEN MINUTES BEFORE, AFTER, OR DURING AN EXAMINATION)

| | | Methodology and Design | | | | Outcome[a] | | |
N	Stress measurement	Baseline measurement	Control for circadian effects?	Verification of stress response	Control group?	S-IgA concentration	S-IgA output	References
42	Immediately after and $1\frac{3}{4}$ hours after exam	Days after exam (unspecified)	No	Norepinephrine	No	↑ ($p < 10$)	—	McClelland et al., 1985
7	Immediately after oral presentation	1 week before and day of presentation	Yes	Questionnaire Cortisol	No	—	↑ ($p < 10$)	Evans et al., 1994
28	30 min before exam	2 and 6 weeks after exam	Yes	Questionnaire	No	↑	↑	Bosch et al., 1996, 1998
16	30 min before, during, and 30 min after oral presentation	Same day, before stress-measurement (unspecified)	No	Questionnaire	No	↑ Before and following presentation	—	Bristow et al., 1997
22	15 min before, and 5 and 15 min after exam	4 weeks before exam	Yes	Questionnaire Cardiovascular Cortisol	No	↑ Before and after exam	—	Spangler, 1997
29	Immediately before and immediately after exam	4 weeks after exam	Yes	Questionnaire Cardiovascular Cortisol	No	↑ During all measurements	↑ During all measurements	Huwe et al., 1998

[a] ↑, Increases; —, not determined.

parameters). In the sections that follow, all the tabulated studies will be discussed with special emphasis on differences in the effects of chronic (which includes academic stress) versus acute stress (which includes examination stress).

B. S-IgA AND CHRONIC STRESS

Chronic stress has been shown to lower the levels of many blood-based functional and enumerative immune measures, including antigen-specific antibodies (i.e., immunoglobulins) following vaccination (Cohen *et al.*, 2001), and, accordingly, it may be hypothesized that chronic stress would also reduce S-IgA. The literature concerning the effects of long-term stress on S-IgA can be divided into studies of academic stress and self-reported stress exposure.

1. *Academic Stress*

Of the studies that measured salivary S-IgA during prolonged academic examination periods, two reported null findings and five reported a decrease in S-IgA (see Table I). However, the studies that reported positive effects suffer from methodological shortcomings; two of the studies collected samples over a period of up to 2 years (Jemmott *et al.*, 1983; Mouton *et al.*, 1989) and thus the stress-associated differences may reflect circannual effects; and two other studies only measured S-IgA concentration, and therefore the stress-associated differences may reflect the effects of stress on salivary flow (Kiecolt-Glaser *et al.*, 1984; Li *et al.*, 1997). A limitation common to all studies is that little information is provided on whether, or how, they controlled for variations in health behaviors, such as exercise, sleep, diet, alcohol, and smoking, that may affect the immune system. It is also worth noting that the only study to include a control group (Deinzer *et al.*, 2000) reported a null finding (see the following for further discussion). Thus, although the balance of evidence suggests that periods of prolonged academic stress are associated with reduced S-IgA, methodological weaknesses constrain any inferences about underlying mechanisms.

Two studies by Deinzer and colleagues (Deinzer and Schuller, 1998; Deinzer *et al.*, 2000) suggested that the effects of a protracted academic stressor may last for days, or even weeks, beyond the formal examination period. In the first study (Deinzer and Schuller, 1998), S-IgA levels, which were reduced during two consecutive examination weeks compared to initial baseline, had not returned to prestress baseline values up to 6 days after the last examination (see Table I). The second study (Deinzer *et al.*,

TABLE III
Questionnaire Studies

N	Methodology and Design		Number of measurements	Outcome		References
	Stress measures	Comments		S-IgA concentration	S-IgA output	
27	Trait anxiety	Undergraduates were preselected on personality trait "high in need of power"	1	Null finding	—	McClelland et al., 1980
29	Daily hassles checklist	Undergraduates	1	Null finding	—	Kubitz et al., 1986
113	Daily hassles checklist Life events Neuroticism/distress questionnaire Frequency of depressive episodes Frequency of anxious episodes	Female nurses	1	Lower levels with higher frequency of anxious episodes No relations with other questionnaires	—	Graham et al., 1988
40	Daily hassles checklist	University students	two, 6 weeks apart	Daily hassles correlated −.32 with S-IgA 6 weeks later	—	Martin et al., 1988

24	Daily hassles checklist	University students	three, 1 month apart	Increase in daily hassles between measurement 2–3 correlate with S-IgA decrease over same period (−.46)	—	Farne *et al.*, 1992
12	Daily hassles checklist Daily uplifts checklist Mood checklist	University students Net desirable events were computed as desirable events minus undesirable events	4 to 11 per subject (mean = 8)	Net desirable events $r = .56$	Undesirable events $r = -.59$ Net desirable events $r = .58$	Evans *et al.*, 1993
48	One item stress scale One item mood scale	Two equal groups of young (20–30 years) and older (60–80) adults	14 (seven consecutive days, twice a day)	—	Stress $r = .26$	Miletic *et al.*, 1996
124	Perceived stress over past month (SAS) Currently perceived stress (LPS) Job stress	Female nurses Much incomplete data (35%)	1	—	$r = .18$ with SAS $r = .25$ with LPS ($N=92$) Lower levels in high-stress hospital units	Ng *et al.*, 1999

[a] —, Not determined.

2000) found no difference in S-IgA concentration or output during the examination period between the experimental and control groups. However, a significant group difference in S-IgA concentration, but not in S-IgA output, was observed 1 and 2 weeks after the examinations. It should be noted, however, that the time-by-group-interaction effects were not significant, indicating that changes in the experimental group were identical to those in the control group. A methodological concern with both studies is that participants collected their own saliva samples immediately after awakening when S-IgA levels are most variable (Hucklebridge *et al.,* 1998). Since the wake-up time was not recorded, it is possible that variations between the two groups in awakening during the study may account for the postexamination differences, if any, between the experimental and control groups. Clearly, further research is needed to substantiate these reports.

2. *S-IgA and Self-reported Chronic Stress*

A number of studies have investigated chronic stress using self-report inventories to determine frequency of exposure to major life events (e.g., death of a spouse, divorce, loss of job) and minor daily hassles (e.g., minor conflicts and annoyances at home or work) and to assess the perceived impact of these exposures. Prior research has established that individuals scoring highly on such questionnaires exhibit increased susceptibility to a variety of infectious and inflammatory conditions, including respiratory infections (Cohen and Herbert, 1996) and oral diseases (da Silva *et al.,* 1995). The evidence provided by questionnaire-based studies, summarized in Table III, tends to support the hypothesis that higher levels of stress exposure and perceived stress are associated with lower levels of salivary S-IgA. However, many of the studies suffer from methodological weaknesses that temper this interpretation. For example, the two studies that reported null findings used sample sizes that were probably insufficient for capturing the intended effect (McClelland *et al.,* 1980; Kubitz *et al.,* 1986); other researchers solved the issue of small sample size by repeated sampling of the same subjects (e.g., Evans *et al.,* 1993; Miletic *et al.,* 1996). In addition, several studies measured stress using instruments with doubtful (e.g., single-item questionnaires) or unknown psychometric properties (Miletic *et al.,* 1996; Ng *et al.,* 1999) or used questionnaire data that were collected months before (Graham *et al.,* 1988). Moreover, none controlled for potential sources of confounding, such as health behaviors, periodontal health, or medication usage (see Evans *et al.,* 2000). Thus, although the data agree with the hypothesis that chronic stress lowers salivary S-IgA, the literature would be improved by large-scale, well-controlled studies to assess the impact of chronic stress on salivary S-IgA.

C. S-IgA AND ACUTE STRESS

1. *Examination Stress and Other Naturalistic Stressors*

Two examination studies (McClelland *et al.*, 1985; Evans *et al.*, 1994), summarized in Table II, provided preliminary indications that acute episodes of stress may be associated with *increases* rather than decreases in S-IgA. However, the S-IgA increases failed to reach conventional statistical significance, and there were also methodological weaknesses, such as no control for salivary flow rate and circadian effects (McClelland *et al.*, 1985) and a very small sample size (Evans *et al.*, 1994). Moreover, both studies collected saliva immediately *after* the stressor, and therefore, the nonsignificant increases may have resulted from experiencing the relief from stress rather than the stress itself. However, four studies were published in 1996 that confirmed these early observations (Bosch *et al.*, 1996; Carroll *et al.*, 1996; Kugler *et al.*, 1996b; Zeier *et al.*, 1996), and they established that acute stress can produce increases in salivary S-IgA concentration and secretion rate: the three that involved naturalistic stressors (Bosch *et al.*, 1996; Kugler *et al.*, 1996b; Zeier *et al.*, 1996) are discussed in this section (see Tables II and IV), and the one that employed a laboratory stressor (Carroll *et al.*, 1996) is considered in the next section (see Table V). Bosch *et al.* (1996) examined 28 undergraduates just *before* an academic examination, and again 2 and 6 weeks later at times of low academic pressure. The increases in S-IgA concentration and output observed before the examination indicated that stress rather than relief was responsible for increased S-IgA levels. Examination stress was also associated with increases in other salivary proteins, such as α-amylase, suggesting that the mechanism behind the increased salivary S-IgA involved greater activity of the salivary glandular cells. Subsequent examination studies (Bristow *et al.*, 1997; Spangler, 1997; Huwe *et al.*, 1998) confirmed that acute examination stress increases salivary S-IgA.

The naturalistic stressor studies are summarized in Table IV. Zeier *et al.* (1996) reported that both S-IgA concentration and output were elevated in a group of 126 air traffic controllers after a work shift and that no S-IgA changes were seen in a group of 10 control participants. It is noted that samples were taken *after* the work sessions (see previous comments). Kugler *et al.* (1996b) measured S-IgA in 17 coaches before, during, and after a football match and found that S-IgA concentration was elevated during the match and at halftime. S-IgA did not change in a control group. Although the S-IgA output data were not presented, salivary flow rate did not change during the match, and therefore, the increases in S-IgA concentration elicited by the competition stress was unlikely to be an artifact or reduced saliva production. It is also worth noting that the acute increases were transient; S-IgA had returned to baseline 1 h after the final whistle.

TABLE IV

S-IgA AND ACUTE NATURALISTIC STRESSORS

	Methodology and Design				Outcome[a]		
N	Stressor	Duration of stressor	Timing of stress measurements	Verification of stress response	S-IgA concentration	S-IgA output	References
126 10 control	Work shift	100 min	After work shift	Questionnaire Cortisol	↑	↑	Zeier et al., 1996
17 8 control	Coaches during a match	120 min	Various time points during match	Questionnaire Cortisol	↑During match and halftime break	↑During match and halftime break	Kugler et al., 1996a
17 17 control	Dental treatment	20 min	During and after treatment	Questionnaire Cortisol	↓During and 10 min after	↓ During and 10 min after	Kugler et al., 1996b

[a] ↑, Increases; ↓, decreases.

2. *Laboratory Stressors*

Further support for the hypothesis that acute stress increases salivary S-IgA was provided by a laboratory study by Carroll and co-workers (Carroll *et al.*, 1996) who found an increase in salivary S-IgA output during a 30-min computer game. Subsequent laboratory studies have replicated this finding of increased S-IgA. Studies using active, effortful coping challenges, such as mental arithmetic and time-paced memory tests, that elicit sympathetic nervous system activation have, with a few exceptions, found increases in S-IgA concentration, although they have been somewhat less consistent regarding increases in S-IgA output (see Table V).

Inconsistencies concerning S-IgA output probably result from differences among studies in both timing of saliva collection and sample size. All studies in which the stress sample was taken *during* the stressor reported a similar modest increase in S-IgA output of approximately 25%. With this effect size, however, only the larger more powerful studies generated statistically significant increases in S-IgA (e.g., Carroll *et al.*, 1996; Willemsen *et al.*, 2000; Bosch *et al.*, 2001). On the other hand, studies in which the stress sample was taken immediately after the stress task generally show a larger increase in S-IgA ouput, which resulted in statistically significant effects irrespective of sample size (e.g., Willemsen *et al.*, 1998; Ohira *et al.*, 1999; Harrison *et al.*, 2000). Support for this interpretation came from a study by Bosch *et al.* (2002b), in which the S-IgA response to a memory test took some time to develop and reached its highest level *immediately after* the stressor. It is again unclear to what extent the increases measured immediately after a stressful task reflect stress or relief, such as parasympathetic rebound (e.g., see Mezzacappa *et al.*, 2001).

Another potential source of discrepant findings is unintentional stimulation of salivary flow rate: In two studies participants had to recall numbers aloud as part of the experimental task (Willemsen *et al.*, 1998, 2000). It is possible that this procedure, in combination with simultaneously having cotton rolls in the mouth as collecting devices, stimulated salivary flow (Navazesh and Christensen, 1982). Stimulation of salivary flow rate reduces S-IgA concentration but enhances S-IgA output (Brandtzaeg, 1998; Proctor and Carpenter, 2001). One of these studies indeed found an increase in S-IgA output (and saliva secretion) while S-IgA concentration did not change (Willemsen *et al.*, 1998).

In sum, research on laboratory stressors generally confirms the findings of other acute-stress studies by showing an increase in salivary S-IgA. These observations are in line with the finding that activation of the sympathetic nerves enhances the secretion of S-IgA into saliva (Carpenter *et al.*, 1998, 2000). Studies tended to find smaller S-IgA effects when samples were collected during rather than immediately after a laboratory stress task. As S-IgA

TABLE V

LABORATORY STRESSORS

		Methodology and Design				Outcome[a]		
N	Stressor	Duration (min)	Timing of stress saliva sample	Verification of stress, response	Control condition	S-IgA concentration	S-IgA output	References
28	Computer game	30	During ($t_{min} = 6$ and 24) and after ($t_{min} = 18$)	Cardiovascular	—	—	↑During game at 24 min	Carroll et al., 1996
22	Memory test	10	5 and 15 min after	Cardiovascular Cortisol Questionnaire	—	↑ 5 and 15 min after	—	Spangler, 1997
16	Mental arithmetic Cold pressor (10°C)	9 4	After (speaking) After	Cardiovascular	—	MA ↔ CP↑	MA↑ CP↑	Willemsen et al., 1998
20	Mental arithmetic	8	During	Cardiovascular	—	↑	↔	Ring et al., 1999
17	Mental arithmetic Cold pressor (10°C)	8 8	During	Cardiovascular	—	MA↑ CP↔	MA↔ CP↔	Winzer et al., 1999 (reanalysis of placebo-only data)
38	Mental arithmetic	30	After	Cardiovascular Questionnaire	—	↑	↑	Ohira et al., 1999
21	Mental arithmetic Cold pressor (10°C)	8 8	During	Cardiovascular	—	MA↑ CP↓	MA↑ CP↓	Ring et al., 2000 (reanalysis of placebo-only data)
29	Video of national team losing penalty shootout	10	After	Cardiovascular Questionnaire	Didactic video	↑	↑	Harrison et al., 2000

#	Task	n	Timing	Measure				Reference
27	Mental arithmetic	8	During (while speaking)	Cardiovascular	Questionnaire	—	↑	↑	Willemsen et al., 2000
30	Memory test / Surgical video	11 / 11	During and $t_{min} = 9$ after	Cardiovascular	Questionnaire	Didactic video	MT:↑S-IgA ↑IgA1 ↑IgA2 ↑SC SV:↓S-IgA ↓IgA1 ↔IgA2 ↔SC, at $t_{min} = 9$ after	MT:↑S-IgA ↑IgA1 ↑IgA2 ↑SC SV: ↓S-IgA ↓IgA1 ↔ IgA2 ↔ SC, at $t_{min} = 9$ after	Bosch et al., 2001
30	Memory test	8	During ($t_{min} = 3$ and 7) and after ($t_{min} = 0, 4, 8$)	Cardiovascular	Questionnaire	—	Biphasic response during MT, (during task ↔ at $t_{min} = 3$; ↑ at t_{min} 7; after task ↑ $t_{min} = 0$)	Biphasic response during MT (during task ↓ at $t_{min} = 3$; after task ↑ $t_{min} = 0$)	Bosch et al., 2003
127	Mental arithmetic / Cold pressor (10°C)	8 / 4	During	Cardiovascular	—	MA↑ CP↓	MA↑ (men only) CP↓	Willemsen et al., 2002
24	Mental arithmetic	28	After 14 min and after 28 min	Cardiovascular	—	↑ After 14 and 28 min	↑ After 14 and 28 min	Ring et al., 2002

[a]MA, Mental arithmetic; CP, cold pressor; MT, memory test; SV, surgical video; ↑, increases; ↓, decreases; ↔, no change; —, not determined.

secretion is regulated by both branches of the autonomic nervous system, it is conceivable that a parasympathetic withdrawal during acute stress may cause attenuation of sympathetically mediated increases in S-IgA or that a parasympathetic rebound immediately after a stressor augments S-IgA secretion. Finally, differences in the method of saliva collection may also account for some variation in the results.

D. ACUTE STRESS THAT DECREASES S-IgA

It is probable that every topic in science will have articles reporting anomalous results, and it is not always immediately clear whether such reports may lead to new insights or whether they just represent experimental artifacts that should be ignored. This is also the case for the literature on stress and S-IgA. To date, at least 19 experiments have been published that report an increase in salivary S-IgA in response to acute stress; however, there are also five studies that report acute decreases in S-IgA.

Kugler and co-workers (Kugler *et al.*, 1996a) observed a decrease in both S-IgA concentration and output during and directly after a caries excavation in patients who were not anesthetized; no changes were seen during a control session (see Table IV). Ring *et al.* (2000) and Willemsen *et al.* (2002) found that cold pressor also caused a rapid decrease in S-IgA, although other such studies have not observed this effect (see Table V). The similarity with the study by Kugler *et al.* (1996a) may be that both tasks are painful and engender passive coping with the stressor. Other passive stressors, such as watching unpleasant scenes were also found to inhibit S-IgA secretion (Hennig *et al.*, 1996; Bosch *et al.*, 2001). In these studies participants viewed video clips selected to induce feelings of disgust (Hennig *et al.* 1996) or scenes of oral surgery (Bosch *et al.*, 2001). The common denominator in these two studies might be a state of disgust. In addition, the latter study found that the decreased S-IgA was accompanied by a cardiac passive-coping response, characterized by sympathetic–parasympathetic coactivation (Bosch *et al.*, 2001). It is possible that a classification of stressors on the basis of their autonomic nervous system effects might resolve the contradictory acute-stress-induced changes in S-IgA (cf., Ring *et al.*, 1999).

One study investigated whether hedonic tone per se might be a determinant of the direction of the acute changes in S-IgA (Hucklebridge *et al.*, 2000). No such discrimination was found; both positive and negative emotional states (induced by imagery and by listening to music tapes) induce acute rises in S-IgA. This complies with the general finding in the literature; apart from the five studies mentioned earlier, to the best of our knowledge virtually all types of acute states (e.g., stress, relaxation, moods, and motivational states) and virtually all experimental manipulations (e.g., viewing videos, listening to music, imagery, performing effortful tasks) similarly cause acute and transient rises in salivary S-IgA.

One study found that the S-IgA response to acute stress can follow a biphasic pattern starting with an acute decrease in S-IgA, which, again, demonstrates that the timing of measurements is critical in stress research. Using an 8-min active-coping stress task, Bosch *et al.* (2002b) found that S-IgA secretion was significantly reduced (-30%) within the first 3 min of the time-paced memory test, followed by a steady increase that reached its peak ($+60\%$) immediately after the stressor (see Table V).

In sum, although the majority of acute-stress studies report S-IgA increases, rapid decreases may occur under some conditions. These conditions include passive-coping stressors, and it is tentatively proposed that the critical determinants are specific emotionally charged states, namely pain (e.g., dental treatment, cold pressor) and disgust (e.g., videos showing surgery and other unpleasant scenes). Preliminary evidence suggests that acute decreases in S-IgA can also be observed at the onset of an active-coping stressor as part of a rapid biphasic response pattern.

E. STRESS AND ORAL IMMUNIZATION

The literature discussed thus far concerned the effects of stress on total S-IgA levels. One group studied that effects of psychosocial stress on the daily variation in specific S-IgA (i.e., S-IgA that is directed to a specific antigen). Stone and colleagues (1987a, 1994) required participants to swallow a capsule containing rabbit albumin each morning for 8–12 weeks, and then they measured the daily variations in specific S-IgA (i.e., S-IgA directed against the rabbit protein) in saliva, as well as in daily mood (Stone *et al.,* 1987a, 1994) and events (Stone *et al.,* 1994). This methodology allowed within-subject comparisons to be made between antibody levels on high-stress and low-stress days. Antigen-specific S-IgA antibody levels were lower on days with relatively high negative mood and were elevated on days with relatively high positive mood (Stone *et al.,* 1987a). Subsequent research demonstrated a positive association between daily positive events and levels of antigen-specific S-IgA and, conversely, a negative association between negative events and S-IgA (Stone *et al.,* 1994). A reanalysis of this data showed that undesirable events lowered antibody level primarily by increasing negative mood, and that desirable events increased antibody levels by decreasing negative mood. After controlling for negative mood, positive moods appeared to explain little of the variance in daily S-IgA levels (Stone *et al.,* 1996).

At the time these studies were performed, oral immunization still was a developing field, which may explain the remarkable choice for rabbit albumin as an oral antigen; typically such proteins are recognized by the mucosal immune system as "food," and for obvious reasons it is very difficult to mount an immune response against such molecules. Today there are several potent protocols available that researchers may prefer to use instead (Czerkinsky *et al.*, 1999). Considering the pivotal role of the secretory immune defenses in determining whether infection will occur, oral immunization provides an important *in vivo* model to study the effects of psychosocial factors on immune function.

F. Effects on S-IgA Translocation and Subclasses

An unresolved issue is whether the effects of stress on S-IgA secretion reflect alterations in epithelial IgA translocation or alterations in the IgA levels at the basolateral epithelial surface. The latter might, for example, be caused by changes in the secretory activity of the plasma cells or by a drain of plasma cells to the circulation. Our results (Bosch *et al.*, 2001) lead us to conclude that both steps in S-IgA secretion can be involved. That is, the decrease observed during the passive-coping stressor probably reflected decreases in the availability of basolateral IgA, as S-IgA levels decreased in spite of an increased secretion of the transporter molecule SC. Conversely, the acute increase during the active-coping memory test may reflect an increase in IgA translocation. Support for this conclusion comes from studies showing that increased IgA near the basolateral surface of the epithelium does not affect the rate in which IgA is translocated, so increases in S-IgA cannot be caused by an increase in basolateral IgA (Norderhaug *et al.*, 1999).

The results of the same study (Bosch *et al.*, 2001) also showed that the rapid stress-induced changes in salivary S-IgA mainly reflect increases in the IgA1 subclass. A comparable finding has been reported for IgA subclass levels in breast milk in response to physical exercise (Gregory *et al.*, 1997). These findings are surprising given that the transporter molecule SC shows no difference in affinity for the two IgA subclasses (Norderhaug *et al.*, 1999). Perhaps another transporter mechanism, besides SC-mediated translocation, is involved that exhibits a preference for a specific subclass. Alternative IgA receptors with subclass preference have been identified (Schiff *et al.*, 1986; Rifai *et al.*, 2000) and are involved, at least in rodents, in the translocation of S-IgA from the blood into bile, and possibly also into gastrointestinal mucus (Shimada *et al.*, 1999). It is also possible that the selective effects of stress on IgA subclass secretion might reflect a differential responsivity of subclass-specific plasma cells to autonomic messenger substances, causing an altered immunoglobulin release or a selective drain of cells into the circulation. Clearly, these speculations require investigation.

G. Conclusion

In sum, chronic stress is generally associated with decreases in S-IgA, whereas acute stressors mostly increase S-IgA. Although there are no clear guidelines to decide when acute stress ends and chronic stress begins, a comparison of Table I with Table II shows that a distinction between acute and chronic stress represents a useful heuristic. Such a distinction between the effects of brief and prolonged stressors is helpful in resolving some of the confusion in this literature.

Decreased S-IgA levels have been observed in students during prolonged examination periods and in individuals reporting high levels of life-event stress. However, the chronic stress literature is hampered by methodological shortcomings, mostly related to inadequate control for possible confounding factors and small sample sizes. In addition, there is limited experimental research to provide a biological rationale (e.g., neuroendocrine pathways) for the observed decreases.

Acute-stress studies, which have been performed under more controlled conditions, have observed increases in S-IgA. Such findings are compatible with research showing that autonomic nerve stimulation enhances S-IgA output by the salivary glands. These acute increases most probably reflect the effects of stress on S-IgA transcytosis. Finally, additional studies show that S-IgA may also rapidly decrease under some conditions, including specific emotionally charged states (e.g., pain, disgust), and during the initial stages of a stress task. Such rapid decreases may involve effects on the IgA-secreting plasma cells, although possible neuroendocrine mechanisms accounting for such rapid decreases have yet to be elucidated.

One finding is that the secretion of the IgA1 subclass is more variable under acute conditions than IgA2, and that during stress the secretion of S-IgA and its transporter molecule SC can be affected independently (Bosch *et al.*, 2001). Finally, immunization studies have shown that stress also affects the secretion of antigen-specific S-IgA (as opposed to total S-IgA, which is measured in most studies), and thus it is capable of modulating immunological processes such as antigen presentation and homing of B cells to the salivary glands. As immunization provides a means for testing immunocompetence in response to novel antigens, this aspect of secretory immunity deserves more attention in future research.

VI. Stress and Innate Secretory Immunity

The introduction to this review pointed out that S-IgA is only one of many protective proteins secreted in the fluids covering the mucosa. This section summarizes the scant literature on the effects of stress on these antimicrobial proteins that comprise innate secretory immunity.

A. STRESS AND MUCINS

Two human studies have shown that acute stressors (a time-paced memory test and a gruesome surgical video presentation) enhance the secretion in saliva of the mucins MUC5B and MUC7 (Bosch *et al.*, 2000, 2002). These findings agree with evidence of increased gastrointestinal and respiratory mucin secretion in response to water-immersion and restraint stress in animals. These animal experiments also found an effect of stress on other biological processes, such as mucin synthesis and posttranslational processing (e.g., sulfatation, glycosylation) (discussed by Bosch *et al.*, 2000, and Soderholm and Perdue, 2001). Administration of physiological doses of β-adrenergic agonists has also been observed to influence salivary protein synthesis and glycosylation in rats (Kousvelari *et al.*, 1988).

Unlike salivary mucin secretion, colonic mucin secretion during acute stress is not initiated by direct neuroendocrine stimulation of the mucin-secreting goblet cells. Castagliuolo *et al.* (1998) demonstrated a stress-induced (by 30 min of restraint) colonic mucin secretion in normal but not in mast cell–deficient mice, indicating that mast cells regulate colonic mucin release in response to restraint stress. However, local neuroendocrine transmitters (e.g., CRH, neurotensin) were responsible for inducing the release of mast cell factors that, in turn, caused the increased mucin secretion. Although increases in mucin secretion are generally interpreted as reflecting an enhanced mucosal barrier function (Soderholm and Perdue, 2001), the exact immunological meaning of increased mucin secretion may depend on various factors, such as the specific microorganism under consideration. For instance, *Helicobacter pylori* specifically bind certain mucin types, which may contribute to the potential of this bacterium to colonize the host (Nieuw Amerongen *et al.*, 1998; Van den Brink *et al.*, 2000). A stress-induced increase in secretion of MUC5B enhances the adherence (*ex vivo*) of *H. pylori* (Bosch *et al.*, 2000). Thus, increases in salivary mucin may have distinct consequences; some that benefit the host and some that benefit microorganisms.

Repeated or protracted stimulation of mucin release can lead to a reduction in mucin release, possibly by depleting the mucin-secreting goblet cells (Castagliuolo *et al.*, 1996). Qiu and co-workers observed that protracted stress decreases colonic mucin secretion in rats. This decrease in mucin secretion enhanced mucosal permeability, thereby potentiating reactivation of experimental colitis by a chemical irritant (Qiu *et al.*, 1999). This study shows how stress-induced impairments in innate secretory immunity may directly contribute to pathophysiology.

B. STRESS AND α-AMYLASE

Morse and co-workers reported that α-amylase increases during relaxation and decreases with acute stress (for review see Morse *et al.*, 1983) and suggested that these changes were mediated by variations in parasympathetic activity. However, a null finding has been reported (Borgeat *et al.*, 1984), and other studies have shown that acute stressors can increase salivary α-amylase (Bosch *et al.*, 1996, 1998, 2002; Chatterton *et al.*, 1996, 1997). These increases appear to be mediated by sympathetic nerve activation as they were positively correlated with serum norepinephrine (Chatterton *et al.*, 1996). This interpretation is further supported by studies showing that α-amylase secretion increases with sympathetic activation by administration of adrenergic agonists (Mandel *et al.*, 1975), by direct sympathetic nerve stimulation (Garrett, 1987; Schneyer and Hall, 1991), and by physical exercise (Chicharro *et al.*, 1998). Nonetheless, claims that salivary α-amylase is a valid noninvasive measure of adrenergic activity (e.g., Skosnik *et al.*, 2000) should be regarded with caution. First, α-amylase is also secreted in response to nonadrenergic sympathetic transmitters, such as neuropeptides (Ekstrom, 1999). Second, α-amylase secretion is also stimulated by parasympathetic stimulation, either alone or in interaction with sympathetic stimulation (Emmelin, 1987; Asking and Proctor, 1989; Ekstrom *et al.*, 1998). It should also be noted that the immunological relevance of α-amylase is still controversial.

C. STRESS AND OTHER MUCOSAL SECRETORY PROTEINS

In humans, Bosch *et al.* (2002) observed increased secretion of salivary cystatin S and lactoferrin in response to a stressor that exhibited a sympathetic–parasympathetic coactivation (i.e., viewing a surgery video), but not in response to a stressor that evoked a sympathetic activation in conjunction with a vagal withdrawal (a time-paced memory test). The former stressor also produced the largest response for most other innate secretory proteins mentioned earlier (i.e., the mucins MUC5B and MUC7, and

α-amylase). This finding is consistent with the literature showing a synergistic effect of the two branches of the autonomic nervous system on secretory gland activity (Emmelin, 1987; Garrett, 1999b); that is, although sympathetic activation is the main stimulus for glandular protein secretion, such sympathetic effects are strongly augmented by concurrent parasympathetic activity.

Perera and co-workers observed a decrease in salivary lysozyme secretion during a final examination period (Perera *et al.*, 1997). In a second study these authors found a significant negative correlation between salivary lysozyme and a self-report measure of stress and arousal in 39 university students (Perera *et al.*, 1997). However, in a study involving 124 nurses, Ng *et al.* (1999; see Table III) did not find an association between self-reports of work stress and salivary lysozyme secretion.

D. Conclusion

Human and animal research shows that innate secretory immunity is modulated by stress. Stress, as with other forms of neuroendocrine stimulation, clearly affects a broad range of biological processes, including protein synthesis, posttranslational modification, and secretion. Interestingly, available data on the susceptibility to infectious disease in stressed humans suggests a role for such innate secretory immune factors involved in *preventing* infection, rather than immune factors that are involved in *responding* to infection. For example, in a seminal study by Cohen *et al.* (1991), in which volunteers were exposed to cold viruses after a stress assessment, a strong relationship between stress and infection was observed. However, the association between stress and actual disease occurrence was much weaker. The authors concluded that "... the relation between stress and colds was primarily attributable to an increased rate of *infections* among subjects with higher stress-index scores, rather than to an increase in *clinical colds* among infected persons ... " (Cohen *et al.*, 1991, italics added). Preventing infection is the primary function of the innate secretory immune defenses, and the role of this aspect of immunity in mediating the association between stress and infectious disease deserves further research.

VII. Stress and Microbial Colonization Processes

A remaining, and critical, issue is whether the effects of stress on secretory immunity translate into functional changes, such as affecting processes that are involved in microbial colonization. Such effects might explain the alterations in mucosal microflora, seen in both humans and animals, under various forms of psychological strain, such as depression (Antilla *et al.*, 1999), space flight simulations (Brown *et al.*, 1973, 1974), maternal separation (Bailey and Coe, 1999), familial strains (Meyer and Haggerty, 1962), animal fighting and relocation (Marcotte and Lavoie, 1998), and life events (Drake *et al.*, 1995). These effects may also explain the close association between stress and oral diseases with a bacterial etiology, such as periodontal disease and dental caries (see Section I). Examination stress reduces the aggregation by saliva of the oral bacterium *Streptococcus gordonii* (Bosch *et al.*, 1996, 1998). A reduced streptococcal aggregation has also been observed in rats, after 1 week of administration of physiological doses of the β-adrenergic agonist isoproterenol (Kousvelari *et al.*, 1988). Aggregation is a process by which bacteria are clumped together through interactions with the host's secretory proteins, so that they cannot effectively adhere to the mucosal surfaces. A decreased bacterial aggregation thus indicates a decreased capacity for mucosal microbial clearance.

Two studies have investigated the effects of stress in humans on saliva-mediated microbial adherence. Microbial adherence is a process by which microbes gain a stable foothold on the host, and it forms a first and essential step in infection. These studies found that acute stress promoted the adhesion (*ex vivo*) of oral bacteria (viridans group streptococci) and nonoral bacteria (*Helicobacter pylori*) (Bosch *et al.*, 2000, 2002). Stress was also found to affect saliva-mediated microbial coadherence (i.e., the adhesion of microorganisms to each other) (Bosch *et al.*, 2002). Other studies have reported that large doses of glucocorticoids in animals were associated with both increased bacterial adherence to the intestinal mucosa and reduced luminal S-IgA (Alverdy, 1991, Spitz, 1994, 1996). Similarly, in rats, 1 week of physiological doses of the β-adrenergic agonist isoproterenol increased the salivary adherence (*ex vivo*) of *Streptococcus mutans* (Kousvelari *et al.*, 1998). The latter changes were related to an altered salivary protein synthesis and protein glycosylation.

VIII. Future Perspectives

The salivary literature reviewed here presents a myriad of effects of psychosocial stressors on secretory immunity, including altered secretion of adaptive and innate immune factors, reduced bacterial aggregation, increased bacterial adherence, and altered microbial coadherence. Clearly, one way to proceed is to study the effects of stress on secretory immunity at anatomical locations other than the oral cavity (i.e., directly where infection is initiated). The scant literature in this area, mostly performed in rodents, confirms and extends the salivary findings in humans; there are effects on protein secretion, synthesis, and posttranslational modification, as well as functional changes, such as enhanced bacterial adherence and increased intestinal permeability.

As well as emphasizing salivary immunity, the literature concerning studies of humans is also characterized by a disproportional focus on S-IgA. The fact that individuals deficient in S-IgA, the most common form of immunodeficiency, are, for the most part, in perfect health, suggests that other secretory factors may be important. With the predominance of innate factors in mucosal secretions, as well as the strong involvement of neuroendocrine systems in their regulation, further PNI research into this aspect of host defense is warranted.

This is by no means to imply that the field of S-IgA research is redundant. An exciting development is the increased insight in the autonomic mechanisms that underlie acute changes in S-IgA, largely obtained through animal studies (see Chapter 8 in this volume). However, this knowledge also poses new questions. For example, whereas experimental animal research seems to indicate that virtually any form of autonomic stimulation (e.g., pharmacological, electrical nerve stimulation, reflex secretion) essentially results in an enhancement of S-IgA secretion, the human stress literature indicates that some stress-induced patterns of autonomic stimulation may result in acute decreases in S-IgA secretion. As there are fundamental differences between humans and rodents in the regulation of S-IgA translocation, more rigorous experimental studies in humans may be needed as well.

More attention needs to be paid to the effects of protracted forms of stress on secretory immunity. Knowledge of the effects of chronic stress on innate secretory immunity is extremely limited, and the studies on chronic stress and S-IgA, although showing a fairly consistent picture, suffer from methodological shortcomings. Moreover, there is little evidence to provide this literature with a biological rationale. Although there is support for the *assumption* that glucocorticoids mediate decreases in S-IgA during protracted stress, this association has never been demonstrated in humans, rather, it is based on studies administering pharmacological doses of dexamethasone to rodents. A most promising model to study the effects of chronic stress on adaptive secretory immunity is mucosal immunization. Knowledge in this field has accumulated over the years, and is now ready for a more thorough exploration within PNI.

Finally, researchers may extend their horizons to include the study of functional implications of stress-induced changes in mucosal protein secretion. A number of microbiological methods exist to study the effects of immunosecretory changes on microbial colonization processes, including viral invasion, bacterial adherence, bacterial growth, and bacterial aggregation.

This approach, rarely considered in the realm of PNI, may reveal the mechanisms by which immunosecretory changes affect the way in which microorganisms are dealt with. Moreover, this methodology will help to strengthen the proposition that stress, via its effects on secretory immunity, has important health consequences.

References

Alverdy, J., and Aoys, E. (1991). The effect of glucocorticoid administration on bacterial translocation—Evidence for an acquired mucosal immunodeficient state. *Annals of Surgery* **214,** 719–723.

Antilla, S. S., Knuuttila, M. L., and Sakki, T. K. (1999). Depressive symptoms favor abundant growth of salivary lactobacilli. *Psychosom. Med.* **61,** 508–512.

Asking, B., and Proctor, G. B. (1989). Parasympathetic activation of amylase secretion in the intact and sympathetically denervated rat parotid gland. *Q. J. Exp. Physiol.* **74,** 45–52.

Bailey, M. T., and Coe, C. L. (1999). Maternal separation disrupts the integrity of the intestinal microflora in infant rhesus monkeys. *Dev. Psychobiol.* **35,** 146–155.

Baron, S., Singh, I., Chopra, A., Coppenhaver, D., and Pan, J. (2000). Innate antiviral defenses in body fluids and tissues. *Antiviral Res.* **48,** 71–89.

Beck, J. D., Kohout, F. J., Hunt, R. J., and Heckert, D. A. (1987). Root caries: Physical, medical and psychosocial correlates in an elderly population. *Gerodontics* **3,** 242–247.

Berczi, I., Chow, D. A., and Sabbadini, E. R. (1998). Neuroimmunoregulation and natural immunity. *Domest. Anim. Endocrinol.* **15,** 273–281.

Bevins, C. L. (1994). Antimicrobial peptides as agents of mucosal immunity. *Ciba Found. Symp.* **186,** 250–260.

Biesbrock, A. R., Reddy, M. S., and Levine, M. J. (1991). Interaction of a salivary mucin-secretory immunoglobulin A complex with mucosal pathogens. *Infect. Immun.* **59,** 3492–3497.

Blankenvoorde, M. J. F., van't Hof, W., Walgreen-Weterings, E., van Steenbergen, T. J. M., Brand, H. S., Veerman, E. C. I., and Nieuw Amerongen, A. V. (1998). Cystatin and cystatin-derived peptides have antibacterial activity against the pathogen *Porphyromonas gingivalis*. *Biol. Chem.* **379,** 1371–1375.

Bonneau, R. H., Padgett, D. A., and Sheridan, J. F. (2001). Psychoneuroimmune interactions in infectious disease: studies in animals. *In* "Psychoneuroimmunology" (R. Ader, D. L. Felten, and N. Cohen, eds.), Vol. 2, pp. 483–497. Academic Press, San Diego.

Borgeat, F., Chagon, G., and Legault, Y. (1984). Comparison of the salivary changes associated with a relaxing and with a stressful procedure. *Psychophysiology* **21,** 690–698.

Bosch, J. A., Brand, H. S., Ligtenberg, A. J. M., Bermond, B., Hoogstraten, J., and Nieuw Amerongen, A. V. (1996). Psychological stress as a determinant of protein levels and salivary-induced aggregation of *Streptococcus gordonii* in human whole saliva. *Psychosom. Med.* **58,** 374–382.

Bosch, J. A., Brand, H. S., Ligtenberg, A. J. M., Bermond, B., Hoogstraten, J., and Nieuw Amerongen, A. V. (1998). The response of salivary protein levels and S-IgA to an academic examination are associated with daily stress. *J. Psychophysiol.* **4,** 170–178.

Bosch, J. A., de Geus, E. J. C., Ligtenberg, A. J. M., Nazmi, K., Veerman, E. C. I., Hoogstraten, J., and Nieuw Amerongen, A. V. (2000). Salivary MUC5B-mediated adherence (*ex vivo*) of *Helicobacter pylori* during acute stress. *Psychosom. Med.* **62,** 40–49.

Bosch, J. A., de Geus, E. J. C., Kelder, A., Veerman, E. C. I., Hoogstraten, J., and Nieuw Amerongen, A. V. (2001). Differential effects of active versus passive coping on secretory immunity. *Psychophysiology* **38**, 836–846.

Bosch, J. A., de Geus, E. J. C., Veerman, E. C. I., Hoogstraten, J., and Nieuw Amerongen, A. V. (2002). Innate secretory immunity in response to laboratory stressors that evoke distinct patterns of cardiac autonomic activity. *Psychosom. Med.* In press.

Bosch, J. A., Willemsen, G. H., Knol, M., de Geus, E. J. C., and Nieuw Amerongen, A. V. (2002b). Acute stress and secretory immunity; timing is everything. Manuscript submitted for publication.

Bosch, J. A., de Geus, E. J. C., Turkenburg, M., Nazmi, K., Veerman, E. C. I., Hoogstraten, J., and Nieuw Amerongen, A. V. (2003). Stress as a determinant of the saliva-mediated adherence and co-adherence of oral and non-oral microorganisms. *Psychosom. Med.* Accepted for publication.

Brandtzaeg, P. (1998). Synthesis and secretion of human salivary immunoglobulins. *In* "Glandular mechanisms of salivary secretion" (J. R. Garrett, J. Ekstrom, L. C. Anderson, eds.), Vol. 10, pp. 167–199. Karger, Basel.

Brandtzaeg, P., Farstad, I. N., Johansen, F. E., Morton, H. C., Norderhaug, I. N., and Yamanaka, T. (1999). The B-cell system of human mucosae and exocrine glands. *Immunol. Rev.* **171**, 45–87.

Breivik, T., and Thrane, P. S. (2001). Psychoneuroimmune interactions in periodontal disease. *In* "Psychoneuroimmunology" (R. Ader, D. L. Felten, and N. Cohen, eds.), Vol. 2, pp. 627–644. Academic Press, San Diego.

Brink, P. R., Walcott, B., Roemer, E., Grine, E., Pastor, M., Christ, G. J., and Cameron, R. H. (1994). Cholinergic modulation of immunoglobulin secretion from avian plasma cells—The role of calcium. *J. Neuroimmunol.* **51**, 113–121.

Bristow, M., Hucklebridge, F. H., Clow, A., and Evans, D. E. (1997). Modulation of secretory immunoglobulin A in saliva in relation to an acute episode of stress and arousal. *J. Psychophysiol.* **11**, 248–255.

Brown, C. C. (1970). The parotid puzzle: A review of the literature on human salivation and its applications to psychophysiology. *Psychophysiology* **7**, 65–85.

Brown, L. R., Shea, C., Allen, S. S., Handler, S., Wheatcroft, M. G., and Frome, W. J. (1973). Effects of a simulated spacecraft environment on the oral microflora of nonhuman primates. *Oral Surg. Oral Med. Oral Pathol.* **35**, 286–293.

Brown, L. R., Wheatcroft, M. G., Frome, W. J., and Rider, L. J. (1974). Effects of a simulated skylab mission on the oral health of astronauts. *J. Dent. Res.* **53**, 1268–1275.

Carpenter, G. H., Garrett, J. R., Hartley, R. H., and Proctor, G. B. (1998). The influence of nerves on the secretion of immunoglobulin A into submandibular saliva in rats. *J. Physiol. (London)* **512**, 567–573.

Carpenter, G. H., Proctor, G. B., Anderson, L. C., Zhang, X. S., and Garrett, J. R. (2000). Immunoglobulin A secretion into saliva during dual sympathetic and parasympathetic nerve stimulation of rat submandibular glands. *Exp. Physiol.* **85**, 281–286.

Carroll, D., Ring, C., Shrimpton, J., Evans, P., Willemsen, G., and Hucklebridge, F. (1996). Secretory immunoglobulin A and cardiovascular responses to acute psychological challenge. *Int. J. Behav. Med.* **3**, 266–279.

Castagliuolo, I., Leeman, S. E., Bartolak-Suki, E., Nikulasson, S., Qiu, B., Carraway, R. E., and Pothoulakis, C. (1996). A neurotensin antagonist, SR 48692, inhibits colonic responses to immobilization stress in rats. *Proc. Natl. Acad. Sci. USA* **93**, 12611–12615.

Castagliuolo, I., Wershil, B. K., Karalis, K., Pasha, A., Nikulasson, S. T., and Pothoulakis, C. (1998). Colonic mucin release in response to immobilization stress is mast cell dependent. *Am. J. Physiol.* **274**, G1094–G1100.

Chatterton, R. T. Jr., Vogelsong, K. M., Lu, Y. C., Ellman, A. B., and Hudgens, G. A. (1996). Salivary α-amylase as a measure of endogenous adrenergic activity. *Clin. Physiol.* **16,** 433–448.

Chatterton, R. T., Jr., Vogelsong, K. M., Lu, Y. C., and Hudgens, G. A. (1997). Hormonal responses to psychological stress in men preparing for skydiving [see comments]. *J. Clin. Endocrinol. Metab.* **82,** 2503–2509.

Chicharro, J. L., Lucia, A., Perez, M., Vaquero, A. F., and Urena, R. (1998). Saliva composition and exercise. *Sports Med.* **26,** 17–27.

Cohen, S., and Herbert, T. B. (1996). Health psychology: Psychological factors and physical disease from the perspective of human psychoneuroimmunology. *Annu. Rev. Psychol.* **47,** 113–142.

Cohen, S., Tyrrell, D. A., and Smith, A. P. (1991). Psychological stress and susceptibility to the common cold [see comments]. *New Engl. J. Med.* **325,** 606–612.

Cohen, S., Miller, G. E., and Rabin, B. S. (2001). Psychological stress and antibody response to immunization: A critical review of the human literature. *Psychosom. Med.* **63,** 7–18.

Czerkinsky, C., Anjuere, F., McGhee, J. R., George-Chandy, A., Holmgren, J., Kieny, M. P., Fujiyashi, K., Mestecky, J. F., Pierrefite-Carle, V., Rask, C., and Sun, J. B. (1999). Mucosal immunity and tolerance: Relevance to vaccine development. *Immunol. Rev.* **170,** 197–222.

da Silva, A. M., Newman, H. N., and Oakley, D. A. (1995). Psychosocial factors in inflammatory periodontal diseases. *A review. J. Clin. Periodontol.* **22,** 516–526.

Deinzer, R., and Schuller, N. (1998). Dynamics of stress-related decrease of salivary immunoglobulin A (sIgA): Relationship to symptoms of the common cold and studying behavior. *Behav. Med.* **23,** 161–169.

Deinzer, R., Kleineidam, C., Stiller-Winkler, R., Idel, H., and Bachg, D. (2000). Prolonged reduction of salivary immunoglobulin A (sIgA) after a major academic exam. *Int. J. Psychophysiol.* **37,** 219–232.

Dhabhar, F. S., and McEwen, B. S. (2001). Bidirectional effects of stress and glucocorticoid hormones on immune function: Possible explanations for paradoxical observations. *In* "Psychoneuroimmunology" (R. Ader, D. L. Felten, and N. Cohen, eds.), Vol. 1, pp. 301–338. Academic Press, San Diego.

Drake, C. W., Hunt, R. J., and Koch, G. G. (1995). Three-year tooth loss among black and white older adults in North Carolina. *J. Dent. Res.* **74,** 675–680.

Ekstrom, J. (1999). Role of Nonadrenergic, Noncholinergic autonomic transmitters in salivary glandular activities *in vitro.* *In* "Neural Mechanisms of Salivary Secretion" (J. R. Garrett, J. Ekstrom, and L. C. Anderson, eds.), Vol. 11, pp. 94–130. Karger, Basel.

Ekstrom, J., Asztely, A., and Tobin, G. (1998). Parasympathetic non-adrenergic, non-cholinergic mechanisms in salivary glands and their role in reflex secretion. *Eur. J. Morphol.* **36** Suppl, 208–212.

El-Mofty, S., and Schneyer, C. A. (1978). Differential effect of autonomic stimulation on salivary secretion of IgG, IgA and amylase. *Proc. Soc. Exp. Biol. Med.* **158,** 59–62.

Emmelin, N. (1987). Nerve interactions in salivary glands. *J. Dent. Res.* **66,** 509–517.

Evans, P., Bristow, M., Hucklebridge, F., Clow, A., and Walters, N. (1993). The relationship between secretory immunity, mood and life-events. *Br. J. Clin. Psychol.* **32,** 227–236.

Evans, P., Bristow, M., Hucklebridge, F., Clow, A., and Pang, F. Y. (1994). Stress, arousal, cortisol and secretory immunoglobulin A in students undergoing assessment. *Br. J. Clin. Psychol.* **33,** 575–576.

Evans, P., Der, G., Ford, G., Hucklebridge, F., Hunt, K., and Lambert, S. (2000). Social class, sex, and age difference in mucosal immunity in a large community sample. *Brain Behav. Immun.* **14,** 41–48.

Farne, M. A., Boni, P., Corallo, A., Gnognoli, D., and Sacco, F. L. (1992). Personality variables as moderators between hassles and objective indicators of distress (S-IgA). *Stress Medicine* **10,** 15–20.

Finkbeiner, W. E. (1999). Physiology and pathology of tracheobronchial glands. *Respir. Physiol.* **118,** 77–83.

Freier, S., Eran, M., and Alon, I. (1989). A study of stimuli operative in the release of antibodies in the rat intestine. *Immunol. Invest.* **18,** 431–447.

Garrett, J. R. (1987). The proper role of nerves in salivary secretion: A review. *J. Dent. Res.* **66,** 387–397.

Garrett, J. R. (1998). Historical introduction to salivary secretion. *In* "Glandular Mechanisms of Salivary Secretion" (J. R. Garrett, J. Ekstrom, and L. C. Anderson, eds.), Vol. 10, pp. 1–20. Karger, Basel.

Garrett, J. R. (1999a). Nerves in the main salivary glands. *In* "Neural Mechanisms of Salivary Gland Secretion" (J. R. Garrett, J. Ekstrom, and L. C. Anderson, eds.), Vol. 11, pp. 1–25. Karger, Basel.

Garrett, J. R. (1999b). Effects of autonomic nerve stimulations on salivary parynchyma and protein secretion. *In* "Neural Mechanisms of Salivary Gland Secretion" (J. R. Garrett, J. Ekstrom, and L. C. Anderson, eds.), Vol. 11, pp. 59–79. Karger, Basel.

Genco, R. J., Ho, A. W., Kopman, J., Grossi, S. G., Dunford, R. G., and Tedesco, L. A. (1998). Models to evaluate the role of stress in periodontal disease. *Ann. Periodontol.* **3,** 288–302.

Gibbons, R. J. (1989). Bacterial adhesion to oral tissues: a model for infectious diseases. *J. Dent. Res.* **68,** 750–760.

Gleeson, M., Cripps, A. W., and Clancy, R. L. (1995). Modifiers of the human mucosal immune system. *Immunol. Cell Biol.* **73,** 397–404.

Graham, N. M., Bartholomeusz, R. C., Taboonpong, N., and La Brooy, J. T. (1988). Does anxiety reduce the secretion rate of secretory IgA in saliva? *Med. J. Aust.* **148,** 131–133.

Gray, J. A. (1987). "The Psychology of Fear and Stress." Cambridge University Press, Cambridge.

Gregory, R. L., Wallace, J. P., Gfell, L. E., Marks, J., and King, B. A. (1997). Effect of exercise on milk immunoglobulin A. *Med. Sci. Sports Exerc.* **29,** 1596–1601.

Harrison, L. K., Carroll, D., Burns, V. E., Corkill, A. R., Harrison, C. M., Ring, C., and Drayson, M. (2000). Cardiovascular and secretory immunoglobulin A reactions to humorous, exciting, and didactic film presentations. *Biol. Psychol.* **52,** 113–126.

Hecht, G. (1999). Innate mechanisms of epithelial host defense: Spotlight on intestine. *Am. J. Physiol.* **277,** C351–C358.

Hennig, J., Possel, P., and Netter, P. (1996). Sensitivity to disgust as an idicator of neuroticism: A psychobiological approach. *Pres. Indiv. Diff.* **20,** 589–596.

Henskens, Y. M. C., Veerman, E. C. I., and Nieuw Amerongen, A. V. (1996). Cystatins in health and disease. *Biol. Chem.* **377,** 71–86.

Herzberg, M. C. (1996). Platelet-streptococcal interactions in endocarditis. *Crit. Rev. Oral Biol. Med.* **7,** 222–236.

Hucklebridge, F., Clow, A., and Evans, P. (1998). The relationship between salivary secretory immunoglobulin A and cortisol: neuroendocrine response to awakening and the diurnal cycle. *Int. J. Psychophysiol.* **31,** 69–76.

Hucklebridge, F., Lambert, S., Clow, A., Warburton, D. M., Evans, P. D., and Sherwood, N. (2000). Modulation of secretory immunoglobulin A in saliva; response to manipulation of mood. *Biol. Psychol.* **53,** 25–35.

Huwe, S., Hennig, J., and Netter, P. (1998). Biological, emotional, behavioral, and coping reactions to examination stress in high and low state anxious subjects. *Anxiety, Stress, and Coping* **11,** 47–65.

Jemmott, J. B. III, and Magloire, K. (1988). Academic stress, social support, and secretory immunoglobulin-A. *J. Pers. Soc. Psychol.* **55,** 803–810.

Jemmott, J. B. III, and McClelland, D. C. (1989). Secretory IgA as a measure of resistance to infectious disease: Comments on Stone, Cox, Valdimarsdottir, and Neale. *Behav. Med.* **15,** 63–71.

Jemmott, J. B. III, Borysenko, J. Z., Borysenko, M., McClelland, D. C., Chapman, R., Meyer, D., and Benson, H. (1983). Academic stress, power motivation, and decrease in secretion rate of salivary secretory immunoglobulin A. *Lancet* **1,** 1400–1402.

Kaliner, M. A. (1991). Human nasal respiratory secretions and host defense. *Am. Rev. Respir. Dis.* **144,** S52–S56.

Kelleher, R. S., Hann, L. E., Edwards, J. A., and Sullivan, D. A. (1991). Endocrine, neural, and immune control of secretory component output by lacrimal gland acinar cells. *J. Immunol.* **146,** 3405–3412.

Kiecolt-Glaser, J. K., Garner, W., Speicher, C., Penn, G. M., Holliday, J., and Glaser, R. (1984). Psychosocial modifiers of immunocompetence in medical students. *Psychosom. Med.* **46,** 7–14.

Kilian, M., Reinholdt, J., Lomholt, H., Poulsen, K., and Frandsen, E. V. (1996). Biological significance of IgA1 proteases in bacterial colonization and pathogenesis: critical evaluation of experimental evidence. *Apmis* **104,** 321–338.

Kousvelari, E. E., Ciardi, J. E., and Bowers, M. R. (1988). Altered bacterial aggregation and adherence associated with changes in rat parotid-gland salivary proteins induced *in vivo* by β-adrenergic stimulation. *Arch. Oral Biol.* **33,** 341–346.

Kubitz, K. A., Peavey, B. S., and Moore, B. S. (1986). The effect of daily Hassles of humoral immunity: An interaction moderated by locus of control. *Biofeedback Self Regul.* **11,** 115–123.

Kugler, J. (1991). Emotional status and immunoglobulin A in saliva—Review of the literature. *Psychother. Psychosom. Med. Psychol.* **41,** 232–242.

Kugler, J., Breitfeld, I., Tewes, U., and Schedlowski, M. (1996a). Excavation of caries lesions induces transient decrease of total salivary immunoglobulin A concentration. *Eur. J. Oral Sci.* **104,** 17–20.

Kugler, J., Reintjes, F., Tewes, V., and Schedlowski, M. (1996b). Competition stress in soccer coaches increases salivary. Immunoglobin A and salivary cortisol concentrations. *J. Sports Med. Phys. Fitness.* **36,** 117–120.

Kusnecov, A. W., Sved, A., and Rabin, B. S. (2001). Immunological effects of acute versus chronic stress in animals. *In* "Psychoneuroimmunology" (R. Ader, D. L. Felten, and N. Cohen, eds.), Vol. 2, pp. 265–278. Academic Press, San Diego.

Lamm, M. E. (1997). Interaction of antigens and antibodies at mucosal surfaces. *Annu. Rev. Microbiol.* **51,** 311–340.

Levine, M. J. (1993). Salivary macromolecules. A structure/function synopsis. *Ann. N.Y. Acad. Sci.* **694,** 11–16.

Li, S., Xiu, B., Qian, Z., and Tang, P. L. (1997). Changes in salivary cortisol and S-IgA in response to test stress. *Chinese Mental Health J.* **11,** 336–338.

Li, X., Kolltveit, K. M., Tronstad, L., and Olsen, I. (2000). Systemic diseases caused by oral infection. *Clin. Microbiol. Rev.* **13,** 547–558.

Ligtenberg, A. J. M., Walgreen-Weterings, E., Veerman, E. C. I., de Soet, J. J., de Graaff., J., and Nieuw Amerongen, A. V. (1992). Influence of saliva on aggregation and adherence of *Streptococcus gordonii* HG 222. *Infect. Immun.* **60,** 3878–3884.

Liu, B., Rayment, S. A., Gyurko, C., Oppenheim, F. G., Offner, G. D., and Troxler, R. F. (2000). The recombinant N-terminal region of human salivary mucin MG2 (MUC7) contains a binding domain for oral Streptococci and exhibits candidacidal activity. *Biochem. J.* **345** Pt 3, 557–564.

Logan, H. L., Lutgendorf, S., Hartwig, A., Lilly, J., and Berberich, S. L. (1998). Immune, stress, and mood markers related to recurrent oral herpes outbreaks. *Oral Surg. Oral Med. Oral Pathol. Oral Radiol. Endod.* **86,** 48–54.

Mandel, I. D., Zengo, A., Katz, R., and Wotman, S. (1975). Effect of adrenergic agents on salivary composition. *J. Dent. Res.* **54** Spec No B, B27–B33.

Marcotte, H., and Lavoie, M. C. (1998). Oral microbial ecology and the role of salivary immunoglobulin A. *Microbiol. Mol. Biol. Rev.* **62,** 71–109.

Martin, R. A., and Dobbin, J. P. (1988). Sense of humor, hassles, and immunoglobulin A: Evidence for a stress-moderating role of humor. *Intl. J. Psychiatry in Medicine* **18,** 93–105.

Marucha, P. T., Kiecolt-Glaser, J. K., and Favagehi, M. (1998). Mucosal wound healing is impaired by examination stress. *Psychosom. Med.* **60,** 362–365.

Matsuo, R. (1999). Central connections for salivary innervations and efferent impulse formation. *In* "Neural Mechanisms of Salivary Gland Secretion" (J. R. Garrett, J. Ekstrom, and L. C. Anderson, eds.), Vol. 11, pp. 26–43. Karger, Basel.

Matsuo, R., Garrett, J. R., Proctor, G. B., and Carpenter, G. H. (2000). Reflex secretion of proteins into submandibular saliva in conscious rats, before and after preganglionic sympathectomy. *J. Physiol.* **527** Pt 1, 175–184.

McCartan, B. E., Lamey, P. J., and Wallace, A. M. (1996). Salivary cortisol and anxiety in recurrent aphthous stomatitis. *J. Oral. Pathol. Med.* **25,** 357–359.

McClellan, K. A. (1997). Mucosal defense of the outer eye. *Surv. Ophthalmol.* **42,** 233–246.

McClelland, D. C., Floor, E., Davidson, R. J., and Saron, C. (1980). Stressed power motivation, sympathetic activation, immune function, and illness. *J. Human Stress* **6,** 11–19.

McClelland, D. C., Ross, G., and Patel, V. (1985). The effect of an academic examination on salivary norepinephrine and immunoglobulin levels. *J. Human Stress* **11,** 52–59.

McGee, D., Eran, M., McGhee, J. R., and Freier, S. (1995). Substance P accelerates secretory component-mediated transcytosis of IgA in the rat intestine. *Adv. Exp. Med. Biol.* 643–646.

Mestecky, J. (1993). Saliva as a manifestation of the common mucosal immune system. *Ann. N.Y. Acad. Sci.* **694,** 184–194.

Meyer, R. J., and Haggerty, R. J. (1962). Streptococcal infections in families: Factors affecting individual susceptibility. *Pediatrics,* **29,** 539–549.

Mezzacappa, E. S., Kelsey, R. M., Katkin, E. S., and Sloan, R. P. (2001). Vagal rebound and recovery from psychological stress. *Psychosom. Med.* **63,** 650–657.

Miletic, I. D., Schiffman, S. S., Miletic, V. D., and Sattely-Miller, E. A. (1996). Salivary IgA secretion rate in young and elderly persons. *Physiol. Behav.* **60,** 243–248.

Morse, D. R., Schacterle, G. R., Furst, M. L., Esposito, J. V., and Zaydenburg, M. (1983). Stress, relaxation and saliva: relationship to dental caries and its prevention, with a literature review. *Ann. Dent.* **42,** 47–54.

Mouton, C., Fillion, L., Tawadros, E., and Tessier, R. (1989). Salivary IgA is weak stress marker. *Behav. Med.* **15,** 179–185.

Navazesh, M. (1993). Methods for collecting saliva. *Ann. N.Y. Acad. Sci.* **694,** 72–77.

Navazesh, M., and Christensen, C. M. (1982). A comparison of whole mouth resting and stimulated salivary measurement procedures. *J. Dent. Res.* **61,** 1158–1162.

Ng, V., Koh, D., Chan, G., Ong, H. Y., Chia, S. E., and Ong, C. N. (1999). Are salivary immunoglobulin A and lysozyme biomarkers of stress among nurses? *J. Occup. Environ. Med.* **41,** 920–927.

Nieuw Amerongen, A. V., Bolscher, J. G. M., and Veerman, E. C. (1995). Salivary mucins: protective functions in relation to their diversity. *Glycobiology* **5,** 733–740.

Nieuw Amerongen, A. V., Bolscher, J. G. M., Bloemena, E., and Veerman, E. C. I. (1998). Sulfomucins in the human body. *Biol. Chem.* **379,** 1–18.

Norderhaug, I. N., Johansen, F. E., Schjerven, H., and Brandtzaeg, P. (1999). Regulation of the formation and external transport of secretory immunoglobulins. *Crit. Rev. Immunol.* **19**, 481–508.

Ohira, H., Watanabe, Y., Kobayashi, K., and Kawai, M. (1999). The type A behavior pattern and immune reactivity to brief stress: Change of volume of secretory immunoglobulin A in saliva. *Percept. Mot. Skills* **89**, 423–430.

Page, R. C. (1998). The pathobiology of periodontal diseases may affect systemic diseases: Inversion of a paradigm. *Ann. Periodontol.* **3**, 108–120.

Perera, S., Uddin, M., and Hayes, J. A. (1997). Salivary lysozyme: A noninvasive marker for the study of the effects of stress on natural immunity. *Int. J. Behav. Med.* **4**, 170–178.

Park, S. J., and Tokura, H. (1999). Bright light exposure during the daytime affects circadian rhythms of urinary melatonin and salivary immunoglobulin A. *Chronobiol. Int.* **16**, 359–371.

Proctor, G. B. (1998). Secretory protein synthesis and constitutive (vesicular) secretion by the salivary glands. *In* "Glandular Mechanisms of Salivary Secretion" (J. R. Garrett, J. Ekstrom, and L. C. Anderson, eds.), Vol. 10, pp. 73–88. Karger, Basel.

Proctor, G. B., and Carpenter, G. H. (2001). Chewing stimulates secretion of human salivary secretory immunoglobulin A. *J. Dent. Res.* **80**, 909–913.

Proctor, G. B., Carpenter, G. H., Anderson, L. C., and Garrett, J. R. (2000a). Nerve-evoked secretion of immunoglobulin A in relation to other proteins by parotid glands in anaesthetized rat. *Exp. Physiol.* **85**, 511–518.

Proctor, G. B., Carpenter, G. H., and Garrett, J. R. (2000b). Sympathetic decentralization abolishes increased secretion of immunoglobulin A evoked by parasympathetic stimulation of rat submandibular glands. *J. Neuroimmunol.* **109**, 147–154.

Qiu, B. S., Vallance, B. A., Blennerhassett, P. A., and Collins, S. M. (1999). The role of CD4[+] lymphocytes in the susceptibility of mice to stress-induced reactivation of experimental colitis. *Nat. Med.* **5**, 1178–1182.

Rifai, A., Fadden, K., Morrison, S. L., and Chintalacharuvu, K. R. (2000). The N-glycans determine the differential blood clearance and hepatic uptake of human immunoglobulin (Ig) A1 and IgA2 isotypes. *J. Exp. Med.* **191**, 2171–2182.

Ring, C., Carroll, D., Willemsen, G., Cooke, J., Ferraro, A., and Drayson, M. (1999). Secretory immunoglobulin A and cardiovascular activity during mental arithmetic and paced breathing. *Psychophysiology* **36**, 602–609.

Ring, C., Harrison, L. K., Winzer, A., Carroll, D., Drayson, M., and Kendall, M. (2000). Secretory immunoglobulin A and cardiovascular reactions to mental arithmetic, cold pressor, and exercise: effects of α-adrenergic blockade. *Psychophysiology* **37**, 634–643.

Ring, C., Drayson, M., Walkey, D. G., Dale, S., and Carroll, D. (2002). Secretory immunoglobulin A reactions to prolonged mental arithmetic stress: Inter-session and intra-session reliability. *Biol. Psychol.* **59**, 1–13.

Rogers, D. F. (1994). Airway goblet cells: responsive and adaptable front-line defenders. *Eur. Respir. J.* **7**, 1690–1706.

Rood, Y. R., Bogaards, M., Goulmy, E., and Houwelingen, H. C. (1993). The effects of stress and relaxation on the *in vitro* immune response in man: A meta-analytic study. *J. Behav. Med.* **16**, 163–181.

Rudney, J. D. (1995). Does variability in salivary protein concentrations influence oral microbial ecology and oral health? *Crit. Rev. Oral Biol. Med.* **6**, 343–367.

Rugh, J. D., Jacobs, D. T., Taverna, R. D., and Johnson, R. W. (1984). Psychophysiological changes and oral conditions. *In* "Social Sciences and Dentistry." (L. K. Cohen and P. S. Bryant, eds.), pp. 19–83. Quintessence, Kingston upon Thames.

Scannapieco, F. A. (1999). Role of oral bacteria in respiratory infection. *J. Periodontol.* **70**, 793–802.

Scannapieco, F. A., Torres, G., and Levine, M. J. (1993). Salivary α-amylase: Role in dental plaque and caries formation. *Crit. Rev. Oral Biol. Med.* **4**, 301–307.

Schenkels, L. C. P. M., Veerman, E. C. I., and Nieuw Amerongen, A. V. (1995). Biochemical composition of human saliva in relation to other mucosal fluids. *Crit. Rev. Oral Biol. Med.* **6**, 161–175.

Schiff, J. M., Fisher, M. M., Jones, A. L., and Underdown, B. J. (1986). Human IgA as a heterovalent ligand: Switching from the asialoglycoprotein receptor to secretory component during transport across the rat hepatocyte. *J. Cell Biol.* **102**, 920–931.

Schmidt, P. T., Eriksen, L., Loftager, M., Rasmussen, T. N., and Holst, J. J. (1999). Fast-acting nervous regulation of immunoglobulin A secretion from isolated perfused porcine ileum. *Gut* **45**, 679–685.

Schneyer, C. A., and Hall, H. D. (1991). Effects of varying frequency of sympathetic stimulation on chloride and amylase levels of saliva elicited from rat parotid gland with electrical stimulation of both autonomic nerves. *Proc. Soc. Exp. Biol. Med.* **196**, 333–337.

Segawa, A., and Yamashina, S. (1998). The dynamics of exocytosis of preformed secretory granules from acini in rat salivary glands. *In* "Glandular Mechanisms of Salivary Secretion" (J. R. Garrett, J. Ekstrom, and L. C. Anderson, eds.), Vol. 10, pp. 89–100. Karger, Basel.

Shimura, S. (2000). Signal transduction of mucous secretion by bronchial gland cells. *Cell Signal* **12**, 271–277.

Shimada, S., Kawaguchi-Miyashita, M., Kushiro, A., Sato, T., Nanno, M., Sako, T., Matsuoka, Y., Sudo, K., Tagawa, Y., Iwakura, Y., and Ohwaki, M. (1999). Generation of polymeric immunoglobulin receptor-deficient mouse with marked reduction of secretory IgA. *J. Immunol.* **163**, 5367–5373.

Skosnik, P. D., Chatterton, R. T., Jr., Swisher, T., and Park, S. (2000). Modulation of attentional inhibition by norepinephrine and cortisol after psychological stress. *Int. J. Psychophysiol.* **36**, 59–68.

Soderholm, J. D., and Perdue, M. H. (2001). II. Stress and intestinal barrier function. *Am. J. Physiol. Gastrointest. Liver Physiol.* **280**, G7–G13.

Spangler, G. (1997). Psychological and physiological responses during an exam and their relation to personality characteristics. *Psychoneuroendocrinology* **22**, 423–441.

Spitz, J. C., Hecht, G., Taveras, M., Aoys, E., and Alverdy, J. (1994). The effect of dexamethasone administration on rat intestinal permeability—The role of bacterial adherence. *Gastroenterology* **106**, 35–41.

Spitz, J. C., Ghandi, S., Taveras, M., Aoys, E., and Alverdy, J. C. (1996). Characteristics of the intestinal epithelial barrier during dietary manipulation and glucocorticoid stress. *Critical Care Medicine* **24**, 635–641.

Stone, A. A., Cox, D. S., Valdimarsdottir, H., Jandorf, L., and Neale, J. M. (1987a). Evidence that secretory IgA antibody is associated with daily mood. *J. Pers. Soc. Psychol.* **52**, 988–993.

Stone, A. A., Cox, D. S., Valdimarsdottir, H., and Neale, J. M. (1987b). Secretory IgA as a measure of immunocompetence. *J. Human Stress* **13**, 136–140.

Stone, A. A., Neale, J. M., Cox, D. S., Napoli, A., Valdimarsdottir, H., and Kennedy-Moore, E. (1994). Daily events are associated with a secretory immune response to an oral antigen in men. *Health Psychol.* **13**, 440–446.

Stone, A. A., Marco, C. A., Cruise, C. E., Cox, D. S., and Neale, J. M. (1996). Are stress-induced immunological changes mediated by mood? A closer look at how both desirable and undesirable daily events influence sIgA antibody. *Int. J. Behav. Med.* **3**, 1–13.

Sullivan, D. A., Underdown, B. J., and Wira, C. R. (1983). Steroid hormone regulation of free secretory component in the rat uterus. *Immunology* **49**, 379–386.

Sutton, P. R. N. (1996). Stress and dental caries. *In* "Advances in Oral Biology" (P. H. Stable, ed.), Vol. 2, pp. 104–148. Academic Press, New York.

Tenovuo, J., Moldoveanu, Z., Mestecky, J., Pruitt, K. M., and Rahemtulla, B. M. (1982). Interaction of specific and innate factors of immunity: IgA enhances the antimicrobial effect of the lactoperoxidase system against *Streptococcus mutans. J. Immunol.* **128**, 726–731.

Travis, S. M., Singh, P. K., and Welsh, M. J. (2001). Antimicrobial peptides and proteins in the innate defense of the airway surface. *Curr. Opin. Immunol.* **13**, 89–95.

Valdimarsdottir, H. B., and Stone, A. A. (1997). Psychosocial factors and secretory immunoglobulin A. *Crit. Rev. Oral Biol. Med.* **8**, 461–474.

Valtonen, V. V. (1999). Role of infections in atherosclerosis. *Am. Heart J.* **138**, S431–S433.

Van den Brink, G. R., Tytgat, K. M., Van der Hulst, R. W., Van der Loos, C. M., Einerhand, A. W., Buller, H. A., and Dekker, J. (2000). *Helicobacter pylori* colocalises with MUC5AC in the human stomach. *Gut* **46**, 601–607.

Vanderas, A. P., Manetas, C., and Papagiannoulis, L. (1995). Urinary catecholamine levels in children with and without dental caries. *J. Dent. Res.* **74**, 1671–1678.

Vanderas, A. P., Manetas, K., and Papagiannoulis, L. (2000). Caries increment in children and urinary catecholamines: findings at one-year. *ASDC J. Dent. Child.* **67**, 355–359, 304.

Veerman, E. C. I., van den Keybus, P. A. M., Vissink, A., and Nieuw Amerongen, A. V. (1996). Human glandular salivas: Their separate collection and analysis. *Eur. J. Oral Sci.* **104**, 346–352.

Vorland, L. H. (1999). Lactoferrin: A multifunctional glycoprotein. *Apmis* **107**, 971–981.

Wang, S. C. (1962). Central nervous system representation of salivary secretion. *In* "Salivary Glands and Their Secretions" (L. M. Sreebny and J. Meyer, eds.), pp. 145–159. MacMillan, New York.

Willemsen, G., Ring, C., Carroll, D., Evans, P., Clow, A., and Hucklebridge, F. (1998). Secretory immunoglobulin A and cardiovascular reactions to mental arithmetic and cold pressor. *Psychophysiology* **35**, 252–259.

Willemsen, G., Ring, C., McKeever, S., and Carroll, D. (2000). Secretory immunoglobulin A and cardiovascular activity during mental arithmetic: Effects of task difficulty and task order. *Biol. Psychol.* **52**, 127–141.

Willemsen, G., Carroll, D., Ring, C., and Drayson, M. (2002). Cellular and mucosal immune reactions to mental and cold stress: Associations with gender and cardiovascular reactivity. *Psychophysiology* **39**, 222–228.

Wilson, I. D., Soltis, R. D., Olson, R. E., and Erlandsen, S. L. (1982). Cholinergic stimulation of immunoglobulin A secretion in rat intestine. *Gastroenterology* **83**, 881–888.

Winzer, A., Ring, C., Carroll, D., Willemsen, G., Drayson, M., and Kendall, M. (1999). Secretory immunoglobulin A and cardiovascular reactions to mental arithmetic, cold pressor, and exercise: Effects of β-adrenergic blockade. *Psychophysiology* **36**, 591–601.

Wira, C. R., and Colby, E. M. (1985). Regulation of secretory component by glucocorticoids in primary cultures of rat hepatocytes. *J. Immunol.* **134**, 1744–1748.

Wira, C. R., and Rossoll, R. M. (1991). Glucocorticoid regulation of the humoral immune system. Dexamethasone stimulation of secretory component in serum, saliva, and bile. *Endocrinology* **128**, 835–842.

Wira, C. R., Sandoe, C. P., and Steele, M. G. (1990). Glucocorticoid regulation of the humoral immune system. I. *In vivo* effects of dexamethasone on IgA and IgG in serum and at mucosal surfaces. *J. Immunol.* **144**, 142–146.

Wu, T., Trevisan, M., Genco, R. J., Dorn, J. P., Falkner, K. L., and Sempos, C. T. (2000). Periodontal disease and risk of cerebrovascular disease: The first national health and nutrition examination survey and its follow-up study. *Arch. Intern. Med.* **160**, 2749–2755.

Zeier, H., Brauchli, P., and Joller-Jemelka, H. I. (1996). Effects of work demands on immunoglobulin A and cortisol in air traffic controllers. *Biol. Psychol.* **42**, 413–423.

CYTOKINES AND DEPRESSION

Angela Clow

Department of Psychology
University of Westminster
London W1B 2UW, United Kingdom

I. Introduction

The belief that there may be a link between emotional well-being and physical illness (see Sternberg, 1997) is the central tenet of psychophysiology and psychoneuroimmunology (PNI). Although at times controversial, research generally supports the notion that progression of some diseases (e.g., cancer, AIDS, cardiovascular disease) can be accelerated by melancholic depression and that psychological interventions can improve physical health (reviewed in Kiecolt-Glaser *et al.*, 2002). The new interdisciplinary science of PNI has made substantial progress in unraveling the mechanisms by which mood and the immune system may interact and have an impact on health. In particular, major depression has been shown to be characterized not only by reduced central monoamine neurotransmitter availability (which is widely believed to be the cause of melancholia) but also by changes in immune function (reduced NK cell activity and, paradoxically, activation of certain aspects of type 1 immunity inflammatory processes and the acute phase response) and disturbance of neuroendocrine activity [increased corticotropin-releasing factor (CRF) and cortisol]. Under normal circumstances these three systems (brain, immune, and neuroendocrine systems) can communicate in a way that is adaptive for survival; depression is associated with dysfunction in the sensitivity of these relationships.

As described in other parts of this book the brain both influences and is influenced by the immune and neuroendocrine systems. At the same time, the immune and neuroendocrine systems have reciprocal communication pathways. Thus when a "strain" is placed on these interrelated systems the entire network is thrown out of kilter. Evidence, to be presented in this chapter, indicates that whether the initial dysfunction originates in the hypothalamic-pituitary-adrenal (HPA) axis (prolonged activation) or in the immune system (prolonged inflammatory disturbance) one associated manifestation might be the induction of major depression. It is unfortunate that the changes in the immune system associated with major depression can not only reduce immunosurveillance and lower mood but also activate the HPA axis, an effect which compounds the immune system changes and also lowers mood while at the same time increases susceptibility to other types of ill health such as cardiovascular disease.

II. Depression and Immune System Dysregulation

A. EARLY STUDIES

The first report to consider the immune system in depressive disorders was published in 1978 (Cappel *et al.,* 1978). During the following 12 years an additional 20 human studies appeared in print. An array of enumerative techniques and functional assays to examine the effect of depression on peripheral circulating immune cells were reported. No consistent change in the representation of immune cell populations in the peripheral circulation of depressed patients emerged as a result of these early investigations. Of the functional studies, the two that received the most attention in this context were the *in vitro* lymphocyte proliferative response to mitogens and the assay of natural killer (NK) cell activity. Unfortunately, again, no consensus emerged on the impact of depression on these measures. Although several groups reported reduced proliferation and decreased NK cell activity, others did not. Of course these inconsistencies in the early literature, which could be attributed to methodological issues, led to widespread confusion and some skepticism that depression might be associated with measurable alterations in immune system functioning (for review see Stein *et al.*, 1991).

The most important methodological issue was that the most marked immunological changes were found to accompany depression associated with only a subpopulation of patients suffering from depression, namely those with profound melancholia. Melancholic depression (synonymous with unipolar endogenous, or major, depression) is characterized by a lack of reactivity to the environment, a pervasive loss of interest in all or almost all activities, significant weight loss, diurnal variation in mood, and early morning awakening. Melancholic depressed patients are also characterized as having the best response to antidepressant drug therapy; evidence that the etiologies of different types of depression are physiologically different. It is not surprising therefore that the early literature was so inconsistent, when so much of it did not take into account the heterogeneous nature of depressive illness. The appreciation that melancholic (major) depression, rather than other types, is specifically associated with the dysregulated immune and neuroendocrine systems (to be described in this chapter) has facilitated current understanding.

B. IMMUNOLOGICAL CHARACTERISTICS OF MAJOR DEPRESSION

The view that depression is associated with less efficient immune system activity as a whole is too simplistic. It is clear that this normally highly tuned system does become dysregulated in depression. In particular those patients who present with severely depressed mood, as well as being older, male, and hospitalized, show the most marked immune system changes. A broad meta-analysis of the relationship between depression and immune measures, which surveyed more than 180 peer-reviewed studies that satisfied strict inclusion criteria (only unmediated major depression, excluding bipolar disorder), lends wide support for a dysregulated immune system (Zorrilla *et al.*, 2001). The main findings were of lymphopenia (reduced numbers of lymphocytes in the circulation) and neutrophilia (increased numbers of neutrophils), reduced NK cell levels and cytotoxity, and reduced lymphocyte response to mitogen. Although they did not observe consistent changes in the absolute numbers of circulating B or T cells nor in the number of $CD4^+$ or $CD8^+$ subsets, they did find an increased ratio of $CD4^+/CD8^+$ across accumulated subjects. Analysis of the accumulated studies also revealed consistently increased circulating haptoglobin and IL6, findings consistent with activation of the acute phase response to infection.

The earliest and most widely reported of these changes was decreased NK cell activity (e.g., Irwin and Gillin, 1987; Irwin *et al.*, 1987) as well as reduced T-cell proliferative response to mitogen stimulation (e.g., Kronfol *et al.*, 1983, Kronfol and House, 1985). These initial findings did indeed suggest that major depression was accompanied by functional aspects of immunosuppression. However, this was only part of the story as, strangely perhaps, these changes (indicative of downregulated Th1 activity) were accompanied by *activation* of cell-mediated immunity. This hint of a paradoxical immunological state first came to light in the late 1980s when depression was first associated with an increased $CD4^+/CD8^+$ cell ratio (Tondo *et al.*, 1988; Maes *et al.*, 1992a). This has now been replicated many times and the size of the increased ratio (in relation to normal) is significantly and positively related to the severity of melancholia. This change has been attributed to both an increase in the number of $CD4^+$ cells and a reduction in the number of $CD8^+$ cells. Others have found higher numbers of phagocytic cells: neutrophils and monocytes (Maes *et al.*, 1992b). Although it was not easy to interpret this enumerative data, the increase in circulating T helper cells ($CD4^+$) challenged the notion of general immunosuppression.

However, perhaps one of the most striking changes seen in depression was the number of *activated* T cells. This is evidenced by the high level of IL-2 receptor binding sites on T lymphocytes (Maes *et al.*, 1993). There were also high levels of soluble IL-2 receptors in the general circulation, another index of T-cell activation (Maes *et al.*, 1995). In fact, high levels of this soluble receptor have been linked with suicidal tendencies (Nassberger and Traskman-Bendz, 1993).

Consistently it has been shown that cultured peripheral blood mononuclear cells of melancholic depressed patients secrete more of the pro-inflammatory cytokines IL-1β and IL-6 when challenged with mitogen (Maes *et al.*, 1991). Additionally plasma and urinary neopterin, a product of IFN-γ-stimulated monocytes/macrophages and a marker of the bias toward cellular, type 1 immune activity, has repeatedly been shown to be elevated in depression (Dunbar *et al.*, 1992; Maes *et al.*, 1994). Critically, these same cells (T cells) lack normal sensitivity to the glucocorticoid dexamethasone (Maes *et al.*, 1991). Normally the production of both IL-1β and soluble IL-2 receptors is inhibited by glucocorticoids. These peripheral blood mononuclear cells in melancholic depressed patients, however, are less inhibited by glucocorticoids in this respect than are those of healthy control subjects. This characteristic is of fundamental importance for understanding the immune response observed in patients with major depression and is addressed more fully later in this chapter.

Depression is also associated with increased plasma concentration of positive acute-phase proteins, such as haptoglobin, alongside reduced negative acute-phase proteins, such as albumin (Joyce *et al.*, 1992, Maes *et al.*, 1992b). It is known that such elevated levels of acute-phase proteins are important indicators of acute or chronic inflammatory states. Consequently these findings support the notion that some major depression patients (typically those with melancholia) suffer from a continuous activation of the "acute"-phase inflammatory response. In one study it has been observed that cancer patients treated with interleukin-2-based therapy displayed concomitant depressive mood symptoms. The intensity of these depressive symptoms at the endpoint of therapy was positively correlated with increases in serum IL-10 between baseline and endpoint. As IL-10 is a Th2 anti-inflammatory cytokine it was argued that elevation at this point in the treatment regimen is indicative of an underlying inflammatory response. The authors proposed that these results supported the hypothesis of a close relationship between depressive symptoms and activation of the cytokine network (Capuron *et al.*, 2001). This finding has been supported by a further study of hepatitis C patients treated with IFN-α. Over a 24-week period of immunotherapy, patients displayed increases in severity of depression and anxiety (Maes *et al.*, 2001). At the same time classical and atypical antidepressant drugs have been shown to attenuate the sickness behavior induced by pro-inflammatory cytokines suggesting that their antidepressant efficacy may be attributed, at least in part, to their immune effects (van West and Maes, 1999; Castanon *et al.*, 2001).

It has been hypothesized therefore (Maes *et al.*, 1995) that T-cell activation and increased secretion of IL-1β and IL-6, characteristic of melancholic depression, are the core events that mediate the reduction in NK cell activity (so vital in disease susceptibility). The reduction in NK cell activity may be the consequence of a combination of increased numbers of neutrophils, soluble IL-2 receptors (which are thought to sequester circulating IL-2, an important NK cell stimulatory cytokine), prostaglandins, and the positive acute-phase proteins (all under the influence of IL-1β and Il-6). Of significance for clinical practice is the observation that antidepressant drug treatment also augments NK cell activity in subjects with low NK cell activity at baseline (Frank *et al.*, 1999). This finding indicates the importance of efficient diagnosis and treatment of depression in cancer patients, as increased NK cell activity has been associated with slowed disease progression.

III. Neuroendocrine Dysregulation

A. HYPERACTIVITY OF THE HPA AXIS

To understand the origins of this unusual profile of immune activity it is necessary to discuss the status of the HPA axis in depression. Patients have enlarged pituitary (Krishnan et al., 1991) and adrenal glands (Nemeroff et al., 1992, Rubin et al., 1995): the result of hyperactivity. It is not surprising therefore to find that major depression is characterized by excessive circulating levels of both pituitary ACTH (Heuser, 1998) and adrenal cortisol, which cannot be explained by decreased activity of the cortisol-metabolizing enzyme 11-β-hydroxysteroid dehydrogenase (Carroll et al., 1976; Weber et al., 2000). Depression is also associated with a flattening of the circadian pattern of cortisol secretion (Rosmond et al., 1998). In addition the paraventricular nucleus (PVN) of the hypothalamus secretes excessive amounts of corticol releasing factor (CRF) in depression (Nemeroff et al., 1984, Nemeroff, 1988). Thus, in major depression, all three components of the HPA axis are hyperactive but have been shown to return to normal following effective treatment (Nemeroff et al., 1991; Rubin et al., 1995; and reviewed in Arborelius et al., 1999).

B. REGULATION OF THE HPA AXIS

It is believed that the reason for excessive HPA activity during major depression is that the body becomes less efficient at regulating levels of CRF and cortisol (Pariante et al., 1995). In nondepressed subjects acute stress-induced increases in these hormones are short lived, as a result of negative feedback control. Glucocorticoid secretion is negatively regulated at the level of the anterior pituitary (Miller et al., 1992) as well as in limbic structures, such as the hypothalamus, hippocampus, and amygdala, and the mesoprefrontal system (Magarinos et al., 1987; Bradbury et al., 1991; Diorio et al., 1993; Ferrini et al., 1999; Feldman and Weidenfeld; 1999). Depression is characterized by inadequate negative feedback, attributable to desensitized receptors for both CRF and cortisol (Modell et al., 1997).

It has been demonstrated that chronic stress can downregulate the mRNA for these receptors (Sapolsky et al., 1984) and induce a hyposuppressive state for basal glucocorticoid secretion (Mizoguchi et al., 2001). This has led many to speculate that chronic stress may be linked to depression via this mechanism (e.g., Raison and Miller, 2001). Evidence suggests that females may be more prone to glucocorticoid receptor desensitization following an acute stressor (Rohleder *et al.*, 2001), which may be associated with the reported greater incidence of depression in the female population (Kockler and Heun, 2002). Furthermore, antidepressant drugs have been shown to upregulate desensitized glucocorticoid receptors and hence restore normalization of HPA axis activity, the timescale of clinical improvement being inline with the observed changes in glucocorticoid receptor sensitivity (Pepin *et al.*, 1989; Peiffer *et al.*, 1991; Barden, 1996). In summary, depression is associated with desensitized CRF and cortisol receptors, the result of which is overproduction of these hormones and thus dysregulation of the HPA axis—the hallmark of melancholic depression.

C. DEXAMETHASONE SUPPRESSION TEST AS AN INDEX OF GLUCOCORTICOID RECEPTOR FUNCTION IN DEPRESSION

Dexamethasone (Dex) is a synthetic glucocorticoid that acts mainly on type 2 cortisol receptors. Under normal circumstances administration of Dex acts, via these receptors, to suppress the production of endogenous cortisol. Nonsuppression is an indication of poor negative feedback, that is, desensitized receptors. In 1981 it was reported that melancholic depression could be simply diagnosed by nonsuppression on the dexamethasone suppression test (DST) (Carroll *et al.*, 1981). A flood of papers, and controversy, quickly ensued: not all depressed patients were nonsuppressors. Many problems stemmed from methodological considerations—different groups executed the test using different regimes and on a broad spectrum of depressed patients. However, it now seems clear that nonsuppression is (usually) associated with high circulating cortisol levels. This led to the classification of depression according to cortisol status, that is, high cortisol and nonsuppression being associated with greater melancholia (Evans *et al.*, 1983). Others have not demonstrated these relationships. It is clear that nonsuppression appears to be related to severity of major depression. The rank order of nonsuppression is striking: normal patients (7–8%), grief reaction (10%), minor depression (23%), major depression (44%), melancholia (50%), and psychotic affective disorders (69%) (Murphy, 1991).

Thus as many as 50% of depressed patients manifest with a dysfunction of the HPA axis: increased activity and flattened circadian cycles (Carroll *et al.*, 1976; Rosmond *et al.*, 1998). Furthermore, the subjects who show most pronounced dysregulation of this axis also show the most marked changes in immune activation and are thus more susceptible to organic ill health. The idea of identifying suicide-prone individuals by these physiological measures has resurfaced. As long ago as 1965 it was first predicted that high cortisol levels provided a good indication of suicidal intent (Bunney and Fawcett,

1965). Although not universally accepted as a simple tool it is recognized that high cortisol is associated with increased severity of depression. A paper linking suicidal intent with activation of Th1 immunity and pro-inflammatory cytokines may reopen this area for discussion (Mendlovic *et al.*, 1999).

IV. Relationship between Immune and Neuroendocrine System Changes

A. ROLE OF CORTISOL

Cortisol has powerful immunomodulatory influences and plays an important role in the connections between depression and the immune system. The individual cells of the immune system to some extent mimic the entire HPA system as they possess receptors for a wide range of hormones, neurotransmitters, and neuropeptides (Lippman and Barr, 1977). The precise role of many of these receptors has yet to be fully determined; however, it is clear that cortisol receptors on lymphocytes of depressed subjects lack sensitivity to cortisol (Lowry *et al.*, 1984; Spencer *et al.*, 1991). Thus the cortisol receptors on both the HPA axis and the T cells share this same characteristic. As a result Th1 immune system activity and macrophage contribution to the acute-phase response is desensitized to cortisol inhibition. Thus depression is characterized by the unusual milieu of high cortisol (which would normally downregulate Th1 and macrophage activity) along with evidence of heightened Th1 and pro-inflammatory activity. The pro-inflammatory cytokines produced also activate the HPA cascade thus exacerbating the overdrive on this axis (Besodovsky *et al.*, 1991). This cascade of events has the effect of reducing NK cell activity and T-lymphocyte proliferative response to mitogen, changes which could underpin accelerated disease progression in those diseases normally held in check by effective Th1 defenses.

B. ROLE OF CRF

Excessive central CRF production is thought to have important influences, independent of its role as initiator of the HPA axis cascade. Whether CRF affects the immune system directly or indirectly via HPA axis and sympathetic nervous system activation is difficult to determine. Certainly, CRF administration to animals suppresses NK cell activity and lymphocyte proliferative responses while stimulating the release of pro-inflammatory cytokines (see Miller, 1998; Torpy and Chrousos, 1996). In adrenalectomized rats (hence no glucocorticoid production) intracerebral ventricular (ICV) administration of CRF significantly suppressed lymphocyte proliferation and NK cell cytotoxicity (Jain *et al.*, 1991).

C. Origin of the Dysregulations

The described abnormalities of the HPA axis and Th1 branch of the immune system tend to coexist in the same patients and are most obvious in those presenting with major depression. Melancholic mood may be a *result* of these neuroendocrine and immune system changes. The initial perturbation may originate either in the HPA axis or in dysregulation of the immune system (toward pro-inflammatory activity). The evidence is accumulating that excessive psychological stress activation of the HPA axis, in susceptible individuals, may lead to dysregulation of the entire neuroendocrine–immune axis. In particular the impact of early life stress has been implicated in the antecedents of adult onset depression (see Arborelius *et al.*, 1999, for a review). Once the neuroendocrine system has lost sensitivity for negative feedback, the accumulation of cortisol and CRF can induce secondary changes in the immune system.

However, it is also possible that primary overproduction of the Th1 pro-inflammatory cytokines (rather than prolonged psychological stress) can be the cause of prolonged HPA activation with consequent dysregulation. For example, one large prospective study (with over 700 subjects) has shown an association between somatic illness in childhood and increased risk of major depression in adulthood (Cohen *et al.*, 1998). This study revealed an association between excessive drive in the immune system in early life with subsequent onset of major depression. The observed association was independent of prior depressive episodes and demographic covariates. Furthermore, the reciprocal relationship was also observed; depression predicted increased risk of future poor physical health. These associations were observed over a long time period (a 17-year span) suggesting a causal relationship rather than simply indicating short-term behavioral changes mediated by the illness itself. However, regardless of the origin of the dysfunction (the HPA axis or the immune system), once a milieu of high levels of Th1 activity and pro-inflammatory cytokines, CRF, and cortisol, predominate, melancholia seems the likely outcome in terms of disturbance of affect.

V. Effect of Cytokines on Mood

The influence that pro-inflammatory cytokines can exert on the HPA axis has already been discussed; however, they can also have direct effects on the brain and mood. The role of these cytokines is to mobilize whole-body responses, including behavioral changes that help to fight against infection. One of the ways to accomplish this is to conserve energy normally expended on activity not relevant to the fight against infection. As a result these cytokines induce a behavioral syndrome associated with withdrawal from social interaction, reduced sexual activity and appetite, and increased sleep (see Kent *et al.*, 1992, and Danzer, 1999, for review). Healthy people given IL-2 and TNF-α (Th1 and pro-inflammatory cytokines) develop depressed mood, increased somatic concern, cognitive impairment, and difficulties with motivation and flexible thinking (reviewed in Maier and Watkins, 1998). These symptoms occur very quickly after cytokine administration and vanish soon after discontinuation of treatment, suggesting that these cytokines play a causal role in induction of the symptoms. However, the dose of cytokines given in these experiments is very large. Although they demonstrate the principle that a link exists between cytokines and mood the most convincing evidence that this link is relevant to health comes from clinical studies.

The most consistent psychological disturbance seen in infection (associated with increased circulating levels of pro-inflammatory cytokines) is depression. Again, individuals who suffer autoimmune diseases attributed to overactive Th1 activity (e.g., multiple sclerosis and rheumatoid arthritis) display a high incidence of depressive disorders (reviewed in Dickens *et al.*, 2002). Similarly depression is twice as prevalent in females as in males, and females tend to be Th1 dominated (Kockler and Heun, 2002). Furthermore, the increased production of INF-γ in the winter months has been linked with seasonal affective disorder (SAD). Patients suffering from SAD can often be successfully treated with high-intensity light therapy. Wintertime darkness is associated with more prolonged melatonin production (a Th1-promoting agent), which can be reversed by bright light so that the shift toward Th1 can be minimized. Thus, the evidence is accumulating that not only is depression associated with raised levels of Th1 activity and pro-inflammatory cytokines but also that these cytokines can induce depressed mood (see Connor and Leonard, 1998, for review).

VI. Role of Brain Monoamines

A. Cytokines and Monoamines

It is known that both psychological and physiological stressors produce alterations in noradrenaline (NA), serotonin (5HT), and dopamine (DA) levels in the brain. These monoamines regulate mood, reward, and vegetative function. If cytokines influence these behaviors (which are clearly evident), then they must make an impact on central monoamine activity. Certainly pro-inflammatory cytokines, and their receptors, are expressed in the brain (see Danzer, 1999). Changes in monoamine activity induced by a single administration of inflammatory cytokines are very similar to those induced by stress. Increased release and utilization of monoamine levels in specific brain regions coincides with the peak of the immunological response to challenge (Besedovsky *et al.*, 1983). These changes can be blocked by pretreatment with immunosuppressive drugs, indicating that they are the direct result of immune system activation. These early experiments determined the effects following a single administration of cytokines or immunological challenge. In reality inflammatory disease and depression are associated with elevated pro-inflammatory cytokines over protracted periods of time. It seems that an acute stressor or immunological challenge causes increased release and use of the monoamines whereas repeated exposure to either of them leads to depletion (Sunanda Rao *et al.*, 2000; Shuto *et al.*, 1997) which would be consistent with the lowered monoamine availability associated with depression.

Another, more indirect, way in which inflammatory cytokines can influence mood is via reduction in circulating levels of the essential amino acid tryptophan, necessary for the synthesis of serotonin. Tryptophan competes with other amino acids for transport into the brain; these are known as competing amino acids (CAAs). A high-protein diet can result in high circulating levels of CAAs that compete with tryptophan for uptake into the brain and also as a result of stimulation of tissue protein synthesis can result in depletion of circulating tryptophan and reduced availability for the brain transporter per se. In a similar way, experimental administration of a drink rich in CAAs to normal healthy subjects causes depletion of central 5HT levels and depressed mood (Ravindran *et al.*, 1999). Levels of tryptophan in the blood of depressed subjects have consistently been shown to be lower than matched controls (Maes *et al.*, 1996). For a long time the reasons for this were not known but it now seems clear that some pro-inflammatory cytokines can decrease tryptophan availability to the brain by inducing its metabolism before being transported into the brain (Maes *et al.*, 1995). Plasma tryptophan levels are significantly and negatively related to IL-6 secretion, plasma levels of positive acute-phase proteins, and neopterin. Consequently, low tryptophan levels have been used as a marker of immune pro-inflammatory activity during major depression and may provide a link between the peripheral responses and mood (Maes *et al.*, 1997).

Furthermore, the synthesis of serotonin from its immediate precursor, tryptophan, requires a cofactor that is derived from neopterin. Although levels of neopterin are elevated in depression (a consequence of increased Th1 activity and macrophage stimulation) its conversion to the essential cofactor is inhibited, a mechanism that also contributes to the high levels of neopterin seen in depressed patients (Coppen *et al.*, 1989; Bottiglieri *et al.,* 1992). So this is yet another avenue whereby immune activation can directly affect brain monoamine availability and mood (van Amsterdam and Opperhuizen, 1999).

B. EFFECTS OF CORTISOL AND CRF ON MONOAMINES

There is no doubt that brain monoamines play a crucial role in the regulation of mood and reward: The initial theory of depression identified NA and 5HT as being the crucially depleted neurotransmitters (see Ban, 2001 for a historical perspective). Drugs used to control depression increase availability of one or both of these neurotransmitters (reviewed in Trimble 1998). It has been suggested that the alterations in brain NA and 5HT, which result in the lowered mood, may be secondary to high cortisol levels (Dinan, 1994). Certainly corticosteroids can alter neural activity by directly interacting with serotonergic $5HT_{1A}$ receptors (Okuhara and Beck, 1998).

The monoamines NA and 5HT activate the HPA axis as well as regulate mood. Cortisol, in return, restrains HPA activity, not only by its direct negative feedback inhibition at the level of the hypothalamus and pituitary but also by inhibiting NA and 5HT availability. Under normal conditions, when cortisol levels are tightly regulated, this does not have any detrimental effect. The consequence of high cortisol levels over a prolonged time period (as seen in depression) is the substantial reduction of 5HT and NA availability and hence a lower mood (Kvetnansky *et al.,* 1993; Pacak *et al.,* 1995).

The long-term consequences of raised cortisol levels can be observed in Cushing's disease (CD). The most common side effect of CD is mental disturbance and depression (Jeffcoate *et al.,* 1979). About one-third of patients with CD have significant psychiatric morbidity, two-thirds are depressed, and approximately 10% attempt suicide (Murphy, 1991). Direct involvement of cortisol in the induction of these mood changes is implied as the severity of the mood disorder in CD has been shown to correlate with the circulating cortisol level (Starkman and Schteingart, 1981). Furthermore, treatment by adrenalectomy (which removes the source of cortisol) or pharmacological inhibition of cortisol effectively reverses the mood disorder (Zeiger *et al.,* 1993).

The evidence that administration of glucocorticoid drugs to normal subjects can lead to euphoria rather than depression (although there may also be irritability and sleeplessness) (Murphy, 1991; Brown and Suppes, 1998) has led some people to question the hypercortisolemia theory of depression. However, this discrepancy can best be accounted for by the relative affinity of synthetic glucocorticoids for the two subtypes of receptor. Whereas in humans endogenous cortisol binds to both types of cortisol receptor the synthetic glucocorticoids are more or less specific for the type 2 receptors only. This means they have little impact on the type 1 receptors located in the brain's emotional centers (see Dinan, 1994). As cortisol and synthetic glucocorticoid drugs work on different receptors, in different brain regions, it is not surprising that they induce different effects.

In addition it has been shown that direct administration of CRF to the brains of laboratory animals leads to a constellation of behavioral abnormalities reminiscent of intense anxiety and melancholic depression (see Arborelius *et al.*, 1999). These include social withdrawal, reduced appetite and sexual activity, alongside psychomotor agitation or retardation (depending on context). The behavioral effects of CRF can be obtained at doses too low to activate the HPA axis. In fact CRF is widely distributed throughout the brain including the raphe nuclei and locus coeruleus (Owens and Nemeroff, 1991) where it has been suggested it plays a regulatory role in central monoamine function. Thus the high levels of CRF commonly associated with depression are thought to play a direct role in causing the symptoms of depression as well as in acting to initiate the HPA cascade (Ritchie and Nemeroff, 1991).

VII. Conclusions

In summary, mood can be modulated by inflammatory cytokines, CRF, and cortisol. In parallel, inflammatory cytokines can modulate neuroendocrine function and visa versa. Under normal circumstances the homeostatic regulation of the interaction between these factors is finely tuned and adaptive for survival. However, when one system becomes dysregulated the others are inevitably affected. Major depression is characterized by reduced NK cell activity, increased production of Th1 and pro-inflammatory cytokines, and stimulation of the acute-phase response as well as increased CRF and cortisol. It seems that the origin of this physiologically maladaptive milieu may be either prolonged inflammatory illness (or immune dysregulation that mimics inflammatory illness) or chronic uncontrollable stress. This is possible as both pro-inflammatory cytokines and psychological stress activate the HPA axis response system. Repeated activation of this system, by either route, induces desensitization of the regulatory negative feedback receptors in the brain, pituitary, and lymphocytes. Once the capacity for regulation is lost the elevated levels of inflammatory cytokines, CRF, and cortisol can prevail, all of which can have a negative influence on mood. It should be no surprise therefore that disturbed mood (depression) is associated with accelerated progression in certain diseases (reviewed elsewhere in this book)

268   ANGELA CLOW

as the parallel changes in both the immune system and the neuroendocrine activity would mediate this. In the past the treatment of depression has focused almost entirely on raising brain monoamine availability with virtually no attention being paid to the other physiological manifestations of the illness. In a similar way clinicians treating physical illness frequently fail to recognize and treat accompanying depression, even though it has been shown that such treatment improves physical health. An understanding of the intimate relationship between mood and the immune and neuroendocrine systems should inform the development of new treatment strategies.

References

Arborelius, L., Owens, M. J., Plotsky, P. M., and Nemeroff, C. B. (1999). The role of corticotrophin-releasing factor in depression and anxiety disorders. *J. Endocrinol.* **160,** 1–12.

Ban, T. A. (2001). Pharmacotherapy of depression: A historical analysis. *J. Neural. Transm.* **108,** 707–716.

Barden, N. (1996). Modulation of glucocorticoid receptor gene expression by antidepressant drugs. *Pharmacopsychiatry* **29,** 12–22.

Besedovsky, H., Del Rey, A., and Sorkin, E. (1983). The immune response evokes changes in brain noradrenergic neurones. *Science* **221,** 564–566.

Besodovsky, H. O., Del Rey, A., Klusman, I., Furukawa, H., Monge Arditi, G., and Kabiersch, A. (1991). Cytokines as modulators of the hypothalamus-pituitary-adrenal axis. *J. Steroid Biochem. Mol. Biol* **40,** 613–618.

Bottiglieri, T., Hyland, K., Laundy, M., Godfrey, P., Carney, M. W., Toone, B. K., and Reynolds, E. H. (1992). Folate deficiency, biopterin and monoamine metabolism in depression. *Psychol. Med.* **22,** 871–876.

Bradbury, M. J., Akana, S. F., Cascio, C. S., Levin, N., Jacobson, L., and Dallman, M. F. (1991). Regulation of basal ACTH secretion by corticosterone is mediated by both type 1 (MR) and type 2 (glucocorticoid receptor) receptors in rat brain. *J. Steroid Biochem. Mol. Biol.* **40,** 105–111.

Brown, E. S., and Suppes, T. (1998). Mood symptoms during corticosteroid therapy: A review. *Harvard Rev. Psychiatry* **5,** 239–246.

Bunney, W. E., and Fawcett, J. A. (1965). Possibility of a biochemical test for suicidal potential. *Arch. Gen. Psychiatry* **13,** 232–239.

Cappel, R., Gregiore, F., Thiry, L., and Sprecher, S. (1978). Antibody and cell-mediated immunity to herpes simplex virus in psychotic depression. *J. Clin. Psychiatry* **39,** 266–268.

Capuron, L., Ravaud, A., Guald, N., Bosmans, E., Dantzer, R., Maes, M., and Neveu, J. (2001). Association between immune activation and early depressive symptoms in cancer patients treated with interleukin-2-based therapy. *Psychoneuroendocrinology* **26,** 797–808.

Carroll, B. J., Curtis, G. C., Davies, B. M., Mendels, J., and Sugarman, A. A. (1976). Urinary free cortisol excretion in depression. *Psychol. Med.* **6,** 43–40.

Carroll, B. J., Feinberg, M., Greden, J. F., Tarika, J., Albala, A. A., Hasket, R. F., James, N. M., Kronfol, Z., Lohr, N., Steiner, M., de Vigne, M. P., and Young, E. (1981). A specific laboratory test for the diagnosis of melancholia: Standardization, validation and clinical utility. *Arch. Gen. Psychiatry* **38,** 15–22.

Castanon, N., Bluthe, R. M., and Dantzer, R. (2001). Chronic treatment with the atypical antidepressant tianeptine attenuates sickness behaviour induced by peripheral but not central lipopolysaccharide and interleukin-1β in the rat. *Psychopharmacology* **154**(1), 50–60.

Cohen, P., Pine, D. S., Must, A., Kasen, S., and Brook, J. (1998). Prospective associations between somatic illness and mental illness from childhood to adulthood. *Am. J. Epidemiol.* **147,** 232–239.

Connor, T. J., and Leonard, B. E. (1998). Depression, stress and immunological activation: The role of cytokines in depressive disorders. *Life Sci* **62**, 583–606.

Coppen, A., Swade, C., Jones, S. A., Armstrong, R. A., Blair, J. A., and Leeming, R. J. (1989). Depression and tetrahydrobiopterin: The folate connection. *J. Affect. Disord.* **16,** 103–107.

Danzer, R. (1999). Sickness behaviour: A neuroimmune-based response to infections. *In* "Psychoneuroimmunology: An Interdisciplinary Introduction" (M. Schedlowski, and U. Tewes, eds.), Kluwer Academic/Plenum, New York.

Dickens, C., McGowan, L., Clark-Carter, D., and Creed, F. (2002). Depression in rheumatoid arthritis: A systematic review of the literature with meta-analysis. *Psychosom. Med.* **65,** 52–60.

Dinan, T. G. (1994). Glucocorticoids and the genesis of depressive illness: A psychobiological model. *Br. J. Psychiatry* **164,** 365–371.

Diorio, D., Viau, V., and Meaney, M. J. (1993). The role of the medial prefrontal cortex (cingulated gyrus) in the regulation of the hypothalamic-pituitary-adrenal responses to stress. *J. Neurosci.* **13**, 3839–3847.

Dunbar, P. R., Hill, J., Neale, T. J., and Mellsop, G. W. (1992). Neopterin measurement provides evidence of altered cell-mediated immunity in patients with depression, but not with schizophrenia. *Psychol. Med.* **22**, 1051–1057.

Evans, D. L., Burnett, G. B., and Nemeroff, C. B. (1983). The dexamethasone suppression test in the clinical setting. *Am. J. Psychiatry* **140**, 586–589.

Feldman, S., and Weidenfeld, J. (1999). Glucocorticoid receptor antagonists in the hippocampus modify the negative feedback following neural stimuli. *Brain Res.* **821**, 33–37.

Ferrini, M., Piroli, G., Frontera, M., Falbo, A., Lima, A., and De Nicola, A. F. (1999). Estrogens normalize the hypothalamic-pituitary-adrenal axis in response to stress and increase glucocorticoid receptor immunoreactivity in hippocampus of aging male rats. *Neuroendocrinology* **69,** 129–137.

Frank, M. G., Hendricks, S. E., Johnson, D. R., Wieseler, J. L., and Burke, W. J. (1999). Antidepressants augment natural killer cell activity: *In vivo* and *in vitro. Neuropsychobiology* **39,** 18–24.

Heuser, I. (1998). The hypothalamic-pituitary-adrenal system in depression. *Pharmacopsychiatry* **31,** 10–13.

Irwin, M., and Gillin, J. C. (1987). Impaired natural killer cell activity among depression patients. *Psychiatry Res* **20**, 181–182.

Irwin, M., Smith, T. L., and Gillin, J. C. (1987). Low natural killer cell cytotoxicity in major depression. *Life Sci.* **41**, 2127–2133.

Jain, R., Zwicker, D., Hollander, C. S., Brand, H., Saperstein, A., Hutchinson, B., Brown, C., and Audhya, T. (1991). Corticotrophin-releasing factor modulates the immune response to stress in rats. *Endocrinology* **128**, 1329–1336.

Jeffcoate, W. J., Silverstone, J. T., Edwards, C. R., and Besser, G. M. (1979). Psychiatric manifestations of Cushing's Syndrome: Response to lowering plasma cortisol. *Q. J. Med.* **191,** 465–472.

Joyce, P., Hawes, C., Mulder, R., Sellman, J., Wilson, D., and Boswell, D. (1992). Elevated levels of acute phase proteins in major depression. *Biol. Psychiatry* **32**, 1035–1041.

Kent, S., Bluthe, R. M., Kelley, K. W., and Dantzer, R. (1992). Sickness behavior as a new target for drug development. *Trends Pharmacol. Sci.* **13**, 24–28.

Kiecolt-Glaser, J. K., McGuire, L., Robles, T. F., and Glaser, R. (2002). Emotions, morbidity, and mortality: New perspectives in psychoneuroimmunology. *Annu. Rev. Psychol.* **53**, 83–107.

Kockler, M., and Heun, R. (2002). Gender differences of depressive symptoms in depressed and nondepressed elderly persons. *Int. J. Geriatr. Psychiatry* **17**, 65–72.

Krishnan, K. R., Doraiswamy, P. M., Lurie, S. N., Figiel, G. S., Husain, M. M., Boyko, O. B., Ellinwood, E. H., and Nemerof, C. B. (1991). Pituitary size in depression. *J. Clin. Endocrinol. Metab.* **72**(2), 256–259.

Kronfol, Z., and House, J. D. (1985). Depression, hypothalamic-pituitary adrenocortical activity, and lymphocyte function. *Psychopharmacol Bull.* **21**(3), 476–478.

Kronfol, Z., Silva, J., Greden, J., Dembinsky, S., Gardener, R., and Carroll, B. (1983). Impaired lymphocyte function in depressive illness. *Life Sci.* **33**, 241–247.

Kvetnansky, R., Fukuhara, K., Pacak, K., Cizza, G., Goldstein, D. S., and Kopin, I. J. (1993). Endogenous glucocorticoids restrain catecholamine synthesis and release at rest and during immobilisation stress in rats. *Endocrinology* **133**, 1411–1419.

Lippman, M., and Barr, R. (1977). Glucocorticoid receptors in purified subpopulations of human peripheral blood lymphocytes. *J. Immunol.* **118**, 1977–1981.

Lowry, M. T., Reder, A. T., Antel, J. P., and Meltzer, H. Y. (1984). Glucocorticoid resistance in depression: The dexamethasone suppression test and lymphocyte sensitivity to dexamethasone. *Am. J. Psychiatry* **141**, 1365–1370.

Maes, M., Bosmans, E., Suy, E., Vandervorst, C., Dejonckheere, C., Minner, B., and Raus, J. (1991). Depression-related disturbances in mitogen-induced lymphocyte responses, interleukin-1β and soluble interleukin-2–receptor production. *Acta Psychiatr. Scand.* **84**, 379–386.

Maes, M., Stephems, W., DeClerck, L., Bridts, C., Peeters, D., Schotte, C., and Cosyns, P. (1992a). Immune disorders in depression: Higher T Helper/T Suppressor-cytotoxic cell ratio. *Acta Psychiatr. Scand.* **86**, 423–431.

Maes, M., Van der Planken, M., Stephens, W., Peeters, D., DeClerck, L., Bridts, C., Schotte, C., and Cosyns, P. (1992b). Leukocytosis, monocytosis and neutrophilia: Hallmarks of severe depression. *J. Psychiatr. Res.* **26**(2), 125–134.

Maes, M., Stephens, W., DeClerck, L., Bridts, C., Peeters, D., Schotte, C., and Cosyns, P. (1993). Significantly increased expression of T-cell activation markers (interleukin 2 and HLA-DR) in depression. Further evidence for an inflammatory process during illness. *Pog. Neuropsychopharmacol. Biol. Psychiatry* **17**, 241–255.

Maes, M., Scharpe, S., Van Grootel, L., Uyttenbroeck, W., Cooreman, W., Cosyns, P., and Suy, E. (1994). Increased neopterin and interferon-γ secretion and lower availability of L-tryptophan in major depression: Further evidence for an immune response. *J. Psychiatr. Res.* **54**, 143–160.

Maes, M., Meltzer, H., Buckley, P., and Bosmans, E. (1995). Plasma soluble interleukin-2 and transferring receptor in schizophrenia and major depression. *Arch. Psychiatr. Clin. Neurosci.* **244**(6), 325–329.

Maes, M., Wauters, A., Verkerk, R., Demedts, P., Neels, H., van Gastel, A., Cosyns, P., Scharpe, S., and Desnyder, R. (1996). Lower serum tryptophan availability in depression as a marker of a more generalized disorder in protein metabolism. *Neuropsychopharmacology* **15**, 243–251.

Maes, M., Verkerk, R., Vandoolaeghe, E., Van Hunsel, F., Neels, H., Wauters, A., Demedts, P., and Scharpe, S. (1997). Serotonin-immune interactions in major depression: Lower serum tryptophan as a marker of an immune-inflammatory response. *Eur. Arch. Psychiatry Clin. Neurosci.* **247**, 154–161.

Maes, M., Bonaccorso, S., Marino, V., Puzella, A., Pasquini, M., Biondi, M., Artini, M., Almerighi, C., and Meltzer, H. (2001). Treatment with interferon-α (INF-α) of hepatitis C patients induces lower dipeptidyl peptidase IV activity, which is related to INF-α–induced depressive and anxiety symptoms and immune activation. *Mol. Psychiatry* **6**(4) 475–480.

Magarinos, A. M., Somoza, G., and De Nicola, A. F. (1987). Glucocorticoid negative feedback and glucocorticoid receptors after hippocampectomy in rats. *Horm. Metab. Res.* **19**, 105–109.

Maier, S. F., and Watkins, L. R. (1998). Cytokines for psychologists: Implications of bidirectional immune-to-brain communication for understanding behavior, mood and cognition. *Psychol. Rev.* **105**, 83–107.

Mendlovic, S., Mozes, E., Eilat, E., Doron, A., Lereya, J., Zakuth, V., and Spirer, Z. (1999). Immune activation in non-treated suicidal major depression. *Immunol. Lett.* **67**, 105–108.

Miller, A. H. (1998). Neuroendocrine and immune system interactions in stress and depression. *Psychoneuroendocrinology* **21**, 443–463.

Miller, A. H., Spenser, R. L., Pulera, M., Kang, S., McEwan, B. S., and Stein, M. (1992). Adrenal steroid receptor activation in rat brain and pituitary following dexamethasone: Implications for dexamesathone suppression test. *Biol. Psychiatry* **32**, 850–869.

Mizoguchi, K., Yuzurihara, M., Ishige, A., Sasaki, H., Chui, D.-H., and Tabira, T. (2001). Chronic stress differentially regulates glucocorticoid negative feedback response in rats. *Psychoneuroendocrinology* **26**, 443–459.

Modell, S., Yassouridis, A., Huber, J., and Holsboer, F. (1997). Corticosteroid receptor function is decreased in depressed patients. *Neuroendocrinology* **65**, 216–222.

Murphy, B. E. P. (1991). Steroids and depression. *J Steroid Biochem. Mol. Biol.* **38**, 537–559.

Nassberger, L., and Traskman-Bendz, L. (1993). Increased soluble interleukin-2 receptor in suicide attempters. *Acta Psychiatr. Scand.* **88**, 48–52.

Nemeroff, C. B. (1988). The role of corticotrophin-releasing factor in the pathogenesis of major depression. *Pharmacopsychiatry* **21**, 76–82.

Nemeroff, C. B., Widerlov, E., Bisset, G., Walleus, H., Karlsson, I., Eklund, K., Kilts, C. D., Loosen, P. T., and Vale, W. (1984). Elevated concentrations of corticotrophin-releasing factor–like immunoreactivity in depressed patients. *Science* **226**, 1342–1344.

Nemeroff, C. B., Bissette, G., Akil, H., and Fink, M. (1991). Neuropeptide concentrations in the CSF of depressed patients treated with electroconvulsive shock therapy. Corticotrophin-releasing factor, β-endorphin and somatostatin. *Br. J. Psychiatry* **158**, 59–63.

Nemeroff, C. B., Krishnan, K. R., Reed, D., Leder, R., Beam, C., and Dunnick, N. R. (1992). Adrenal gland enlargement in major depression. A computed tomographic study. *Arch. Gen. Psychiatry* **49**(5), 384–387.

Okuhara, D. Y., and Beck, S. G. (1998). Corticosteroids alter 5–hydroxytryptamine$_{1A}$ receptor-effector pathway in hippocampal subfield CA3 pyramidal cells. *J. Pharmacol. Exp. Ther.* **284**, 1227–1233.

Owens, M. J., and Nemeroff, C. B. (1991). Physiology and pharmacology of corticotropin-releasing factor. *Pharmacol. Rev.* **43**, 425–473.

Pacak, K., Palkovits, M., Kvetnansky, R., Matern, P., Hart, C., Kopinm, I. J., and Goldstein, D. S. (1995). Catecholaminergic inhibition by hypercortisolemia in the PVN nucleus of the conscious rat. *Endocrinology* **136**, 4814–4819.

Pariante, C. M., Nemeroff, C. B., and Miller, A. H. (1995). Glucocorticoid receptors in depression. *Isr. J. Med. Sci.* **31**, 705–712.

Peiffer, A., Veilleux, S., and Barden, N. (1991). Antidepressant and other centrally acting drugs regulate glucocorticoid receptor messenger RNA levels in rat brain. *Psychoneuroendocrinology* **16**, 505–515.

Pepin, M. C., Beaulieu, S., and Barden, N. (1989). Antidepressants regulate glucocorticoid receptor messenger RNA concentrations in primary neuronal cultures. *Mol. Brain Res.* **6**, 77–83.

Raison, C. L., and Miller, A. H. (2001). The neurobiology of stress and depression. *Semin. Clin. Neuropsychiatry* **6**, 277–295.

Ravindran, A. V., Griffiths, J., Merali, Z., Knott, V. J., and Anisman, H. (1999). Influences of acute tryptophan depletion on mood and immune measure in healthy males. *Psychoneuroendocrinology* **24**, 99–113.

Ritchie, J. C., and Nemeroff, C. B. (1991). Stress, the hypothalamic-pituitary-adrenal axis and depression. *In* "Stress, Neuropeptides and Systemic Disease". Academic Press, San Diego.

Rohleder, N., Schomer, N. C., Hellhammer, D. H., Engel, R., and Kirschbaum, C. (2001). Sex differences in glucocorticoid sensitivity of proinflammatory cytokine production after psychosocial stress. *Psychosom. Med.* **63**, 966–972.

Rosmond, R., Dallman, M. F., and Bjorntorp, P. (1998). Stress-related cortisol secretion in men: Relationships with abdominal obesity and endocrine, metabolic and hemodynamic abnormalities. *J. Clin. Endocrinol. Metab.* **83**, 1853–1859.

Rubin, R. T., Phillips, J. J., Sadow, T. F., and McCracken, J. T. (1995). Adrenal gland volume in major depression. Increase during the depressive episode and decrease with successful treatment. *Arch. Gen. Psychiatry* **52**(3), 213–218.

Sapolsky, R. M., Krey, L. C., and McEwan, B. S. (1984). Stress downregulates corticosterone receptors in a site-specific manner in the brain. *Endocrinology* **114**, 287–292.

Shekelle, R. B., Raynor, W. J., Ostfield, A. M., Garron, D. C., Bieliauskas, L. A., Liu, S. C., Maliza, C., and Paul, O. (1981). Psychological depression and 17-year risk of death from cancer. *Psychosomat. Med.* **43**, 117–125.

Shuto, H., Kataoka, Y., Horikawa, T., Fujihara, N., and Oishi, R. (1997). Repeated interferon-α administration inhibits dopaminergic neural activity in the mouse brain. *Brain Res.* **747**, 348–351.

Spencer, R. L., Miller, A. H., Stein, M., and McEwen, B. S. (1991). Corticosterone regulation of type I and type II adrenal steroid receptors in brain, pituitary, and immune tissue. *Brain Res.* **549**, 236–246.

Starkman, M. N., and Schteingart, D. E. (1981). Neuropsychiatric manifestations of patients with Cushing's syndrome. *Arch. Intern. Med.* **141**, 215–219.

Stein, M., Miller, A. H., and Trestman, R. L. (1991). Depression and the immune system. *In* "Psychoneuroimmunology" 2nd ed. (Robert Ader, David L. Felter, and Nicholas Cohen, eds.), pp. 897–930. Academic Press, New York.

Sternberg, E. M. (1997). Emotions and disease: From balance of humours to balance of molecules. *Nat. Med.* **3**, 264–267.

Sunanda Rao, B. S., and Raju, T. R. (2000). Restraint stress-induced alterations in the levels of biogenic amines, amino acids, and AChE activity in the hippocampus. *Neurochem. Res.* **25**, 1547–1552.

Tondo, L., Pani, P. P., Pellegrini-Bettoli, R., Milia, G., and Manconi, P. E. (1988). T-Lymphocytes in depressive disorder. *Med. Sci. Res.* **16**, 867–868.

Torpy, D. J., and Chrousos, G. P. (1996). The three-way interactions between the hypothalamic-pituitary-adrenal and gonadal axes and the immune system. *Baillieres Clin. Rheumatol.* **10**, 181–198.

Trimble, M. R. (1998). "Biological Psychiatry," pp. 241–281. Wiley, Chichester.

van Amsterdam, J. G., and Opperhuizen, A. (1999). Nitric oxide and biopterin in depression and stress. *Psychiatry Res.* **18**, 33–38.

Van West, D., and Maes, M. (1999). Activation of the inflammatory response system: A new look at the etiopathogenesis of depression. *Neuroendocrinol. Lett.* **20**(1–2), 11–17.

Weber, B., Lewicka, S., Deuschle, M., Colla, M., Vecsei, P., and Heuser, I. (2000). Increased diurnal plasma concentrations of cortisone in depressed patients. *J. Clin. Endrocrin. Metab.* **85,** 1133–1136.

Zeiger, M. A., Franker, D. L., Pass, H. I., Nieman, L. K., Cutler, G. B., Chrousos, G. P., and Norton, J. A. (1993). Effective reversibility of the signs and symptoms of hypercortisolism by bilateral adrenalectomy. *Surgery* **114,** 1138–1143.

Zorrilla, E. P., Luborsky, L., McKay, J. R., Rosenthal, R., Houldin, A., Tax, A., McCorkle, R., Seligman, D. A., and Schmidt, K. (2001).). The relationship of depression and stressors to immunological assays: A meta-analytic review. *Brain Behav. Immun.* **15,** 199–226.

IMMUNITY AND SCHIZOPHRENIA: AUTOIMMUNITY, CYTOKINES, AND IMMUNE RESPONSES

Fiona Gaughran

Ladywell Unit, University Hospital
Lewisham, London SE13 6LH, United Kingdom

I. Background

Schizophrenia is a psychotic illness that affects almost 1% of the population worldwide. It is commonly diagnosed in late adolescence or young adulthood. Its geographic distribution appears to be uniform (International Pilot Study of Schizophrenia, 1973). The diagnosis of schizophrenia is a clinical one, relying on the presence of clusters of clinical features. The symptoms can be divided into three classes. Positive symptoms are psychotic features such as delusions or hallucinations and thought disorganization. Negative symptoms include poor motivation, social withdrawal, day–night reversal, poor self-care, and poverty of thought. Disturbance in aspects of cognitive function such as attention, executive functions, and working memory are also consistently observed, contributing greatly to the significant functional disability associated with the disorder (Lewis and Lieberman, 2000). Schizophrenia is a chronic illness. Management is thus expensive and life-long, both in terms of human resources and medication.

It is possible that schizophrenia, as we understand it, is more than one disease. It is almost certain that there is more than one cause. The etiology of schizophrenia is as yet unclear. There is undoubtedly a significant genetic component, although this is likely to be related to a combination of many genes rather than to a single "schizophrenia" gene. Other factors linked to the causation of this condition include psychosocial variables, environmental triggers, viral infection, autoimmune activity, endocrinological irregularities, and neurological structural/anatomical abnormalities.

Epidemiological studies have linked schizophrenia with environmental factors. An excess of obstetric complications, particularly in male patients with schizophrenia, has been noted, along with minor physical anomalies and dermatoglyphic (fingerprint) abnormalities, suggesting a possible intrauterine insult (O'Callaghan et al., 1991, 1992; Bracha et al., 1992). The possibility of a viral or nutritional origin to these developmental anomalies has been raised. Schizophrenia is more common in cities and occurs more frequently in lower socioeconomic groups (Kohn, 1968). First trimester famine, occurrence of influenza epidemics during gestation, and early perinatal viral infections are also linked with the later development of schizophrenia (Susser and Lin, 1992; Cooper, 1992; Jones et al., 1998).

By the time a person presents with schizophrenia, brain changes are often evident on an MRI scan (Nopoulos et al., 1995). The most consistent abnormalities are enlargement of the ventricles and dilatation of the cortical sulci, along with reduced volume of the anterior hippocampus, the parahippocampal gyrus, the frontal lobe, and the superior temporal gyrus. These findings are especially marked in the left hemisphere. Postmortem investigation confirms these changes and a relative absence of gliosis, or scar tissue, suggests that these structural changes may date from before the third trimester of intrauterine development (Waddington, 1993).

The theory that the neuropathology predates the onset of symptoms is strengthened by studies of children suggesting that those who go on to develop schizophrenia are more likely to show abnormalities of behavior in childhood (Jones et al., 1994). Men presenting with schizophrenia, in particular, are more likely to have attended a child guidance clinic well before the onset of their schizophrenic illness (Ambelas, 1992).

The monoamine neurotransmitters, dopamine and serotonin, are important variables in the pathogenesis of the disorder. A glutamatergic deficiency model has been proposed (Carlsson and Carlsson, 1990). Dopamine agonists, such as L-dopa and amphetamine, along with glutamatergic antagonists such as ketamine, can produce acute psychosis. All the drugs used for the treatment of schizophrenia over the last four decades of the twentieth century have in common an antagonism of dopamine. The newer "atypical" neuroleptics often exhibit both dopamine- and serotonin-receptor blockade.

Immune abnormalities have been described in schizophrenia for many years and predate the use of neuroleptic medication. There are two possible roles that the immune system could play in the causation of schizophrenia. First, the early neurodevelopmental changes described may be mediated by imbalances in the immune system. Second, the immune system may modulate the ongoing clinical course of the illness. Interestingly, the new "atypical" antipsychotic medications appear to have immunomodulatory effects.

II. Autoimmune Diseases and Schizophrenia

There are many clinical similarities linking schizophrenia and certain forms of autoimmune disease. Schizophrenic-like symptoms can be directly induced by autoimmune illnesses. Such illnesses may be chronic, as in CNS lupus erythematosis, or acute, such as Sydenham's chorea. Sydenham's chorea causes psychosis, obsessional thoughts, and abnormal involuntary movements which wax and wane in line with changes in systemic antibrain antibody titers (Swedo *et al.*, 1993).

Knight *et al.* (1992) considered if autoimmune mechanisms could account for the genetic predisposition to schizophrenia. These authors noted various similarities. For example, the remitting–relapsing course of autoimmune illnesses (e.g., lupus) is similar to that of schizophrenia: There is a variable age of onset in both conditions; both systemic lupus erythematosis (SLE) and thyrotoxicosis behave like schizophrenia in displaying a correlated age of onset within families; both can be precipitated by drugs, physical injury, or infection. Schizophrenia and autoimmune diseases have similar monozygotic twin discordance rates. Patients with insulin-dependent diabetes mellitus (IDDM), like those with schizophrenia, are characterized by a clustering toward winter births.

Both positive and negative epidemiological associations exist between schizophrenia and various autoimmune diseases. Patients with schizophrenia have an increased risk of suffering from autoimmune conditions in general (Ganguli *et al.*, 1987). On the other hand, specific autoimmune conditions such as rheumatoid arthritis and IDDM are less likely to coexist with schizophrenia than would be expected (Eaton *et al.*, 1992; Finney, 1989). Although autoimmune disorders tend to occur together, negative associations can result from genes having opposing effects on the risk of developing different disorders.

It also appears that the inheritance of schizophrenia may be related to that of autoimmune disorders. Not only the sufferers themselves, but also the families of patients with schizophrenia, show an excess of autoimmune conditions. People with a first-degree relative suffering from schizophrenia are more likely to also have a parent or sibling with an autoimmune disease (Wright *et al.*, 1996). Autoimmune thyroid disease (thyrotoxicosis and myxedema) is more common in the first-degree relatives of psychotic patients, as are both insulin-dependent and non-insulin-dependent diabetes mellitus even though IDDM is negatively associated with schizophrenia (see earlier). Mothers of patients with schizophrenia have lower rates of rheumatoid arthritis than controls, mirroring the findings in the patients themselves (Gilvarry *et al.*, 1996; Baldwin, 1979).

It has been suggested that some cases of schizophrenia may have a viral cause and that the effects of the virus on the developing brain may be immunologically mediated. Influenza epidemics have been followed by an increase in births of people who will later go on to develop schizophrenia (Cooper, 1992). It has been proposed that virus-induced maternal antibodies may cross the placenta and the immature blood–brain barrier to cross-react with fetal brain tissues. These antibodies may interfere with neurodevelopment, resulting in schizophrenia in later life. It is known that rabbits inoculated with influenza A virus produce an antibody which cross-reacts with a protein in the human hippocampus, cortex, and cerebellum leading to suggestions that certain mothers, perhaps those with enhanced antibody resistance to viral infections, are immunologically predisposed to such a reaction (Laing *et al.*, 1996). Interestingly, relatives of patients with schizophrenia have been shown to suffer from fewer viral infections than controls (Carter and Watts, 1971).

A. AUTOANTIBODIES IN SCHIZOPHRENIA

Higher than expected levels of various antibodies to both brain and non-brain tissue have been reported in schizophrenia. Antibodies to non-CNS-specific antigens that were reported include antihistone antibodies, anti-ganglioside antibodies, rheumatoid factor, anticardiolipin antibodies, lupus anticoagulant, and antibodies to nicotinic acetylcholine receptors (Yannitsi *et al.*, 1991; Stevens and Weller, 1992; Mukherjee *et al.*, 1994; Chengappa *et al.*, 1991, 1992b). Platelet autoantibodies that inhibit dopamine uptake have also been described, which may be relevant given the dopamine hypothesis of schizophrenia (Kessler and Shinitzky, 1993). Galinowski *et al.* (1992) found decreased IgG, IgM, and IgA autoantibodies against a variety of autoantigens, namely actin, tubulin, myosin, DNA, thyroglobulin, elastin, and albumin. By contrast patients with schizophrenia (including unmedicated patients) along with their well relatives show high levels of anticardiolipin antibodies (Sirota *et al.*, 1993; Chengappa *et al.*, 1991; Firer *et al.*, 1994), implying a possible association with antiphospholipid syndrome, which is also characterized by anticardiolipin antibodies.

Antinuclear antibodies, strongly associated with SLE, have classically been linked to antipsychotic medication, but high levels have also been found in drug-free patients, and so it is reasonable to conclude that the elevation cannot be entirely explained by neuroleptic medication. Increased antinuclear antibodies (ANA), anti-double-strand DNA, and anti-single-strand DNA titers have been reported not only in patients with schizophrenia but also in their healthy first-degree relatives, with no significant difference in autoantibody systems between the patients and their relatives (Sirota *et al.*, 1991).

Non-right-handedness is especially prevalent in schizophrenia (Gur, 1977). Chengappa *et al.* (1992) found that non-right-handed, neuroleptic-naive patients with schizophrenia have more autoantibodies. They suggested that non-right-handedness is a clinical marker of predisposition to autoimmunity in affected men. Handedness was not a distinguishing factor in women. If a fetus was subjected to an insult during neurodevelopment affecting the dominant hemisphere, the immune system could be exposed to unusual or intracellular antigens, precipitating an antibody response. The association of non-CNS antibodies with a past history of obstetric complications in schizophrenia (Chengappa *et al.*, 1995) may also be explained in this way.

B. ANTINEURONAL ANTIBODIES

Lehmann-Facius first found antibrain antibodies in schizophrenia as far back as 1939. In 1963, Fessel replicated this work followed shortly afterward by Heath *et al.* in 1967. DeLisi *et al.*, in 1985, noted that 18% of patients with schizophrenia had antibrain membrane-binding antibodies significantly outside the control range. However, not all researchers have reported these positive associations (Kelly *et al.*, 1987).

Antibodies to various specific brain areas have been described in schizophrenia. These regions include the hippocampus, the septal region, the cingulate gyrus, the amygdala, and the frontal cortex, as well as neurons, glia, and blood vessels (Ganguli *et al.*, 1987; Kuznetsova and Semenov, 1961). Antibodies to the septal region of the brain, in particular, have been found by a number of researchers (Heath *et al.*, 1989; Henneberg *et al.*, 1994). Increased antihippocampal antibody concentration in schizophrenia is associated with decreased lymphocyte production of IL-2 (Yang *et al.*, 1994).

Antibodies to the 60 and 70 kilodalton (kDa) human heat-shock proteins (HSP) have been described. Especially high anti-HSP70 titers are seen in never-medicated patients. High anti-HSP60 titers are mainly found in patients being treated with neuroleptics. Since heat-shock proteins are involved in diverse neuroprotective mechanisms, antibodies against them may inhibit neuroprotective processes (Schwarz *et al.*, 1999).

Heath *et al.* (1967) reported that schizophrenic patients exhibit abnormal brain waves in electroencephalograph (EEG) recordings from the caudate nucleus and septal area. He also recorded these abnormal waves from similar sites in monkey brains after injections of IgG isolated from the blood of acutely ill schizophrenic patients into the monkey lateral ventricle cerebrospinal fluid (CSF). In 1980, Bergen *et al.* prepared IgG fractions from control subjects and acutely ill patients with schizophrenia and tested them in rhesus monkeys under double-blind conditions. Of 107 serums tested from 24 schizophrenic patients, 29 produced positive EEG recordings in the monkeys. From 30 control subjects they tested 80 samples and found only 6 to be positive. This amounted to more than 1 positive reaction for every 4 tested in the schizophrenia group and approximately 1 in 13 positive from control serum fractions. This suggested a substance present in serum from patients with schizophrenia that causes an effect on brain function.

In 1989 Heath and co-workers used crossed-immunoelectrophoresis to evaluate reactivity of an IgG fraction from serum of schizophrenic patients and controls with homogenates of tissues of the septal region, hippocampus, vermal cerebellum, frontal cortex, and liver of rhesus monkeys. When IgG fractions of unmedicated schizophrenic patients and schizophrenic patients who had received neuroleptic medication for less than 24 h were tested against the septal region homogenate, a precipitin arc was identified, indicating a positive result, with more than 95% of the fractions. In contrast, IgG fractions of schizophrenic patients who had received neuroleptic medication for more than 24 h were rarely positive. Fractions of all control subjects tested negatively. Potentially this illustrates a remarkable relationship between neuroleptic medication and the immune system's behavior in producing and sustaining IgG specific for antigens present in critical regions of the brain.

Pandey *et al.* (1981), using a hemagglutination technique, found high titers of antibrain antibodies in serum and CSF, not just from patients but also from their relatives. These antibrain antibodies were more frequent in those with a family history of schizophrenia and with a past history of illness.

C. Immunoglobulin Isotype Representation in Schizophrenia

Many studies have examined differential immunoglobulin isotype levels in schizophrenia but the results have been inconsistent. Four separate groups found no difference in serum IgG, IgM, IgA, and IgE levels overall between patients with schizophrenia and controls (Roos *et al.*, 1985; Rao *et al.*, 1988; Stevens *et al.*, 1990; Cazzullo *et al.*, 1998).

DeLisi's team measured immunoglobulin levels in schizophrenia on two occasions. In 1981 they reported decreased CSF and serum IgG, IgA, and IgM in patients with chronic schizophrenia, but when they repeated the work in acutely ill patients in 1984 only the IgM finding persisted. Bhatia's team again reported decreased serum IgA and IgG levels in 1992, although they found normal IgM levels (Bhatia *et al.*, 1992).

In marked contrast, Legros *et al.* (1985) reported increased serum IgM levels in psychiatric illness overall, whereas in China, Chong-Thim *et al.* (1993) have also found high IgM levels in acutely ill patients along with increased IgG levels in male patients. Overall, therefore, no consistent direction of change in the representation of serum immunoglobulin isotype levels has been clearly demonstrated in schizophrenia.

Centrally, however, it appears that CSF IgG is positively correlated with negative symptoms in schizophrenia and, conversely, patients with these CSF alterations may be at higher risk of developing negative symptoms (Muller and Ackenheil, 1995).

Reichelt and Landmark (1995) described increased specific IgA antibodies in schizophrenia. More patients than controls showed serum IgA antibody levels above the upper normal limit to gliadin, β-lactoglobulin, and casein. Interestingly, earlier discharge from the hospital has been reported after a milk-free, gluten-free diet, implying some association with mucosal antigen challenge and the nature of the immune response thus engendered (Dohan and Grasberger, 1973).

III. T and B Lymphocytes in Schizophrenia

Morphologically atypical lymphocytes ("P cells"), similar to those found in mononucleosis and other viral diseases, have been reported in patients with schizophrenia, including a subgroup of patients who had never taken antipsychotic medication (Lahdelma *et al.*, 1995).

Many studies have since examined the distribution of peripheral blood lymphocytes in schizophrenia. Such studies tend to reveal abnormalities although the direction of change is inconsistent. The total T cell number is low although the percentage of T cells in the blood is increased (Coffey *et al.*, 1983; DeLisi *et al.*, 1982).

One of the most consistent findings is of an increase in CD4$^+$ (T helper cell) subpopulations (Muller *et al.*, 1991). T suppressor cell activity is low (Muller *et al.*, 1993) although both increased and decreased T suppressor cell numbers have been reported (Achiron *et al.*, 1994; DeLisi *et al.*, 1982). Muller *et al.* (1993) quoted increased $CD3^+$ (total T cells) and $CD4^+$ cell numbers along with an increased ratio of $CD4^+/CD8^+$ cells in schizophrenia. The distribution of T lymphocyte subsets in the CSF of patients with schizophrenia is also abnormal, with 74% of patients having a $CD4^+$ and/or $CD8^+$ level outside the normal range (Nikkila *et al.*, 1995).

In Cazzullo *et al.*'s study (1998), however, when compared to healthy controls and their own relatives, patients with schizophrenia showed a lower level of $CD4^+$ cells, with the $CD4^+$ $45RA^+$ (naive) subset being significantly higher. Conversely, the number of $CD4^+$ $45RA^-$ (memory) lymphocytes was lower in the patients in comparison to their relatives and controls, while the $CD8^+$ cytotoxic T cell percentage was significantly higher.

Decreased natural killer cell activity has been reported (DeLisi *et al.*, 1983), although again, there have been inconsistent findings (McDaniel *et al.*, 1992). Theodoropoulou *et al.* (2001) described increased percentages of activated $CD4^+$ and $CD16^+$ natural killer cells, as well as cells expressing ICAM-1 adhesion molecules and IL-2 specific receptors.

Patients with schizophrenia have an increased number and percentage of B lymphocytes (DeLisi *et al.*, 1982; Masserini *et al.*, 1990). McAllister *et al.* (1989) found that $CD5^+$ B cells, a subset of B cells with limited immunoglobulin diversity of low affinity, were elevated to a similar degree as in rheumatoid arthritis. However, Ganguli and Rabin (1993) were unable to replicate these findings.

Zorilla *et al.* (1996) examined leukocyte counts in patients with schizophrenia and their siblings and concluded that familial vulnerability for schizophrenia was characterized by a relative lymphopenia in the context of a relative granulocytosis, that is, a general reduction in peripheral blood lymphocytes in relation to granulocytes.

A. THE ACUTE-PHASE RESPONSE IN SCHIZOPHRENIA

Acute schizophrenia is accompanied by an acute-phase response as indicated by increased serum and/or plasma concentrations of a number of acute-phase proteins. Increased levels of serum α1-antitrypsin, α2-macroglobulin, haptoglobin, ceruloplasmin, thyroxine-binding globulin, fibrinogen, complement component 3, C4, α1-acid-glycoprotein, and hemopexin have been observed in patients with schizophrenia when compared to controls. Albumin, a negative acute-phase protein, transferrin, and retinol-binding protein levels are reduced. Hemopexin levels are increased only in acutely ill patients while complement C3 and complement hemolytic activity is decreased in chronically ill patients (Maes *et al.*, 1997; Wong *et al.*, 1996; Spivak *et al.*, 1993).

Interleukin (IL)-1 synergizes strongly with IL-6 in generating the acute-phase response and both these cytokines have also been found in increased quantities in schizophrenia (El-Mallakh *et al.*, 1993; Ganguli *et al.*, 1994). IL-6 is particularly high in the acute illness.

Maes *et al.* (1997) found that differences in acute-phase reactants between normal volunteers, and schizophrenic, manic, and depressed patients disappeared following chronic treatment with psychotropic drugs. Their results suggested that schizophrenia, along with depression (see Chapter 10 in this volume) and mania, is accompanied by an acute-phase response, which may be normalized by treatment with psychotropic medication.

B. CYTOKINES AND SCHIZOPHRENIA

Other chapters in this book have emphasized the close links between the CNS, the endocrine system, and the immune system. Cytokines in the CNS are involved in various regulatory mechanisms including:

- Initiation of an inflammatory immune response in the CNS
- Regulation of the blood–brain barrier
- Developmental mechanisms
- Repair after injury
- Regulation of the endocrine system in the HPA axis
- Different stimulatory and inhibitory influences on dopaminergic, serotinergic, noradrenergic, and cholinergic neurotransmission.

This communication between the brain and the immune system is a two-way process. Both microglia and astroctyes can produce and release cytokines once activated (Muller and Ackenheil, 1998). Conversely, IL-2 activates T cells, which in turn trigger the proliferation of oligodendrocytes and affect the differentiation of brain cells. Cytokines also play an important role in the development of the CNS which may be strongly altered by their over- or underproduction (Merrill, 1992).

Various cytokines influence the release of a number of neurotransmitters, including glutamate, dopamine, and serotonin, and therefore may be important in the pathogenesis of schizophrenia. The subtle neuropathological abnormalities in schizophrenia such as alterations in neuronal number and density are consistent with the actions of cytokines on neuronal survival and programmed cell death.

1. Effect of Therapy with Cytokines

Currently, recombinant IL-2 is used with some success in cancer immunotherapy. Its administration, however, has been reported to induce symptoms similar to the positive and negative features of schizophrenia, such as visual and auditory hallucinations, paranoia, delusions, agitation, irritability, cognitive impairment, and fatigue. In one study, 65% of patients treated with high-dose r-IL-2 developed symptoms such as delusions and severe cognitive impairment (Denicoff *et al.*, 1987). Evidence for enhanced IL-2 activity in schizophrenia is summarized in Section III.B.2.b.

Interestingly, in one study, mice treated with r-IL-2 on a schedule similar to that used in humans were tested for acquisition of a passive-avoidance task and then sacrificed for histological examination. Older treated mice showed impairment of acquisition of the task, along with hippocampal degenerative changes (Nemni *et al.*, 1992). The damage was maximal in the cA3 region. In the older treated mice, there was a loss of neurons, without the accompanying reactive astrocytosis or increases in numbers of oligodendroglia, in the cA1, cA3, and cA4 hippocampal regions. These psychological and neuropathological changes are remarkably similar to those described in schizophrenia.

2. Specific Cytokines and Schizophrenia

a. Interleukin 1. IL-1 is a cytokine implicated in a variety of central activities, including fever, sleep, ischemic injury, and neuromodulatory responses, such as neuroimmune and neuroendocrine interactions. Brief application of IL-1β causes a profound decrease of glutamate transmission in hippocampal cA1 pyramidal neurons (Luk *et al.*, 1999). IL-1 enhances dopaminergic sprouting and can regulate astroglia-derived dopaminergic neurotrophic factors, such as acidic and basic fibroblast growth factor or glial cell line–derived neurotrophic factor. Moreover, dopaminergic neurons themselves express IL-1 receptors (Ho and Blum, 1998). IL-1β, synergistically with IL-6, also modulates the serotonin response in the rat anterior hypothalamus (Wu *et al.*, 1999).

IL-1β serum levels are raised in acute schizophrenia (Katila *et al.*, 1994; Theodoropoulou *et al.*, 2001). It has been suggested that this increase of IL-1β in schizophrenia may be as a result of decreased REM sleep (Appelberg *et al.*, 1997).

b. Interleukin 2. IL-2 influences neurotransmitter release in the hippocampus and striatum (Plata-Salaman and ffrench-Mullen, 1993; Lapchak, 1992). IL-2 potentiates dopamine release evoked by a number of different stimuli in mesencephalic cell cultures. It is active at very low concentrations, a finding that indicates a potent effect of IL-2 on dopaminergic neurons and implicates a physiological role for this cytokine in the modulation of dopamine release (Alonso *et al.*, 1993). Peripheral application of IL-2 also causes an increase of catacholaminergic neurotransmission in the hippocampus and frontal cortex (Zalcman *et al.*, 1994). Moreover, it is a potent modulator of hippocampal acetylcholine release (Hanisch *et al.*, 1993).

Serum soluble IL-2 receptor (sIL-2R) levels, a marker of immune activation, are increased in schizophrenia, although not all laboratories concur (Gaughran *et al.*, 1998; Barak *et al.*, 1995). Increases in sIL-2Rα (a single-chain component of the trimeric receptor molecule) in schizophrenia are present before treatment with neuroleptics and are higher in patients with tardive dyskinesia, negative symptoms, and soft neurological signs (Rapaport *et al.*, 1994; Hornberg *et al.*, 1995). These findings are not limited to the patients alone. Rapaport *et al.* (1993) found increased sIL-2R in discordant monozygotic twins. Unaffected siblings of patients with schizophrenia also have higher serum sIL-2R levels than controls (Gaughran *et al.*, 2002). Hornberg *et al.* (1995) noted that patients with increased levels of sIL-2Rα along with decreased mitogen-stimulated IL-2 production in cell culture, had a worse clinical course.

Gattaz *et al.* (1992) found no abnormalities in serum concentrations of IL-2 itself in schizophrenia. Theodoropoulou *et al.* (2001), however, found significantly lower IL-2 serum levels in schizophrenic patients than in controls with a higher percentage of cells expressing IL-2 receptors in medicated chronic schizophrenic patients compared with drug-naive patients.

A reduction in IL-2 production in an *in vitro* assay from stimulated peripheral blood mononuclear cells (PBMCs), a standard test of cellular immunocompetence, in the disorder has been repeatedly reported, although, again, there have been some dissenters (Kim *et al.*, 1998; Cazzullo *et al.*, 2001). Decreased IL-2 production, in this assay system, in schizophrenia is associated with serum antihippocampal antibodies (Yang *et al.*, 1994). In one patient, followed over 33 weeks, Ganguli *et al.* (1995a), found that relapse could be predicted with a great degree of certainty by the IL-2 production levels in the *in vitro* assay of the previous week.

Licinio *et al.* (1991) found elevated IL-2 in the CSF of drug-free schizophrenic patients although other groups did not concur, with Barak *et al.* (1995) reporting low sIL-2R CSF levels. High CSF levels of IL-2 have been suggested to predict schizophrenic relapse (McAllister *et al.*, 1995).

Muller *et al.* (1997) observed that sIL-2Rα levels increased in patients after clinical improvement, although this has not been a uniform finding (Gaughran *et al.*, 2001). He attributed the symptoms of schizophrenia to excess intracranial IL-2 and suggested that the symptoms decrease because of a neutralization of this effect with treatment, mediated by sIL-2Rα. Plasma sIL-2Rα binds its ligand, IL-2, thereby competing for the binding of free

IL-2 to cellular IL-2R on the responding cells. Increased sIL-2Rα therefore mediates an immunosuppressive effect due to its capacity to reduce IL-2 availability.

Muller postulated that once bound to sIL-2Rα, less IL-2 crosses the blood–brain barrier (BBB), as it is then a larger molecule. Alternatively, increased serum sIL-2Rα could result in increased CSF sIL-2Rα levels, as the diffusion of proteins across the BBB depends on serum concentrations. Higher CSF sIL-2Rα will then bind more IL-2 and mediate a decrease of free CSF IL-2 levels. Another proposed mechanism was that, as neuroleptics seem to increase the permeability of the BBB for certain compounds, more sIL-2Rα cross the BBB during treatment and inhibit further central activation by IL-2.

The reduced IL-2 production by *in vitro* assay from sampled PBMCs reported in schizophrenia is also seen in certain autoimmune disease (Caruso *et al.*, 1993) and has been explained by *in vivo* overproduction of IL-2 causing T-cell exhaustion and the consequent *in vitro* hyporesponsiveness of stimulated peripheral blood mononuclear cells. Consistent with this explanation is the finding of low serum cholesterol levels in schizophrenia (Boston *et al.*, 1996). IL-2 causes a decrease in cholesterol, especially HDL, along with an increase in serum triglycerides (Penttinen, 1995).

c. Interleukin 6. IL-6 affects the levels of monoamine neurotransmitters in the brain. It modulates the serotonin response in the rat hypothalamus (Wu *et al.*, 1999). Systemic administration of IL-6 induces reductions of interstitial dopamine levels (Song *et al.*, 1999).

Elevated serum IL-6 and sIL-6R have been reported and have been associated with treatment resistance (Lin *et al.*, 1998). Levels of IL-6 are maximal during acute episodes of schizophrenia (Frommberger *et al.*, 1997) and are associated with duration of illness (Ganguli *et al.*, 1994). High CSF levels of sIL-6R are found in patients with more marked positive symptoms of the illness (Muller *et al.*, 1997). It is of interest to compare the evidence of increased cellular immune activity and activation of the acute-phase response outlined earlier with the parallel literature evidencing similar associations with immune dysregulation in relation to depression (see Chapter 10 in this volume).

d. Interferons. Interferon-γ (INF-γ) is lowered in the acute phase of schizophrenia (Arolt *et al.*, 1997) although Becker *et al.* (1990) found no difference in serum interferon in first psychotic attacks. Decreased production of interferon-α (INF-α) and INF-γ has been reported (Moises *et al.*, 1985), although Cazzullo *et al.* (2001) described higher production of INF-γ in drug-free and drug-naive patients. Inglot *et al.* (1994) found that patients with chronic schizophrenia with a high interferon response had predominately positive symptoms, while those with a low interferon response had mainly negative symptoms. In Preble and Torrey's 1985 study, high titers of interferon were found in serum from 24.4% of patients with psychosis and from 3.1% of controls. Interferon-positive patients were more likely to have had a recent onset or exacerbation of their illness and to be on low-dose or no medication. No interferon was detected in the CSF of 65 patients or 20 control subjects. Patients with decreased INF-γ production have a worse clinical course (Hornberg et al., 1995).

e. Other cytokines. TNF-α levels and IL-3 like activity are high in schizophrenia (Theodoropoulou et al., 2001; Sirota et al., 1995). There is no difference in IL-4 and IL-10 production in drug-free patients (Cazzullo et al., 2001). Serum IL-18, however, a Th1 pro-inflammatory cytokine produced by macrophage-like cells, is increased (Tanaka et al., 2000). Increased plasma IL-1 receptor antagonist (IL-1Ra) and lower levels of Clara cell protein (CC16), a natural anti-cytokine, were described by Maes et al. (1996), who went on to suggest that schizophrenia may be accompanied by in vivo activation of the monocytic arm of cell-mediated immunity.

The regulated expression of neural cell adhesion molecule (NCAM) isoforms in the brain is critical for many neurodevelopmental processes including axonal outgrowth, and the establishment of neuronal connectivity. Barbeau et al. (1995) investigated the expression of the major adult isoforms of NCAM and its embryonic isoform (PSA-NCAM) in the hippocampal region of postmortem brains from 10 schizophrenic and 11 control individuals. They observed a 20–95% reduction in the number of hilar PSA-NCAM immunoreactive cells in the dentate gyrus of the great majority of schizophrenic brains. The expression of this embryonic form of NCAM appears to be related to synaptic rearrangement and plasticity. Therefore, the decrease in hippocampal PSA-NCAM immunoreactivity in schizophrenia may suggest altered plasticity of the hippocampus in a large proportion of people with schizophrenia.

Schizophrenia is associated with both obstetric complications and maternal exposure to infection (O'Callaghan et al., 1991, 1992). Cytokine levels, including levels of IL-1β, IL-6, and TNF-α, are altered in amniotic fluid or neonatal cord blood in pregnancies complicated by infection (Gilmore and Jarskog, 1997). Given that maternal cytokines can cross the placenta, birth trauma, associated with a disturbed BBB, could facilitate invasion of immune-activating agents into the developing CNS (Muller and Ackenheil, 1998). This could, in theory, result in neurodevelopmental changes, ultimately presenting as schizophrenia.

As a note of caution Haack and colleagues (1999) cast doubt on the cytokine abnormalities reported in schizophrenia. They measured circulating

IL-1Ra, sIL-2R, TNF-α, soluble TNF receptors (sTNF-R p55, sTNF-R p75), and IL-6 and found that the levels were affected by age, body-mass index (BMI), gender, smoking habits, ongoing or recent infectious diseases, or prior medication. Cytokine or cytokine receptor levels were significantly increased in patients treated with clozapine (sIL-2R, sTNF-R p75), lithium (TNF-α, sTNF-R p75, IL-6), and benzodiazepines (TNF-α, sTNF-R p75). Taking these confounding factors into account, they found no evidence for disease-related alterations in the levels of IL-1Ra, sIL-2R, sTNF-R p75, and IL-6, whereas levels of sTNF-R p55 were slightly decreased in schizophrenia compared to healthy controls. They concluded that, if confounding factors are carefully taken into account, plasma levels of the previously mentioned cytokines and cytokine receptors yield little evidence for immunopathology in schizophrenia.

IV. HLA Antigens in Schizophrenia

HLA antigens have been linked to several immunologically mediated disorders, and there have been a number of studies examining their distribution in patients with schizophrenia. According to Wright et al.'s (2001) review of the subject, early studies were limited by their use of unstandardized diagnostic criteria. Although the earlier studies focused mostly on class I antigens in schizophrenia, more recent ones have examined the frequencies of class II antigens. Two or more groups have reported associations of HLA A9 (or its A24 subspecificity), A28, A10, DRB1*01, and DRw6 with schizophrenia.

The same authors have previously reported negative associations of schizophrenia with HLA DQB1*0602, an allele which may protect against insulin-dependent diabetes mellitus (IDDM), which itself is negatively associated with schizophrenia. They also examined the HLA DRB1 gene locus on chromosome 6p21.3 in 94 patients with schizophrenia and 92 unrelated mothers of different schizophrenic patients. The HLA DRB1*04 gene is positively associated with rheumatoid arthritis. They found decreased HLA DRB1*04 gene frequency both in the patients and in the unrelated mothers. This is consistent with the epidemiological findings of a negative association between schizophrenia and rheumatoid arthritis. This provides some evidence that the HLA DQB1*0602 and DRB1*04 alleles (or alleles at linked loci) may protect against schizophrenia.

Wright et al. (2001) suggest possible mechanisms for this. First, some property of the HLA molecules encoded by these alleles may directly protect against schizophrenia. There is already evidence that the HLA DQ beta chain with aspartate at position 57 protects against IDDM. Second, a negative genetic association with HLA DQB1*0602 and DRB1*04 may be secondary to a positive association at a linked locus such as DPB1 or DQA1. Third, pedigrees with a schizophrenic member may have a propensity for autoimmune diseases and the presence or absence of DQB1*0602 and DRB1*04 alleles in some members of such a pedigree may influence whether IDDM, rheumatoid arthritis, or schizophrenia develop.

They acknowledge that two well-designed studies have failed to find associations between HLA types and schizophrenia so the cumulative evidence thus far remains weak. Genetic linkage studies have found limited evidence of a susceptibility locus for schizophrenia on the short arm of chromosome 6 near the HLA region at 6p21.3. Although there have been negative studies, linkage has been found for D6S274, D6S296, and D6S291, which map close to the HLA region.

V. HPA Axis in Schizophrenia

The precise relationship of the HPA axis to schizophrenia is, as yet, uncertain. Marx and Lieberman (1998) have reviewed the subject extensively and note that studies have focused on basal cortisol secretion, cortisol or ACTH responses to stressors, dexamethasone suppression tests (DST), and cortisol responses to pharmacological probes.

Basal cortisol level studies in patients with schizophrenia have shown an inconsistent pattern, with most finding no differences between patients with schizophrenia and controls, although there have been individual reports of high basal cortisol secretion in schizophrenia, with others of low levels in medicated schizophrenic patients. No differences in basal ACTH secretion in CSF or serum have been reported. Following neuroleptic treatment of schizophrenia, plasma cortisol levels decrease. It is speculated that this inhibition could be related to reduced noradrenergic activity (Wik, 1995).

Patients with schizophrenia have reduced cortisol, ACTH, and growth hormone responses to stressors such as lumbar puncture, suggesting a disturbance of the flexibility of the HPA axis in the disorder. Increased severity of psychosis is associated with a blunted ACTH response to lumbar puncture, although cortisol and ACTH responses to hypoglycemia appear normal (Breier et al., 1988; Kathol et al., 1992). Patients with schizophrenia, however, have an exaggerated ACTH response to acute metabolic stress induced by 2-deoxy-D-glucose (Elman et al., 1998).

The rates of dexamethasone suppression test (DST) nonsuppression cited in schizophrenia vary greatly, with figures between 5 and 44% reported.

Some groups have found an association between DST nonsuppression and negative symptoms, but others found no such link. Cognitive impairment is correlated with 8 a.m. postdexamethasone cortisol concentrations in unmedicated schizophrenic patients (Newcomer *et al.*, 1991). Depressed patients with schizophrenia, do not have higher rates of DST nonsuppression compared to nondepressed patients (Garyfallos *et al.*, 1993). HPA axis activation is elicited by cytokines such as IL-1, IL-6, and TNF-α. These cytokines are increased in schizophrenia and therefore the hypercortisolemia recorded could be induced by their pro-inflammatory activity (Altamura *et al.*, 1999).

ACTH and cortisol responses to apomorphine are reduced in schizophrenia, suggesting dysfunction in the dopaminergic systems (Mokrani *et al.*, 1995). There is no difference in the cortisol response to D-fenfluramine challenge however (Abel *et al.*, 1996).

VI. Effects of Antipsychotic Medication

Antipsychotic medications have long been associated with the production of autoantibodies. Antinuclear antibodies have classically been linked to antipsychotic medication. Other antibodies linked to the prescription of neuroleptics include rheumatoid factor, antihistone antibodies and anticardiolipin antibodies (Gallien *et al.*, 1977; Chengappa *et al.*, 1991, 1992b; Canoso *et al.*, 1990) mimicking to some extent the autoantibody profiles characteristic of SLE and antiphospholipid syndrome.

Some authors have described immunosuppressive effects of neuroleptics (Saunders and Muchmore, 1964; Baker *et al.*, 1977). Others found no suppression of the immune system (Muller *et al.*, 1991) and some *in vitro* investigations have even demonstrated an immune-activating function of antipsychotic drugs (Zarrabi *et al.*, 1979). These contradictory results suggest that both *in vitro* and *in vivo* effects, as well as short- and long-term effects, must be considered when interpreting studies.

There appears to be, following short-term treatment at least, a difference in effects on the immune system between "typical" and "atypical" antipsychotics. Taking IL-2 as an example, haloperidol, a typical neuroleptic, does not alter sIL-2Rα levels (Pollmacher *et al.*, 1997), although clozapine does (Pollmacher *et al.*, 1995). The findings of increased sIL-2Rα levels more with atypical neuroleptics may be related to their mixed D2 and 5HT2a receptor blockade or, as previously stated, their ability to modulate the HPA axis. No difference in sIL-2Rα levels is found, however, between patients on typical and atypical neuroleptics after 6 months of treatment (Ganguli *et al.*, 1995b).

There have also been several reports that antipsychotic therapy in schizophrenia is associated with a decrease of IL-6 or SIL-6R levels. Overall it appears that the signs of immune activation seen in schizophrenia and depression normalize after 1–2 months of neuroleptic, mood stabilizer, or antidepressant therapy (Muller and Ackenheil, 1998).

Clozapine, the drug used for treatment-resistant schizophrenia, is widely accepted to have immunomodulatory properties. It increases the plasma levels of TNF-α, soluble TNF receptors p55 and p75, and sIL-2r. Increased TNF-α and sIL-2r levels are more pronounced in patients with clozapine-induced fever who also have increased plasma IL-6 levels and granulocyte counts. Plasma IL-1 receptor antagonist levels and monocyte and lymphocyte counts are not affected by clozapine treatment (Pollmacher *et al.,* 1996).

VII. Treatment of Schizophrenia with Immunosuppressants

Given the links with autoimmune disease and the evidence of T-cell activation in the condition, some researchers have been curious about the effects of immunosuppressive therapy in the treatment of schizophrenia. Smidt *et al.,* in 1985, treated twelve patients with chronic schizophrenia with glucocorticoids in combination with neuroleptic drugs. Two were discontinued due to aggravation of psychiatric illness. Seven had a greater than 50% reduction in first-rank symptoms. The patients who responded to steroid treatment were younger, with a shorter history of illness, and were more likely to have a positive family history of schizophrenia. Immunological investigations, including α-1-antitrypsin levels, haptoglobins, orosomucoid, and serum IgA, IgM, and IgG levels failed to differentiate between glucocorticoid responders and nonresponders.

Levine *et al.* (1997) treated fourteen patients with schizophrenia who had high antiplatelet antibodies with azothiaprine. Only two of eleven compliant patients showed a reduction in psychiatric symptomatology but the treatment was generally well tolerated.

VIII. Vaccine Response

There is evidence of impaired responsiveness to routine vaccination schedules in schizophrenia. In 1942, Molholm described decreased delayed hypersensitivity to guinea pig serum in schizophrenia and Vaughan *et al.*

(1949) recorded decreased responsivity to pertussus vaccine in the disorder. Ozec *et al.* (1971) found lower antibody production after vaccination against salmonella in unmedicated patients with schizophrenia. Psychiatric patients also demonstrate a reduced antibody response to hepatitis B vaccine (Russo *et al.*, 1994). The response to influenza vaccination in institutionalized elderly patients with schizophrenia and dementia is likewise significantly less than in healthy community populations (Edgar *et al.*, 2000).

IX. Th1/Th2 Imbalance in Schizophrenia

Muller *et al.* (1999) proposed a Th1/Th2 imbalance with a shift to the Th2 system in schizophrenia. Both the unspecific "innate" and the specific arm of the immune system are involved in the dysfunction of the immune system in the illness. The "innate" immune system shows signs of overactivation in unmedicated schizophrenic patients, with increased monocytes and $\gamma\delta$T-cells (a primitive form of T cell which contributes toward innate defense). The increased levels of IL-6 and the activation of the IL-6 system seen in schizophrenia might also be the result of the activation of monocytes/macrophages.

In addition, several parameters of the specific cellular immune system are blunted, for example, the decreased Th1-related immune parameters in schizophrenic patients both *in vitro* and *in vivo*. Muller suggests that the role of neuroleptic therapy may not have been taken seriously into consideration when interpreting studies showing increased antibody levels in schizophrenia. He proposes that neuroleptic therapy results in activation of both arms of the acquired immune system, both the cellular (Th1) and the humoral (Th2). He cites papers showing normalization of IFN-γ production, the increase of IL-2 receptors with treatment, and the fact that the blunted antibody response to vaccination with salmonella was not observed in patients that were treated with neuroleptics (Wilke *et al.*, 1996; Ozek *et al.*, 1971).

The reduced *in vitro* production of IL-2 by stimulated PBMCs, well described in schizophrenia, has been traditionally interpreted as due to cell exhaustion. However, another possibility is that it may reflect the reduced capacity of lymphocytes to produce IL-2, consistent with a shift away from Th1-mediated immunity. Furthermore, the response to stimulation with tuberculin, which is mediated through Th1 activity, is blunted in schizophrenia (Muller *et al.*, 1991).

Neopterin levels, another indicator of the activity in the Th1 cellular immune system, are lowest at baseline and increase significantly during the first week of treatment of acute schizophrenia. The baseline levels are significantly lower in comparison to healthy controls, suggesting normalization of cellular immunity during treatment (Sperner-Unterweger *et al.*, 1989).

The soluble intercellular adhesion molecule-1 (sICAM-1) is also a marker for the activation of the Th1 cellular immune system. Patients with schizophrenia showed significantly decreased serum levels of sICAM-1 both without antipsychotic medication immediately after admission to the hospital, and after clinical improvement (Schwarz *et al.*, 2000). Muller and Ackenheil (1998) postulate that the negative relationship between schizophrenia and rheumatoid arthritis mentioned earlier may be an illustration of this Th1/Th2 divide. Rheumatoid arthritis is associated with high sICAM-1 levels and schizophrenia has low sICAM-1 levels. Rheumatoid arthritis is a disorder that is primarily mediated by the Th1-related immune system.

Functionally, in 1942, before neuroleptics were invented, Molholm demonstrated hyposensitivity to foreign protein (guinea pig serum) in patients with schizophrenia, also implying blunting of Th1 immune reactions.

Schwarz *et al.* (2001) also hypothesize a shift to Th2-like immune reactivity in a subgroup of schizophrenic patients. Besides the immunological abnormalities, this subgroup appears to suffer with more pronounced negative symptoms and a poorer therapy outcome. It is suggested that the type of schizophrenia affecting this subgroup may be related to a prenatal viral infection. The increase in $CD3^+$ and $CD4^+$ cells seen in schizophrenia is also consistent with a Th2 shift in the illness.

Therefore, a pattern of aberrations in the T-cell cytokine system that is typical for Th2-skewed autoimmune disorders, such as SLE, and consistent with a shift to Th2 domination, namely decreased IL-2 production and increased IL-2 receptors along with lowered production of IFN-γ has been reported in patients with schizophrenia. These findings are not affected by clinical variables but may be associated with acute exacerbations of the illness (Arolt *et al.*, 2000).

Against the theory of a Th1/Th2 imbalance with a shift to the Th2 system in schizophrenia is that increased sIL-2Rα levels are described not just in the patients but also in their first-degree relatives. The increased lymphocyte proliferation and the increased levels of IL-2 in the CSF also suggest activation of the Th1 system. Rothermundt *et al.* (2000) examined IL-2 and IFN-γ production in unmedicated schizophrenia patients before and during 4 weeks of neuroleptic treatment and found no change in cytokine production. These results do not support the notion that neuroleptic medication *in vivo* might influence Th1 cytokine production in schizophrenia.

X. Summary

As is evident from the present account, there is no single or persuasive argument that signals emanating from the immune system are directly involved in the etiology of schizophrenia. We do not even know if we are dealing with a single disorder with a single causality; almost certainly we are not. The precise etiology of schizophrenia, as with so many neurological disorders, remains obscure. However, there is abundant evidence in schizophrenia of mutual dysregulation of neuronal function and immune system activity. Although this evidence is not always consistent, a pattern emerges suggesting aspects of immune activity being involved in the pathology of neuronal development that characterizes schizophrenia. Exposure to infective agents, HLA associations, autoimmune associations, disturbances in lymphocyte populations, and cytokine imbalances with a skew toward Th2 activity are supportive of this view. That the evidence is not always consistent is a testament to the complexity and heterogeneity of the disorder, to confounding by antipsychotics that themselves are immunomodulatory, and to the multifaceted nature, with all its checks and balances, of the immune system itself.

References

Abel, K., O'Keane, V., and Murray, R. (1996). Enhancement of the prolactin response to d-fenfluramine in drug-naive schizophrenic patients. *Br. J. Psychiatry* **168,** 57.

Achiron, A., Noy, S., Pras, E., Lereya, J., Hermesh, H., and Laor, N. (1994). T-cell subsets in acute psychotic schizophrenic patients. *Biol. Psychiatry* **35,** 27–31.

Alonso, R., Chadieu, I., Diorio, J., Krishnamurthy, A., Quirion, R., and Boksa, P. (1993). Interleukin modulates evoked release of [^{3}H]Dopamine in rat cultured mesencephalic cells. *J. Neurochem.* **61**(4), 1284–1290.

Altamura, A. C., Boin, F., and Maes, M. (1999). HPA axis and cytokines dysregulation in schizophrenia: Potential implications for the antipsychotic treatment. *Eur. Neuropsychopharmacol.* **10**(1), 1–4.

Ambelas, A. (1992). Schizophrenia in an evolutionary perspective. *Br. J. Psychiatry* **1609,** 401–404.

Appelberg, B., Katila, H., and Rimon, R. (1997). Plasma Interleukin 1β and sleep architecture in schizophrenia and other non-affective psychoses. *Psychosom. Med.* **59**(5), 529–532.

Arolt, V., Weitzsch, C., Wilke, I., Nolte, A., Pinnow, M., Rothermundt, M., and Kirchner, H. (1997). Production of interferon γ in families with multiple occurrence of schizophrenia. *Psychiatry Res.* **66**(2–3), 145–152.

Arolt, V., Rothermundt, M., Wandinger, K. P., and Kirchner, H. (2000). Decreased *in vitro* production of interferon γ and interleukin 2 in whole blood of patients with schizophrenia during treatment. *Mol. Psychiatry* **5**(2), 150–158.

Baker, G. A., Santalo, R., and Blumenstein, J. (1977). Effect of psychotropic agents on the blastogenic response of human T-lymphocytes. *Biol. Psychiatry* 12(2), 159–169.
Baldwin, J. A. (1979). Schizophrenia and physical disease. *Psychol. Med.* 9, 611–618.
Barak, V., Barak, Y., Levine, J., Nisman, B., and Roisman, I. (1995). Changes in interleukin 1B and soluble interleukin 2 receptor levels in CSF and serum of schizophrenic patients. *J. Basic Clin. Physiol. Pharmacol.* 6(1), 61–69.
Barbeau, D., Liang, J. J., Robitalille, Y., Quirion, R., and Srivastava, L. K. (1995). Decreased expression of the embryonic form of the neural cell adhesion molecule in schizophrenic brains. *Proc. Nat. Acad. Sci. USA* 92(7), 2785–2789.
Becker, D., Kritschmann, E., Floru, S., Shlomo-David, Y., and Gatlieb-Stematxky, T. (1990). Serum interferon in first psychotic attacks. *Br. J. Psychiatry* 157, 136–138.
Bergen, J. R., Grinspoon, L., Pyle, H. M., Martinez, J. L. J., and Pennell, R. B. (1980). Immunologic studies in schizophrenia and control subjects. *Biol. Psychiatry* 15, 369–379.
Bhatia, M. S., Dhar, N. K., Agrawal, P., Khurana, S. K., Bohra, N., and Malik, S. C. (1992). Immunoglobulin profile in schizophrenia. *Indian J. Med. Sci.* 46(8), 239–242.
Boston, P. F., Dursun, S. M., and Reverly, M. A. (1996). Cholesterol and mental disorder. *Br. J. Psychiatry* 169(6), 682–689.
Bracha, H. S., Torrey, E. F., Gottesman, I. I., Bigelow, L. B., and Cunniff, C. (1992). Second-trimester markers of fetal size in schizophrenia: A study of monozygotic twins. *Am. J. Psychiatry* 149(10), 1355–61.
Breier, A., Wolkowitz, O., Doran, A., Bellar, S., and Pickar, D. (1988). Neurobiological effects of lumbar puncture stress in psychiatric patients and healthy volunteers. *Psychiatric Research* 25, 187.
Canoso, R. T., de Oliveira, R. M., and Nixon, R. A. (1990). Neuroleptic-associated autoantibodies. A prevalence study. *Biol. Psychiatry* 27(8), 863–870.
Carlsson, M., and Carlsson, A. (1990). Interactions between glutamatergic and monoaminergic systems within the basal ganglia: Implications for schizophrenia and Parkinson's disease. *Trends Neurosci.* 13, 272–276.
Carter, M., and Watts, C. A. H. (1971). Possible biological advantages among schizophrenic relatives. *Br. J. Psychiatry* 118, 453–460.
Caruso, C., Candore, G., Cigna, D., Collucci, A. T., and Modica, M. A. (1993). Biological significance of soluble IL-2 receptor. *Mediators Inflamm.* 2, 3–21.
Cazzullo, C. L., Saresella, M., Roda, K., Calvo, M. G., Bertrando, P., Doria, S., Clerici, M., Salvaggio, A., and Ferrante, P. (1998). Increased levels of CD8$^+$ and CD4$^+$ 45RA$^+$ lymphocytes in schizophrenic patients. *Schizophr. Res.* 31(1), 49–55.
Cazzullo, C. L., Sacchetti, E., Galluzzo, A., Panariello, A., Colombo, F., Zagliani, A., and Clerici, M. (2001). Cytokine profiles in drug-naive schizophrenic patients. *Schizophr. Res.* 47(2–3), 293–298.
Chengappa, K. N., Ganguli, R., Yang, Z. W., Brar, J. S., Li, L., and Rabin, B. S. (1992a). Left-handed first-episode, neuroleptic-naive schizophrenic patients have a higher prevalence of autoantibodies. *Schizophr. Res.* 8(1), 75–80.
Chengappa, K. N., Carpenten, A. B., Keshavan, M. S., Yang, Z. W., Kelly, R. H., Rabin, B. S., and Ganguli, R. (1991). Elevated IgG and IgM anticardiolipin antibodies in a subgroup of medicated and unmedicated schizophrenic patients. *Biol. Psychiatry* 30(7), 731–735.
Chengappa, K. N., Nimgaonkar, V. L., Bachert, C., Yang, Z. W., Rabin, B. S., and Ganguli, R. (1995). Obstetric complications and autoantibodies in schizophrenia. *Acta Psychiatr. Scand.* 92(4), 270–273.
Chengappa, K. N., Betts Carpenter, A, Yang, Z. W., Brar, J. S., Rabin, B. S., and Ganguli, R. (1992b). Elevated IgG anti-histone antibodies in a subgroup of medicated schizophrenic patients. *Schizophr. Res.* 7, 49–54.

Chong-Thim, W., Wing-Foo, T., and Nilmani, S. (1993). Serum immunoglobulin levels in Chinese male schizophrenics. *Schizophr. Res.* **10**, 61–66.

Coffey, C. E., Sullivan, J., and Rice, J. (1983). T lymphocytes in schizophrenia. *Biol. Psychiatry* **18**(1), 113–119.

Cooper, S. J. (1992). Schizophrenia after prenatal exposure to 1957 A2 influenza epidemic. *Br. J. Psychiatry* **161**, 394–396.

DeLisi, L. E., Weinberger, D., Potkin, S., Neckers, L., Shiling, D., and Wyatt, R. (1981). Quantitative determination of immunoglobulins in CSF and plasma of chronic schizophrenic patients. *Br. J. Psychiatry* **139**, 513–518.

DeLisi, L. E., Goodman, S., Neckers, L., and Wyatt, R. (1982). An analysis of lymphocyte subpopulations in schizophrenic patients. *Biol. Psychiatry* **17**(9), 1003–1009.

DeLisi, L. E., Ortaldo, J. R., Maluish, A. E., and Wyatt, R. J. (1983). Deficient natural killer cell activity and macrophage functioning in schizophrenic patients. *J. Neural Transm.* **58**, 99–106.

DeLisi, L. E., King, C., and Targum, S. (1984). Serum immunoglobulin concentrations in patients admitted to an acute psychiatric inpatient service. *Br. J. Psychiatry* **145**, 661–666.

DeLisi, L. E., Weber, R. J., and Pert, C. B. (1985). Are there antibodies against brain in sera from schizophrenic patients? Review and prospectus. *Biol. Psychiatry* **20**, 94–115.

Denicoff, K. D., Rubinoff, D., Papa, M. Z., Simpson, C., Seipp, C. A., Lotze, M. T., Chang, A. E., Rosenstein, D., and Rosenberg, S. A. (1987). The neuropsychiatric effects of treatment with Interleukin 2 and lymphokine activated killer cells. *Ann. Intern. Med.* **107**, 293–300.

Dohan, F. C., and Grasberger, J. C. (1973). Relapsed schizophrenics: earlier discharge from the hospital after cereal-free, milk-free diet. *Am. J. Psychiatry* **130**(6), 685–688.

Eaton, W. W., Hayward, C., and Ram, R. (1992). Schizophrenia and rheumatoid arthritis: A review. *Schizophr. Res.* **6**, 181–192.

Edgar, S., Gaughran, F., Macdonald, A., Oxford, J. S., and Lambkin, R. (2000). Response to influenza vaccine in an elderly institutionalised population with schizophrenia or dementia. Royal College of Psychiatrists Annual Meeting, Edinburgh.

El-Mallakh, R. S., Suddath, R. L., and Wyatt, R. J. (1993). Interleukin 1-α and interleukin 2 in cerebrospinal fluid of schizophrenic subjects. *Prog. Neuro-Psychopharmacol. Biol. Psychiatry* **17**(3), 383–391.

Elman, I., Adler, C. M., Malhotra, A. K., Bir, C., Pickar, D., and Breier, A. (1998). Effect of acute metabolic stress on pituitary-adrenal axis activation in patients with schizophrenia. *Am. J. Psychiatry* **155**(7), 979–981.

Fessel, W. J. (1963). Antibrain factor in psychiatric patients' sera: Further studies with haemagglutinin technique. *Arch. Gen. Psychiatry* **8**, 614.

Finney, G. O. H. (1989). Juvenile onset diabetes and schizophrenia? *Lancet* **2**(8673), 1214–1215.

Firer, M., Sirota, P., Schild, K., Elizur, A., and Slor, H. (1994). Anticardiolipin antibodies are elevated in drug-free, multiply affected families with schizophrenia. *J. Clin. Immunol.* **14**(1), 73–78.

Frommberger, U. H., Bauer, J., Haselbauer, P., Fraeulin, A., Riemann, D., and Berger, M. (1997). Interleukin 6 plasma levels in depression and schizophrenia: Comparison between the acute state and after remission. *Eur. Arch. Psychiatry Clin. Neurosci.* **247**(4), 228–233.

Galinowski, A., Barbouche, R., Truffinet, P., Louzir, H., Poirrier, M. F., Bouvet, O., Loo, H., and Avrameas, S. (1992). Natural autoantibodies in schizophrenia. *Acta Psychiatr. Scand.* **85**(3), 240–242.

Gallien, M., Schnetzler, J. P., and Morin, J. (1977). Antinuclear antibodies and lupus cells in 600 hospitalised, phenothiazine treated patients. *Ann. Med. Psychol. Med.* **1**, 237–248.

Ganguli, R., and Rabin, B. S. (1993). CD5$^+$ B lymphocytes in Schizophrenia: No alteration in numbers or percentage as compared with control subjects. *Psychiatry Res.* **48**, 69–78.

Ganguli, R., Brar, J., Chengappa, K. N. R., Deleo, M., Yang, Z., and Rabin, B. (1995a). Mitogen-stimulated interleukin 2 production in never-medicated, first episode schizophrenic patients: The influence of age of onset and negative symptoms. *Arch. Gen. Psychiatry* **52**(8), 668–672.

Ganguli, R., Brar, J. S., and Rabin, B. S., Reply to Pollmacher, T., Hinze-Selch, D., Mullington, J., and Holsboer, F. (1995b). Clozapine induced increase in plasma levels of soluble inter-leukin 2 receptors. *Arch. Gen. Psychiatry* **52**, 878.

Ganguli, R., Rabin, B. S., Kelly, R. H., Lyte, M., and Ragu, U. (1987). Clinical and labora-tory evidence of autoimmunity in acute schizophrenia. *Ann. N. Y. Acad. Sci.* **496**, 676–685.

Ganguli, R., Yang, Z. W., Shurin, G., Chengappa, R., Brar, J. S., Gubbi, A. B., and Rabin, B. S. (1994). Serum Interleukin 6 concentration in schizophrenia: Elevation associated with duration of illness. *Psychiatry Res.* **51**(1), 1–10.

Garyfallos, G., Lavrentiadis, G., Amoutzias, D., Monas, K., and Monos, N. (1993). Negative symptoms of schizophrenia and the dexamethasone suppression test. *Acta Psychiatr. Scand.* **88**, 425.

Gattaz, W. F., Dalgalarrondo, P., and Schroder, H. C. (1992). Abnormalities in serum con-centration of Interleukin 2, Interferon-α and Interferon-γ in schizophrenia not detected. *Schizophr. Res.* **6**, 237–241.

Gaughran, F., O'Neill, E., Cole, M., Collins, K., Daly, R. J., and Shanahan, F. (1998). Increased soluble interleukin 2 receptor levels in schizophrenia. *Schizophr. Res.* **29**, 263–267.

Gaughran, F., O'Neill, E., Sham, P., Daly, R. J., and Shanahan, F. (2001). Soluble interleukin 2 receptor levels in acute and stable schizophrenia. *Schizophr. Res.* **52**(1–2), 143–144.

Gaughran, F., O'Neill, E., Sham, P., and Daly, R. J. (2002). Shanahan F. Soluble interleukin 2 receptor levels in families of people with schizophrenia. *Schizophr. Res.* **56**(3), 235–239.

Gilmore, J. H., and Jarskog, L. F. (1997). Exposure to infection and brain development: Cytokines in the pathogenesis of schizophrenia. *Schizophr. Res.* **24**(3), 365–367.

Gilvarry, C. M., Sham, P. C., Jones, P. B., Cannon, M., Wright, P., Lewis, S., Bebbington, P., Toone, B. K., and Murray, R. M. (1996). Family history of autoimmune disease in psychosis. *Schizophr. Res.* **19**, 33–40.

Gur, R. E. (1977). Motoric laterality imbalance in schizophrenia. *Arch. Gen. Psychiatry* **34**, 33–37.

Haack, M., Hinze-Selch, D., Fenzel, T., Kraus, T., Kuhn, M., Schuld, A., and Pollmacher, T. (1999). Plasma levels of cytokines and soluble cytokine receptors in psychiatric patients upon hospital admission: Effects of confounding factors and diagnosis. *J. Psychiatr. Res.* **33**(5), 407–418.

Hanisch, U. K., Seto, D., and Quirion, R. (1993). Modulation of hippocampal acetylcholine release: A potent central action of interleukin-2. *J. Neurosci.* **13**(8), 3368–3374.

Heath, R. G., Krupp, I. M., Byers, L. W., and Lijekvist, J. I. (1967). Schizophrenia as an immuno-logic disorder. 3. Effects of antimonkey and antihuman brain antibody on brain function. *Arch. Gen. Psychiatry* **16**(1), 24–33.

Heath, R. G., McCarron, K. L., and O'Neil, C. E. (1989). Antiseptal brain antibody in IgG of schizophrenic patients. *Biol. Psychiatry* **25**, 725–733.

Henneberg, A. E., Horter, S., and Ruffert, S. (1994). Increased prevalence of antibrain anti-bodies in the sera from schizophrenic patients. *Schizophr. Res.* **14**, 15–22.

Ho, A., and Blum, M. (1998). Induction of interleukin-1 associated with compensator dopami-nergic sprouting in the denervated striatum of young mice: Model of aging and neurode-generative disease. *J. Neurosci.* **18**(15), 5614–5629.

Hornberg, M., Arolt, V., Wilke, I., Kruse, A., and Kirchner, H. (1995). Production of interferons and lymphokines in leukocyte cultures of patients with schizophr. *Schizophr. Res.* **15**(3), 237–242.

Inglot, A. D., Leszek, J., Piasecki, E., and Sypula, A. (1994). Interferon responses in schizophrenia and major depressive disorders. *Biol. Psychiatry* **35,** 464–473.

International Pilot Study of Schizophrenia. (1973). World Health Organization, Geneva.

Jones, P., Rodgers, B., and Murray, R. (1994). Child development risk factors for adult schizophrenia. *Lancet* **344,** 1398–1402.

Jones, P., Rantakallio, P., Hartikainen, A. L., Isohanni, M., and Sipila, P. (1998). Does schizophrenia result from pregnancy, delivery, and perinatal complications? A 28-year study in the l966 North Finland Birth Cohort. Presented at the 8th Congress of the Association of Psychopharmacology.

Kathol, R. G., Gehris, T. L., Carroll, B. T., Samuelson, S. D., Pitts, A. F., Meller, W. H., and Carter, J. L. (1992). Blunted ACTH response to hypoglycemic stress in depressed patients but not in patients with schizophrenia. *J. Psychiatr. Res.* **26**(2), 103–116.

Katila, H., Appelberg, B., Hurme, M., and Rimon, R. (1994). Plasma levels of interleukin-1B and Interleukin-6 in schizophrenia, other psychoses, and affective disorders. *Schizophr. Res.* **12,** 29–34.

Kelly, R. H., Ganguli, R., and Rabin, B. S. (1987). Antibody to discrete areas of the brain in normal individuals and patients with schizophrenia. *Biol. Psychiatry* **22**(12), 1488–1491.

Kessler, A., and Shinitzky, M. (1993). Platelets from schizophrenic patients bear autoimmune antibodies that inhibit dopamine uptake. *Psychobiology* **21**(4), 299–306.

Kim, C. E., Lee, M. S., and Suh, K. Y. (1998). Decreased interleukin 2 production in Korean schizophrenic patients. *Biol. Psychiatry* **43**(9), 701–704.

Knight, J., Knight, A., and Ungvari, G. (1992). Can Autoimmune mechanisms account for the genetic predisposition to schizophrenia? *Br. J. Psychiatry* **160,** 533–540.

Kohn, M. L. (1968). Social class and schizophrenia: A review. *J. Psychiatr. Res.* **6,** 155–173.

Kuznetsova, Ni., and Semenov, S. F. (1961). Determining antibodies to the brain in the serum of patients with neuropsychiatric disease. *Zh. Nevropatol. Psikhiatr.* **61,** 869–874.

Lahdelma, R. L., Katila, H., Hirata-Hibi, M., Andersson, L., Appelberg, B., and Rimon, R. (1995). Atypical lymphocytes in schizophrenia. *Eur. Psychiatry* **10**(2), 92.

Laing, P., Knight, J. G., Wright, P., and Irving, W. L. (1996). Disruption of fetal brain development by maternal antibodies as an aetiological factor in schizophrenia. *In* "Neural Development in Schizophrenia: Theory and Research" (S. A. Mednick, ed.). Plenum, New York.

Lapchak, P. A. (1992). A role for interleukin 2 in the regulation of striatal dopaminergic function. *Neuroreport* **3,** 165–168.

Legros, S., Mendlewicz, J., and Wybran, J. (1985). Immunoglobulins, autoantibodies and other serum protein fractions in psychiatric disorders. *Eur. Arch. Psychiatry Neurol. Sci.* **235**(1), 9–11.

Lehmann-Facius, H. (1939). Serologischanalytische veruche it liqouren und seren von schizophrenien. *Alg. Z. Psychiatr.* **110,** 232–243.

Levine, J., Gutman, J., Feraro, R., Levy, P., Kihmi, R., Leykin, I., Deckmann, M., Handzel, Z. T., and Shinitzky, M. (1997). Side effect profile of azothioprine in the treatment of chronic schizophrenic patients. *Neuropsychobiology* **36**(4), 172–176.

Lewis, D. A., and Lieberman, J. A. (2000). Catching up on schizophrenia: Natural History and Neurobiology. *Neuron* **28**(2), 325–334.

Licinio, J., Krystal, J., Seibyl, J., Altemus, M., and Charney, D. (1991). Elevated central levels of interleukin 2 in drug free schizophrenic patients. *Schizophr. Res.* **4,** 372.

Lin, A., Kenis, G., Bignotti, S., Tura, G. J., De Jong, R., Bosmans, E., Pioli, R., Altamura, C., Scharpe, S., and Maes, M. (1998). The inflammatory response system in treatment resistant schizophrenia increased serum interleukin 6. *Schizophr. Res.* **32**(1), 9–15.

Luk, W. P., Zhang, Y., White, T. D., Lue, F. A., Wu, C., Jiang, C. G., Zhang, L., and Moldofsky, H. (1999). Adenosine: A mediator of interleukin-1β-induced hippocampal synaptic inhibition. *J. Neurosci.* **19**(11), 4238–4244.

Maes, M., Bosmans, E., Ranjan, R., Vandoolaeghe, E., Meltzer, H. Y., De Ley, M., Berghmans, R., Stans, G., and Desnyder, R. (1996). Lower plasma CC16, a natural anti-inflammatory protein, and increased plasma interleukin 1 receptor antagonist in schizophrenia: Effects of antipsychotic drugs. *Schizophr. Res.* **21**(1), 39–50.

Maes, M., Delange, J., Ranjan, R., Meltzer, H. Y., Desnyder, R., Cooremans, W., and Scharpe, S. (1997). The acute phase protein response in schizophrenia, mania and major depression: Effects of psychotropic drugs. *Psychiatry Res.* **66**(1), 1–11.

Marx, C. E., and Lieberman, J. A. (1998). Psychoneuroendocrinology of schizophrenia. *Psychiatr. Clin. North Am.* **21**(2), 413–434.

Masserini, C., Vita, A., Basile, R., Morselli, R., Boato, P., Peruzzi, C., Pugnetti, L., Ferrante, P., and Cazzullo, C. L. (1990). Lymphocyte subsets in schizophrenic disorders. Relationship with clinical, neuromorphological and treatment variables. *Schizophr. Res.* **3**(4), 269–275.

McAllister, C. G., Rapaport, M. H., Pickar, D., Podruchny, T. A., Christison, G., Alphs, L. D., and Paul, S. M. (1989). Increased numbers of CD5$^+$ B lymphocytes in schizophrenic patients. *Arch. Gen. Psychiatry* **46**(10), 890–894.

McAllister, C. G., van Kammen, D. P., Rehn, T. J., Miller, A. L., Gurklis, J., Kelley, M. E., Yao, J., and Peters, J. L. (1995). Increases in CSF levels of interleukin 2 in schizophrenia: Effects of recurrence of psychosis and medication status. *Am. J. Psychiatry* **152**, 1291–1297.

McDaniel, J. S., Jewart, R. D., Eccard, M. B., Pollard, W. E., Caudle, J., Stipetic, M., Risby, E. D., Lewine, R., and Risch, S. C. (1992). Natural killer cell activity in schizophrenia and schizoaffective disorder: A pilot study. *Schizophr. Res.* **8**(2), 125–128.

Merrill, J. E. (1992). Tumor necrosis factor α, interleukin 1 and related cytokines in brain development: Normal and pathological. *Dev. Neurosci.* **14**, 1–10.

Moises, H. W., Schindler, L., Leroux, M., and Kirchner, H. (1985). Decreased production of interferon α and interferon γ in leucocytes cultures of schizophrenic patients. *Acta Psychiatr. Scand.* **72**(1), 45–50.

Mokrani, M., Duval, F., Crocq, M., Bailey, P. E., and Macher, J. P. (1995). Multihormonal responses to apomorphine in mental illness. *Psychoneuroendocrinology* **20**, 365.

Molholm, H. B. (1942). Hyposensitivity to foreign protein in schizophrenic patients. *Psychiatr. Q.* **16**, 565–571.

Mukherjee, S., Mahadik, S. P., Korenovsky, A., Laev, H., Schnur, D. B., and Reddy, R. (1994). Serum antibodies to nicotinic acetylcholine receptors in schizophrenic patients. *Schizophr. Res.* **12**, 131–136.

Muller, N., and Ackenheil, M. (1995). Immunoglobulin and albumin content of cerebrospinal fluid in schizophrenic patients: Relationship to negative symptomatology. *Schizophr. Res.* **14**, 223–228.

Muller, N., and Ackenheil, M. (1998). Psycho-neuro-immunology and the cytokine action in the CNS: Implications for psychiatric disorders. *Prog. Neuro-Psychopharmacol. Biol. Psychiatry* **22**(1), 1–33.

Muller, N., Ackenheil, M., Hofschuster, E., Mempel, W., and Eckstein, R. (1991). Cellular immunity in schizophrenic patients before and during neuroleptic treatment. *Psychiatry Res.* **37**, 147–160.

Muller, N., Hofschuster, E., Ackenheil, M., and Eckstein, R. (1993). T-cells and psychopathology in schizophrenia: Relationship to the outcome of neuroleptic therapy. *Acta Psychiatr. Scand.* **87**(1), 66–71.

Muller, N., Empl, M., Riedel, M., Schwartz, M., and Ackenheil, M. (1997). Neuroleptic treatment increases soluble IL2 receptors and decreases soluble IL6 receptors in schizophrenia. *Eur. Arch. Psychiatry Clin. Neurosci.* **247**(6), 308–313.

Muller, N., Riedel, M., Ackenheil, M., and Schwarz, M. J. (1999). The role of immune function in schizophrenia: An overview. *Eur. Arch. Psychiatry Clin. Neurosci.* **249**(Supply 4), 62–68.

Nemni, R., Iannaccone, S., Quattrini, A., and Smirne, S. (1992). Effect of chronic treatment with recombinant interleukin 2 on the central nervous system of adult and old mice. *Brain Res.* **591**(2), 248–252.

Newcomer, J. W., Faustman, W. O., Whiteford, H. A., Moses, J. A. Jr., and Csernansky, J. G. (1991). Symptomatology and cognitive impairment associate independently with post-dexamethasone cortisol concentrations in unmedicated schizophrenic patients. *Biol. Psychiatry* **29**(9), 855–864.

Nikkila, H., Muller, K., Ahokas, A., Miettinen, K., Andersson, L. C., and Rimon, R. (1995). Abnormal distributions of T lymphocyte subsets in the cerebrospinal fluid of patients with schizophrenia. *Schizophr. Res.* **14**, 215–221.

Nopoulos, P., Torres, I., Flaum, M., Andreasen, N., Ehrhardt, J. C., and Yuh, W. T. C. (1995). Brain morphology in first episode schizophrenia. *Am. J. Psychiatry* **152**, 1721–1723.

O'Callaghan, E., Larkin, C., Kinsella, A., and Waddington, J. (1991). Familial, obstetric and other clinical correlates of minor physical anomalies in schizophrenia. *Am. J. Psychiatry* **148**, 479–483.

O'Callaghan, E., Gibson, T., Colohan, H. A., Buckley, P., Walshe, D. G., Larkin, C., and Waddington, J. L. (1992). Risk of schizophrenia in adults born after obstetric complications and their association with early onset of illness: A controlled study. *BMJ* **305**, 1256–1259.

Ozek, M., Toreci, K., Akkok, I., and Guvener, Z. (1971). [Influence of therapy on antibody-formation.] [German] *Psychopharmacologia* **21**(4), 401–412.

Pandey, R. S., Gupta, A. K., and Chaturvedi, V. C. (1981). Autoimmune model of schizophrenia with special reference to antibrain antibodies. *Biol. Psychiatry* **16**, 1123–1136.

Penttinen, J. (1995). Hypothesis: Low serum cholesterol, suicide and interleukin 2. *Am. J. Epidemiol.* **141**, 716–718.

Plata-Salaman, C. R., and ffrench-Mullen, J. M. (1993). Interleukin-2 modulates calcium currents in dissociated hippocampal CA1 neurons. *Neuroreport* **4**(5), 579–581.

Pollmacher, T., Hinze-Selch, D., Mullington, J., and Holsboer, F. (1995). Clozapine induced increase in plasma levels of soluble interleukin 2 receptors. *Arch. Gen. Psychiatry* **52**, 877–878.

Pollmacher, T., Hinze-Selch, D., and Mullington, J. (1996). Effects of clozapine on plasma cytokine and soluble cytokine receptor levels. *J. Clin. Psychopharmacol.* **16**(5), 403–409.

Pollmacher, T., Hinze-Selch, D., Fenzel, T., Kraus, T., Schuld, A., and Mullington, J. (1997). Plasma levels of cytokines and soluble cytokine receptors during treatment with haloperidol. *Am. J. Psychiatry* **154**(12), 1763–1765.

Preble, O. T., and Torrey, E. F. (1985). Serum interferon in patients with psychosis. *Am. J. Psychiatry* **142**(10), 1184–1186.

Rao, N. IV, Gopinath, P. S., Jayasimha, N., and Subhash, M. N. (1988). Immunochemical profiles in schizophrenia. *NIMHANS J.* **6**(1), 27–28.

Rapaport, M., and Lohr, J. B. (1994). Serum interleukin 2 receptors in neuroleptic naive schizophrenic subjects with and without tardive dyskinesia. *Acta Psychiatr. Scand.* **90**, 311–315.

Rapaport, M. H., Fuller Torrey, E., McAllister, C., Nelson, D. L., Pickar, D., and Paul, S. M. (1993). Increased serum soluble interleukin 2 receptors in schizophrenic monozygotic twins. *Eur. Arch. Psychiatry Clin. Neurosci.* **243**, 7–10.

Reichelt, K. L., and Landmark, J. (1995). Specific IgA antibody increases in schizophrenia. *Biol. Psychiatry* **37**, 410–413.

Roos, R. P., Davis, K., and Meltzner, H. Y. (1985). Immunoglobulin studies in patients with psychiatric diseases. *Arch. Gen. Psychiatry* **42**, 124–128.

Rothermundt, M., Arolt, V., Leadbeater, J., Peters, M., Rudolf, S., and Kirchner, H. (2000). Cytokine in unmedicated and treated schizophrenic patients. *Neuroreport* **11**(15), 3385–3388.

Russo, R., Ciminale, M., Ditommaso, S., Siliquini, R., Zotti, C., and Ruggenini, A. M. (1994). Hepatitis B vaccination in psychiatric patients. *Lancet* **343**, 346.

Saunders, J. C., and Muchmore, E. (1964). Phenothiazine effect on human antibody synthesis. *Br. J. Psychiatry* **110**, 84–89.

Schwarz, M. J., Riedel, M., Gruber, R., Ackenheil, M., and Muller, N. (1999). Antibodies to heat shock proteins in schizophrenic patients: Implications for the mechanism of the disease. *Am. J. Psychiatry* **156**(7), 1103–1104.

Schwarz, M. J., Riedel, M., Ackenheil, M., and Muller, N. (2000). Decreased levels of soluble intercellular adhesion molecule-1 (sICAM-1) in unmedicated and medicated schizophrenic patients. *Biol. Psychiatry* **47**(1), 29–33.

Schwarz, M. J., Muller, N., Riedel, M., and Ackenheil, M. (2001). The Th2-hypothesis of schizophrenia: A strategy to identify a subgroup of schizophrenia caused by immune mechanisms. *Med. Hypotheses* **56**(4), 483–486.

Sirota, P., Schild, K., Firer, M., Tanay, A., Elizur, A., Meytes, D., and Slor, H. (1991). Autoantibodies to DNA in multicase families with schizophrenia. *Biol. Psychiatry* **29**, 131A.

Sirota, P., Firer, M., Schild, K., Zurgil, N., Barak, Y., Elizur, A., and Slor, H. (1993). Increased anti-Sm antibodies in schizophrenic patients and their families. *Prog. Neuro-Psychopharmacol. Biol. Psychiatry* **17**, 793–800.

Sirota, P., Schild, K., Elizur, A., Djaldetti, M., and Fishman, P. (1995). Increased interleukin 1 and interleukin 3 like activity in schizophrenic patients. *Prog. Neuropsychopharmacol. Biol. Psychiatry* **19**, 75–83.

Smidt, E., Axelsson, R., and Steen, G. (1985). Treatment of chronic schizophrenia with glucocorticoids in combination with neuroleptic drugs: A pilot study. *Curr. Therapeut. Res.* **43**, 842–850.

Song, C., Merali, Z., and Anisman, H. (1999). Variations of nucleus accumbens dopamine and serotonin following systemic interleukin-1, interleukin-2 or interleukin-6 treatment. *Neuroscience* **88**(3), 823–836.

Sperner-Unterweger, B., Barnas, C., Fleischhacker, W. W., Fuchs, D., Meise, U., Reibnegger, G., and Wachter, H. (1989). Is schizophrenia linked to alteration in cellular immunity? *Schizophr. Res.* **2**(4–5), 417–421.

Spivak, B., Radwan, M., Brandon, J., Baruch, Y., Stawski, M., Tyano, S., and Weizman, A. (1993). Reduced total complement haemolytic activity in schizophrenic patients. *Psychol. Med.* **23**(2), 315–318.

Stevens, A., and Weller, M. (1992). Ganglioside antibodies in schizophrenia and major depression. *Biol. Psychiatry* **32**(8), 728–730.

Stevens, J., Papadopoulos, N. M., and Resnick, M. (1990). Oligoclonal bands in acute schizophrenia: A negative search. *Acta Psychiatr. Scand.* **81**(3), 262–264.

Susser, E. S., and Lin, S. P. (1992). Schizophrenia after prenatal exposure to the Dutch Hunger Winter of 1944–1945. *Arch. Gen. Psychiatry* **49**, 983–988.

Swedo, S., Leonard, H., Schipiro, M., Casey, B. J., Mannheim, G. B., Lenane, M., and Rettew, D. C. (1993). Sydenham's Chorea: Physical and psychological symptoms of St. Vitus Dance. *Pediatrics* **91**, 706–713.

Tanaka, K. F., Shintani, F., Fujii, Y., Yagi, G., and Asai, M. (2000). Serum interleukin-18 levels are elevated in schizophrenia. *Psychiatry Res.* **96**(1), 75–80.

Theodoropoulou, S., Spanakos, G., Baxevanis, C. N., Economou, M., Gritzapis, A. D., Papamichail, M. P., and Stefanis, C. N. (2001). Cytokine serum levels, autologous mixed lymphocyte reaction and surface marker analysis in never medicated and chronically medicated schizophrenic patients. *Schizophr. Res.* **47**(1), 13–25.

Vaughan, W. T., Sullivan, J. C., and Elmadjian, F. (1949). Immunity and schizophrenia. *Psychosom. Med.* **11**, 327–333.

Waddington, J. (1993). Schizophrenia: Developmental neuroscience and pathobiology. *Lancet* **341,** 531–536.

Wik, G. (1995). Effects of neuroleptic treatment on cortisol and 3-methoxy-4-hydroxy-phenylethyl glycol levels in blood. *J. Endocrinol.* **144**(3), 425–429.

Wilke, I., Arolt, V., Rothermundt, M., Weizsch, C. H., Hornberg, M., and Kirchner, H. (1996). Investigations of cytokine production in whole blood cultures of paranoid and residual schizophrenic patients. *Eur. Arch. Psychiatry Clin. Neurosci.* **246,** 279–284.

Wong, C. T., Tsoi, W. F., and Saha, N. (1996). Acute phase proteins in male Chinese schizophrenic patients in Singapore. *Schizophr. Res.* **22**(2), 165–171.

Wright, P., Sham, P. C., Gilvarry, C. M., Jones, P. B., Cannon, M., Sharma, T., and Murray, R. M. (1996). Autoimmune diseases in the first degree relatives of schizophrenic and control subjects. *Schizophr. Res.* **20,** 261–267.

Wright, P., Nimgaonkar, V. L., Donaldson, P. T., and Murray, R. M. (2001). Schizophrenia and HLA: A review. *Schizophr. Res.* **47**(1), 1–12.

Wu, Y., Shaghaghi, E. K., Jacquot, C., Pallardy, M., and Gardier, A. M. (1999). Synergism between interleukin-6 and interleukin-1β in hypothalamic serotonin release: A reverse *in vivo* microdialysis study in F344 rats. *Eur. Cytokine Netw.* **10**(1), 57–64.

Yang, Z. W., Chengappa, R., Shurin, G., Brar, J. S., Rabin, B. S., Gubbi, A. B., and Ganguli, R. (1994). An association between anti-hippocampal antibody concentration and lymphocyte production of IL-2 in patients with schizophrenia. *Psychol. Med.* **24,** 449–455.

Yannitsi, S. G., Manoussakis, M. N., Mavridis, A. K., Tzioufas, A. G., Loukas, S. B., and Plataris, G. K. (1991). Factors related to the presence of auto-antibodies in patients with chronic mental disorders. *Biol. Psychiatry* **30,** 731–735.

Zalcman, S., Green-Johnson, J. M., Murray, L., Nance, D. M., Dyck, D., Anizman, H., and Greenberg, A. H. (1994). Cytokine specific central monoamine alterations induced by interleukin-1, -2 and -6. *Brain Res.* **643,** 40–49.

Zarrabi, M. H., Zucker, S., Miller, T., Derman, R. M., Romeno, G. S., and Hartnett, J. A. (1979). Immunologic and coagulation disorders in chlorpromazine treated patients. *Ann. Intern. Med.* **91,** 194–199.

Zorilla, E. P., Cannon, T. D., Gur, R. E., and Kessler, J. (1996). Leukocyte and organ-non-specific auto-antibodies in schizophrenics and their siblings: Markers of vulnerability or disease? *Biol. Psychiatry* **40,** 825–833.

CEREBRAL LATERALIZATION AND THE IMMUNE SYSTEM

Pierre J. Neveu

Neurobiologie Intégrative
INSERM U394
Institut François Magendie
33077 Bordeaux, France

I. Introduction

The immune system has classically been considered as a sophisticated autoregulated system. However, clinicians have observed since long ago that susceptibility to illness may be influenced by psychological factors. Conversely, in humans, immune reactivity may be impaired in neurological and psychiatric diseases. Direct evidence of brain–immune interactions comes from experimental data. Indeed, immune responses may be influenced by lesion and stimulation of various brain areas (Keller *et al.*, 1980; Nance *et al.*, 1987; Roszman *et al.*, 1982; Lambert *et al.*, 1981; for review see Deleplanque and Neveu, 1994), physical and psychological stressors (Khansari *et al.*, 1990; Laudenslager *et al.*, 1983), and classically conditioned stimuli associated with an immunomodulatory stimulus (Ader and Cohen, 1985). The autonomic nervous system and the neuroendocrine system are considered major candidates for the mediation of brain influences on the immune system (Ader *et al.*, 1990). Lymphoid tissues (thymus, spleen, lymph nodes, mesenteric patches) receive direct innervation from sympathetic and parasympathetic components of the autonomic nervous system (Felten *et al.*, 1985). In addition, a wide body of data involves different components of the neuroendocrine system in the modulation of immune function (Blalock, 1989; Bateman *et al.*, 1989). Conversely, modulation of the nervous and neuroendocrine systems by immunomodulators, especially cytokines, has also been demonstrated (Plata-Salaman, 1991). Since the 1980s, the neuroimmune interactions have been known to be based on a bidirectional communication, and the existence of complex networks between the two systems is being increasingly characterized (Kroemer *et al.*, 1988).

Cerebral lateralization is evidenced on the basis of functional, anatomical, and neurochemical data. There is widespread agreement that the two cerebral hemispheres contribute differentially to the regulation of human behavior (Levy, 1974). In right-handers, the left hemisphere is specialized for speech and handedness and the right hemisphere is specialized for spatial abilities and the expression of affect (Bradshaw and Nettleton, 1981). The individual cerebral organization probably depends on hormonal and genetic factors during brain development. Morphological brain asymmetries have been described in man for a long time. These asymmetries mainly involve the cortical areas located around the posterior end of the Sylvius fissure, essentially the planum temporale. In most instances, cortical asymmetries favor the left hemisphere (Habib, 1989). However, the functional significance of these asymmetries as well as the relationship between callosal size and cerebral dominance remain controversial. By contrast, correlations between functional lateralization and asymmetrical distribution of neurotransmitters have been established, at least in experimental animals. Furthermore, the asymmetry in the distribution of neurotransmitters is impaired in several neurological and psychiatric diseases and could explain the fact that some drugs act predominantly on one hemisphere.

As the central nervous system can modulate the activity of the immune system and a number of brain functions are lateralized, the question arises whether the brain can regulate immune responses asymmetrically. In the early 1980s, two different paradigms were used to answer this question. First, using a model of cortical ablations in mice, Renoux *et al.* (1983) showed that left and right hemispheres have opposite effects on various immune parameters. Second, Geschwind and Behan (1982) have described an association between left-handedness and the incidence of immune disorders in humans. From these first observations, the following studies on the role of cerebral lateralization in the modulation of immune responses have given new insights into brain–immune networks.

II. Opposite Effects of Left- and Right-Cortical Ablations on Immunity

The opposite effects of right and left hemispheres on immune responses were first described in the 1980s in laboratory animals subjected to unilateral neocortex ablations. Ten years later, clinical data extend this phenomenon to humans and provide information for the management of patients suffering from brain lesions.

A. DESCRIPTION OF THE PHENOMENON IN RODENTS

In laboratory animals, the opposite effects of right- and left-cortical lesions on immune reactivity are obvious when measured a long time, 6 to 10 weeks, after surgery. In female C3H mice, mitogen-induced T-cell proliferation is decreased after ablation of the left fronto-parieto-occipital cortex. After a symmetrical right ablation, mitogenesis is enhanced (Neveu et al., 1986; Renoux et al., 1983). Differences are statistically significant only when left- and right-hemisphere-lesioned groups are compared. Usually, neither of these groups differ from the non-operated control group. These modifications in T-cell proliferation parallel IL-2 production (Neveu et al., 1989b). Additionally, natural killer (NK) cell activity is impaired after left cortical ablation but is unaffected by right lesion (Bardos et al., 1981; Betancur et al., 1991a). Cortical lesions also modified, but only slightly and nonsignificantly, mitogen-induced proliferation of B cells (Neveu et al., 1986). Production of antibodies of the IgG isotype is depressed after left lesions, whereas IgM antibody production remains unaffected (Renoux et al., 1983). Brain modulation of macrophage activation also depends on asymmetry. The intraperitoneal injection of bacillus Calmette-Guérin is known to induce an accumulation of activated macrophages in the peritoneum. Such an accumulation is not observed after a left-cortical ablation. Moreover, the oxidative metabolism of macrophages is decreased after left but not after right-cortical lesions (Neveu et al., 1989a). Damaging the cortex does not alter the function of nonactivated resident macrophages. Likewise, IL-1 production by macrophages stimulated by lipopolysaccharide (LPS) is depressed after left lesions, but enhanced after right lesions (Li and Yang, 1987). Our results obtained in female C3H mice have been replicated for NK cell activity in males of the same strain of mice (Betancur et al., 1991b) and for lymphoproliferation in female Sprague-Dawley and in male Wistar rats (Barnéoud et al., 1988a; LaHoste et al., 1989). This suggests that brain asymmetry control of immune reactivity, which was first described in female mice, is probably a generalized phenomenon among mammals.

The respective role of each hemisphere in neocortex-mediated immuno-modulation is still controversial. Renoux and Bizière (1986) have postulated that the right hemicortex modulates the activity of the left, which in turn controls the immune system. According to this hypothesis, the effects of bilateral cortical ablation should be similar to that of a left lesion alone. Indeed, they have shown that both bilateral and left lesions depress NK cell activity to the same extent. However, in our experiments, bilateral lesions (two unilateral lesions performed within 3 weeks to avoid mortality observed after one stage of bilateral damage) do not modify mitogen-induced lymphoproliferation (Neveu et al., 1988a). Suppression of the asymmetrical immunoregulatory effects by bilateral cortical ablation suggests that each hemicortex may be acting oppositely on the immune system. The right hemisphere may decrease, while the left increases, certain T-cell functions.

Each hemicortex appears to be heterogeneous in regard to its immuno-regulatory functions. The immune effects of unilateral lesions restricted to the parieto-occipital areas are different from those observed after lesions involving all of the fronto-parieto-occipital cortex (Barnéoud et al., 1987). Furthermore, small electrolytic lesions of the right or left frontal cortex have no effect on immune reactivity (Barnéoud, 1988, unpublished data). The fact that different immunological parameters may be modulated in various ways depending on the size and location of the lesions suggests that each hemicortex may contain both activating and suppressing areas, which may interact within and between hemispheres.

The immunoregulatory role of each hemisphere is consistent when studied a long time after surgery. To study the possible involvement of neuronal reorganization during the postoperative period, we have tested the immune system as early as 2 weeks after left- or right-cortical ablation (Barnéoud et al., 1988b). Comparison between early and late effects of right- and left-neocortical lesions shows that the immunoregulatory functions of each hemisphere evolve differently with time. The immunosuppressive effects of right lesions are transitory and can be completely reversed. On the contrary, after left lesions, the early depression of immune reactivity persists with time. This suggests that neural reorganization following surgery can be different according to the side of the cortex lesions. However, no more immune alterations can be observed 6 months after either a right- or a left-cortical ablation (Barnéoud and Neveu, 1989, unpublished data).

B. IMMUNE TARGETS SUBJECTED TO BRAIN CONTROL

Although it has been shown that the neocortex modulates the activity of both lymphocytes (T and B) and macrophages, the cellular targets of

brain cortex immunomodulation are not yet known. In addition, the phys-
iopathological effects of immune perturbations induced by cortical lesions
remain to be clarified. Until now, the effects of cortical ablation on infection have been studied in just one experiment in which mice were infected
with *Trypanosoma musculi*. Even though parasitemia is slightly increased in
the brain lesioned mice, asymmetrical effects are not observed (Neveu *et al.*,
1989a). Similar experiments, using other infectious diseases or tumor grafts
have yet to be reported.

   Cortical lesions do not induce a lymphocyte redistribution similar to
that observed during stress (Dhabhar *et al.*, 1995). Furthermore, cortical
lesions modulate mitogen-induced proliferation of lymphocytes from both
lymph nodes or spleen in a similar way (Barnéoud *et al.*, 1990). Therefore,
the cortex modulation of the immune system may not directly occur at the
level of secondary lymphoid organs but rather at the level of the bone marrow or thymus. If this is the case, neocortical damage may affect immune
functions in one of two ways. As the cortex affects both lymphocytes and
macrophages, it is possible that it acts on a hematopoietic stem cell at the
bone marrow level as previously postulated (Neveu *et al.*, 1989a). Alternatively, the neocortex may first act on T lymphocytes in the thymus and only
secondarily modulate B-cell and macrophage functions through the production of cytokines produced by T lymphocytes. In fact, it has been shown
that the left-cortical lesions decrease production and/or release of serum
factor(s) involved in T-cell maturation (Renoux and Bizière, 1987). Further
experiments are needed to address these issues.

C. Brain Structures Involved in Neocortex Immunomodulation

   Subcortical structures implicated in cortical immunomodulation are not
yet known. Among the subcortical structures known to send projections to
the neocortex, the nucleus basalis magnocellularis, which provides the major source of cortical cholinergic innervation, can modulate cortical activity.
Bilateral lesions of this nucleus strongly increase lymphoproliferation and
NK cell activity, though, unilateral lesions have no effects on these immune
parameters (Cherkaoui *et al.*, 1990). However, the conclusion that cholinergic systems are involved in neuroimmunomodulation must be considered
cautiously because the lesion of the nucleus basalis may affect neurotransmitters other than acetylcholine. Other experimental data confirm that catecholaminergic pathways may be involved in neocortex immunomodulation.
One series of experiments has used a pharmacological approach. Sodium
diethyldithiocarbamate is an inhibitor of the dopamine β-hydroxylase and
an immune stimulator. The immune effects of this compound, which are

related to the inhibition of dopamine β-hydroxylase, depend on the presence of intact cortex (Renoux and Bizière, 1987). In the same vein, there is evidence to suggest that unilateral cortical lesions asymmetrically modify the concentrations of catecholamines in various subcortical structures (Barnéoud et al., 1991). This might suggest that catecholamines are modified by cortical lesions and that they, in turn, modulate the immune system. Even though these suggestions are speculative, we have demonstrated that unilateral lesions of the right or left substantia nigra had opposite effects on lymphoproliferation (Neveu et al., 1992).

Brain modulation of immune reactivity is well known to be mediated by the HPA axis and the sympathetic nervous system. However, the role of these systems in the immune alterations induced by cortical lesions is not known. Very few experiments have questioned the possible role of hormonal mediators in brain cortex immunomodulation. To our knowledge, only one hormone has been shown to be involved in asymmetrical brain modulation of the immune system. Plasma levels of prolactin are negatively correlated with lymphocyte proliferation after asymmetrical cortical lesions (LaHoste et al., 1989).

After cortex damage, there are many secondary functional changes involving neural plasticity as well as alterations in several subcortical structures. These secondary modifications may mask the respective role of each hemisphere in neuroimmunomodulation. Additionally, the immune system is known to send information to the central nervous system. For example, interleukin-1 (IL-1) produced by macrophages stimulates corticosteroid production by acting at the hypothalamic and/or the pituitary level. During antibody production, the firing rate increases and the noradrenaline turnover decreases in the hypothalamus (Besedovsky et al., 1977). In mice, after stimulation of the immune system induced by bacillus Calmette-Guérin, noradrenaline levels increase in both hemispheres but only significantly in the right one. This increase was correlated with lymphoproliferation (Barnéoud et al., 1988c). These results suggest that the communication pathways from the immune system toward the brain may also be lateralized. Feedback mechanisms could modify the immune effects of cortical lesions and therefore make analysis of these effects difficult.

D. Immune Effects of Asymmetrical Brain Lesions in Humans

Data first obtained in laboratory animals appear to be sufficiently reliable to be extended to humans. Indeed, immune defects observed after brain lesions in humans depend on the side of the lesion. However, literature on clinical data is still sparse. The perturbations of blood lymphocyte subsets observed 3 weeks after a stroke in humans (Czolonkowska et al.,

1987) are in agreement with the modifications of mitogenesis observed in mice 2 weeks after brain cortex ablation. One month poststroke, patients displayed enhanced antigen-specific T-cell reactivity on the paretic side of the body when the lesions were on the right but not the left side (Tarkowski *et al.*, 1998). These authors have further suggested that the frontal cortex-putamen is a key brain structure in regulating the magnitude of immune responses. Their results indicate a similar specialization of the regulation of immune responses between the right and left brain hemispheres in humans as that described in mice. In patients with stroke or cerebral palsy, immune alterations are more important after left than after right hemispheric lesions (Rogers *et al.*, 1998). When studying the effects of resections for epilepsy surgery on T-cell indices, Meador *et al.* (1999) show differential immunological responses to focal cerebral lesions as a function of cerebral lateralization. Preliminary data on elderly patients with cerebrovascular diseases show a greater incidence of severe infection, mainly pneumonia, after left lesions (Kawaharda and Urasawa, 1992).

E. CONCLUSION

The experimental evidence of brain asymmetry in immunomodulation, obtained in animals, has been extended to humans and may have important clinical incidence. In animals the immune effects of unilateral neocortex brain lesions are well established after a lag period of 6–10 weeks. As compared to right lesions, the left ones depress T-cell mitogenesis, IL-2 production, T-cell-dependent antibody production, NK cell activity, and macrophage activation. By contrast, T-cell-independent antibody production is not altered by left or right lesions. In humans, immune alterations induced mainly by left lesions must be better established in order to predict possible infectious complications. New techniques including neuroimaging would be useful to further characterize the cortex areas and the subcortical structures responsible for immune alterations and the mechanisms involved in such a brain–immune regulation. They would also allow the analysis of neural plasticity following lesioning and the effects of immune alterations on brain functioning.

III. Association between Behavioral Lateralization and Immune Reactivity

A. CLINICAL DATA

Geschwind and Behan (1982) were the first to describe a higher incidence of immune diseases, developmental learning disorders, and migraine in left-handers as compared to right-handers. To account for these results, Geschwind and Galaburda (1985) have suggested that, during fetal life, either a high level of testosterone or a higher sensitivity to this hormone could result in the development of left-handedness, due to delayed growth of the left hemisphere. Additionally, testosterone could also affect the development of the thymus, favoring immune disorders later in life. According to Geschwind and Galaburda, males should display a higher incidence of left-handedness and immune disorders, as they have higher testosterone levels. Following this first observation, an association between left-handedness and autoimmune diseases has been found by other authors for certain disorders such as Crohn's disease, ulcerative colitis, and type I diabetes (Searleman and Fugali, 1987), but not for others like lupus erythematosus (Salcedo et al., 1985). Similarly, for allergic diseases whose ethiopathogenesis is better known, the association remains controversial. Strong associations are found in some studies. Geschwind and Behan (1982) have reported a significant increase in allergies in strongly left-handed individuals, without specifying what type of allergic disorders are included. Smith (1987) has found that among patients attending an allergy clinic, there are significantly more left-handers when compared to control subjects; the proportion of left-handedness was largest in patients suffering from rhinitis, asthma, eczema, and urticaria. Lelong et al. (1986) have reported that the frequency of non-right-handedness is significantly greater in children allergic to mites than in a control population of school children of the same age. In a study of mathematically precocious youths at the Johns Hopkins, Benbow and Benbow (1984) have found that there is a marked predominance of males in this group and that 20% of these students are left-handed or ambidextrous and 56% had symptomatic atopic disease. However, other authors have been unable to demonstrate an association between left-handedness and allergies. Bishop (1986), in a study of a population of British children, does not find that allergy (usually hay fever), eczema, asthma, or psoriasis are associated with left-handedness. Although Pennington et al. (1987) have found a significant elevation of both autoimmune and allergic disorders in dyslexic patients, there is no association between left-handedness and immune disorders. In our study, we have compared the distribution of right- and left-handers in a population of patients consulting an allergy clinic and a control group with a similar sex and age distribution (Betancur et al., 1990). There is no overall association between left-handedness and allergy. However, in accordance with Geschwind's theory, we have found a tendency toward left-handedness in patients whose allergic symptoms start before puberty, suggesting that left-handers may have an increased predisposition to allergic disease that manifests itself during early life.

The discrepancies between the different human studies probably result from differences among the patient populations tested. Furthermore, in these studies only clinical signs are taken into account, and the immune status of patients has not been studied. These contradictions in the human studies are difficult to resolve because of both theoretical and methodological problems (Satz and Soper, 1986; Smith, 1987). To overcome the difficulties encountered in human studies for demonstrating a possible association between lateralization and immune responses, an experimental approach in laboratory animals has been developed.

B. ANIMAL OBSERVATIONS

In rodents, postural/motor asymmetries have been assessed using a variety of tests, including paw preference (Collins 1985; Betancur *et al.*, 1991a) or lever-pressing (Glick and Jerussi, 1974) in a food-reaching task, side bias in an open field (Camp *et al.*, 1984), side preference in a T-maze (Camp *et al.*, 1984; Castellano *et al.*, 1989), tail pinch–induced asymmetries (Camp *et al.*, 1984; Myslobodsky and Braun, 1980), tail suspension (Castellano *et al.*, 1989; Myslobodsky and Braun, 1980), tail posture in neonates (Denenberg *et al.*, 1982), swimming rotation (Collins, 1985), and spontaneous or D-amphetamine-induced rotational behavior (Glick *et al.*, 1986). Animals usually display a consistent lateralized response when tested with the same protocol, but they do not show a consistent bias across the different tests. Indeed, studies on the relationship between different measures of behavioral asymmetries have shown that the direction of one postural/motor asymmetry does not predict the direction of another one (Camp *et al.*, 1984; Castellano *et al.*, 1989; Myslobodsky and Braun, 1980). This suggests that in rodents as in humans, different measures of behavioral lateralization may reflect different brain asymmetries. In animals, some studies have found population biases for the right side in amphetamine-induced rotation (Glick and Ross, 1981), electrified T-maze (Alonso *et al.*, 1991; Castellano *et al.*, 1987, 1989), tail suspension (Castellano *et al.*, 1989), or paw preference (Waters and Denenberg, 1994). Usually, these population biases are found only after repetition of the behavioral tasks that increase individual and population laterality with practice. However, it is generally held that although animals exhibited behavioral asymmetries at the individual level, they do not display a strong asymmetry at the population level, that is, side preferences are nearly distributed equally between left and right. Furthermore, lateralization tested by paw preference in a food-reaching task, as previously described by Collins (1985), is stable over time and results obtained in test sessions separated by intervals of several weeks are highly correlated (Collins, 1985;

Neveu *et al.,* 1988b). According to paw preference distributions, a mouse population may be divided into subpopulations of right-pawed, left-pawed, and ambidextrous animals. These subpopulations may be compared when studying the possible differences in the activity of any physiological system supposed to be linked to brain lateralization. Furthermore, each animal can be characterized by a lateralization score, which enables correlation analyses to be performed.

An association between paw preference and immune reactivity was first described in C3H mice. In females, left-pawed animals exhibit higher mitogen-induced T-lymphocyte proliferation than right-pawed mice. However, B-lymphocyte mitogenesis is not dependent on behavioral lateralization (Neveu *et al.,* 1988b). Such an association between paw preference and T-cell mitogenesis is not found in males of the same strain. In these males, but not in females, NK cell activity is associated with paw preference. Left-pawed males exhibit lower NK cell activity than right-pawed animals (Betancur *et al.,* 1991b). Furthermore, production of cytokines, especially IL-1, in response to restraint (Merlot *et al.,* 2002) or to lipopolysaccharide (LPS) (Dong *et al.,* 2002) is higher in right-pawed than in left-pawed mice. In Balb/c mice, the production of cytokines in response to LPS has also been shown to depend on lateralization (Gao *et al.,* 2000). LPS, which is known to induce a depression of T-cell-induced mitogenesis, depresses lymphoproliferation in right-pawed mice but not in left-pawed animals (Delrue *et al.,* 1994). An association between lateralization and immune reactivity has also been demonstrated for the production of autoantibodies (Neveu *et al.,* 1989c). New Zealand Black mice spontaneously develop autoimmune disorders such as lupuslike glomerulonephritis and hemolytic anemia related to anti-DNA and antierythrocyte antibodies, respectively. No correlation between paw preference and autoantibody production was observed in males. By contrast, in females, both anti-erythrocyte and anti-DNA antibodies of the IgG isotype appear earlier in left-pawed animals. The production of autoantibody of the IgM isotype is not influenced by lateralization. Since isotype switch to IgG is strongly T-cell dependent, it may be suggested that paw preference is mainly associated with T-cell functions. Kim *et al.* (1999) show in Balb/c mice that animals with right-turning preference have a higher host resistance to the intracellular bacteria *Listeria monocytogenes.* Mice with right-turning preference also exhibit a higher primary antibody response and a higher delayed-type hypersensitivity to a protein antigen. However, the secondary humoral response is similar in right- and left-turners. By contrast, the production of cytokines (IL-6, IFN-γ) in response to *L. monocytogenes* is lower in right-turners.

The data already available concerning the association between the immune system and behavioral lateralization clearly show that this association depends on the immune parameters tested. This suggests that not all the components of the immune system are under the control of lateralization. The T-cell lineage appears to be more susceptible than the B-cell lineage. No experiments have yet tested the possible influence of brain lateralization on macrophage functions. The immune target(s) of brain immunomodulation remains to be elucidated. The hypothetical impact of behavioral lateralization on immune cell production in the bone marrow or on T-cell maturation in the thymus remains to be studied. The association between lateralization and immune reactivity also depends on the sex of the animals, suggesting that the sex hormones are involved. This point is not surprising because sex hormones are known to be involved both in brain development (McLusky and Naftolm, 1981) and in immune reactivity (Comsa et al., 1982). Finally, the fact that the importance of the link between behavioral lateralization and immunity depends on the strains tested (Fride et al., 1990; Neveu et al., 1991) suggests that this association could be influenced by genetic factors which are still unknown. This point has to be specifically questioned using a genetic approach.

C. BRAIN–IMMUNE PATHWAYS INVOLVED IN LATERALIZED BRAIN CONTROL OF IMMUNE REACTIVITY

Because the communication pathways between the brain and the immune system involve the neuroendocrine system and, especially, the hypothalamic-pituitary-adrenal (HPA) axis (Bateman et al., 1989) as well as the autonomic nervous system (Ader et al., 1990), it may be hypothesized that the asymmetrical brain organization influences the reactivity of these systems.

Only a few differences have been observed in the activity of the HPA axis in basal conditions. The in vitro production of ACTH by the pituitary and of corticosterone by the adrenals is similar whether tissues are taken from right- or left-pawed mice (Betancur et al., 1992). As the activity of the HPA axis is regulated by the hippocampus, we looked for a possible difference in the distribution of glucocorticoid receptors in relation to behavioral lateralization. In the hippocampus, the binding capacity of glucocorticoid receptors was distributed symmetrically and does not depend on the behavioral lateralization of animals. By contrast a greater expression of mineralocorticoid receptors (MR) in the right hippocampus, as a ratio of the total hippocampal MR, is associated with left-paw preference (Neveu et al., 1998b). The inhibitory function of the hippocampus on the HPA axis has been shown to be mainly mediated by the mineralocorticoid receptor and differences in hippocampal mineralocorticoid receptor binding activity may explain interindividual differences in the corticoid stress response (Schöbitz *et al.*, 1994; Oitzl *et al.*, 1995). To our knowledge, no experiment has been performed to look for differential sympathetic innervation of lymphoid tissues in relation to brain lateralization.

The role of lateralization in the modification of immune reactivity may be easily studied during the stress response. Indeed, it represents an integrated response of the organism to environmental demands that involves the endocrine system, especially the HPA axis, and the sympathetic nervous system. Different stressors have been used in mice to demonstrate the role of brain lateralization on various parameters of the stress response. The influence of lateralization on the alterations of the dopaminergic metabolism induced by restraint, a stressor known to have immune effects, has been studied in female C3H mice. The asymmetry of dopamine metabolism in the nucleus accumbens, which is correlated with paw preference scores in controls (Cabib *et al.*, 1995) disappears during stress. Conversely, asymmetry in dopamine and its metabolite (DOPAC) contents in the prefrontal cortex, which appear in stressed animals are correlated with paw preference (Neveu *et al.*, 1994). These experiments show that a stressful event known to affect immune reactivity, induces asymmetrical modifications of the dopaminergic metabolism that are related to behavioral lateralization. However, the immunomodulating role of dopamine in the nucleus accumbens is not known, A stresslike response can also be induced by an injection of LPS. This response includes a decrease in T-lymphocyte proliferation, an increase in ACTH and corticosterone plasma levels, and augmentation of central serotoninergic and dopaminergic metabolisms (Dunn and Welch, 1991). Interestingly, modifications of brain monoaminergic metabolism observed 2 and 4 h after injection of LPS are asymmetrically expressed and depend on behavioral lateralization (Delrue *et al.*, 1994). Moreover, plasma corticosterone levels are increased in both left- and right-pawed animals, but plasma ACTH levels were increased only in right-pawed and ambidextrous animals. Immune and neuroendocrine modifications induced by an injection of LPS could have been mediated by the sympathetic system. Indeed, immune depression as well as activation of the HPA axis induced by LPS could be partly or totally suppressed by a pharmacological depletion of peripheral catecholamines using 6-hydroxydopamine (Delrue-Perollet *et al.*, 1995). These results suggest that the activity of the sympathetic nervous system can be, at least partly, under the control of lateralized brain functions. In accordance with this hypothesis, we have shown that an $\alpha1/\alpha2$-adrenergic receptor antagonist, prazosin, suppresses more efficiently an LPS-induced increase of plasma IL-1 in right- than in left-pawed animals (Dong *et al.*, 2002).

Finally, it can be hypothesized that the effects of immune messengers at the brain level may differ according to lateralization. Indeed, sickness behavior (referred to as depressed social behavior, increased immobility, loss of body weight, reduced food and water intake, and increased non-REM sleep) induced by an intraperitoneal injection of IL-1 is more severe in right-pawed mice than in left-pawed animals (Neveu *et al.*, 1998a).

D. Biobehavioral Significance of the Association between Lateralization and Immune Reactivity

The role of a functional brain asymmetry in the modulation of the immune system is well established in rodents, and it is probable in humans. If so, the question arises why no difference has been observed already in the incidence of infectious diseases or in survival time in relation to immune responsiveness despite differences in immune reactivity between left- and right-handed humans. Several hypotheses may be proposed. A first possibility is that differences in immune reactivity between left- and right-handers are in fact very low and therefore physiologically irrelevant. A second possibility may be that, even though these differences have a physiological significance, the immune response of an individual exhibiting a specific depression of a single immune parameter may be effective because it is achieved by alternative mechanisms. For example, in left-pawed animals with low NK cell activity, tumor killing may be assumed by cytotoxic T cells. A third possibility, not exclusive to the second one, is that some immune parameters may be positively correlated, while others may be inversely correlated with lateralization scores. According to this hypothesis, in a mouse population, left- as compared to right-pawed animals should exhibit a higher value for one immune parameter but a lower value for another one. Therefore, strict left- as compared to strict right-pawed animals should be considered as high responders in relation to some specific immune components, but as low responders for another type of response. According to this hypothesis, discussed elsewhere (Neveu, 1992), the resistance to various diseases (intracellular vs extracellular infections, for example), which involves different types of immune responses (Th1/type 1 as opposed to Th2/type 2), may differ according to functional asymmetry. Left-handers should be more sensitive to some diseases but more resistant to others compared to right-handers. This point, which needs experimental confirmation, may explain why no difference in survival time is observed between right- and left-handers in a population naturally subjected to numerous pathogenic agents.

Lateralization, which can be considered as an individual behavioral trait, is not only related to the activity of the neuroendocrine and immune systems. In fact, lateralization has been correlated with various behavioral and personality traits in humans and experimental animals. Emotionality, which is defined as the capacity to perceive and react to potentially anxiogenic situations (Boissy, 1995), has been associated with lateralization and with an asymmetric organization of motor behavior (Sackeim *et al.*, 1982). In humans and primates, hemispheric specialization underlies interindividual differences in affective behavior and behavioral reactivity (Davidson, 1995). It has also been shown that in children, behavioral inhibition—thought in its extreme form to be an early marker of fearful temperament—is associated with relative prefrontal asymmetric brain activation (Davidson, 1992). The relationship between hemispheric lateralization and affectivity has been further illustrated in studies showing that monkeys with extreme right frontal activity exhibited greater levels of traitlike fearful behavior than do monkeys with extreme left frontal activity (Kalin *et al.*, 1998). In rats, asymmetries in the distribution of brain monoamines, especially serotonin in the amygdala and dopamine in the frontal cortex are predictive of interindividual differences in the elevated plus-maze (Andersen and Teicher, 1999). The degree of side preferences has been correlated to performances in various tasks including active- and passive-avoidance responses. A low degree of laterality is associated with an optimal learning ability, suggesting for some authors that laterality should have some adaptative significance (Glick *et al.*, 1977). A relationship between rat behavioral despair in the forced-swimming test and motor lateralization in the T-maze test has been related to dopaminergic innervation of the nucleus accumbens (Alonso *et al.*, 1991). We have previously shown that the neuroendocrine and immune responses to a psychological stressor (i.e., restraint) depend on motor lateralization in mice (Neveu and Moya, 1997; Merlot *et al.*, 2001). It could therefore be hypothesized that lateralization may be related to anxiety which is known to influence stress reactivity and immune responses (Haas and Schauenstein, 1997). According to this hypothesis, the relation between lateralization and immune reactivity would involve individual behavioral patterns as they appear in exploratory-based anxiety models. In accordance with this hypothesis, we have reported that exploratory behavior in the open field, but not in the elevated plus-maze, is influenced by the interactive effect of gender and behavioral lateralization (Mrabet *et al.*, 2000). Westergaard *et al.* (2002) reported that in monkeys, there is a correlation among immune functioning (percentage of $CD8^+$ and $CD4^+$ cells), behavioral responses to a threat (human-directed aggression in response to an invasive threat), and hand preference in a food-reaching task. These findings have important implications since they could point out key factors allowing a clear characterization of subpopulations according to both anxiety and physiological responses to stress. Co-variations between changes in exploratory behaviors and neuroendocrine correlates of stress response should be evidenced in future experiments. Other kinds of behavioral traits, like social status which is known to have immune correlates (Koolhaas *et al.*, 1999), may be putatively linked to lat- eralization. Further investigation in this direction is needed to confirm this hypothesis. Finally, behavioral investigations should benefit from etholog- ically based approaches to evidence not only possible discrete differences relative to lateralization status of subjects but also the putative functions of lateralization. More generally, lateralization may be considered as one of the behavioral traits, which characterized the personality of an individual, the latter including the functioning of physiological systems such as the immune system.

IV. General Conclusion and Perspectives

The data concerning the immune effects of asymmetrical brain lesions clearly demonstrate that each hemisphere plays a particular role in im- munomodulation. First obtained in animals, they have been extended by human observations. Further work is needed, especially in humans, to de- lineate the intermediary mechanisms involved and to predict the possible immune defects consecutive to brain lesions.

The consistent observation of an association between lateralization and immune reactivity has led to important notions. Lateralization may repre- sent a neurobehavioral trait linked to the activity of physiological systems in- volved in the response to external challenges (either intra- or interspecific). Therefore, the description of these associations should be a first step for delineating the personality of an individual (taken as the whole of his or her characteristics), including various aspects of behavior, of brain metabolism, and of neuroendocrine and immune responsiveness, in order to predict the responsiveness to stressful conditions.

These experimental observations might have clinical relevance. Groups of subjects may be better characterized in terms of the various components of the immune response in addition to the clinical course of disease. In fact, experimental models have shown that this association may be observed for some immune parameters but not for others. As the immune processes in- volved in pathogenesis of various diseases may be different, the association may, or may not be pertinent for the predictability of the disease process. The association between lateralization and immune reactivity may also exist in clinical situations not already expected. For example, depressive symp- tomatology has been shown to depend on lateralization (Flor-Henry, 1976). Studies show that depression is accompanied by an increased production of cytokines (Capuron *et al.*, 2001), which may have a pathogenic role in the activation of the HPA axis and in the depression of brain serotonin

(Capuron *et al.*, 2002). As animal studies have shown that the production of cytokines in response to an inflammatory stimulus depends on lateralization, the latter could be used as a behavioral marker to predict severity and evolution of depression. This example may be generalized to include inflammatory diseases such as multiple sclerosis or arthritis where stressful situations have been involved in the relapse of these chronic diseases as lateralization is involved both in stress response and cytokine production. These considerations open a large field of new investigations.

References

Ader, R., and Cohen, N. (1985). CNS-immune system interactions: Conditioning phenomena. *Behav. Brain Sci.* **8,** 379–394.

Ader, R., Felten, D., and Cohen, N. (1990). Interactions between the brain and the immune system. *Ann. Rev. Pharmacol. Toxicol.* **30,** 561–602.

Alonso, J., Castellano, M. A., and Rodriguez, M. (1991). Behavioral lateralization in rats: Prenatal stress effects on sex differences. *Brain Res.* **539,** 45–50.

Andersen, S. L., and Teicher, M. H. (1999). Serotonin laterality in amygdala predicts performance in the elevated plus maze in rats. *Neuroreport* **10,** 3497–3500.

Bardos, P., Degenne, D., Lebranchu, Y., Bizière, K., and Renoux, G. (1981). Neocortical lateralization of NK activity in mice. *Scand. J. Immunol.* **13,** 609–611.

Barnéoud, P., Neveu, P. J., Vitiello, S., and Le Moal, M. (1987). Functional heterogeneity of the right and left cerebral neocortex in the modulation of the immune system. *Physiol. Behav.* **41,** 525–530.

Barnéoud, P., Neveu, P. J., Vitiello, S., and Le Moal, M. (1988a). Brain neocortex immunomodulation in rats. *Brain Res.* **474,** 394–398.

Barnéoud, P., Neveu, P. J., Vitiello, S., and Le Moal, M. (1988b). Early effects of right and left cerebral cortex ablation on mitogen-induced spleen lymphocyte DNA synthesis. *Neurosci. Lett.* **90,** 302–307.

Barnéoud, P., Rivet, J. M., Vitiello, S., Le Moal, M., and Neveu, P. J. (1988c). Brain norepinephrine levels after BCG-stimulation of the immune system. *Immunol. Lett.* **18,** 201–204.

Barnéoud, P., Neveu, P. J., Vitiello, S., and Le Moal, M. (1990). Lymphocyte homing after left or right brain neocortex ablation. *Immunol. Lett.* **24,** 31–36.

Barnéoud, P., Le Moal, M., and Neveu, P. J. (1991). Asymmetrical effects of cortical ablation on brain monoamines in mice. *Int. J. Neurosci.* **56,** 283–294.

Bateman, A., Singh, A., Kral, T., and Solomon, S. (1989). The immune-hypothalamic-pituitary-adrenal axis. *Endocrine Rev.* **10,** 92–112.

Benbow, C. P., and Benbow, R. M. (1984). Biological correlates of high mathematical reasoning ability. *Prog. Brain Res.* **61,** 469–490.

Besedovsky, H., Sorkin, E., Felix, D., and Haas, H. (1977). Hypothalamic changes during the immune response. *Eur. J. Immunol.* **7,** 323–325.

Betancur, C., Velez, A., Cabanieu, G., Le Moal, M., and Neveu, P. J. (1990). Association between left-handedness and allergy: A reappraisal. *Neuropsychologia* **28,** 223–227.

Betancur, C., Neveu, P. J., and Le Moal, M. (1991a). Strain and sex differences in the degree of paw preference in mice. *Behav. Brain Res.* **45,** 97–101.

Betancur, C., Neveu, P. J., Vitiello, S., and Le Moal, M. (1991b). Natural killer cell activity is associated with brain asymmetry in male mice. *Brain Behav. Immunol.* **5**, 162–169.

Betancur, C., Sandi, C., Vitiello, S., Borrell, J., Guaza, C., and Neveu, P. J. (1992). Activity of the hypothalamo-pituitary-adrenal axis in mice selected for left- or right-handedness. *Brain Res.* **589**, 302–306.

Bishop, D. V. M. (1986). Is there a link between handedness and hypersensitivity? *Cortex* **22**, 289–296.

Blalock, J. E. (1989). A molecular basis for bidirectional communication between the immune and neuroendocrine systems. *Physiol. Rev.* **69**, 1–32.

Boissy, A. (1995). Fear and fearfulness in animals. *Q. Rev. Biol.* **70**, 165–191.

Bradshaw, C., and Nettleton, N. C. (1981). The nature of hemispheric specialization in man. *Behav. Brain Sci.* **4**, 51–91.

Cabib, S., D'Amato, F., Neveu, P. J., Deleplanque, B., and Puglisi-Allegra, S. (1995). Paw preference and brain dopamine asymmetries. *Neuroscience* **64**, 427–432.

Camp, D. M., Robinson, T. E., and Becker, J. B. (1984). Sex differences in the effects of early experience on the development of behavioral asymmetries in rats. *Physiol. Behav.* **33**, 433–439.

Capuron, L., Ravaud, A., Gualde, N., Bosmans, E., Dantzer, R., Maes, M., and Neveu, P. J. (2001). Association between immune activation and early depressive symptoms in cancer patients treated with interleukin-2-based therapy. *Psychoneuroendocrinology* **26**, 797–808.

Capuron, L., Ravaud, A., Neveu, P. J., Miller, A. H., Maes, M., and Dantzer, R. (2002). Association between decreased serum tryptophan concentrations and depressive symptoms in cancer patients undergoing cytokine therapy. *Mol. Psychiatr.* **7**, 468–473.

Castellano, M. A., Diaz-Palarea, M. D., Rodriguez, M., and Barroso, J. (1987). Lateralization in male rats and dopaminergic system: Evidence of right-side population bias. *Physiol. Behav.* **40**, 607–612.

Castellano, M. A., Diaz-Palarea, M. D., Barroso, J., and Rodriguez, M. (1989). Behavioral lateralization in rats and dopaminergic system: Individual and population laterality. *Behav. Neurosci.* **103**, 46–53.

Cherkaoui, J., Mayo, W., Neveu, P. J., Kelley, K. W., Vitiello, S., Le Moal, M., and Simon, H. (1990). The nucleus basalis is involved in brain modulation of the immune system in rats. *Brain Res.* **516**, 345–348.

Collins, R. L. (1985). On the inheritance of direction and degree of asymmetry. *In* "Cerebral Lateralization in Non-human Species" (S. D. Glick, ed.), pp. 41–71. Academic Press, New York.

Comsa, J., Leonhardt, H., and Wekerle, H. (1982). Hormonal coordination of the immune response. *Rev. Physiol. Biochem. Pharmacol.* **92**, 115–191.

Czolonkowska, A., Korlak, J., and Kuczynska-Zarzewialy, A. (1987). Lymphocyte subsets after stroke (abstract). *J. Neuroimmunol.* **16**, 40.

Davidson, R. J. (1992). Anterior cerebral asymmetry and the nature of emotion. *Brain Cogn.* **20**, 125–151.

Davidson, R. J. (1995). Cerebral asymmetry, emotion and affective style. *In* "Brain Asymmetry" (R. J. Davidson and K. Hugdahl, eds.), pp. 361–387. MIT Press, Cambridge, MA.

Deleplanque, B., and Neveu, P. J. (1994). Brain regions involved in modulation of immune responses. *In* "Brain Control of Responses To Trauma" (N. Rothwell and F. Berkenbosch, eds.), pp. 108–122. Cambridge University Press, Cambridge, U.K.

Delrue, C., Deleplanque, B., Rouge-Pont, F., Vitiello, S., and Neveu, P. J. (1994). Brain monoaminergic, neuroendocrine and immune responses to an immune challenge in relation to brain and behavioral lateralization. *Brain Behav. Immun.* **8**, 137–152.

Delrue-Perollet, C., Li, K. S., and Neveu, P. J. (1995). Peripheral catecholamines are involved in the neuroendocrine and immune effects of LPS. *Brain Behav. Immun.* **9**, 149–162.

Denenberg, V. H., Rosen, G. D., Hofmann, M., Gall, J., Stockler, J., and Yutzey, D. A. (1982). Neonatal postural asymmetry and sex differences in the rat. *Dev. Brain Res.* **2**, 417–419.

Dhabhar, F. S., Miller, A. H., McEwen, B. S., and Spencer, R. L. (1995). Effects of stress on immune cell distribution. Dynamics and hormonal mechanisms. *J. Immunol.* **154**, 5511–5527.

Dong, J., Mrabet, O., Moze, E, Li, K. S., and Neveu, P. J. (2002). Lateralization and catecholaminergic neuroimmunomodulation: Prazosin, an $\alpha1/\alpha2$ adrenergic receptor antagonist, suppresses IL-1 and increases IL-10 production induced by LPS. *Neuroimmunomodulation* (in press).

Dunn, A. J., and Welch, J. (1991). Stress and endotoxin-induced increases in brain tryptophan and serotonin depend on sympathetic nervous system activity. *J. Neurochem.* **57**, 1615–1622.

Felten, D. L., Felten, S. Y., Carlson, S. L., Olschowka, J. A., and Livnat, S. (1985). Noradrenergic and peptidergic innervation of lymphoid tissue. *J. Immunol.* **135**, 755s–765s.

Flor-Henry, P. (1976). Lateralized temporal-limbic dysfunction and psychopathology. *Ann. N. Y. Acad. Sci.* **280**, 777–795.

Fride, E., Collins, R. L., Skolnick, P., and Arora, P. K. (1990). Strain-dependent association between immune function and paw preference in mice. *Brain Res.* **522**, 246–250.

Gao, M. X., Li, K. S., Dong, J., Liège, S., Jiang, B., and Neveu, P. J. (2000). Strain-dependent association between lateralization and LPS-induced IL-1β and IL-6 production in mice. *Neuroimmunomodulation* **8**, 78–82.

Geschwind, N., and Behan, P. (1982). Left-handedness: Association with immune disease, migraine and developmental learning disorders. *Proc. Natl. Acad. Sci. USA* **79**, 5097–5100.

Geschwind, N., and Galaburda, A. M. (1985). Cerebral lateralization. Biological mechanisms, associations and pathology. III. A hypothesis and a program for research. *Arch. Neurol.* **42**, 634–654.

Glick, S. D., and Jerussi, T. P. (1974). Spatial and paw preferences in rats: Their relationship to rate-dependent effects of *d*-amphetamine. *J. Pharmacol. Exp. Ther.* **188**, 714–725.

Glick, S. D., and Ross, D. A. (1981). Right-sided population bias and lateralization of activity in normal rats. *Brain Res.* **205**, 222–225.

Glick, S. D., Zimmerberg, B., and Jerussi, T. P. (1977). Adaptive significance of laterality in the rodent. *Ann. N. Y. Acad. Sci.* **299**, 180–185.

Glick, S. D., Shapiro, R. M., Drew, K. L., Hinds, P. A., and Carlson, J. N. (1986). Differences in spontaneous and amphetamine-induced rotational behavior, and in sensitization to amphetamine, among Sprague-Dawley–derived rats from different sources. *Physiol. Behav.* **38**, 67–70.

Haas, H. S., and Schauenstein, K. (1997). Neuroimmunomodulation via limbic structures—the neuroanatomy of psychoneuroimmunology. *Prog. Neurobiol.* **51**, 195–222.

Habib, M. (1989). Anatomical asymmetries of the human cerebral cortex. *Int. J. Neurosci.* **47**, 67–79.

Kalin, N. H., Larson, C., Shelton, S. E., and Davidson, R. J. (1998). Asymmetric frontal brain activity, cortisol, and behavior associated with fearful temperament in rhesus monkeys. *Behav. Neurosci.* **112**, 286–292.

Kawaharda, M., and Urasawa, K. (1992). Immunological functions and clinical course of elderly patients with cerebrovascular diseases. *Jpn. J. Geriatr.* **29**, 652–660.

Keller, S. E., Stein, M., Caminero, M. S., Schleifer, S. J., and Sherman, J. (1980). Suppression of lymphocyte stimulation by anterior hypothalamic lesions in guinea pigs. *Cell. Immunol.* **52**, 334–340.

Khansari, D. N., Murgo, A. J., and Faith, R. E. (1990). Effects of stress on the immune system. *Immunol. Today* **11**, 170–175.

Kim, D., Carlson, J. N., Seegal, R. F., and Lawrence, D. A. (1999). Differential immune responses in mice with left- and right-turning preference. *J. Neuroimmunol.* **93**, 164–171.

Koolhaas, J. M., Korte, S. M., DeBoer, S. F., VanDerVegt, B. J., VanReenen, C. G., Hopster, H., DeJong, I. C., Ruis, M. A. W., and Blokhuis, H. J. (1999). Coping styles in animals: Current status in behavior and stress physiology. *Neurosci. Biobehav. Rev.* **23**, 925–935.

Kroemer, G., Brezinschek, H. P., Faessler, R., Schauenstein, K., and Wick, G. (1988). Physiology and pathology of an immunoendocrine feedback loop. *Immunol. Today* **9**, 163–165.

LaHoste, G. J., Neveu, P. J., Mormède, P., and Le Moal, M. (1989). Hemispheric asymmetry in the effects of cerebral cortical ablations on mitogen-induced lymphoproliferation and plasma prolactin levels in female rats. *Brain Res.* **483**, 123–129.

Lambert, P. L., Harrell, E. H., and Achterberg, J. (1981). Medial hypothalamic stimulation decreases the phagocytic activity of the reticuloendothelial system. *Physiol. Psychol.* **9**, 911–914.

Laudenslager, M. L., Ryan, S. M., Drugan, R. C., Hyson, R. L., and Maier, S. F. (1983). Coping and immunosuppression: Inescapable but not escapable shock suppresses lymphocyte proliferation. *Science* **221**, 568–570.

Lelong, M., Thelliez, F., and Thelliez, P. (1986). Les gauchers sont-ils plus souvent des allergiques? *Allergie et Immunologie* **18**, 10–13.

Levy, J. (1974). Psychobiological implications of bilateral asymmetry. In "Hemispheric Function in the Human Brain" (S. J. Dimond and J. G. Beaumont, eds.), pp. 121–183. P. Elek, London.

Li, Q. S., and Yang, G. S. (1987). Immunoregulatory effect of neocortex in mice. *Immunol. Invest.* **16**, 87–96.

McLusky, N. J., and Naftolm, F. (1981). Sexual differentiation of the central nervous system. *Science* **211**, 1294–1303.

Meador, K. J., DeLecuona, J. M., Helman, S. W., and Loring, D. W. (1999). Differential immunologic effects of language-dominant and nondominant cerebral resections. *Neurology* **52**, 1183–1187.

Merlot, E., Moze, E., Dantzer, R., and Neveu, P. J. (2002). Suppression of restraint-induced plasma cytokines in mice pretreated with LPS. *Stress* **5**, 131–135.

Mrabet, O., Es-Salah, Z., Telhiq, A., Aubert, A., Liège, S., Choulli, K., and Neveu, P. J. (2000). Influence of gender and behavioral lateralization on two exploratory models of anxiety in C3H mice. *Behav. Processes* **52**, 35–42.

Myslobodsky, M. S., and Braun, H. (1980). Postural asymmetry and directionality of rotation in rats. *Pharmacol. Biochem. Behav.* **13**, 743–745.

Nance, D. M., Rayson, D., and Carr, R. I. (1987). The effects of lesions in the septal and hippocampal areas on the humoral immune response of adult female rats. *Brain Behav. Immun.* **1**, 292–305.

Neveu, P. J. (1992). Asymmetrical brain modulation of immune reactivity in mice: A model for studying interindividual differences and physiological population heterogeneity. *Life Sci.* **50**, 1–6.

Neveu, P. J., and Moya, S. (1997). In the mouse, the corticoid stress response depends on lateralization. *Brain Res.* **749**, 344–346.

Neveu, P. J., Taghzouti, K., Dantzer, R., Simon, H., and Le Moal, M. (1986). Modulation of mitogen-induced lymphoproliferation by cerebral neocortex. *Life Sci.* **38**, 1907–1913.

Neveu, P. J., Barnéoud, P., Vitiello, S., and Le Moal, M. (1988a). Immune functions after bilateral neocortex ablation in mice. *Neurosci. Res. Commun.* **3**, 183–190.

Neveu, P. J., Barnéoud, P., Vitiello, S., Betancur, C., and Le Moal, M. (1988b). Brain modulation of the immune system: Association between lymphocyte responsiveness and paw preference in mice. *Brain Res.* **457**, 392–394.

Neveu, P. J., Barnéoud, P., Georgiades, O., Vitiello, S., Vincendeau, P., and Le Moal, M. (1989a). Brain neocortex influence of the mononuclear phagocytic system. *J. Neurosci. Res.* **22**, 188–193.

Neveu, P. J., Barnéoud, P., Vitiello, S., Kelley, K. W., and Le Moal, M. (1989b). Brain neocortex modulation of mitogen-induced interleukin-2 but not interleukin-1 production. *Immunol. Lett.* **21,** 307–310.

Neveu, P. J., Betancur, C., Barnéoud, P., Preud'homme, J. L., Aucouturier, P., Le Moal, M., and Vitiello, S. (1989c). Functional brain asymmetry and murine systemic lupus erythematosus. *Brain Res.* **498,** 159–162.

Neveu, P. J., Betancur, C., Vitiello, S., and Le Moal, M. (1991). Sex-dependent association between immune function and paw preference in two substrains of C3H mice. *Brain Res.* **559,** 347–351.

Neveu, P. J., Deleplanque, B., Vitiello, S., Rouge-Pont, F., and Le Moal, M. (1992). Hemispheric asymmetry in the effects of substantia nigra lesioning on lymphocyte reactivity in mice. *Int. J. Neurosci.* **64,** 267–273.

Neveu, P. J., Delrue, C., Deleplanque, B., D'Amato, F. R., Pugli-Allegra, S., and Cabib, S. (1994). Influence of brain and behavioral lateralization in brain monoaminergic, neuroendocrine, and immune stress responses. *Ann. N. Y. Acad. Sci.* **741,** 271–282.

Neveu, P. J., Bluthé, R. M., Liège, S., Michaud, B., Moya, S., and Dantzer, R. (1998a). Interleukin-1 induced sickness behavior depends on behavioral lateralization in mice. *Physiol. Behav.* **63,** 587–590.

Neveu, P. J., Liège, S., and Sarrieau, A. (1998b). The asymmetrical distribution of hippocampal mineralocorticoid receptors depends on lateralization in mice. *Neuroimmunomodulation* **5,** 16–21.

Oitzl, M. S., VanHaarst, A. D., Sutanto, W., and DeKloet, E. R. (1995). Corticosterone, brain mineralocorticoid receptors (MRs) and the activity of the hypothalamic-pituitary-adrenal (HPA) axis: The Lewis rat is an example of increased central MR capacity and a hyporesponsive HPA axis. *Psychoneuroendocrinology* **20,** 655–675.

Pennington, B. F., Smith, S. D., Kimberling, W. J., Green, P. A., and Haith, M. M. (1987). Handedness and immune disorders in familial dyslexics. *Arch. Neurol.* **44,** 634–639.

Plata-Salaman, C. R. (1991). Immunoregulators in the nervous system. *Neurosci. Biobehav. Rev.* **15,** 185–215.

Renoux, G., and Bizière, K. (1986). Brain neocortex lateralized control of immune recognition. *Integr. Psych.* **4,** 32–40.

Renoux, G., and Bizière, K. (1987). Asymmetrical involvement of the cerebral neocortex on the response to an immunopotentiator, sodium diethyldithiocarbamate. *J. Neurosci. Res.* **18,** 230–238.

Renoux, G., Bizière, K., Renoux, M., Guillaumin, J. M., and Degenne, D. (1983). A balanced brain asymmetry modulates T-cell-mediated events. *J. Neuroimmunol.* **5,** 227–238.

Rogers, S. L., Coe, C. L., and Karaszewski, J. W. (1998). Immune consequences of stroke and cerebral palsy in adults. *J. Neuroimmunol.* **91,** 113–120.

Roszman, T. L., Cross, R. J., Brooks, W. H., and Markesbery, W. R. (1982). Hypothalamic-immune interactions. II. The effect of hypothalamic lesions on the ability of adherent spleen cells to limit lymphocyte blastogenesis. *Immunology* **45,** 737–743.

Sackeim, R. J., Decina, P., and Malitz, S. (1982). Functional brain asymmetry and affective disorders. *Adolesc. Psychiatry* **10,** 320–335.

Salcedo, J. R., Spiegler, B. J., Gibson, E., and Magilavy, D. B. (1985). The autoimmune disease systemic lupus erythematosus is not associated with left-handedness. *Cortex* **21,** 645–647.

Satz, P., and Soper, H. V. (1986). Left-handedness, dyslexia and autoimmune disorder: A critique. *J. Clin. Exp. Neuropsychol.* **8,** 453–458.

Schöbitz, B., Sutanto, W., Carey, M. P., Holsboer, F., and DeKloet, E. R. (1994). Endotoxin and interleukin-1 decrease the affinity of hippocampal mineralocorticoid (type I) receptor in parallel to activation of the hypothalamic-pituitary-adrenal axis. *Neuroendocrinology* **60,** 124–133.

CEREBRAL LATERALIZATION AND THE IMMUNE SYSTEM   323

Searleman, A., and Fugali, A. K. (1987). Suspected autoimmune disorders and left-handedness. Evidence for individuals with diabetes, Crohn's disease and ulcerative colitis? *Neuropsychologia* **25,** 367–374.

Smith, J. (1987). Left-handedness: Its association with allergic disease. *Psychopharmacologica* **25,** 665–674.

Tarkowski, E., Jensen, C., Ekholm, S., Ekelund, P., and Blomstrand, C. (1998). Localization of the brain lesion affects the lateralization of T-lymphocyte-dependent cutaneous inflammation. Evidence for an immunoregulatory role of the right frontal cortex-putamen region. *Scand. J. Immunol.* **47,** 30–36.

Waters, N. S., and Denenberg, V. H. (1994). Analysis of two measures of paw preference in a large population of inbred mice. *Brain Res.* **63,** 195–204.

Westergaard, G. C., Lussier, I. D., Suomi, S. J., and Higley, J. D. (2002). Handedness is associated with immune functioning and temperament in rhesus macaques. *Laterality* In press.

BEHAVIORAL CONDITIONING OF THE IMMUNE SYSTEM

Frank Hucklebridge

Department of Biomedical Sciences
University of Westminster
London W1M 8JS, United Kingdom

I. Introduction

The various demonstrations that the workings of the immune system are
subject to behavioral conditioning offer some of the most compelling evidence of functional links between the brain and the immune system, and
they challenge most closely and intriguingly the caveat of clinical relevance.
Many, perhaps most, immunologists have traditionally viewed the immune
system as autonomous, in a sense their own preserve. The main reason for
this is that the immune system is geared to respond adaptively to antigen,
as the primary stimulus, in essence, foreign pathogenic microorganisms.
This primary stimulus was viewed as distinct from other sensory modali-
ties, gustatory, olfactory, visual, and so on; that is, the seven senses. This
distinction may be artificial. The immune system is embedded within the

physiological organization of the body as a whole and its organization is deeply physiological; it is linked to the neural and neuroendocrine regulators of physiological stasis. It should therefore be no surprise that elements of immune system activity are subject to the same learning processes as apply to, say, cardiovascular activity and something as apparently mundane as salivation. That this is the case and that immune activity can be manipulated by classical conditioning is important, not only to our appreciation of physiology, and the immune systems integrated position in the way our bodies work, but also to our understanding of novel avenues of immunotherapeutic intervention.

Behavioral conditioning was first systematically explored by the Russian physiologist Pavlov early in the twentieth century. Pavlov's approach is well known. If a physiologically relevant stimulus, one that reliably produces a physiological response, is paired with a distinct but physiologically neutral stimulus, the response may be transferred to the formally neutral stimulus alone. In Pavlov's classical studies the physiologically relevant stimulus, termed the unconditioned stimulus (US), was the presentation of meat powder to the tongue of a dog whereas the response induced by this stimulus was salivation. An auditory stimulus, sounding of a tone was used as the neutral, conditioned stimulus (CS). After a number of trials temporally pairing the CS with the US, salivation could be induced by the sounding of the tone alone. The conditioning effect was most robust if the CS briefly preceded the US—forward conditioning. The response becomes a conditioned response (CR). This form of associated learning is sometimes referred to as Pavlovian, or classical, conditioning.

A. EARLY STUDIES OF IMMUNE CONDITIONING

Remarkably, demonstrations that aspects of immune system functioning could be conditioned in a similar manner soon followed Pavlov's pioneering work. In a series of studies carried out in the early 1920s at the Pasteur Institute in Paris, Metal'nikoff (a student of Pavlov) and colleagues began to explore immune system activity as a conditioned response. The approach of these workers was to use a challenge to the immune system as the US. In a first series of experiments using guinea pigs the US was intraperitoneal (ip) injection of tapioca. This induces an easily measurable immune response—the increase in circulating leukocytes (leukocytosis). In the conditioning paradigm this response could be transferred to dermal stimulation, placing a warm bar on the subjects skin or lightly scratching it. In further studies, injection of killed *Bacillus anthracoides* (anthrax) was used as the US and the response monitored was again leukocytosis, with the CS dermal stimulation as before. After multiple pairings of CS with US the CS alone could induce leukocytosis. The potential clinical significance of these findings was illustrated by the demonstration that conditioned immune protection to anthrax could extend to another disease, cholera. The conditioned leukocytosis response to anthrax was induced, then animals were infected with cholera and further presented with the CS. Survival was enhanced in animals so treated compared with nonconditioned controls.

Leukocytosis is part of the acute-phase response to immunological challenge and hence represents an aspect of natural or innate immunity. This facet is revealed by the nonspecificity of conditioned protection in these studies; the conditioned response to anthrax confers protection against cholera.

The Metal'nikoff experiments were somewhat overlooked in the West although a number of workers in Russia continued to extend these studies during the 1950s (see Exton *et al.*, 1999a, also Spector, 1987). Wider interest in behavioral conditioning was rekindled in the 1970s by what was originally a chance observation reported by Ader and Cohen (1975). Ader and Cohen were conducting studies into conditioned taste aversion (CTA) in the rat. This is one of the main conditioning paradigms (see Section I.C). Most commonly, presentation of a novel taste (CS) is paired with injection of a toxic drug (US), which aggravates the gastrointestinal tract and induces nausea. Commonly used nausea-inducing drugs are cyclophosphamide, lithium chloride, and apomorphine. A commonly employed novel taste is made by adding saccharin to drinking water. So that this is a distinct stimulus, animals are water deprived and allowed to drink saccharine-flavored water for restricted periods each day. After just one trial in pairing the novel drinking solution (CS) with nausea-inducing injection (US), rats acquire aversion to the saccharine-flavored water and restrict their intake. The degree of restriction indicates the strength of the CTA. Extinction of conditioning can also be determined by monitoring the volume of saccharine-flavored water consumed at each re-presentation of the CS. Ader and Cohen noted that the strength of conditioning and the resistance to extinction was directly related to the volume of saccharine solution consumed during the original conditioning trial. An unexpected finding was that some of the conditioned rats died (of common animal-house pathogens), and mortality was related to the volume of saccharine solution originally consumed. They reasoned that in addition to CTA they had induced conditioned suppression of the immune system, which might account for the increased pathogen susceptibility. Cyclophosphamide (CY) is a drug that not only induces nausea but also is an antimitotic agent used in cancer therapy, and it has immunosuppressive properties.

B. Conditioning of the Immune Response Rediscovered

Ader and Cohen set out to test this hypothesis directly. They replicated the saccharine CY–CTA paradigm on rats restricted to water consumption for just a single 15-min period every 24 h between 9 and 10 a.m. This was the active phase for nocturnal rats maintained on a reversed light cycle. Three days after the single-trial conditioning, rats were challenged with an antigen by ip injection of sheep red blood cells (srbc). Thirty minutes following this a subgroup were given their 15-min saccharine-solution exposure. On Day 9, 6 days after antigen challenge, animals were sacrificed and serum tested for anti-srbc antibodies and plasma for corticosterone. Compared with relevant controls the conditioned animals that had been given saccharine solution 30 min following antigen challenge showed a marked reduction in specific anti-srbc antibody titers. These authors found no association between immunosuppression and plasma corticosterone levels, which argued against the interpretation that immunosuppression might be a straightforward stress response to the CTA with corticosterone as immunosuppressive mediator.

To further test that the immunosuppression observed in relation to the CTA was not simply a stress response, Ader and Cohen repeated the study but using lithium chloride (LiCl) as the CS. LiCl also induces gastrointestinal nausea but has no immunoregulatory influence [this notion has since been challenged (Gauci et al., 1992) but the findings reported here are clear cut]. A single trial successfully induced CTA as evidenced by reduction of saccharine water consumption. LiCl did not directly suppress the antibody response to sheep red blood cell injection nor was there any evidence that the induction of CTA per se was immunosuppressive (Ader and Cohen, 1975). A new chapter in the appreciation of the relationships between the brain and the immune system had begun. Not only were these studies important because they revealed that the acquired immune response was subject to behavioral conditioning (antibody response to a T-dependent antigen) but also they formed the foundation to the nascent interdisciplinary science of psychoneuroimmunology.

C. Conditioning Paradigms

It is useful at this point to briefly describe the paradigms that have commonly been used to explore various aspects of immune function as the conditioned response (CR). The application of a conditioned taste aversion (CTA) paradigm by Ader and Cohen was serendipitous. In retrospect it was fortunate that the aversive drug CY also had immunosuppressive properties. CTA has been widely adopted. Usually a novel tasting substance with hedonic properties (sweet or sour, e.g., saccharine) is added to drinking water to provide the CS. Water deprivation and restricted supply of the flavored water focuses the CS. Conditioning is initiated by pairing the CS with injection of a drug that causes gastrointestinal aggravation and nausea. This induces CTA, the strength of which, and latency to extinction, can be determined by monitoring the volume of "saccharine"-flavored water consumed. Animals can be protected from dehydration by allowing them access to normal drinking water for a limited period at some other time during the 24-h cycle. A major advantage to the CTA paradigm is that, in the rat, acquisition is very rapid, often this can be achieved in a single trial. If the US is also an immunomodulatory drug, immunomodulation can be shown to be acquired in parallel with taste aversion.

Odor conditioning is a variant of CTA but in which the US is an immunomodulatory agent that has no aversive properties in relation to the CS. The CS is a novel odor such as peppermint or camphor. The odor is usually presented under the nose of the animal using a cotton bud, or exposure is in an enclosed chamber. Odors are potent stimuli in rodents that are olfactorally dominant. In contrast to CTA the animals do not restrict their exposure to the CS so there is no behavioral indicator of the effectiveness of conditioning. In studies designed to demonstrate behavioral conditioning of the immune response in human participants, a novel-tasting substance, such as sherbet, has similarly been used as CS (see Section V.B). Finally, fear conditioning has been employed in relation to animal studies. In this paradigm the animal is placed into a chamber with novel environmental cues such as color, light, or noise. The chamber also is equipped to provide an aversive stimulus such as a small electric shock to induce a fear response. Induction of a fear response modulates various aspects of immune function hence this is used as the US. It can be demonstrated that the exposure to the chamber itself with strong environmental cues can become a CS in relation to the immune modulation.

D. FURTHER DEVELOPMENTS AND CLINICAL RELEVANCE

The demonstration that CY immunosuppression could be behaviorally conditioned proved to be a robust phenomenon and was replicated in a number of laboratories. In most studies reexposure to the conditioned stimulus (saccharin, in the taste aversion paradigm) followed an antigen challenge to the immune system, thus demonstrating that the CS was effective in downregulating an activated immune response. It was also demonstrated, again using CY as the US, that presentation of the CS prior to the antigen challenge could suppress an antibody response, thus showing that the

CS could exert an inhibitory influence over the immune system even before specific activation by antigen (as is the case with CY itself). Further developments employed more sensitive assays of an antibody response than the hemagglutination titration originally employed by Ader and Cohen. In particular, splenic plaque-forming cell assay has been utilized, which identifies and quantifies an immune response at the level of antibody-secreting plasma cells harvested from the spleen (for review see Cohen *et al.*, 1994). T cells, which play a central regulatory role in the immune system, seem to be the main target of conditioned immunosuppression. In rodent studies conditioned immunosuppression can be demonstrated by testing the capacity of lymphocytes, usually harvested from the spleen, to respond *in vitro* to various lymphocyte mitogens. This offers some indication of the functional capacity of lymphocyte populations. Conditioned immunosuppression of the lymphoproliferative response to T-cell mitogens is more robust than to B-cell mitogens. Additionally, T-cell immunosuppression has been demonstrated by classical adoptive-transfer experiments. In this system, transfer of syngeneic immunocyte populations functionally reconstitutes a mouse whose own immune system has been disabled, such that different characteristics of immunological function can be attributed to different cell populations. This approach revealed that conditioned immunosuppression is transferable from donor to recipient by T cells (see Ader and Cohen, 1993).

Fear conditioning has been used in a study to explore aspects of central regulation. Electric foot shock was employed as the US paired with auditory or visual cues as the CS. The US induced downregulation of various aspects of cellular immunity, reduced splenic NK cell activity and splenic T-cell proliferation in response to mitogens, and diminished levels of IFN-γ production in an *in vitro* assay. These responses can be conditioned to the CS. The acquisition of the conditioned response in this system was, however, blocked by intracerebroventricular administration of opioid receptor antagonists (Perez and Lysle, 1997). Hence this study revealed the role of central opioid-mediated circuits in the acquisition of an immunosuppressive CS.

A number of studies began to explore the possibility that the CY–CTA paradigm might be effective in downregulating aggressive immune activity in a number of animal models of human autoimmune diseases. One such model is adjuvant-induced arthritis in the rat. In this model the arthritis-like inflammatory joint condition is induced in the paw of a rat by injection of Freund's complete adjuvant, a potent inflammatory agent (for detailed description of the adjuvant arthritis model see Chapter 6 of this volume). Usually just one hind paw is so treated but on about the 12th postinjection day, the inflammatory swelling also begins to manifest in the other uninjected hind paw. Klosterhalfen and Klosterhalfen (1983) were able to demonstrate that the CS developed by CY–CTA conditioning (using saccharine/vanilla-flavored water as the CS in a single-pairing trial) was effective in preventing this secondary inflammatory process. There was no amelioration of the primary inflammation but this was attributed to subsensitive quantitative indicators for distinguishing degrees of extreme inflammation. The less severe secondary inflammation was thought more sensitive to both the US–CY and the CS. The authors dismissed the possibility of corticosterone stress responses being responsible since previous work has shown that exposure to social stress or noise, stimuli known to elevate circulating corticosterone, before and after adjuvant treatment does not reliably influence the progress of inflammation.

Ader and Cohen (1982) reported studies of the (NZB × NZW) FI mouse, a hybrid strain, which has a genetic predisposition to develop a systemic-lupus-erythematosus-like autoimmune disease. Weekly treatment with CY can delay onset of the disease. Substitution of the CY–CTA CS for half of the active pharmacological treatments was effective in delaying onset. Similarly in the same mouse model after saccharine–CY–CTA conditioning, presentation of the CS alone after discontinuation of the drug regimen significantly prolonged survival (Ader, 1985). In contrast to the beneficial effects of conditioned immunosuppression on the progress of autoimmune disorders, susceptibility to cancer can be adversely influenced. In a mouse model, animals conditioned to CY as the US were more susceptible to an artificially induced cancer (inoculation with syngeneic plasmacytoma cells) when represented with the CS compared with relevant controls (Gorczynski *et al.*, 1985). This study is of particular interest since it reveals that the CS might have physiological influences somewhat different to the pharmacological properties of the US. Recall CY is an anticancer drug.

The study of Gorczynski and co-workers cited earlier revealed that conditioned immunosuppression can adversely influence the course of neoplastic disease, defense against which is largely dependent on Th1-mediated cellular immunity. The delayed-type hypersensitivity (DTH) response, dermal injection of an antigen to which an individual has already been sensitized, which results in a delayed (by several days) local inflammatory wheel and flare (erythema and induration) is a very good indicator of the strength of Th1 immunity. Roudebush and Bryant (1991) showed, using a murine model of the DTH response and the CY–CTA paradigm, that presentation of the CS alone, saccharin, was effective in inhibiting the response. This influence was independent of serum corticosterone and could not be replicated when the synthetic steroid dexamethasone was used as the conditioning agent, suggesting again that the conditioning effect is genuine and not a manifestation of a stress–corticosterone response to the conditioned stimulus.

II. Conditioned Immunosuppression Using Cyclosporin A

CY is an antimitotic agent and therefore inhibits any rapidly dividing cell population including all subsets of activated lymphocytes. Although it is effective in the CTA–immunosuppression paradigm it is a broad-acting "dirty" drug. Cyclosporin A (CsA), by contrast, selectively inhibits IL-2 signaling within the immune system and therefore primarily inhibits cellular cytotoxic activation (CD8[+] cytotoxic T cells and NK cell activity). It has found wide and important application in inhibiting cytotoxic rejection of organ allografts in transplantation surgery. Importantly, ip injection of CsA is also effective as the US in the CTA paradigm in the rat, and this has allowed detailed investigation of behaviorally conditioned immunosuppression with possible clinical validity.

A. PROMOTION OF ALLOGRAFT SURVIVAL

Grochowicz and co-workers (1991) first investigated the possibility that a CTA paradigm using CsA as the immunosuppressive US might be effective in promoting allograft survival thus mimicking the main clinical application of CsA. In this they were following a logical development since it had already been demonstrated that the CY–CTA paradigm could sustain skin allograft survival (across inbred mouse strains) on a schedule of presentation of the CS alone (Gorczynski, 1990). In the first study employing CsA, heterotopic heart allograft survival in the rat, which is a well-established animal model to study heart rejection processes, was examined. The heart of the donor rat, which is of different inbred strain than the recipient, is transplanted heterotopically into the peritoneal cavity of the recipient. Functional survival of the allograft can be monitored by palpation and also ECG monitoring. Fifteen days prior to transplantation the recipient rats were subjected to an habituation regimen of water deprivation except for a 10-min period each day. On day 10 and day 6 prior to transplantation the CTA group were given saccharine solution (CS) followed immediately by ip injection of an immunosuppressive preparation (20 mg/kg) of CsA (US). Relevant controls were included in the experimental design. Heart transplantations were performed at Day 0. One day prior to transplantation and also three days following transplantation rats were exposed to the CS alone. Heart allograft survival was significantly prolonged by exposure to the CS compared with controls and in parallel with the acquisition of taste aversion. One animal so conditioned retained a functional allograft for 100 days. This is considered to be long-term survival.

The reproducibility of this phenomenon was demonstrated in a later study employing a different donor–recipient inbred strain pairing and a modified conditioning schedule. Again exposure to the CS was effective in prolonging functional heart allograft survival compared with nonconditioned controls and equivalent to that of animals maintained on a parallel CsA treatment schedule (Exton *et al.*, 1998a).

Exton *et al.*, (1998b) replicated the findings in relation to heart allograft survival outlined earlier above but also explored the mechanisms. These workers found evidence of cellular immunosuppression in splenocytes induced by the CS and paralleling allograft survival. In conditioned animals the *in vitro* splenic lymphocyte proliferation response was significantly suppressed. In addition splenocyte production of IL-2 and IFN-γ was inhibited. These changes represent a downregulation of Th1 cellular immune capability. Neuronal pathways seem to mediate these immune changes in response to the CS since they are completely abrogated by prior sympathetic denervation of the spleen.

In a further study Exton *et al.* (1999b) superimposed conditioned CsA immunosuppression of heart allograft rejection on a schedule of subtherapeutic CsA administration. The subtherapeutic CsA administration alone, a treatment schedule employing 2 mg/kg administrations (10 × less than a pharmacologically effective dose), did not prolong allograft survival beyond that of a non-CsA treatment group. However, the addition of the behavioral conditioning CTA paradigm to the subtherapeutic CsA administration schedule markedly enhanced allograft survival, including long-term survival (in excess of 100 days) in 20% of the animals. In this study Exton and coworkers again probed the mechanisms by which behavioral conditioning might influence the immune system in relation to allograft rejection. The conditioning effect was completely blocked by prior sympathetic denervation of the spleen. Taking these latter two studies together it can be argued that behavioral conditioning prolongs allograft survival by suppressing production of those Th1 cytokines that are important in promoting cytotoxic activity. Sympathetic innervation to lymphoid organs seems to provide the efferent pathway.

B. Mechanisms of CsA-Conditioned Immunosuppression: Further Studies

The mechanisms by which the CS in the CsA–CTA paradigm initiates immunosuppression have been studied in a series of experiments separate from the allograft transplantation studies. In one such study, three pairings of saccharin with ip CsA (either on consecutive or alternate days) was sufficient to induce CTA for up to seven re-presentations of the CS alone (Exton *et al.*, 1998c). Immune functions were assessed after the third presentation of the CS alone. Thymus and splenic weight was significantly reduced and the *in vitro* proliferative response of splenocytes to the mitogen Concanavalin-A (Con-A) was reduced. This is generally regarded as an index of the strength of cellular immunity.

Exton *et al.* (2000a) explored the role of sympathetic splenic innervation in a different model of immunosuppression. In this study CsA–CTA was induced in the rat as in previous studies to explore heart allograft survival. Subsequently dermal contact hypersensitivity (CHS) was induced by application of 2,4-dinitrochlorobenzene (DNCB) to abdominal skin. Four days later the animals were challenged with DNCB solution applied to the ear. The strength of the CHS response, which is a DTH response—a measure of macrophage activation and Th1 cellular immunity—was determined by measuring the degree of ear swelling. Re-presentation of the CS was as effective in reducing the magnitude of ear swelling as was the US, CsA, itself. However, in this model the CS effect was not abrogated by splenic denervation. Thus in the CsA–CTA paradigm the CS may influence cellular immunity by different efferent channels relative to the nature of the immunological challenge.

Interestingly there may be differences in the way that the immune system responds to the CS compared with CsA in the CTA paradigm. In terms of immune cell trafficking to the peripheral circulation differences can be distinguished. After CsA–CTA induction in the rat, presentation of the CS alone induces leukopenia (reduction in leukocytes) in peripheral blood. This is broadly based including reduction in B cells, CD8$^+$ T cells, CD4$^+$ T cells (both naive and memory cells), and granulocytes. Granulocyte functional tests also reveal impairment. By contrast CsA treatment alone did not affect granulocyte function, and lymphocyte numbers in the peripheral circulation are actually increased (von Horsten *et al.*, 1998). Whatever the signals are that control the redistribution of immune cells and their representation in peripheral blood, the conditioned stimulatory pathway seems to distinguish drug-induced changes, although, as we have seen, the final immunological outcome may be comparable.

In the CsA–CTA paradigm Exton *et al.* (2000c) have capitalized on the conditioned leukopenia that follows presentation of the CS (as a diagnostic window) to determine latency to extinction of the CR. They were able to show in the rat that this association is very resistant to extinction and can be sustained (in the same group of rats for up to a year) either by reinstating the conditioning paradigm or by reexposure to the saccharine solution only.

Taken as a whole the studies described in this section show, in the rat at least, how potent might be conditioned immunosuppression induced by the CTA paradigm.

Experimental protocol to examine CTA conditioned prolongation of heart transplant survival

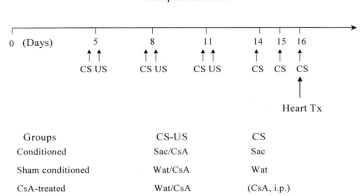

Groups	CS-US	CS
Conditioned	Sac/CsA	Sac
Sham conditioned	Wat/CsA	Wat
CsA-treated	Wat/CsA	(CsA, i.p.)
Untreated	–	–

Fig. 1. Following water deprivation animals are subjected to three CS–US training trials receiving either saccharin (Sac) or water (Wat) as the CS, with Cyclosporin A (CsA) as the US. On Day 14, animals are administered the CS only. On Day 16, rats undergo heart allograft transplantation. The CS is presented every day thereafter. Heart survival in the conditioned group is comparable to that in the CsA-treated group (normal drug regimen) and greater than in the untreated or sham-conditioned groups. The sham-conditioned group controls for any residual effect of CsA. Cs, Conditioned stimulus; US, unconditioned stimulus. (Adapted from Exton *et al.*, 1999a).

A typical CTA conditioning schedule as used in the studies described, together with appropriate control procedures, is illustrated in Fig 1.

III. Conditioned Immunopotentiation

The chance observation of Ader and Cohen, reported in 1975, lead to the exploration of behavioral conditioning of immunosuppression using variants of the CTA paradigm. But recall, the pioneering studies of Metal'nikoff and colleagues were of behavioral conditioning of immunopotentiation. They demonstrated the capacity of a CS to mobilize aspects of the innate response to antigen challenge or infection, namely leukocytosis. Contemporary research has also revisited this aspect of behavioral conditioning of immune system activity. Buske-Kirschbaum *et al.* (1996) exploited the pharmacological properties of nicotine in relation to the immune system. Nicotine influences immune cell trafficking to induce an increase in peripheral blood mononuclear cells (PBMC), that is, lymphocytes and monocytes. Male rats were exposed to four conditioning trials in an odor chamber whence exposure to peppermint odor was followed 5 min later with infusion, by indwelling catheter, of nicotine bitartrate. The use of nicotine as US was interesting since it is also known to stimulate the hypothalamic-pituitary-adrenocortical (HPA) axis thus resulting in a corticosterone response. Conditioned animals exhibited both a corticosterone response and an increase in PBMCs on first presentation of the CS alone. The PBMC response was resistant to extinction and remained over five unreinforced CS trials but the corticosterone response was lost after the initial test trial. This study showed not only that odor conditioning could be utilized but also that there was a distinction between a corticosterone response and the immune response. This added weight to the argument that conditioned immune responses are not merely corticosteroid-mediated responses to stress, associated with the conditioning procedure.

A. NK CELLS AND NK CELL ACTIVITY

NK cells are an important arm of our cytotoxic defenses. They are called natural killer cells since they can recognize and kill tumor or virally infected cells without the need for priming and cellular expansion in response to initial antigenic contact. Mobilization of NK cells to the peripheral circulation and enhancement of NK cell activity (against tumor cells) is a well-established indicator of cellular immune upregulation in PNI research and has more physiological validity than simply changes in PBMC trafficking as described earlier. Reductions in NK cell numbers and activity similarly reveal downregulation of an aspect of immune cell activity. NK cell activity is sensitive to behavioral conditioning. Solvason *et al.* (1991) explore NK cell conditionability by using poly-inosinic:poly-cytidylic acid (poly I:C), which is a strong inducer of NK cell activity as the US. In this paradigm, conditioned enhancement of NK cell activity can be demonstrated if the CS is presented together with a suboptimal dose of poly I:C. Conditioning is established by pairing odor presentation with an NK cell stimulatory dose of poly I:C. In fact two odors were used, camphor and citronella, oil in an experimental design that allowed the testing of odor discrimination. The conditioned response (increase in NK cell activity) was only elicited by the odor-CS used in the formation of the conditioned association. Exposure of conditioned mice to the nonassociated stimulus (camphor if conditioning had been with citronella oil or vice versa) as CS did not increase NK cell responsiveness to a suboptimal dose of poly I:C. This odor discrimination supports the view that enhancement of NK cell activity in this paradigm is as a result of Pavlovian conditioning and dependent on central nervous system associated processes rather than on a generalized physiological response to odor presentation per se.

Odor conditioning of the poly I:C NK cell activity response can be achieved after just one conditioning trial but only by forward conditioning; that is, the odor (CS) precedes the US (poly I:C). Reversing the order, backward conditioning is not effective in this protocol as was the conditioning of salivation in Pavlov's dogs (Solvason *et al.*, 1992).

1. *Mechanisms*

Poly I:C is also a pyrogen (fever inducer) and similar behavioral paradigms have shown that the pyrogenic response can also be conditioned to the odor of camphor as CS after just one conditioning trial (Hiromoto *et al.*, 1991). Interestingly the pathway by which conditioned learning of the NK cell activation response occurs seems to be distinguishable from that by which fever is induced. Injection of sodium carbonate could block the CS–US learning of the NK cell response but left the fever response intact. Additionally indomethacin, which blocks prostaglandin-induced fever, interfered with the conditioned learning of the fever response but had no adverse effect on acquisition of the NK cell response (Rogers *et al.*, 1992). This also implies that during acquisition of the conditioning association the brain must receive afferent signals from conditioned response, in the present context immune cell activity. There is evidence that in the case of odor conditioning of the NK cell response to poly I:C these afferent signals might be provided by the cytokine interferon (IFN)-β (Hiramoto *et al.*, 1993). In fact IFN-β can be substituted for poly I:C as the US to condition an increase in NK cell activity in response to camphor as the CS (Solvason *et al.*, 1988).

The efferent pathway has also been explored in this system. In conditioned animals plasma ACTH and β-endorphin and IFN-α message expression in the spleen were determined in relation to presentation of the CS (Hsueh *et al.*, 1994). Plasma ACTH and splenic IFN-α message expression were elevated on presentation of the CS indicating that the efferent pathway might involve ACTH induced upregulation of splenic IFN-α, which in turn stimulates NK cell activity.

Expression of the conditioned response in the camphor odor poly I:C paradigm seems to be dependent on opioid signaling. Intraperitoneal injection of naltrexone, an opioid receptor antagonist, blocks the CR (increase in NK cell activity) if given after the conditioning trial but immediately prior to the test trial, that is, exposure to the CS alone. Central opioid pathways are thought to be involved since ip injection of the less potent opioid receptor antagonist quaternery naltrexone (which is thought not to block central receptors) was without effect. By contrast acquisition of conditioning does not seem to involve opiotergic pathways since ip naltrexone, given prior to the conditioning trial itself, did not interfere with the formation of the conditioned association, which could be revealed by subsequent demonstration of the CR (Solvason *et al.*, 1989).

Various aspects of immune function deteriorate with age. Aged mice have decreased splenic NK cell activity. Spector *et al.* (1994) have demonstrated that camphor odor–poly I:C conditioning of upregulation of splenic NK cell activity is also applicable to old mice with impaired NK cell function. This raises the intriguing possibilities that behavioral conditioning might be applied to enhance the effectiveness of compromised immune systems.

B. OTHER UNCONDITIONED STIMULI TO ENHANCE NK CELL ACTIVITY AND CELLULAR IMMUNITY

Poly I:C is not the only stimulator of NK cells that has been employed as the US in conditioning trials. Arecoline, a muscarinic cholinergic agonist also elevates NK cell activity and by comparison with poly I:C depresses body temperature. Both responses can be conditioned to camphor odor as the CS (Ghanta *et al.*, 1996). As will be described in due course (Section V.B) in studies of conditioning of NK cell activity in humans, injection of the hormone adrenalin has been utilized as an ethically acceptable means of inducing the physiological–immunophysiological response of interest.

Finally in this section, not only can NK cell activity be enhanced by presentation of a CS but also $CD8^+$ cytotoxic lymphocyte (CTL) activity can be augmented. In animals stimulated with an alloantigen to induce a CTL response and conditioned to either poly I:C or arecoline as US with exposure to camphor odor as CS, recall presentation with camphor odor was effective in enhancing CTL activity, thus paralleling the influence over NK cell activity (Demissie *et al.*, 2000).

IV. Antigen as the Unconditioned Stimulus

What has thus far been described as behavioral conditioning of the immune system is in reality a series of studies that have demonstrated immune conditioning by pharmacological manipulation of the immune system— immunopharmacological conditioning, with both up- and downregulation. The primary stimulus to the immune system however is antigen challenge.

A central question therefore is whether antigen can serve as US to induce a physiological response, the characteristic immune response to antigen, and this response be retrieved by re-presentation of a conditioned stimulus. In a sense this question was originally posed and answered by the early studies of Metal'nikoff and colleagues when they paired dermal stimulation as CS with injection of killed *Bacillus anthracis* (anthrax) as US although the leukocytosis they observed as the CR was a manifestation of the innate immune response to challenge. Nonetheless it seems that physiological responses associated with antigen challenge could serve as afferent signals to the brain for conditioned associated learning. Ader and colleagues pioneered a reexamination of this issue.

Mice were given repeated immunizations with a potent soluble protein antigen, keyhole limpet hemocyanin (KLH) as the US. KLH is often used in studies in mice to stimulate an antibody response. The injections were paired with a gustatory stimulus as the neutral CS. They were able to observe significant enhancement of an antibody response inducible by the CS in comparison with relevant controls (Ader *et al.*, 1993), although to observe the effect they paired re-presentation of the CS with injection of a minimally immunogenic dose of KLH.

This demonstration that antigen itself might serve as the US in behavioral conditioning was soon replicated in studies of the rat. Alvarez-Borda *et al.* (1995) again used a gustatory conditioning paradigm pairing injection of protein antigen with exposure to saccharin dissolved in drinking water as the CS. This is not CTA since the antigen did not induce gastrointestinal upset, and there was no development of aversion to the saccharine-flavored water. In this study the protein antigen was hen egg lysozyme (HEL), commonly used in antibody response studies. The conditioning trial was a single pairing of saccharine water exposure for 10 min on a water-restricted schedule followed immediately by an ip injection of an immunogenic concentration of HEL. Twenty-five days later when the primary antibody response to the HEL challenge is no longer detectable conditioned animals were reexposed to the CS alone. A marked antigen-specific (anti-HEL) antibody response was induced to the CS comparable to that induced by a secondary antigen challenge. In this case immunological memory was invoked entirely by the CS. The antibody response to the CS was characteristic of a secondary response to this kind of antigen challenge, with isotype switch to IgG and a pronounced IgG response. It is remarkable, considering the complexity of T cell–B cell communication involved in elaborating the secondary response to antigen challenge, that these signals can be simulated by the nervous system as a result of central nervous system learning. This response is illustrated in Fig 2.

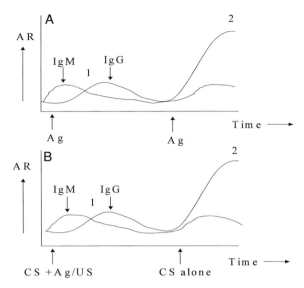

FIG. 2. A primary challenge with soluble protein antigen (Ag) induces a primary antibody response (1). If paired with a CS, either gustatory or olfactory, presentation of the CS alone induces a characteristic secondary antibody response, with predominance of the IgG isotype (2), comparable to that induced in response to secondary challenge with the Ag. AR, Antibody response. See text for further details.

Further studies explored the neural substrates of this association. In a replication of the study (Ramirez-Amaya and Bermudez-Rattoni, 1999), the influence of bilateral lesions to the insular cortex, the amygdala, and the dorsal hippocampus was investigated. Lesioning was excitotoxic by microinjection of N-methyl-D-aspartate (NMDA). The surgical procedure was performed 15 days prior to the conditioning trial. Previously it had been shown that bilateral lesions to the insular cortex and amygdala disrupt the acquisition and evocation of immunosuppression in the CTA paradigm and these regions might be important in the learning process (Ramirez-Amaya et al., 1998). The insular cortex is related to various aspects of learning and memory as, of course, is the hippocampus. Both the insular cortex and the amygdala have connections to autonomic regions such as the nucleus of the solitary tract, and the autonomic nervous system has been implicated as a signaling pathway associated with conditioned immunomodulation (as described in Section II.A). In addition the hippocampus expresses IL-1 receptors and is intimately involved in modulating afferent signals from the immune system to the brain (see Rothwell and Hopkins, 1995; also Chapter 4 of this volume).

Using saccharin as a taste-conditioning stimulus this study replicated the original finding that presentation of the CS alone resulted in a marked antigen-specific IgG response comparable to the secondary response to protein antigen (HEL). This secondary antibody response was almost completely abrogated by bilateral lesioning to the insular cortex and amygdala, whereas lesion to the hippocampus was without effect. These findings were replicated in every respect in a second experiment in which odor conditioning (almond odor) was used instead of taste conditioning. Thus it could not be argued that lesions to the insular cortex or amygdala had simply interfered with taste discrimination since the same effect could be produced by odor and suggests that involvement of these brain regions in conditioned stimulation of a secondary antibody response is not attributable only to a sensory mediation. These authors also demonstrated that lesions to these brain areas did not adversely influence the normal immune response. Lesioned animals produced a perfectly normal antibody response to antigen challenge per se.

Histological analysis revealed that the insular cortex lesions were located in the medial part of the anteroposterior axis of the insular cortex. Amygdala lesions were mainly located in the basolateral nucleus with some damage to the central nucleus. Lesions into the hippocampus destroyed mainly the CA1 region. The insular cortex, amygdala, and hippocampus are all limbic structures that have been implicated in brain–immune system interactions, mediating both afferent and efferent communication. These studies reveal the involvement of the insular cortex and the amygdala in conditioned immunoenhancement of an antibody response but no involvement of the hippocampal CA1 region.

These remarkable experiments suggest that the panoply of immunological defenses can be informed by external, environmental cues, in addition to molecular challenge by antigen, and that there is a representation of immunological memory imprinted within the nervous system. Strong association of environmental stimuli that originally accompany antigen challenge (exposure to pathogen), it seems, can alert the immune system to renewed danger. It is remarkable that this should be the case since the acquired immune response is characterized by its specificity and selectivity. There may be adaptive advantage that memory acquired in the nervous system should facilitate that "acquired" by the immune system itself. There can be no selective advantage if the resources of the immune system are mobilized by neuronal processes inappropriately, that is, by environmental sensory cues of impending infection that are in fact inaccurate. These are teleological arguments but the various demonstrations of conditioned immunoenhancement invite these considerations.

And now, to quote Auden "what of man, who can whistle tunes by heart."

V. Human Studies

The degree to which behavioral conditioning of immune activity and response, clearly established in rodent studies, can be generalized to humans is of more than passing interest. Allusion has been made several times to clinical relevance and if conditioning paradigms can influence the human immune system significantly there are important therapeutic implications. Anecdotal clinical observation of such conditioning first appeared in the literature over 100 years ago. A paper rose was observed to induce an allergic reaction in a susceptible individual (Mackenzie, 1896). Similarly, it had been reported that viewing a picture of a hay field was sufficient to elicit a hay fever attack in very sensitive subjects (Hill, 1930). Subsequent laboratory studies in both humans and animals confirmed that exposure to symbolic, non-allergenic environmental stimuli (CS) previously associated with allergen exposure (US) could induce asthmatic symptoms in some subjects (Khan, 1977; Ottenberg *et al.*, 1958).

A. A CLASSICAL STUDY

One of the first truly systematic explorations of behavioral conditioning of the immune response in humans examined conditioning on the delayed-type hypersensitivity (DTH) response to tuberculin injection. Conditioned immunosuppression of the DTH response in the mouse using the CY–CTA paradigm has already been described (Section I.D) although this was an example of behavioral immunopharmacological conditioning (conditioning of the pharmacological effect of an immunosuppressive drug). The present study on humans took a direct approach to see if conditioning might influence the relationship between antigen challenge and the immunological response per se. The study merits detailed description. A small group of nine hospital employees were recruited to participate in what they were told was a study of the reproducibility of the tuberculin reaction. Dermal injection of tuberculin stimulates a local DTH response since most individuals have been sensitized as a result of BCG vaccination. All participants tested positively in the tuberculin test. As previously mentioned, DTH response is a reliable indicator of the strength of Th1 cellular immunity and involves macrophage activation and recruitment to the local dermal site of antigen challenge. The participants were told that they would be given a tuberculin skin test on each arm monthly for 6 months, that the test would be the same each month, and that the concentrations of tuberculin would be different between arms. They were paid $ 25 for each test. The testing conditions were identical each month: the same office, day of the week, time of day, nurse, and arrangement of office furniture. When a subject arrived for testing he or she could see a red vial and a green vial on the desk. The contents of each vial were drawn into a syringe so that the subject could see that the content of the red vial was always applied to the right arm and the content of the green vial to the left arm. Subjects returned to the same office 24 and 48 h after every administration, where the same nurse measured the extent of their reaction to the injection, the diameter of the inflammatory erythema and induration, always reading the right arm first. In fact the subjects were given an active dose of tuberculin to the right arm (red vial) and saline to the left arm (green vial).

This identical protocol was followed consecutively for 5 months but on the 6th month (the experimental trial), without the knowledge of either the nurse or the patient, the content of the vials was switched. Saline (now in the red vial) was applied to the arm that for the previous 5 months had been given tuberculin, and tuberculin (now in the green vial) was applied to the arm that had previously been given saline. As a control, on the 7th month the subjects were debriefed and a final dose of either tuberculin or saline (in the original colored vials) was administered to the original arm. The results showed that the tuberculin injection caused a stable and measurable reaction for the first 5 weeks whereas the saline injection caused no reaction. However, in the experimental month the arm that was given the tuberculin (misleadingly colored) showed a very marked reduction in the intensity of its skin reaction. Saline injection to the previously exposed tuberculin arm caused no skin reaction. The final (control) test demonstrated that the reaction to tuberculin was as it had been in the initial 5-month period. Thus, this experiment demonstrated that the delayed hypersensitivity response to tuberculin could be significantly diminished by psychological mediation, that is, the expectation that there would be no response in that arm (Smith and McDaniel, 1983).

The findings can be explained in terms of behaviorally conditioned suppression of the immune response, and direct conditioning of immunological challenge as opposed to conditioning of immunopharmacological influence. In the experiment, the vials, the room, the day of the week, and the nurse as well as the idea that they would always be positive or negative (depending on arm) served as conditioned stimuli. The study is elegant because it explored, in humans, behavioral conditioning of immunosuppression and immunopotentiation at the same time. There were two conditioned responses (one in each arm): the positive response to tuberculin and its associated stimuli and the negative response to saline with its associated stimuli. The experiment showed that in this paradigm suppression of the immune response was subject to behavioral conditioning, but not activation.

The response to antigen (US) was sensitive to conditioned inhibition but it was not possible to demonstrate conditioning of a response in the absence of antigen. Since it has been shown that the skin DTH response can be modulated by neural regulation over leukocyte trafficking to the skin (Dhabhar and McEwan, 1997, 1999) the efferent pathway to explain the conditioned association would seem to be in place. However neural influences over cell trafficking clearly could not result in a DTH response in the absence of the provoking antigen. In this sense, the demonstration that a secondary immune response following antigen-mediated sensitization of the immune system can be initiated by a CS, described in studies of the rat (Section IV), was not transferable to the human. There are, however, numerous differences between the two procedures, not the least of which is that in the rat studies behavioral conditioning of an antibody response was demonstrated whereas in the human studies aspects of the cellular immune response were investigated.

B. Human Studies: Conditioning of NK Cell Activity

Conditioning of immunoenhancement has in fact been demonstrated in studies on human participants in which the model of immunopharmacological stimulation of NK cell trafficking and activity has been employed. An injection of adrenaline, capable of causing a transient increase in NK cell activity and circulating numbers in the periphery, was used as the US. It is known that adrenaline and/or stimulation of the sympathetic nervous system has this influence over NK cell activity and it can be blocked by β-adrenergic antagonists (Schedlowski et al., 1996). Studies using human participants and adrenalin as the US to stimulate NK cells effectively parallel the pharmacological studies of NK cell conditioned activation in rodents previously described (Section III.A) Buske-Kirschbaum et al. (1992) first described how in humans, repeated trials of pairing of a taste CS, sherbet sweet, could be conditioned to the NK cell response to subcutaneous injection of adrenalin (US). These authors explored this paradigm further, using a discriminative learning protocol, and were able to demonstrate classical stimulus discrimination. Sherbet sweet coupled with white noise was used as the CS, paired in repeated trials with adrenalin injection. A second stimulus complex, herbal sweet/auditory stimulus was not so reinforced. Stimulus discrimination, which is important in demonstrating the classical nature of conditioning, was revealed in as much as conditioning of increase in NK cell number and NK cell activity was assigned to the reinforced stimulus (CS) but not to the unreinforced stimulus (Buske-Kirschbaum et al., 1994).

Further studies revealed that indeed sympathetic β-adrenergic mechanisms may be responsible for the efferent pathway in the conditioned association. Injection of the β-adrenergic antagonist propanolol 1 h prior to presentation of the CS in this paradigm blocked the conditioned response (see Exton *et al.*, 2000b). Hence in studies using human participants it has been demonstrated that it is possible to directly condition recruitment of NK cells to the periphery and also perhaps to augment their contribution to immunological defense. As a cautionary note, however, this conditioning effect is not always easily reproduced. Although the Buske-Kirschbaum group could successfully replicate their findings (in Trier, Germany) a collaborative Dutch group could not (Kirschbaum *et al.*, 1992). This emphasizes the complexity of human information processing and the difficulty, given all the background setting and environmental cues, of establishing a clear conditioned stimulus. In fact the ease with which CTA can be established in the rat may be a function of the fact that the rat is very taste oriented since it cannot regurgitate.

C. IMMUNOSTIMULATORY CYTOKINE CONDITIONING

The Th1 immunostimulatory cytokine IFN-γ has a role in cancer immunotherapy as a result of its capacity to promote cellular immune defenses. It has also been used to investigate the possibility of conditioning of immunopharmacological stimulation in normal human participants. In a placebo-controlled, double-blind study of 31 normal participants; IFN-γ (the US) was paired with an immunologically neutral, gustatory stimulus, oral propylene glycol (CS), in a typical classical conditioning paradigm. During a 4-week period the experimental group received progressively fewer injections of IFN-γ. By the final week these participants received only the neutral stimulus (CS). At the end of this regime the behaviorally conditioned group had higher levels of serum quinolinic acid and neopterin (marker for cellular immune activation) than positive controls that had actually received the immunostimulatory agent throughout the trial and into the 4th week (Longo *et al.*, 1999). Similarly, in comparison to a relevant control, the conditioned group exhibited greater expression of Fc receptors (CD46) on PBMCs at the end of the trial, a direct cellular index of immune activation. These data strongly support the notion that the human immune system can be activated as well as downregulated by classical conditioning and despite the caveat of failure to reproduce the conditioned effect previously cited (Kirschbaum *et al.*, 1992) clearly immunopharmacological manipulation can be enhanced in human participants by classical conditioning procedures.

D. Conditioned Immunosuppression: Clinical Implications

Behavioral conditioning can occur with pairings between physiologically neutral stimuli, that is in terms of the response of interest (nothing that is perceived as a stimulus is neutral) and the active agent. Examples of neutral stimuli would be sherbet sweet, as shown earlier, or procedures such as those relating to the color of a vial. The more specific and rapidly acting the physiologically active agent the more likely it is that environmental cues or stimuli will be associated with that activity and hence induce a conditioned response. The reason for this is that afferent pathways are involved in establishing the association. A clearly defined physiological/pharmacological response will ring "bells" in the central nervous system. This suggests that as clinical medicine progresses with the development of more accurately focused drugs the opportunity for conditioning will be increased. Research into the significance of the clinical setting in the treatment of cancer illustrates the point. It has been appreciated for many years that patients repeatedly exposed to nauseating cytotoxic drugs (such as CY in cancer chemotherapy) can develop "anticipatory" nausea prior to drug exposure. Since CY and comparable drugs are powerful immunosuppressive agents, the possibility for anticipatory (or conditioned) immunosuppression presents itself.

An early study to explore this phenomenon investigated ovarian cancer patients after they had received at least three chemotherapy infusions. In the 4-week interval between treatments they were contacted by phone to arrange a home assessment at least 3 days before their forthcoming treatment. In this session psychological tests were administered and a blood sample taken for determining proliferative responses to mitogens and NK cell activity as well as complete blood counts and cell subset enumeration. A similar blood sample was taken, and analyzed, from the same patients on the occasion of their visits to the hospital, but before exposure to chemotherapy. The results showed that the *in vitro* proliferative responses to the T-cell mitogens phytohemagglutinin (PHA) and Con A (generally regarded as indexes of the effectiveness of cellular immune defenses as previously described) were lower for the cells isolated at the hospital visit than from the blood collected in the patients' homes a few days earlier. NK cell activity was not significantly lower in the hospital than at home nor were the number of white blood cells, the percentage of lymphocytes, or the absolute number of lymphocytes. Although patients scored higher on measures of anxiety while in the hospital setting the reduced proliferative response to mitogen could not be accounted for by anxiety alone. The authors of this study argued that their findings were a consequence of repeated pairing of the hospital stimuli (the conditioned stimuli) with immunosuppressive chemotherapy (the unconditioned stimulus) such that patients showed conditioned immunosuppression in the hospital setting (Bovbjerg *et al.*, 1990).

A subsequent study investigated breast cancer patients undergoing a form of chemotherapy that was less aggressive than that used in the ovarian cancer study described earlier (Fredrikson *et al.*, 1993). As before, blood was sampled in the patients' homes a few days before their 4th or 5th course of chemotherapy and in the hospital immediately before treatment. This time patients displayed *higher* numbers of white blood cells in the hospital, immediately prior to treatment compared to 2 days earlier in their own home. This was due to an increased number of granulocytes in the peripheral blood. These authors were unable to replicate the reduced *in vitro* proliferative responses to T-cell mitogens reported by Bovbjerg and co-workers (1990). The discrepancy in the results was explained by subtle differences in procedure between the two studies. In the original study, patients arrived at the hospital the night before treatment whereas in the latter study they arrived only several hours prior to treatment. It was hypothesized that conditioned immunosuppression may occur several hours after exposure to the conditioned stimuli and that the observed effects of the second study represented only the first stage in the treatment setting–induced effect. For example an epinephrine-mediated increase in granulocytes may later be followed by the immunosuppressive effect of catecholamines (see Chapter 2 in this volume).

VI. Conclusions

Both animal and human studies reveal that immune system activity can be modulated by behavioral conditioning. Unconditioned stimuli are pharmacological regulators of immune system activity, which are variously stimulatory or inhibitory to aspects of immune function. This might properly be termed "behavioral immunopharmacological conditioning." Antigen challenges to the immune system can also serve as the US such that on re-presentation of a CS (otherwise immunologically neutral) an enhanced immunological response is observed. In fact in studies of the rat a gustatory or odor CS can induce, in terms of specific antibody titer and isotype the characteristics of a secondary response to a protein antigen challenge.

The mechanism by which conditioning of the immune system occurs has only been partly explored although it is clear that no single mechanism could possibly be responsible for all of the observed effects. This merely reflects the complexity of both the nervous and the immune systems and their interrelatedness. A primary efferent pathway from the brain to the immune system is the neuroendocrine HPA axis, and it is clear that activation of the HPA axis and secretion of corticosterone can be conditioned (DeVries *et al.*, 1998) and that some of the conditioning paradigms described in this chapter condition its activation in conjunction with conditioning of some aspect of immunity (as in Buske-Kirschbaum *et al.*, 1996). This does not, however, imply that HPA axis conditioning alone is causally related to the observed immune conditioning, as some of these effects are not under the regulatory influence of corticosterone and other factors have been shown to be important. Central nervous system structures have been implicated. Experiments have demonstrated the role of certain brain regions (the insular cortex and amygdala) in acquiring a conditioned antibody response to protein antigen as the US. Other studies have implicated activity of endogenous opioid systems within the brain. Direct autonomic innervation to lymphoid organs of the immune system is also considered important and might carry both efferent and afferent signaling traffic involved in establishing conditioned immunological memory.

On the immune system side of the conditioning equation T helper cells almost certainly play a central role, and modulation of T helper cell cytokine production (such as Il-2) has been demonstrated in some rodent studies. Immunocyte trafficking is also implicated in conditioned immunomodulation. Examples of conditioned changes in immune function are found in both the Th1 cellular and Th2 humoral arms of the immune system.

In the human context the more potent, selective, and well defined both the CS and the US, the greater is likely to be the impact of conditioning on the relationship between immunomodulatory drug and immune system function. This might have a beneficial or an adverse influence on health, depending on the direction of manipulation. In various animal studies and now in some human studies, observed conditioned responses have been shown to be of sufficient magnitude to impact disease progression. As medicine produces more selective and potent immunopharmacological agents the potential for conditioning of immune system activity to cues within the clinical setting becomes of greater relevance, and some appreciation of this on the part the medical establishment is merited.

References

Ader, R. (1985). Conditioned immunopharmacologic effects in animals: implications for conditioning model of pharmacotherapy. *In* "Placebo: Theory Research and Mechanisms" (L. White, B. Tursky, and G. Schwartz, eds.), pp. 306–323. Guilford Press, New York.
Ader, R., and Cohen, N. (1975). Behaviourally conditioned immunosuppression. *Psychosom. Med.* **37**, 333–340.

Ader, R., and Cohen, N. (1982). Behaviourally conditioned immunosuppression and murine systemic lupus erythematosus. *Science* **215,** 1534–1536.

Ader, R., and Cohen, N. (1993). Psychoneuroimmunology: Conditioning and stress. *Ann. Rev. Psychol.* **44,** 53–85.

Ader, R., Kelly, K., Moynihan, J. A., Grota, L. J., and Cohen, N. (1993). Conditioned enhancement of antibody production using antigen as the unconditioned stimulus. *Brain. Behav. Immun.* **7,** 334–343.

Alvarez-Borda, B., Ramirez-Amaya, V., Perez-Montfort, R., and Bermudez-Rattoni, F. (1995). Enhancement of antibody production by a learning paradigm. *Neurobiol. Learn. Mem.* **64,** 103–105.

Bovbjerg, D. H., Redd, W. H., Maier, L. A., Holland, J. C., Lesko, L. M., Niedzwiecki, D., Rubin, S. C., and Hakes, T. B. (1990). Anticipatory immune suppression and nausea in women receiving cyclic chemotherapy for ovarian cancer. *J. Consult. Clin. Psychol.* **58,** 153–157.

Buske-Kirschbaum, A., Kirschbaum, C., Stierle, H., Lhnert, H., and Hellhammer, D. (1992). Conditioned increase of natural killer cell activity (NKC) in humans. *Psychosom. Med.* **54,** 123–132.

Buske-Kirschbaum, A., Kirschbaum, C., Stierle, H., Jabaij, L., and Hellhammer, D. (1994). Conditioned manipulation of natural killer cells in humans using discriminative learning protocol. *Biol. Psychol.* **38,** 143–155.

Buske-Kirschbaum, A., Grota, L., Kirschbaum, C., Bienen, T., Moynihan, J., Ader, R., Blair, M. L., Hellhammer, D. H., and Felton, D. L. (1996). Conditioned increases in peripheral blood mononuclear cell (PBMC) number and corticosterone secretion in the rat. *Pharmacol. Biochem. Behav.* **55,** 27–32.

Cohen, N., Moynihan, J. A., and Ader, R. (1994). Pavlovian conditioning of the immune system. *Int. Arch. Allergy Immunol.* **105,** 101–106.

Demissie, S., Ghanta, V. K., Hiramoto, N. S., and Hiramoto, R. N. (2000). NK cell and CTL activities can be raised via conditioning of the CNS with unrelated unconditioned stimuli. *Int. J. Neurosci.* **103,** 79–89.

DeVries, A. C., Taymans, S. E., Sundstrom, J. M., and Pert, A. (1998). Conditioned release of corticosterone by contextual stimuli associated with cocaine is mediated by corticotrophin-releasing factor. *Brain Res.* **789,** 39–46.

Dhabhar, F. S., and McEwan, B. S. (1997). Acute stress enhances while chronic stress suppresses cell-mediated immunity *in vivo:* A potential role for leukocyte trafficking. *Brain Behav. Immun.* **11,** 286–306.

Dhabhar, F. S., and McEwan, B. S. (1999). Enhancing versus suppressive effects of stress hormones on skin immune function. *Proc. Nat. Acad. Sci. USA.* **96,** 1059–1064.

Exton, M. S., von Auer, A. K., Buske-Kischbaum, A., Stockhorst, U., Gobel, U., and Schedlowski, M. (2000b). Pavlovian conditioning of immune function: Animal investigation and the challenge of human application. *Behav. Brain Res.* **110,** 129–141.

Exton, M. S., Shult, M., Donath, S., Strubel, T., Nagel, E., Westerman, J., and Schedlowski, M. (1998a). Behavioural conditioning prolongs heart allograft survival in rats. *Transpl. Proc.* **30,** 2033.

Exton, M. S., von Horsten, S., Voge, J., Westermann, J., Schult, M., Nagel, E., and Schedlowski, M. (1998c). Conditioned taste aversion produced by cyclosporine A: Concomitant reduction in lymphoid organ weight and splenocyte proliferation. *Physiol. Behav.* **63,** 241–247.

Exton, M. S., King, M. G., and Husband, A. J. (1999a). Behavioural conditioning of immunity. *In* "Psychoneuroimmunology: An Interdisciplinary Introduction" (M. Schedlowski, and U. Tewes, eds.). Kluwer Academic/Plenum, New York/London.

Exton, M. S., von Horston, S., Schult, M., Voge, J., Strubel, T., Donath, S., Steinmuller, C., Seeliger, H., Nagel, E., Westermann, J., and Schedlowski, M. (1998b). Behaviourally conditioned immunosuppression using cyclosporine A: Central nervous system reduces IL-2 production via splenic innervation. *J. Neuroimmunol.* **88,** 182–191.

Exton, M. S., Schultz, M., Donath, S., Strubel, T., Del Rey, A., Westerman, J., and Shedlowski, M. (1999b). Conditioned immunosuppression makes subtherapeutic cyclosporin effective via splenic innervation. *Am. J. Physiol.–Regul. Integr. Comp. Physiol.* **276,** r1710–r1717.

Exton, M. S., Elfers, A., Jeong, W. Y., Bull, D. F., Westermann, J., and Schedlowski, M. (2000a). Conditioned suppression of contact sensitivity is independent of sympathetic splenic innervation. *Am. J. Physiol.–Regul. Integr. Comp. Physiol.* **279,** r1310–r1315.

Exton, M. S., von Horstoen, S., Strubel, T., Donath, S., Schedlowski, M., and Westerman, J. (2000c). Conditioned alterations of specific blood leukocyte subsets are reconditionable. *Neuroimmunomodulation* **7,** 106–114.

Fredrikson, M., Furst, C. J., Lekander, M., Rotstein, S., and Blomgren, H. (1993). Trait anxiety and anticipatory immune reactions in women receiving adjuvant chemotherapy for breast cancer. *Brain, Behav. Immun.* **7,** 79–90.

Gauci, M., Bull, D. F., Schedlowski, M., Husband, A. J., and King, M. G. (1992). Lithium chloride and immunomodulation in taste aversion conditioning. *Physiol. Behav.* **51,** 207–210.

Ghanta, V. K., Demissie, S., Hiramoto, N. S., and Hiramoto, R. N. (1996). Conditioning of body temperature and natural killer cell activity with arecoline, a muscarinic cholinergic agonist. *Neuroimmunomodulation* **3,** 233–238.

Gorczynski, R. M. (1990). Conditioned enhancement of skin allografts in mice. *Behav. Immun.* **4,** 85–92.

Gorczynski, R. M., Kennedy, M., and Ciampi, A. (1985). Cimetidine reverses tumour growth enhancement of plasmacytoma tumours in mice demonstrating conditioned immunosuppression. *J. Immunol.* **134,** 4261–4266.

Grochowicz, P. M., Schedlowski, M., Husband, A. J., Kingm, M. G., Hibberd, A. D., and Bowen, K. M. (1991). Behavioral conditioning prolongs heart allograft survival in rats. *Brain Behav. Immun.* **5,** 349–356.

Hill, L. E. (1930). "Philosophy of a Biologist". Arnold, London.

Hiramoto, R., Ghanta, V., Solvason, B., Lorden, J., Hsueh, C. M., Rogers, C., Demissie, S., and Hiromoto, N. (1993). Identification of specific pathways of communication between the CNS and NK cell system. *Life Sci.* **53,** 527–540.

Hiramoto, R. N., Ghanta, V. K., Rogers, C. F., and Hiromoto, N. S. (1991). Conditioning the elevation of body temperature, a host defensive reflex response. *Life Sci.* **49,** 93–99.

Hsueh, C. M., Tyring., S. K., Hiramoto, R. N., and Ghanta, V. K. (1994). Efferent signal(s) responsible for the conditioned augmentation of natural killer cell activity. *Neuroimmunomodulation* **1,** 74–81.

Khan, A. U. (1977). Effectiveness of biofeedback and counterconditioning in the treatment of bronchial asthma. *J. Psychosom. Res.* **21,** 97–104.

Kirschbaum, C., Jabaaij, L., Buske-Kirschbaum, A., Hennig, J., Blom, M., Dorst, K., Bauch, J., DiPauli, R., Schmodt, G., Ballieux, R., and Hellhammer, D. (1992). Conditioning of drug-induced immunomodulation of human volunteers: A European collaborative study. *Br. J. Clin. Psychol.* **31,** 459–472.

Klosterhalfen, W., and Klosterhalfen, S. (1983). Pavlovian conditioning of immunosuppression modifies adjuvant arthritis in rats. *Behav. Neurosci.* **97,** 663–666.

Longo, D. L., Duffey, P. L., Kopp, W. C., Heyes, M. P., Alvord, W. G., Sharfman, W. H., Schmidt, P. J., Rubinow, D. R., and Rosenstein, D. L. (1999). Conditioned immune response to interferon-γ in humans. *Clin. Immunol.* **90,** 173–181.

Mackenzie, J. N. (1896). The production of the so-called "rose cold" by means of an artificial rose. *Am. J. Med. Sci.* **91,** 45–57.

Ottenberg, P., Stein, M., Lewis, J., and Hamilton, C. (1958). Learned asthma in the guinea pig. *Psychosom. Med.* **20,** 395–400.

Perez, L., and Lysle, D. T. (1997). Conditioned immunomodulation: Investigations of the role of endogenous activity at μ, κ, and δ opioid receptor subtypes. *J. Neuroimmunol.* **79,** 101–112.

Ramirez-Amaya, V., and Bermudez-Rattoni, F. (1999). Conditioned enhancement of antibody production is disrupted by insular cortex and amygdala but not hippocampal lesions. *Brain Behav. Immum.* **13,** 46–60.

Ramirez-Amaya, V., Alvarez-Borda, B., and Bermudez-Rattoni, F. (1998). Differential effects of NMDA-induced lesions into the insular cortex and amygdala on the acquisition and evocation of conditioned immunosuppression. *Brain Behav. Immun.* **12,** 149–160.

Rogers, C., Ghanta, V., Hseuh, C. M., Hiramoto, N., and Hiramoto, R. (1992). Indomethacin and sodium carbonate effects on conditioned fever and NK cell activity. *Pharmacol. Biochem. Behav.* **43,** 417–422.

Rothwell, N. J., and Hopkins, S. J. (1995). Cytokines and the nervous system. II. Actions and mechanisms of action. *Trends Neurosci.* **18,** 130–136.

Roudebush, R E., and Bryant, H. U. (1991). Conditioned immunosuppression of a murine delayed type hypersensitivity response: Dissociation from corticosterone elevation. *Brain Behav. Immun.* **5,** 308–317.

Schedlowski, M., Hosch, W., Oberdeck, R., Benschop, R. J., Jacobs, R., Raab, H. R., and Schmidt, R. E. (1996). Catecholamines modulate human NK cell circulation and function via splenic-independent β_2-adrenergic mechanisms. *J. Immunol.* **156,** 93–99.

Smith, G. R., and McDaniel, P. (1983). Psychologically mediated effect on the delayed hypersensitivity reaction to tuberculin in humans. *Psychosom. Med.* **45,** 65–70.

Solvason, H. B., Ghanta, V., and Hiramoto, R. N. (1988). Conditioned augmentation of natural killer cell activity: Independence from nociceptive effects and dependence on Interferon-β. *J. Immunol.* **140,** 661–665.

Solvason, H. B., Hiamoto, R. N., and Ghanta, V. K. (1989). Naltrexone blocks the expression of the conditioned elevation of natural killer cell activity in BALB/c mice. *Brain Behav. Immun.* **3,** 247–262.

Solvason, H. B., Ghanta, V. K., Lorden, J. F., Soong, S. J., and Hiramoto, R. N. (1991). A behavioral augmentation of natural immunity: Odor specificity supports a Pavlovian conditioning model. *Int. J. Neurosci.* **61,** 277–288.

Solvason, H. B., Ghanta, V. K., Soong, S. J., Rogers, C. F., Hsueh, C. M., Hiramoto, N. S., and Hiramoto, R. N. (1992). A simple, single trial-learning paradigm for conditioned increase in natural killer cell activity. *Proc. Soc. Biol. Med.* **199,** 199–203.

Spector, N. H. (1987). Old and new strategies in the conditioning of immune responses. *Ann. N.Y. Acad. Sci.* **496,** 522–531.

Spector, N. H., Provinciali, M., Di Stefano, G., Muzzioli, M., Bulian, D., Viticchi, C., Rossano, F., and Fabris, N. (1994). Immune enhancement by conditioning of senescent mice. *Ann. N.Y. Acad. Sci.* **741,** 283–291.

von Horsten, S., Exton, M. S., Schult, M., Nagel, E., Stalp, M., Schweitzer, G., Voge, J., del Rey, A., Schedlowski, M., and Westermann, J. (1998). Behaviourally conditioned effects of Cyclosporine A on the immune system of rats: Specific alterations of blood leukocyte numbers and decrease of granulocyte function. *J. Neuroimmunol.* **85,** 193–201.

PSYCHOLOGICAL AND NEUROENDOCRINE CORRELATES OF DISEASE PROGRESSION

Julie M. Turner-Cobb

Department of Psychology
University of Kent at Canterbury
Canterbury, Kent CT2 7NP
United Kingdom

I. Introduction

Evidence is rapidly accumulating that dispels the idea of biological influences on disease progression existing in isolation from psychological influences. This chapter discusses the scientific evidence which elucidates the relationships between psychological factors and disease progression by posing the following three questions: (1) What are the most salient psychological correlates of disease progression and how do these relate across different diseases? (2) How might intervention studies facilitate psychological resources to elicit a reduction in the rate of disease progression? (3) What neuroendocrine correlates of disease progression may be mediating these effects and what immune responses do they utilize?

The chronic illnesses of cancer and human immunodeficiency virus/acquired immunodeficiency syndrome (HIV/AIDS) are the main focus, with evidence also drawn from other immune-related diseases including autoimmunity. Psychological correlates discussed include stress and depression, the influence of social relations, coping responses, and emotional repression, as well as the influence of socioeconomic status. The relevance of the theory of allostasis and the concept of allostatic load and the importance of including a developmental perspective are emphasized. Neuroendocrine–immune correlates that have been focused on include cortisol and its relationship to circadian rhythmicity and corresponding T helper 1/T helper 2 cell (Th1/Th2) immune balance, CD4 activity (T helper cell marker), and functional and enumerative activity of natural killer (NK) cells. Evidence abounds of the impact of psychological correlates of disease progression as determined by symptomatology, endocrine–immune alteration, and survival in chronic illness— in particular, cancer and HIV. Future work needs to concentrate on further extrapolating the details of effective psychological interventions in mapping neuroendocrine and immune responses to disease resistance and thereby enabling maximization of mainstream medical interventions.

The importance of psychological factors, indeed that "mind matters" (Spiegel, 1999) in physical illness to the extent of altering survival via neuroendocrine and immune variables in disease progression is becoming increasingly evident. In this chapter the scientific evidence that elucidates the relationships between psychological factors and disease progression will be discussed. Putative neuroendocrine mediators of such processes and their possible links to immune functioning will be examined.

To investigate the three questions posed at the start of this section, the chronic illnesses of cancer and human immunodeficiency virus/acquired immunodeficiency syndrome (HIV/AIDS) will be the main focus. However, I will also be drawing from evidence relating to other immune-related diseases including autoimmune disease as well as infectious illness in otherwise healthy individuals, to elucidate the psychoimmune connections in illness. One of the consistent themes that runs throughout such research, but is often less well attended to, is the developmental perspective involved in disease progression. In looking at the involvement of psychological correlates of disease status, it is important to consider the starting point of the organism's physiological status. While an individual's immune status prior to becoming ill is generally not known for obvious methodological reasons, the importance of early development on an individual's ability to cope both psychologically and physiologically should not be underestimated or overlooked.

We can now say with some certainty that psychological factors do appear to be associated with disease progression across a number of diseases. Twenty years ago this may have been a dubious notion purported by a handful of seemingly misdirected interdisciplinary scientists. The link between psychological factors and disease unleashes a much greater potential in attenuating disease progression than would have been available through biological approaches alone. Furthermore, an understanding of disease progression contributes in making the logical leap to evaluating strength of resistance to disease in healthy individuals. However, the specificity of this link in regard to the combination of psychological factors and relevant physiological mediators that would best determine a positive outcome in relation to disease progression remains to be determined and is the subject of this chapter.

II. Psychological Correlates Associated with Disease Progression

The most important psychological influences on disease in relation to progression include stress, depression or distress, coping response, and social support. However, the direction of effect and the relative importance of these influences varies both across diseases and within diseases (for example, due to differences in sex or stage of illness). As Garssen and Goodkin (1999) point out in their comprehensive review of cancer initiation and progression, although a total of 20 out of 26 studies found evidence in support of there being some effect of psychological factors on cancer progression (compared to 6 out of 12 for cancer initiation) the degree of involvement by any one factor is as yet difficult to determine. They point to the three psychological correlates of a lack of supportive social relations, a helplessness coping response, and a repressive style of dealing with emotions as the most significant determinants of disease progression. Each of these psychological correlates will now be considered in conjunction with the broader terms of stress, coping, and social relations.

A. STRESS

There is much research on healthy populations that attests to the influence of stress in reducing an individual's resistance to disease (for reviews, see Cohen and Herbert, 1996; Kemeny and Gruenewald, 1999; Biondi and Picardi, 1999; Baum and Posluszny, 1999). For example, in studies of other diseases including upper respiratory infections and autoimmune diseases, significant effects of stress on both the onset and the severity of disease have been noted in human (Cohen and Herbert, 1996; Turner-Cobb and Steptoe, 1996; Turner-Cobb et al., 1998) and in animal studies (Sheridan et al., 1998).

In regard to the acceleration of cancer progression in response to stress, results have been mixed. For example, in studies examining the relapse rate for breast cancer in women, some studies have reported an association between major life stressors such as bereavement and job loss and progression of the disease (Ramirez et al., 1989), while others have not found evidence in support of such a connection (Barraclough et al., 1992).

Stress pervades the progression of HIV/AIDS from a number of sources, from onset throughout disease progression, particularly due to the social stigma associated with HIV (for reviews, see Ironson et al., 1994b; Kalichman, 1998; Balbin et al., 1999). Some very convincing evidence of this relationship is provided by Evans and colleagues (Evans et al., 1997) in a prospective study of 93 asymptomatic HIV-positive homosexual men, assessing life events using a semistructured interview measured every 6 months over the course of a maximum of 42 months (depending on survival rate). As severity of life stress experienced by these men increased so did the risk of disease progression. The authors report that for each severe stress over a 6-month period, the risk of disease progression doubled and furthermore, in a subset of men who remained in the study for 2 years, the chances of disease progression were almost quadrupled if they had experienced higher severe life stress. A later study by this group (Leserman et al., 1999) examined the relative influences of high stress, low social support, and greater depressive symptoms on the course of HIV as measured by the progression to AIDS. Each of these factors was significantly related to disease acceleration and in combination the factors of stress and social support remained significant. In fact, the authors, following up at 5.5 years, found that progression to AIDS was 2–3 times more likely in individuals with high stress or low social support. Some earlier studies, however, failed to find evidence in support of the relationships between stress and disease progression (Kessler et al., 1991; Perry et al., 1992). From a developmental perspective, the influence of life-event stress in children with HIV has also been linked to survival (Moss et al., 1998).

Disease progression and exacerbation in autoimmune illnesses have also been strongly associated with stress, for example, in rheumatoid arthritis (Koehler, 1985; Zautra et al., 1997) and multiple sclerosis (Schwartz et al., 1999). Interestingly, it is not always simply a case of a higher level of stress or negative events which correlate with such illnesses but a prevalence of fewer positive events (Schwartz *et al.*, 1999; Turner-Cobb *et al.*, 1998).

B. Depression

The role of depression in mediating stress effects on disease progression has been highlighted, particularly in HIV/AIDS. Some evidence for the role of depression in HIV has linked shorter survival time rather than more severe symptomatology in a study of HIV-positive heterosexual and gay men (Patterson *et al.*, 1996). The authors interpret this as depressive symptoms having greater association with the later stages of illness via the mechanism of an indirect influence on health behaviors rather than a direct immune effect from neuroendocrine alteration, since no link was found between depression and $CD4^+$ (T helper cells) decline. Positive associations have also been found in a number of other studies, particularly with HIV-positive gay men, between depression and immune functioning (most notably $CD4^+$ T cell level) as a marker of disease progression (for example, Burack *et al.*, 1993; Kemeny *et al.*, 1994). Other studies however, have failed to show such effects with depression and immune measures or survival (for example, Lyketsos *et al.*, 1993; Perry *et al.*, 1992). Using a more generalized assessment of distress in homosexual men with HIV, Vedhara and colleagues (1997) also report a relationship with $CD4^+$ cell decline, as a disease progression indicator. However, in a sample of women with advanced HIV, studied over a 2-year period (Vedhara *et al.*, 1999), no comparable differences were found between depressed vs nondepressed women in terms of CD4 count as a marker of depression. Although not all results for men have found significant associations, this study highlights the fact that psychological correlates of disease progression for women with HIV may reveal marked differences compared to those in men. Indeed, the importance of psychological factors in HIV for women's health is also strongly indicated by the findings of Goodkin and colleagues of a relationship between psychological factors and the link between human papillomavirus (HPV) infection and HIV status (Goodkin *et al.*, 1993).

Similarly, depression has been associated with disease progression in cancer (McDaniel *et al.*, 1995; Spiegel, 1996) not least because of its role in delaying treatment seeking, although the physiological mechanisms mediating these effects are still under investigation (Spiegel, 1996).

As Spiegel (1999) points out, it is likely that the manner of responding to stress may either reduce or amplify its effects, both psychologically and physiologically. From this perspective it seems appropriate to turn to how individuals cope with stressful events including the "event" of the disease itself.

C. COPING

Psychological responses to the disease itself can influence the course of the disease and this has been well documented in chronic illness, particularly for cancer and HIV. Of studies addressing the relationship between cancer progression and coping style, a program of research conducted by Greer and colleagues (Greer *et al.*, 1979, 1994; Greer and Watson, 1985; Morris *et al.*, 1992; Pettingale *et al.*, 1985) has been particularly influential. These authors found strong evidence for an association between coping and the stress of cancer via a fighting-spirit response and having a more positive prognosis, while "stoic acceptance" and "helplessness" were found to correlate with relapse and hence disease progression (Greer, 1991; Greer *et al.*, 1979); this further supports the influence of an attitude of helplessness/hopelessness being associated with disease relapse or death in a sample of women with early-stage breast cancer at 5-year follow-up (Watson *et al.*, 1999). They also found depression to be associated with survival at follow-up. Yet this study failed to find a relationship between fighting spirit and either recurrence or survival. One study prospectively examined relationships between coping style and survival in a 5-year study of adults undergoing bone marrow transplantation for leukemia (Tschuschke *et al.*, 2001). They report a significantly higher rate of survival at 5-years posttransplant in patients who exhibited a low level of coping by distraction and similarly in patients with strong fighting spirit pretransplant. When subgroups of high and low distraction and fighting spirit were compared the combination of high fighting spirit and low distraction had a pronounced effect on survival rate. The authors attribute their results to the sensitivity of the semistructured interview technique used and view the use of less-detailed questionnaire methods as the failure of some studies (for example, Watson *et al.*, 1999).

In general, in line with other chronic illnesses such as cancer, as described earlier, the HIV literature reveals a pattern of problem-focused coping being inversely related to distress and disease progression, while more passive-defensive coping has been linked with less positive outcomes (Vassend and Eskild, 1998). In HIV, particularly because of the social stigma relating to the illness, the effect of responses to the illness itself have been well documented. The majority of studies examining coping responses to HIV infection have found links between type of coping and disease progression as measured by symptomatology, immune response, and survival. Symptomatology and survival will be examined first, and immune responses will be considered later in the chapter. For example, denial and a lack of fighting spirit have been associated with increased symptom onset in HIV-positive men and women (Ironson *et al.*, 1994a; Solano *et al.*, 1993). Another study (Thornton *et al.*, 2000), however, investigating a number of coping styles, failed to find a relationship between disengagement or denial and disease progression in a group of gay men infected with HIV for between 6 and 11 years. They did, however, find a significant association between acceptance coping and faster progression to AIDS-related complex or AIDS diagnosis.

The issue of positive versus negative expectancies concerning the future appear to be different for HIV/AIDS than for some aspects of cancer progression and survival. For example, as evidenced by Spiegel *et al.*'s studies of women with metastatic breast cancer (Spiegel, 2001; Spiegel *et al.*, 1989), a realistic acceptance of the situation has been linked to better survival outcome. Indeed Spiegel (Spiegel, 2001) refers to the "prison of positive thinking" as opposed to realistic acceptance. Yet in other studies, positive illusions have been reported as adaptive (Taylor, 1983) indicating instead the power of positive illusions in conferring health protective effects (Taylor *et al.*, 2000). This latter response appears to be particularly so in the case of HIV disease, which has been utilized as a model for examining the effects of psychological resources in relation to health outcome. In a modified version of the Lazarus and Folkman (1984) model of coping, Folkman (1997) incorporates positive states of mind with the underlying theme of searching for and finding meaning in the situation, which appears to be the key element to potentially influence disease progression. Interestingly the proposal of this model is based on a study of caregiving by spouses/partners of men in the end stages of AIDS. As these authors (Taylor *et al.*, 2000) point out, the use of HIV disease as a model of illness is becoming less viable due to the effectiveness of recently accessible HIV medications. In children with HIV, a link between parental ratings of the child's "resiliency" has been linked to survival (Moss *et al.*, 1998).

D. EMOTIONAL REPRESSION AND PSYCHOLOGICAL INHIBITION

Repression and psychological inhibition are theoretically strong candidates as correlates of disease progression since both are thought to raise the levels of the stress response via sympathetic activation and to result in an increase in immune suppression (Pennebaker and Chew, 1985; Pennebaker *et al.*, 1987, 1988; Petrie *et al.*, 1998). Repressive coping has consistently been linked to more rapid progression of disease in cancer patients (Kneier and Temoshok, 1984; Temoshok *et al.*, 1985). One particularly important element of cancer progression appears to be the expression of negative emotion or distress associated with fear of the disease (Spiegel, 1993). In suppressing negative emotion, there may also occur a reduction in the expression of positive emotion since expression of both types of emotion are closely linked (Lane *et al.*, 1997). In fact, it has been found that attempts to suppress negative affect not only seem to suppress positive emotion as well, but instead increase rather than decrease dysphoria (Koopman *et al.*, 1998; Lane *et al.*, 1997). Furthermore, although some evidence has been found in support of a link between extroversion and breast cancer progression (Hislop *et al.*, 1987) in general personality correlates of cancer progression have been limited. Indeed the implications of personality factors within medically related research frequently simultaneously inform and inflame lay opinion producing a counteractive effect.

One aspect of disease management that could be viewed in relation to repression versus expression or psychological inhibition is that of disclosure, particularly salient to HIV infection largely because of its frequent asymptomatic nature in the early stages. Psychological inhibition in HIV has taken the form of concealment of either homosexual identity or of HIV status itself. There is evidence that such concealment may be related to subsequent progression of the disease through the various stages and eventually to AIDS. In a study of 88 HIV-positive gay men, Cole and colleagues (Cole *et al.*, 1996b) found a more rapid progression of infection (as measured by a decrease in CD4 count, AIDS diagnosis, and mortality) in men who concealed their sexual identity compared to those who were open about their sexuality. This effect was significant only for men whose concealment was of a moderate to high level. Interestingly, such differences in acceleration were not related to individual differences in mood, coping style, or social support, or in health behaviors or medical treatment. This relationship of increased susceptibility to disease in relation to concealment of HIV status has been further evidenced in relation to infectious and neoplastic diseases in a group of gay and bisexual seropositive men. (Cole *et al.*, 1996a). From a developmental perspective, a study examining childrens' self-disclosure of their HIV status to friends reports a greater increase in CD4 percentage at 1-year follow-up in children who disclosed their status (Sherman *et al.*, 2000). This study opens up a wealth of potential research opportunities for examining the underlying neuroendocrine alterations and resulting immune alterations in children living with HIV. The study of children with known immunosuppressive diseases provides a rich window of opportunity for the study of the concept of allostasis or allostatic load (accumulated lifetime stress) (McEwen, 1998) and resulting influences on health outcome.

Furthermore, in a sample of functional bowel disease and fibromyalgia patients, classified according to social inhibition characteristics, conditions of high social engagement were compared with a baseline control condition (Cole *et al.*, 1999). These authors found delayed-type hypersensitivity (DTH) induration responses to be greater for individuals who were more socially inhibited, indicating a link between an inflammatory immune response and coping in relation the social environment.

E. SOCIAL RELATIONS

The psychological domain, most convincingly linked to disease progression in cancer is that of social support, or its converse, social isolation. A number of large-scale studies have revealed the impact of social relations on health in general (for reviews, see Cohen, 1988; House *et al.*, 1988; Uchino *et al.*, 1996). In fact, there is evidence (House *et al.*, 1988) that the strength of the relationship between social integration and age-adjusted mortality is comparable to the effect of such standard health behavior risk factors as smoking and serum cholesterol levels. Studies examining disease outcome have found that positive, supportive social relations from a variety of sources, such as marriage (Goodwin *et al.*, 1987; Maunsell *et al.*, 1993), daily contact with others (Arnetz *et al.*, 1983), and the presence of confidants, reduces mortality risk from cancer as well as from other diseases. For example, in a study of the associations between social support and survival for patients diagnosed with breast, colorectal, or lung cancer (Ell *et al.*, 1992), psychosocial factors were found to operate differently depending on the site and severity of the cancer. For patients with earlier disease stages and particularly among women with breast cancer, emotional support from primary network members was found to have a protective influence in terms of survival (Ell *et al.*, 1992). An effect was not found, however, with more advanced disease or among lung or colorectal cancer patients only. Results concerning the buffering of emotional support have been differential across diseases (Penninx *et al.*, 1998). Emotional support was not found to buffer depression in cancer, diabetes, or lung disease but it did act as a buffer for cardiac or arthritis patients.

Although not all studies have found evidence for the positive influence of social support on disease progression (Barraclough *et al.*, 1992; Marshall and Funch, 1983), others have found positive effects of supportive relationships in breast cancer patients in contexts such as the marital relationship, supportive friendships, confiding in a confidant, size of social network, and employment status to be associated with longer survival (Hislop *et al.*, 1987; Maunsell *et al.*, 1993; Waxler-Morrison *et al.*, 1991).

Inconsistencies in the findings of prospective studies examining social support in relation to symptoms, immune changes, and mortality in HIV have revealed a time course pattern of effects for the influence of social relations on disease progression. Immune changes, particularly in CD4 count are discussed later (Section IV.A). In the study of HIV-positive heterosexual gay men (Patterson *et al.*, 1996) mentioned earlier, larger social networks were found to be associated with symptom onset and survival but only in those men who had progressed to AIDS status. The authors interpret this as the importance of support particularly for those in the later stages of HIV.

F. IMPORTANCE OF HEALTH BEHAVIORS

The importance of health behaviors in disease progression cannot be underestimated since the action of stress and other psychological factors may occur indirectly, via altered health-related behaviors. For example, during stressful periods, alcohol consumption may rise, sleep quality may be diminished, there may be poorer nutritional value in the diet, and exercise may be avoided. That the weakening of such factors may be associated with reduced resistance to advancing disease needs to be given greater attention to help elucidate some of the mediating pathways through which neuroendocrine mechanisms may operate. This is applicable not only to cancer (Garssen and Goodkin, 1999) but also to HIV and other immune-related diseases with variable progression.

G. SOCIOECONOMIC STATUS

An important group of variables not be overlooked in this relationship between psychosocial influences on cancer survival is the impact of social inequalities and related factors such as socioeconomic status (SES) and race. For example, an inverse relationship has been reported between SES and overall pattern of cancer mortality (for review, see Balfour and Kaplan, 1998). In a study of breast, cervix, and uterine corpus cancers, SES and race were found to be independent predictors of survival (Greenwald *et al.*, 1996). SES predicted survival from breast and uterine corpus cancers while race predicted survival in each of the three types of cancer examined (Greenwald *et al.*, 1996).

While a number of health behavior and health care access issues are obviously implicated in the survival relationship in physical illness, endocrine factors may play an important role in the link between differential SES utilization of psychosocial resistance factors and health outcome. That social hierarchy may play a role in influencing hormonal responses has been demonstrated in an ongoing series of naturalistic baboon studies which have consistently found evidence in support of instability or demotion in social rank being reflected in endocrine responses (Sapolsky, 1989, 1995; Sapolsky *et al.*, 1997; Sapolsky and Spencer, 1997). In relation to the mechanism by which social support enhances coping with life stress, a link to studies where religion and church attendance have been associated with positive physiological health outcomes is worth making (Koenig *et al.*, 1997; Matthews *et al.*, 1998). It is possible that at least a part of the physiological enhancement that is experienced through religion and church attendance, is related to the increased acceptance and elevation in social hierarchy frequently sought and received through such a support system. (Greenwald *et al.*, 1996).

III. Intervention Studies

Before discussing findings relating to intervention studies and disease progression, as Kiecolt-Glaser and Glaser (1992) point out, the enhancement of immune function is most beneficial for immunocompromised populations, while such an enhancement in those whose immune functioning is already optimum may be impossible or even harmful. The diseases focused on in this chapter meet this criteria, since cancer is associated with immunosuppression both directly through metabolic, metastatic, and other mechanisms and indirectly through the effects of chemotherapy and radiotherapy, and HIV/AIDS is a progressive assault on the immune system.

A. CANCER AND SURVIVAL

Given the previously mentioned literature reporting that stress can increase the rate of cancer progression and that psychological resources such as coping responses and supportive social relations can act to buffer against these stress effects, it is plausible that appropriate interventions utilizing such psychological factors may offer positive outcomes for disease progression and mortality. Indeed in many cases psychological interventions, whose primary aim was to improve psychological adjustment to illness, have reported positive physical health correlates (for reviews, see Blake-Mortimer et al., 1999; Bottomley, 1997; Cwikel et al., 1997; Fawzy et al., 1995; Loscalzo, 1998). The theory behind such interventions is that social support acts as a buffer between stress and illness and in doing so limits the physiological response to the stress of the disease such as cancer (Cohen and Wills, 1985; Levine et al., 1989; Spiegel, 1999). A number of intervention studies have focused specifically on survival as an outcome of receiving psychotherapeutic intervention. One of the early studies (Grossarth-Maticek et al., 1984) reported evidence for such a survival effect. However, the results are less than conclusive as the methodology generated a number of serious criticisms. One very influential study in this area reported a few years later (Spiegel, 1992; Spiegel et al., 1989) found that Supportive–Expressive group psychotherapy was effective in increasing the length of life of women with metastatic breast cancer by an average of 18 months. Similarly, increased survival following supportive interventions in other cancer populations have subsequently been reported, including positive findings for lymphoma and leukemia patients (Richardson et al., 1990) and with newly diagnosed malignant melanoma patients (Fawzy et al., 1990a, 1993). Other cancer intervention studies, however, have not found such a survival effect with supportive interventions (Gellert et al., 1993; Ilnyckyj et al., 1994; Linn et al.,

1982; Cunningham *et al.,* 1998). Evidence also exists in support of medita-
tional interventions in diseases such as prostate cancer (Coker, 1999; Kaplan
et al., 1993).

B. HIV AND SURVIVAL

The necessity for including social support in intervention studies has
been evidenced earlier and there is an obvious, strong theoretical rationale
for including it as a component in HIV interventions as in other chronic
illnesses (Mulder, 1994). Hence many intervention studies examining HIV
and associated infection are currently underway but initial results have frequently been inconclusive (Mulder, 1994) and as yet are less clear than
for cancer; however, that interventions have potential in attenuating HIV
progression is evidenced by the literature reviewed earlier. There is some evidence of psychological interventions using stress management techniques
slowing the decline in immune defenses via endocrine and immune markers
of HIV progression, and these are discussed below (Section IV.B). However,
in relation specifically to survival outcomes in HIV, no large-scale prospective
studies have been reported to date.

Furthermore, in relation to other diseases, convincing evidence is provided by Smyth and colleagues (1999) in a study examining the influence
of emotional expression through writing about stressful events. Comparing
between experimental and control groups, asthma patients receiving the intervention revealed improvements in lung function and rheumatoid arthritis patients revealed a reduction in disease activity at the 4-month follow-up.
These improvements were clinically significant for almost half (47.1%) of
patients compared to clinical improvement in only one quarter (24.3%) of
the control group (Smyth *et al.,* 1999).

IV. Neuroendocrine and Immunological Correlates That Underpin Disease Progression

On the basis that psychological factors can upregulate immune mechanisms that defend against disease progression, interventions utilizing such
psychological mechanisms may also have the potential to retard the progression of disease. However, a substantial amount of further research is needed
to determine the combinations of psychosocial factors and diseases in which
such mechanisms may operate.

As Spiegel (1995) rightly points out in relation to cancer, that psychological interventions can be effective in altering disease progression should

no longer be surprising, rather the focus of attention is now at liberty to turn to how such processes work by elucidating the mechanisms that operate to bring about such positive effects. To do this it is necessary to turn to some of the studies that have linked psychological factors with neuroendocrine and immune mediators.

A. CORTISOL

There is growing interest in the cascade of hormones produced by the stress-response system of the HPA axis. Of particular interest in the progression of disease is the production of the glucocorticoid cortisol, not least because of its ease of measurement in the unbound state in saliva. An inverse relationship between social support and cortisol levels has been found in a number of studies in healthy populations (for example, Arnetz et al., 1985; Seeman et al., 1994; Wadhwa et al., 1996) but other studies found no association (for example, Groer et al., 1994; Wadhwa et al., 1996). In women with early stage breast cancer, a significant decrease in plasma cortisol levels was related to receiving group therapy intervention. Likewise in a sample of homosexual men attending a bereavement intervention group, plasma cortisol levels decreased significantly in comparison to the control group (Goodkin et al., 1998).

As well as an examination of the relationship between disease progression and overall mean levels of cortisol, studies have also explored the relationships between psychological factors and health outcome with regard to the potential mediating effect of diurnal cortisol rhythm. In a replication trial of Spiegel's survival finding (Spiegel et al., 1989), diurnal cortisol rhythm evaluated in women with metastatic breast cancer was assessed from twelve measures taken four times a day across three consecutive days. Baseline diurnal cortisol measures were found to predict survival up to 7-years later such that a marked diurnal declining pattern was associated with longer survival (Sephton et al., 2000). The authors suggest that cortisol rhythmicity is a long-term indicator of disease progression (Sephton et al., 2000). In an assessment of this data at baseline, overall mean salivary cortisol levels were found to be inversely associated with three specific types of social support—appraisal, belonging, and tangible support (Turner-Cobb et al., 2000). Sephton et al. (2000) argue that NK cells, which are sensitive to the immunosuppressive influence of cortisol (discussed in Section V.B) mediate the relationship between diurnal cortisol dysregulation and shorter survival. Furthermore, specific links have been found between androgen receptors that are sensitive to cortisol, leading to an enhancement of tumor progression in prostate cancer (Zhao et al., 2000). Thus increased production of cortisol under stressful conditions can be strongly linked to cancer progression.

Other possible hormonal candidates that may play a role in disease progression or may provide a marker of progression and which are also influenced by psychological factors include prolactin (Bhatavdekar *et al.*, 1990; van der Pompe *et al.*, 1997) and oxytocin (Ito *et al.*, 1995; Murrell, 1995; Uvnas-Moberg, 1997).

B. Alteration in CD4 Cell Populations

One important marker of disease progression in HIV disease is the shift in balance from the Th1 toward the Th2 response, reflecting a shift in balance away from cellular immunity and toward humoral dominance. (This marker has also been referred to in relation to depression in Section II.B). Given that this is a similar effect to that seen in chronic stress, the importance of the stress response in HIV becomes evident. For example, the active-coping response has been positively associated with natural killer (NK) cytotoxicity, CD4 counts, and total T-cell count in gay men (Goodkin *et al.*, 1992a,b). In the study of life-event stress and survival in children with HIV, mentioned earlier in Section II.A, while CD4 percentage was also linked with survival, no link was found to implicate it as a physiological mediator of the life-event stress (Moss *et al.*, 1998). Conversely, in autoimmune diseases it is Th1 imbalance that is associated with the hormonal profile of disease exacerbation. This is particularly well evidenced in the remission from certain autoimmune diseases during pregnancy (Mastorakos and Ilias, 2000; Wilder, 1998), described as a "transient state of relative hypercortisolism" (Mastorakos and Ilias, 2000).

While some studies have yielded evidence of the positive effect of social support on HIV progression in relation to CD4 counts (Theorell *et al.*, 1995) others have found the opposite effect whereby individuals with a high level of network support have been found to have a more rapid decline in CD4 (Persson *et al.*, 1994). In addition to the study by Patterson *et al.* (1996) mentioned earlier, a further finding by Miller *et al.* (1997) emphasizes the importance of stage of disease. Miller *et al.* report an association between low scores on loneliness and a decline in the CD4 level, yet as they point out, the differences across these studies can be at least partly explained by the differences in the HIV subpopulations studied, variability in disease stage, and differences in measures. These authors (Miller *et al.*, 1997) speculate that HIV-positive gay men with greater social support may have a greater decline in CD4 cells because of what they term the "cumulative exposure" to the stressors associated with the course of AIDS, including those relating to their own diagnosis and illness and to those in their community (e.g., caregiving, bereavement). This fits with the recurring theme of the association between allostatic load and health outcomes including disease progression. Another study cited earlier which also focused on this cumulative experience of stress (Leserman *et al.,* 1999) reports satisfaction with support (but not network size) as being significantly linked to AIDS progression (as defined by $CD4^+$ cell decline and the presence of an AIDS-indicator condition). Interestingly, a further vital point made by Miller *et al.* (1997) is that social relationships do not in themselves always confer positive effects on health.

C. NATURAL KILLER (NK) CELL NUMBER AND ACTIVITY

1. *Chronic Stress*

In general, research findings suggest that rather than adaptation in response to chronic stress, a process of downregulation occurs (Kiecolt-Glaser and Glaser, 1992). A large body of literature has accumulated in support of the notion that psychosocial factors influence the relationship between a stressor and its converse, relaxation, and immunity (Snyder *et al.,* 1993; Bergsma, 1994; Kiecolt-Glaser and Glaser, 1992; Van Rood *et al.,* 1993). In a meta- analysis of 23 human studies of exam or experimental stress and relaxation on immune response, stress was found to increase white blood cells, Epstein-Barr virus, and herpes simplex virus titers (Van Rood *et al.,* 1993). Similarly, evidence was found for decreases in IL-2 receptor expression and T-cell proliferation in response to the T-cell stimulator phytohemagglutinin under stress. However, inconsistency in the direction of change was found for monocytes, B cells, $CD4^+$ T helper cells, $CD8^+$ T cytotoxic cells, the $CD4^+/CD8^+$ ratio, and NK cells (Van Rood *et al.,* 1993). Conversely, consistency in direction of change was found with relaxation such that white blood cell concentration decreased and salivary immunoglobulin A and NK cell activity increased during or after it. A lower lymphocyte proliferation response 3 weeks following exposure to a novel antigen in subjects who had experienced more negative stress or psychological distress has also been reported (Snyder *et al.,* 1993). Hence evidence exists in support of a model in which psychosocial processes act as mediators between stressful events and immunity (Snyder *et al.,* 1993).

Reactivity of natural killer cells in response to stressful stimuli has been reported in a number of populations (Herbert and Cohen, 1993). Measures taken immediately following an acute stress have revealed an increase in NK cell numbers and/or NK cell activity, for example, following cognitive conflict tasks, mental arithmetic, and parachute jumping (Landmann *et al.*, 1984; Naliboff *et al.*, 1991; Schedlowski *et al.*, 1993). Such acute-stress-induced increases in NK cell numbers and function may be followed by decreases below baseline shortly afterward (Schedlowski *et al.*, 1993). Furthermore, NK cell decreases have been observed following such periods of enduring or chronic stress as those involved in interpersonal difficulties or challenges, for example, examinations, caring for a chronically ill spouse, divorce, separation, and bereavement (Esterling *et al.*, 1994; Irwin *et al.*, 1987; Kiecolt-Glaser *et al.*, 1984, 1987a,b). Alteration in NK number and function has also been linked to mood, for example, lower NK cell counts have been evidenced in depressed mood (Kemeny *et al.*, 1989; Stein *et al.*, 1991) and lower cytotoxicity is seen with anxiety (Ironson *et al.*, 1990). A decline in NK cells and a rise in cortisol have also been reported in advanced-stage cancer patients with depression (Lechin *et al.*, 1990). In women with metastatic breast cancer, in addition to the association between diurnal cortisol variability and survival mentioned previously, a relationship was also found between flattened profiles and low number and activity of NK cells (Sephton *et al.*, 2000; Spiegel *et al.*, 1998).

2. *Social Support and Emotional Expression*

Although some studies have not found evidence of a relationship between social support and changes in immune function (for example, Kiecolt-Glaser *et al.*, 1985; Schlesinger and Yodfat, 1991), in general, supportive relationships seem to modulate stress-induced immunosuppression, as seen, for example, in medical students undergoing examination stress (Kiecolt-Glaser *et al.*, 1984a). Psychosocial factors such as social support and interpersonal relationships seem to influence the effect of stress on NK cells, even among cancer patients. In a sample of stage I and II breast cancer patients, Levy (1990) reported an association between NK cell activity and perceived support from their spouse or partner. A relationship was also found between NK cell activity and perceived support from their physician, as well as with the coping strategy of active social support seeking. These findings are especially noteworthy since among patients who were immunologically assessed approximately 1 week postsurgically and then followed 5 years later, NK cell activity was a strong predictor of disease recurrence, revealing higher cytotoxicity to be related to a longer disease-free interval (Levy *et al.*, 1991). Thus quality of family support and overall mood may potentially predict disease progression in patients whose cancer does recur (Levy *et al.*, 1991).

In a study of immune function in women after surgical treatment for regional breast cancer, higher stress levels were a significant predictor of decreased NK cytotoxicity, NK cell response to IFN-γ, and T-cell responses to mitogen stimulation (Andersen *et al.*, 1998). Social support has been reported to be related to aspects of immune status in women treated with adjuvant chemotherapy for breast cancer (Lekander *et al.*, 1996). They report that perceived attachment significantly influenced white blood cell levels and the number and percentages of granulocytes in women after, but not during, chemotherapy. As these authors point out, although causality is not discernible from their study, since the overall white blood cell count can be positively affected by social support, this has important implications for the recovery process following treatment.

Evidence for a relationship between social support and blastogenic responses of lymphocytes to mitogens has also been reported (Baron *et al.*, 1990; Glaser *et al.*, 1992; Linn *et al.*, 1988; Snyder *et al.*, 1993; Theorell *et al.*, 1990; Thomas *et al.*, 1985). An interaction between chronic stress and helpful support in relation to blastogenic responses has been found by Kiecolt-Glaser and colleagues (1991), such that caregivers who were low in helpful social support showed greater negative changes in functional immune response even after statistical controls for age, income, and depression.

Further support is provided for psychological influences on disease progression in a series of studies investigating biopsychosocial aspects of cutaneous malignant melanoma by Temoshok and colleagues (1985). They report that expression of negative emotions (sadness and anger) was positively related to the disease outcome and the number of lymphocytes at the tumor site and negatively related to the mitotic rate of the tumor. In addition, animal studies have provided some evidence that stress-associated increases in tumor growth may occur concomitant with NK cell activity suppression (Ben- Eliyahu *et al.*, 1991; Rowse *et al.*, 1992; Shavit *et al.*, 1985). For example, in rats implanted with a mammary adenocarcinoma, stress in the form of shock resulted in reduced NK activity and survival time (Shavit *et al.*, 1985). To elucidate the underlying mechanism involved in this relationship between stress and tumor growth, Ben-Eliyahu and his colleagues (1991) injected rats with lung tumor cells prior to being subjected to an acute stressor. Stressed animals revealed a significant reduction in NK cell activity at a critical point in the process of NK activity in defense against tumor cells. This resulted in a twofold increase in tumor metastases for stressed rats compared to control rats (Ben-Eliyahu *et al.*, 1991).

V. Neuroimmune Alterations Associated with Psychological Interventions

A. HIV

There is some evidence for interventions using stress management techniques to slow the decline in immune defenses to a greater extent than in a comparison control group (for review, see Cole and Kemeny, 1997;

Kiecolt-Glaser and Glaser, 1992). For example, some comparatively early intervention studies using an aerobic exercise or cognitive-behavioral intervention for HIV-positive homosexual men reported a positive influence on the attenuation of NK cell numbers, CD4 T lymphocytes, and mitogenic response (Antoni *et al.*, 1991; LaPerriere, 1990 & 1991; Esterling, 1990; Taylor, 1995). One early study in this area failed to find a link between stress-reduction intervention and immune alteration in HIV- infected individuals (Coates *et al.*, 1989). However, as Kiecolt-Glaser and Glaser (1992) point out, the lack of findings may have been due to a number of methodological issues. Mulder *et al.* (1995) also failed to find changes in CD4 cell decline or T-cell responses in a group of asymptomatic HIV-infected homosexual men who participated in either cognitive-behavioral or existential group therapy, although they did find that decreases in distress were related to increases in CD4 cell counts. One particular stressor associated with HIV is that of AIDS-related bereavement in a partner of close friends. In a study of a 10-week bereavement support group intervention for HIV-1-seronegative and -seropositive homosexual men (Goodkin *et al.*, 1998), several immunoenhancing long-term (6-month follow-up) changes were found for those in the intervention group. An increase was found in CD4 cells, total T lymphocytes, and total lymphocyte counts but no differences were found for the CD4/CD8 ratio or CD8 cell count. In addition, plasma cortisol decreased significantly in the intervention group and CD4 cell changes were associated with changes in plasma cortisol, revealing neuroimmune alterations.

In a study using a stress management intervention, Cruess and colleagues measured the adrenal steroid dehydroepiandrosterone (DHEA) in addition to cortisol. Whereas cortisol favors the shift away from Th1 cellular immunity, DHEA exerts the opposite influence. Evidence was found that a cognitive-behavioral stress management intervention provided a buffering effect on DHEA-S (the circulating sulfated form of the steroid) and on the ratio between cortisol/DHEA-S in men with HIV. Both of these endocrine measures are potential markers of disease progression. While a control group showed changes in these hormones indicative of disease progression, no such deleterious hormone changes were revealed in the intervention group. Reductions in distress in the intervention group were also found and the authors stress the importance of this alteration in relation to the cortisol/DHEA-S balance rather than the DHEA-S level per se (Cruess *et al.*, 1999). This balance between DHEA-S and cortisol provides a further link in understanding the underlying mechanisms involved in neuroimmune alterations under stress and their attenuation via interventions, in relation to disease progression. Hence the counterbalance of DHEA-S and its relationship to cortisol regulation offers potential for further research as the authors suggest (Cruess *et al.*, 1999). A later study by these authors adds further support to the effectiveness of intervention in influencing cortisol responses in relation to relaxation (Cruess *et al.*, 2000). The necessity for including social support in intervention studies has been shown earlier and there is an obvious, strong theoretical rationale for including it as a component in HIV interventions as in other chronic illnesses (Mulder, 1994).

B. CANCER

Physiological mediators of intervention effects that have been implicated in cancer include an increase in immune cell function. For example, an increase in interleukin-2 levels and T-cell mitogens has been reported following psychosocial intervention for metastatic cancer patients using guided imagery and relaxation (Gruber *et al.*, 1988). The clinical relevance of this study is not discernable, however, given the lack of a control group and the lack of statistical power due to the low participant number ($n = 10$). In a more detailed study, van der Pompe and associates investigated the effectiveness of intervention employing a randomized, controlled trial consisting of a 13-week experiential–existential group psychotherapy program with breast cancer patients (van der Pompe *et al.*, 1997). After adjusting for baseline endocrine and immune levels, the authors observed significant differences between those receiving group therapy and those in the wait list control. Patients in the therapy condition who also had high endocrine profiles at baseline (as measured by prolactin and cortisol) revealed a decrease in these endocrine levels at follow-up, when compared to controls. In the same manner, patients with a comparatively high baseline immune level showed lower percentages of natural killer cells, CD8 cells, CD4 cells, and a lower proliferative response to pokeweed mitogen than the cancer control group. The authors explain the endocrine changes as reflecting a decrease in HPA axis activity following the group therapy intervention. While the immune effects appear to be to some extent contradictory, the authors point to previous research which indicates that lower percentages of some lymphocyte populations may in fact be associated with a more advantageous outcome.

As mentioned earlier, Spiegel and colleagues have for some years, been conducting a replication trial of the earlier intervention study where a survival effect was found (Spiegel *et al.*, 1989) incorporating immune and endocrine measures (Sephton *et al.*, 2000; Turner-Cobb *et al.*, 2000). However, only one study to date has been reported which directly examines the influence of immune mediators of psychosocial intervention in conjunction with cancer progression or survival. That is a 6-week structured psychiatric group intervention study by Fawzy and his colleagues (Fawzy *et al.*, 1990b). This evaluation of the effects of a 6-week group intervention for stage I or II malignant melanoma patients reported a significant increase in $CD57^+$ large granular lymphocytes (LGLs) at the end of the 6-week intervention. This increase in LGLs in the intervention group was found to take place in the $CD8^+$ T-cell subpopulation rather than in NK cells. At a 6-month follow-up, the presumed immunoenhancement among those who had been through group treatment was evident in INF-α augmented $CD56^+$ NK cell cytotoxicity rather than CD8 cell counts. Interestingly, while this higher NK cell activity was predictive of a lower rate of recurrence, it did not predict survival time (Fawzy, 1994; Fawzy *et al.*, 1990b). The authors suggest that the NK cell system may be especially responsive to psychological or behavioral changes. In addition, CD4 helper T cells were unexpectedly reduced at 6 months, explained by the authors as due to selective redistribution (Fawzy *et al.*, 1990b) or possibly due to immune cell trafficking (Dhabhar *et al.*, 1994, 1995).

VI. Other Explanatory Mechanisms in Disease Progression

There are of course many other mechanisms that are known to have an influence on disease progression. For example, in cancer, other biological mechanisms include the process of angiogenesis and associated influencing cytokines involved in this process (Garssen and Goodkin, 1999) as well as the processes of DNA repair and apoptosis (cell death). The influence of stress on these biological processes remains to be explored.

To summarize, disease progression is associated with a number of psychological and neuroendocrine correlates which frequently although not solely influence disease course via immune alteration, reducing host resistance to external antigens and internal disease processes. In this chapter, it has been seen that psychological influences, particularly enduring stress, emotional repression, the ability to find meaning under difficult circumstances, and the broad realm of social relations play not just an important but an integral role in disease progression with variation in effect across a number of diseases. The interplay of health behaviors and SES within this multifactorial pattern of influences on disease progression should not be underestimated. Evidence abounds of the impact of psychological correlates of disease progression as determined by symptomatology, endocrine–immune alteration, and survival in chronic illness, particularly in cancer and HIV. Although in the comparatively early stages, mapping of known neuroendocrine–immune correlates of disease progression with these psychological influences has begun, including the stress hormone cortisol and its relationship to circadian rhythmicity and corresponding Th1/Th2 immune balance, the link with

CD4 activity, and functional and enumerative activity of NK cells. Future work needs to concentrate on further extrapolating the details of effective psychological interventions in mapping neuroendocrine and immune responses to disease resistance and thereby enabling maximization of mainstream medical interventions.

References

Andersen, B. L., Farrar, W. B., Golden-Kreutz, D., Kutz, L. A., MacCallum, R., Courtney, M. E., and Glaser, R. (1998). Stress and immune responses after surgical treatment for regional breast cancer. *J. Nat. Cancer Inst.* **90**(1), 30–36.

Antoni, M. H., Baggett, L., Ironson, G., LaPerriere, A., August, S., Klimas, N., Schneiderman, N., and Fletcher, M. A. (1991). Cognitive-behavioral stress management intervention buffers distress responses and immunologic changes following notification of HIV-1 seropositivity. *J. Cons. Clin. Psychol.* **59**(6), 906–915.

Arnetz, B. B., Theorell, T., Levi, L., Kallner, A., and Eneroth, P. (1983). An experimental study of social isolation of elderly people: Psychoendocrine and metabolic effects. *Psychosom. Med.* **45**(5), 395–406.

Arnetz, B. B., Edgren, B., Levi, L., and Otto, U. (1985). Behavioural and endocrine reactions in boys scoring high on Sennton neurotic scale viewing an exciting and partly violent movie and the importance of social support. *Soc. Sci. Med.* **20**(7), 731–736.

Balbin, E. G., Ironson, G. H., and Solomon, G. F. (1999). Stress and coping: The psychoneuroimmunology of HIV/AIDS. *Baillieres best Pract. Res. Clin. Endocrinol. Metab.* **13**(4), 615–633.

Balfour, J. L., and Kaplan, G. A. (1998). Social class/Socioeconomic factors. *In* "Psycho-Oncology" (J. Holland, ed.). Oxford Univ. Press, New York.

Baron, R. B., Cutrona, C. E., Hicklin, D., Russel, D. W., and Lubaroff, D. M. (1990). Social support and immune function among spouses of cancer patients. *Journal of Personality and Social Psychology,* **59**(2), 344–352.

Barraclough, J., Pinder, P., Cruddas, M., Osmond, C., Taylor, I., and Perry, M. (1992). Life events and breast cancer prognosis. *Br. Med. J.* **304**, 1078–1081.

Baum, A., and Posluszny, D. M. (1999). Health psychology: Mapping biobehavioral contributions to health and illness. *Annu. Rev. Psychol.* **50**, 137–163.

Ben-Eliyahu, S., Yirmiya, R., Liebeskind, J. C., Taylor, A. N., and Gale, R. P. (1991). Stress increases metastatic spread of a mammary tumor in rats: Evidence for mediation by the immune system. *Brain Behav. Immun.* **5**(2), 193–205.

Bergsma, J. (1994). Illness, the mind, and the body—Cancer and immunology: An introduction. *Theor. Med.* **15**(4), 337–347.

Bhatavdekar, J. M., Shah, N. G., Balar, D. B., Patel, D. D., Bhaduri, A., Trivedi, S. N., Karelia, N. H., Ghosh, N., Shukla, M. K., and Giri, D. D. (1990). Plasma prolactin as an indicator of disease progression in advanced breast cancer. *Cancer* **65**, 2028–2032.

Biondi, M., and Picardi, A. (1999). Psychological stress and neuroendocrine function in humans: The last two decades of research. *Psychother. Psychosom.* **68**, 114–150.

Blake-Mortimer, J. S., Gore-Felton, C., Kimerling, R., and Turner-Cobb, J. M. (1999). Improving the quality and quantity of life among patients with cancer: A review of the effectiveness of group psychotherapy. *Eur. J. Cancer* **35**(11), 1581–1586.

Bottomley, A. (1997). Where are we now? Evaluating two decades of group interventions with adult cancer patients. *J. Psychiatr. Mental Health Nursing* **4**, 251–265.

Burack, J. H., Barrett, D. C., Stall, R. D., Chesney, M. A., Ekstrand, M. L., and Coates, T. J. (1993). Depressive symptoms and CD4 lymphocyte decline among HIV-infected men. *JAMA* **270**, 2568–2573.

Coates, T. J., McKusick, L., Kuno, R., and Stites, D. P. (1989). Stress reduction training changed number of sexual partners but not immune function in men with HIV. *Am. J. Publ. Health* **79**(7), 885–887.

Cohen, S. (1988). Psychosocial models of the role of social support in the etiology of physical disease. *Health Psychol.* **7**(3), 269–297.

Cohen, S., and Herbert, T. B. (1996). Health psychology: Psychological factors and physical disease from the perspective of human psychoneuroimmunology. *Annu. Rev. Psychol.* **47**, 113–142.

Cohen, S., and Wills, T. A. (1985). Stress, social support, and the buffering hypothesis. *Psychol. Bull.* **98**(2), 310–357.

Coker, K. H. (1999). Meditation and prostate cancer: Integrating a mind/body intervention with traditional therapies. *Semin. Urol. Oncol.* **17**(2), 111–118.

Cole, S. W., and Kemeny, M. E. (1997). Psychobiology of HIV infection. *Crit. Rev. Neurobiol.* **11**(4), 289–321.

Cole, S. W., Kemeny, M. E., Taylor, S. E., and Visscher, B. R. (1996a). Elevated physical health risk among gay men who conceal their homosexual identity. *Health Psychol.* **15**(4), 243–251.

Cole, S. W., Kemeny, M. E., Taylor, S. E., Visscher, B. R., and Fahey, J. L. (1996b). Accelerated course of human immunodeficiency virus infection in gay men who conceal their homosexual identity. *Psychosom. Med.* **58**(3), 219–231.

Cole, S. W., Kemeny, M. E., Weitzman, O. B., Schoen, M., and Anton, P. A. (1999). Socially inhibited individuals show heightened DTH response during intense social engagement. *Brain Behav. Immun.* **13**, 187–200.

Cruess, D. G., Antoni, M. H., Kumar, M., Ironson, G., McCabe, P., Fernandez, J. B., Fletcher, M., and Schneiderman, N. (1999). Cognitive-behavioral stress management buffers decreases in dehydroepiandrosterone sulfate (DHEA-S) and increases in the cortisol/DHEA-S ratio and reduces mood disturbance and perceived stress among HIV-seropositive men. *Psychoneuroendocrinology* **24**(5), 537–549.

Cruess, D. G., Antoni, M. H., Kumar, M., and Schneiderman, N. (2000). Reductions in salivary cortisol are associated with mood improvement during relaxation training among HIV-seropositive men. *J. Behav. Med.* **23**(2), 107–122.

Cunningham, A. J., Edmonds, C. V., Jenkins, G. P., Pollack, H., Lockwood, G. A., and Warr, D. (1998). A randomized controlled trial of the effects of group psychological therapy on survival in women with metastatic breast cancer. *Psychooncology* **7**(6), 508–517.

Cwikel, J. G., Behar, L. C., and Zabora, J. R. (1997). Psychosocial factors that affect the survival of adult cancer patients: A review of research. *J. Psychosoc. Oncology* **15**(3/4), 1–34.

Dhabhar, F. S., Miller, A. H., Stein, M., McEwen, B. S., and Spencer, R. L. (1994). Diurnal and acute stress-induced changes in distribution of peripheral blood leukocyte subpopulations. *Brain Behav. Immun.* **8**(1), 66–79.

Dhabhar, F. S., Miller, A. H., McEwen, B. S., and Spencer, R. L. (1995). Effects of stress on immune cell distribution. Dynamics and hormonal mechanisms. *J. Immunol.* **154**(10), 5511–5527.

Ell, K., Nishimoto, R., Mediansky, L., Mantell, J., and Hamovirchm, M. (1992). Social relations, social support and survival among patients with cancer. *J. Psychosom. Res.* **36**(6), 531–541.

Esterling, B. A., Kiecolt-Glaser, J. K., Bodnar, J. C., and Glaser, R. (1994). Chronic stress, social support, and persistent alterations in the natural killer cell response to cytokines in older adults. *Health Psychol.* **13**(4), 291–298.

Evans, D. L., Leserman, J., Perkins, D. O., Stern, R. A., Murphy, C., Zheng, B., Gerres, G.,

Longmare, J. A., Silva, S. G., van der Horst, C. M., Hall, C. D., Folds, J. D., Folden, R. N., and Petirro, J. M. (1997). Severe life stress as a predictor of early disease progression in HIV infection. *Am. J. Psychiatry* **154**(5), 630–634.

Fawzy, F. I. (1994). Immune effects of a short-term intervention for cancer patients. *Advances* **10**(4), 32–33.

Fawzy, F. I., Cousins, N., Fawzy, N. W., Kemeny, M. E., Elashoff, R., and Morton, D. (1990a). A structured psychiatric intervention for cancer patients. I. Changes over time in methods of coping and affective disturbance. *Arch. Gen. Psychiatry* **52**, 100–113.

Fawzy, F. I., Kemeny, M. E., Fawzy, N. W., Elashoff, R., Morton, D., Cousins, N., and Fahey, J. L. (1990b). A structured psychiatric intervention for cancer patients. II. Changes over time in immunological measures. *Arch. Gen. Psychiatry* **47**(8), 729–735.

Fawzy, F. I., Fawzy, N. W., Hyun, C. S., Elashoff, R., Guthrie, D., Fahey, J. L., and Morton, D. L. (1993). Malignant melanoma: Effects of an early structured psychiatric intervention, coping, and affective state on recurrence and survival 6 years later. *Arch. Gen. Psychiatry* **50**, 681–689.

Fawzy, F. I., Fawzy, N. W., Arndt, L. A., and Pasnau, R. O. (1995). Critical review of psychosocial interventions in cancer care. *Arch. Gen. Psychiatry* **52**(2), 100–113.

Folkman, S. (1997). Positive psychological states and coping with severe stress. *Soc. Sci. Med.* **45**(8), 1207–1221.

Garssen, B., and Goodkin, K. (1999). On the role of immunological factors as mediators between psychosocial factors and cancer progression. *Psychiatry Res.* **85**, 51–61.

Gellert, G. A., Maxell, R. M., and Siegel, B. S. (1993). Survival of breast cancer patients receiving adjunctive psychosocial support therapy: a 10-year follow-up study. *J. Clin. Oncol.* **11**, 66–69.

Glaser, R., Kiecolt-Glaser, J. K., Bonneau, R. H., Malarky, W., Kennedy, S., and Hughes, J. (1992). Stress-induced modulation of the immune response to recombinant hepatitis B vaccine. *Psychosomatic Medicine,* **54**, 22–29.

Goodkin, K., Antoni, M. H., Helder, L., and Sevin, B. (1993). Psychoneuroimmunological aspects of disease progression among women with human papillomavirus-associated cervical dysplasia and human immunodeficiency virus type 1 co-infection. *Int. J. Psychiatry Med.* **23**(2), 119–148.

Goodkin, K., Blaney, N. T., Feaster, D., Fletcher, M. A., Baum, M. K., Mantero-Atienza, E., Klimas, N. G., Millon, C., Szapocznik, J., and Eisdorfer, C. (1992a). Active coping style is associated with natural killer cell cytotoxicity in asymptomatic HIV-1 seropositive homosexual men. *J. Psychosom. Res.* **36**(7), 635–650.

Goodkin, K., Fuchs, I., Feaster, D., Leeka, J., and Rishel, D. (1992b). Life stressors and coping style are associated with immune measures in HIV-1 infection—A preliminary report. *Int. J. Psychiatry Med.* **22**(2), 155–172.

Goodkin, K., Feaster, D. J., Asthana, D., Blaney, N. T., Kumar, M., Baldewicz, T., Tuttle, R. S., Maher, K. J., Baum, M. K., Shapshak, P., and Fletcher, M. A. (1998). A bereavement support group intervention is longitudinally associated with salutary effects on the CD4 cell count and number of physician visits. *Clin. Diagnostic Laboratory Immunol.* **5**(3), 382–391.

Goodwin, J. S., Hunt, W. C., Key, C. R., and Samet, J. M. (1987). The effect of marital status on stage, treatment, and survival of cancer patients. *JAMA* **258**(21), 3125–3130.

Greenwald, H. P., Polissar, N. L., and Dayal, H. H. (1996). Race, socioeconomic status and survival in three female cancers. *Ethnicity and Health* **1**(1), 65–75.

Greer, S. (1991). Psychological response to cancer and survival. *Psychol. Med.* **21**(1), 43–49.

Greer, S., and Watson, M. (1985). Towards a psychobiological model of cancer: Psychological considerations. *Soc. Sci. Med.* **20**(8), 773–777.

Greer, S., Morris, T., and Pettingale, K. W. (1979). Psychological response to breast cancer: Effect on outcome. *Lancet* **2**(8146), 785–787.

Greer, S., Morris, T., and Pettingale, K. W. (1994). *In* Psychological response to breast cancer: Effect on outcome. *In* "Psychosocial Processes in Health: A Reader" (A. Steptoe and J. Wardle, eds.), Cambridge University Press, U.K.

Groer, M. W., Humenick, S., and Hill, P. D. (1994). Characterizations and psychoneuroimmunologic implications of secretory immunoglobulin A and cortisol in preterm and term breast milk. *J. Perinatal Neonatal Nursing* **7**(4), 42–51.

Grossarth-Maticek, R., Schmidt, P., Vetter, H., and Arndt, S. (1984). Psychotherapy research in oncology. *In* "Health Care and Human Behavior" (A. Steptoe and A. Mattews, eds.), pp. 325–341. Academic Press, London.

Gruber, B. L., Hall, N. R., Hersh, S. P., and Dubois, P. (1988). Immune system and psychological changes in metastatic cancer patients using relaxation and guided imagery: A pilot study. *Scand. J. Behav. Ther.* **17**(1), 25–46.

Herbert, T. B., and Cohen, S. (1993). Stress and immunity in humans: A meta-analytic review. *Psychosom. Med.* **55**(4), 364–379.

Hislop, T. G., Waxler, N. E., Coldman, A. J., Elwood, J. M., and Kan, L. (1987). The prognostic significance of psychosocial factors in women with breast cancer. *J. Chronic Dis.* **40**(7), 729–735.

House, J. S., Landis, K. R., and Umberson, D. (1988). Social relationships and health. *Science* **241**(4186), 540–545.

Ilnyckyj, A., Farber, J., Cheang, M., and Weinerman, B. (1994). A randomized controlled trial of psychotherapeutic intervention in cancer patients. *Annals of the Royal College of Physicians and Surgeons of Canada,* **27**(2), 93–96.

Ironson, G., Lapierre, A., Antoni, M., O'Hearn, P., Schneiderman, N., Klimas, N., and Fletcher, M. A. (1990). Changes in immune and psychological measures as a function of anticipation and reaction to news of HIV-1 antibody status. *Psychosom. Med.* **52**(3), 247–270.

Ironson, G., Friedman, A., Klimas, N., Antoni, M., Fletcher, M.A., LaPerriere, A., Simoneau, J., and Schneiderman, N. (1994a). Distress, denial, and low adherence to behavioral interventions predict faster disease progression in gay men infected with human immunodeficiency virus. *Int. J. Behav. Med.* **1,** 90–105.

Ironson, G., Schneiderman, H., Kumar, M., Antoni, M., LaPerriere, A., Klimas, N., and Fletcher, M. A. (1994b). Psychosocial stress, endocrine and immune response in HIV-1 disease. *Homeostasis in Health and Disease* **35**(3), 137–148.

Irwin, M., Daniels, M., Smith, T. L., Bloom, E., and Weiner, H. (1987). Impaired natural killer cell activity during bereavement. *Brain Behav. Immun.* **1,** 98–104.

Ito, I., Kimura, T., Wakasugi, E., Takeda, T., Kobayashi, T., Shimano, T., Kubota, Y., Makino, Y., Azuma, C., and Saji, F. *et al.* (1995). Expression of the oxytocin receptor in clinical human breast cancer tissues. *Adv. Exp. Med. Biol.* **395**, 555–556.

Kalichman, S. C. (1998). "Understanding AIDS. Advances in Treatment and Research." American Psychological Association, Washington, D.C.

Kaplan, K. H., Goldenberg, D. L., and Galvin-Nadeau, M. (1993). The impact of a meditation-based stress reduction program on fibromyalgia. *Gen. Hospital Psychiatry* **15**(5), 284–289.

Kemeny, M. E., and Gruenewald, T. L. (1999). Psychoneuroimmunology update. *Semin. Gastrointest. Dis.* **10**(1), 20–29.

Kemeny, M. E., Cohen, F., Zegans, L. S., and Conant, M. A. (1989). Psychological and immunological predictors of genital herpes recurrence. *Psychosom. Med.* **51**(2), 195–208.

Kemeny, M. E., Weiner, H., Taylor, S. E., Schneider, S., Visscher, B., and Fahey, J. L. (1994). Repeated bereavement, depressed mood, and immune parameters in HIV seropositive and seronegative gay men. *Health Psychol.* **13**(1), 14–24.

Kessler, R. C., Foster, C., Joseph, J., Ostrow, D., Wortman, C., Phair, J., and Chmiel, J. (1991). Stressful life events and symptom onset in HIV infection. *Am. J. Psychiatry* **148**(6), 733–738.

Kiecolt-Glaser, J. K., and Glaser, R. (1992). Psychoneuroimmunology: Can psychological interventions modulate immunity? Special issue—Behavioral medicine: An update for the 1990s. *J. Cons. Clin. Psychol.* **60**(4), 569–575.

Kiecolt-Glaser, J. K., Garner, W., Speicher, C. E., Penn, G., and Glaser, R. (1984a). Psychosocial modifiers of immunocompetence in medical students. *Psychosom. Med.* **46**, 7–14.

Kiecolt-Glaser, J. K., Ricker, D., George, J., Messick, G., Speicher, C. E., Garner, W., and Glaser, R. (1984b). Urinary cortisol levels, cellular immunocompetency, and loneliness in psychiatric inpatients. *Psychosom. Med.* **46**(1), 15–23.

Kiecolt-Glaser, J. K., Glaser, R., Williger, D., Stout, J., Messick, G., Sheppard, S., Ricker, D., Romisher, S. C., Briner, W., and Bonnell, G. (1985). Psychosocial enhancement of immunocompetence in a geriatric population. *Health Psychol.* **4**(1), 25–41.

Kiecolt-Glaser, J. K., Fisher, L. D., Ogrocki, P., Stout, J. C., Speicher, C. E., and Glaser, R. (1987a). Marital quality, marital disruption, and immune function. *Psychosom. Med.* **49**(1), 13–34.

Kiecolt-Glaser, J. K., Glaser, R., Shuttleworth, E. C., Dyer, C. S., Ogrocki, P., and Speicher, C. E. (1987b). Chronic stress and immunity in family caregivers of Alzheimer's disease victims. *Psychosom. Med.* **49**(5), 523–535.

Kiecolt-Glaser, J. K., Dura, J. R., Speicher, C. E., Trask, O. J., and Glaser, R. (1991). Spousal caregivers of dementia victims: Longitudinal changes in immunity and health. *Psychosom. Med.* **53**(4), 345–362.

Kneier, A. W., and Temoshok, L. (1984). Repressive coping reactions in patients with malignant melanoma as compared to cardiovascular disease patients. *J. Psychosom. Res.* **28**(2), 145–155.

Koehler, T. (1985). Stress and rheumatoid arthritis: A survey of empirical evidence in human and animal studies. *J. Psychosom. Res.* **29**, 655–663.

Koenig, H. G., Cohen, H. J., George, L. K., Hays, J. C., Larson, D. B., and Blazer, D. G. (1997). Attendance at religious services, interleukin-6, and other biological parameters of immune function in older adults. *Int. J. Psychiatry Med.* **27**(3), 233–250.

Koopman, C., Hermanson, K., Diamond, S., Angell, K., and Spiegel, D. (1998). Social support, life stress, pain and emotional adjustment to advanced breast cancer. *Psycho-Oncology* **7**, 101–111.

Landmann, R. M., Muller, F. B., Perini, C., Wesp, M., Erne, P., and Buhler, F. R. (1984). Changes of immunoregulatory cells induced by psychological and physical stress: relationship to plasma catecholamines. *Clin. Exp. Immunol.* **58**(1), 127–135.

Lane, R. D., Reiman, E. M., Bradley, M. M., Lang, P. J., Ahern, G. L., Davidson, R. J., and Schwartz, G. E. (1997). Neuroanatomical correlates of pleasant and unpleasant emotion. *Neuropsychologia* **35**(11), 1437–1444.

Lazarus, R. S., and Folkman, S. (1984). "Stress, Appraisal and Coping." Springer-Verlag, New York.

Lechin, F., van der Dijs, B., Vitelli-Florez, G., Lechin-Baez, S., Azocar, J., Cabrera, A., Lechin, A., Jara, H., Lechin, M., Gomez, F., *et al.* (1990). Psychoneuroendocrinological and immunological parameters in cancer patients: Involvement of stress and depression. *Psychoneuroendocrinology* **15**(5-6), 435–451.

Lekander, M., Furst, C. J., Rotstein, S., Blomgren, H., and Fredrikson, M. (1996). Social support and immune status during and after chemotherapy for breast cancer. *Acta Oncologica* **35**(1), 31–37.

Leserman, J., Jackson, E. D., Petitto, J. M., Golden, R. N., Silva, S. G., Perkins, D. O., Cai, J., Folds, J. D., and Evans, D. L. (1999). Progression to AIDS: The effects of stress, depressive symptoms, and social support. *Psychosom. Med.* **61**(3), 397–406.

Levine, S., Coe, C., and Wiener, S. G. (1989). Psychoneuroendocrinology of Stress: A Psychobiological Perspective. *In* "Psychoendocrinology" (F. R. Brush and S. Levine, eds.), pp. 341–377, Academic Press, Inc., San Diego.

Levy, S. M. (1990). Perceived social support and tumor estrogen/progesterone receptor status as predictors of natural killer cell activity in breast cancer patients. *Psychosom. Med.* **52**(1), 73–85.

Levy, S. M., Herberman, R. B., Lippman, M., D'Angelo, T., and Lee, J. (1991). Immunological and psychosocial predictors of disease recurrence in patients with early-stage breast cancer. *Behav. Med.* **17**(2), 67–75.

Linn, B. S., Linn, M. W., and Kilmas, N. G. (1988). Effects of psychosocial stress on surgical outcomes. *Psychosom. Med.* **50**, 230–244.

Linn, M. W., Linn, B. S., and Harris, R. (1982). Effects of counseling for late stage cancer patients. *Cancer* **49**(5), 1048–1055.

Loscalzo, M. (1998). Interventions. *In* "Psycho-Oncology" (J. C. Holland, ed.). OUP, New York.

Lyketsos, C. G., Hoover, D. R., Guccione, M., Senterfitt, W., Dew, M. A., Wesch, J., VanRaden, M. J., Treisman, G. J., and Morgenstern, H. (1993). Depressive symptoms as predictors of medical outcomes in HIV infection. Multicenter AIDS Cohort Study. *JAMA* **270**(21), 2563–2567.

Marshall, J. R., and Funch, D. P. (1983). Social environment and breast cancer: A cohort analysis of patient survival. *Cancer* **52**, 1546–1550.

Mastorakos, G., and Ilias, I. (2000). Maternal hypothalamic-pituitary-adrenal axis in pregnancy and the postpartum period, Postpartum-related disorders. *Ann. N.Y. Acad. Sci.* **900**, 95–106.

Matthews, D. A., McCullough, M. E., Larson, D. B., Koenig, H. G., Swyers, J. P., and Milano, M. G. (1998). Religious commitment and health status: A review of the research and implications for family medicine. *Arch. Family Med.* **7**(2), 118–124.

Maunsell, E. B., Jacques, B., and Deschenes, L. (1993). Social support and survival among women with breast cancer. *Cancer* **76**(4), 631–637.

McDaniel, J. S., Musselman, D. L., Porter, M. R., Reed, D. A., and Nemeroff, C. B. (1995). Depression in patients with cancer. Diagnosis, biology, and treatment. *Arch. Gen. Psychiatry* **52**(2), 89–99.

McEwen, B. S. (1998). Stress, adaptation, and disease. Allostasis and allostatic load. *Ann. N.Y. Acad. Sci.* **840**, 33–44.

Miller, G. E., Kemeny, M. E., Taylor, S. E., Cole, S. W., and Visscher, B. R. (1997). Social relationships and immune processes in HIV seropositive gay and bisexual men. *Ann. Behav. Med.* **19**(2), 139–151.

Morris, T., Pettingale, K. W., and Haybittle, J. (1992). Psychological response to cancer diagnosis and disease outcome in patients with breast cancer and lymphoma. *Psycho-Oncology* **1**, 105–114.

Moss, H., Bose, S., Wolters, P., and Brouwers, P. (1998). A preliminary study of factors associated with psychological adjustment and disease course in school-age children infected with the human immunodeficiency virus. *J. Dev. Behav. Pediatr.* **19**(1), 18–25.

Mulder, C. L. (1994). Psychosocial correlates and the effects of behavioral interventions on the course of human immunodeficiency virus infection in homosexual men. *Patient Education & Counselling* **23**(3), 237–247.

Mulder, C. L., Antoni, M. H., Emmelkamp, P. M., Veugelers, P. J., Sandfort, T. G., van de Vijver, F. A., and de Vries, M. J. (1995). Psychosocial group intervention and the rate of decline of immunological parameters in asymptomatic HIV-infected homosexual men. *Psychother. Psychosom.* **63**(3-4), 185–192.

Murrell, T. G. (1995). The potential for oxytocin (OT) to prevent breast cancer: A hypothesis. *Breast Cancer Res. Treat* **35**(2), 225–229.

Naliboff, B., Benton, D., Solomon, G., Morley, J., Fahey, J., Bloom, E., Makinodan, T., and Gilmore, S. (1991). Immunological change in young and old adults during brief laboratory stress. *Psychosomatic Medicine* **53**, 121–132.

Patterson, T. L., Shaw, W. S., Semple, S. J., Cherner, M., *et al.* (1996). Relationship of psychosocial factors to HIV disease progression. *Ann. Behav. Med.* **18**(1), 30–39.

Pennebaker, J. W., and Chew, C. H. (1985). Behavioral inhibition and electrodermal activity during deception. *J. Personality and Soc. Psychol.* **49**(5), 1427–1433.

Pennebaker, J. W., Hughes, C. F., and O'Heeron, R. C. (1987). The psychophysiology of confession: Linking inhibitory and psychosomatic processes. *J. Personality Soc. Psychol.* **52**(4), 781–793.

Pennebaker, J. W., Kiecolt-Glaser, J. K., and Glaser, R. (1988). Disclosure of traumas and immune function: Health implications for psychotherapy. *J. Cons. Clin. Psychol.* **56**(2), 239–245.

Penninx, B. W. J. H., van Tilburg, T., Boeke, A. J. P., Deeg, D. J. H., Kriegsman, D. M. W., and van Eijk, J. T. M. (1998). Effects of social support and personal coping resources on depressive symptoms: Different for various chronic diseases? *Health Psychol.* **17**(6), 551–558.

Perry, S., Fishman, B., Jacobsberg, L., and Frances, A. (1992). Relationships over 1 year between lymphocyte subsets and psychosocial variables among adults with infection by human immunodeficiency virus. *Arch. Gen. Psychiatry* **49**(5), 396–401.

Persson, L., Gullberg, B., Hanson, B. S., Moestrup, T., and Ostergren, P. O. (1994). HIV infection: Social network, social support, and CD4 lymphocyte values in infected homosexual men in Malmo, Sweden. *J. Epidemiol. Community Health* **48**(6), 580–585.

Petrie, K. J., Booth, R. J., and Pennebaker, J. W. (1998). The immunological effects of thought suppression. *J. Personality Soc. Psychol.* **75**(5), 1264–1272.

Pettingale, K. W., Morris, T., Greer, S., and Haybittle, J. (1985). Mental attitudes to cancer: An additional prognostic factor. *Lancet* **1**, 750.

Ramirez, A. J., Craig, T. K. J., Watson, J. P., Fentiman, I. S., North, W. R. S., and Rubens, R. D. (1989). Stress and relapse of breast cancer. *Br. Med. J.* **298**, 291–293.

Richardson, J. L., Zarnegar, Z., Bisno, B., and Levine, A. (1990). Psychosocial status at initiation of cancer treatment and survival. *J. Psychosom. Res.* **34**(2), 189–201.

Rowse, G. L., Weinberg, J., Bellward, G. D., and Emerman, J. T. (1992). Endocrine mediation of psychosocial stressor effects on mouse mammary tumor growth. *Cancer Lett.* **65**, 85–93.

Sapolsky, R. M. (1989). Hypercortisolism among socially subordinate wild baboons originates at CNS level. *Arch. Gen. Psychiatry* **46**, 1047–1051.

Sapolsky, R. M. (1995). Social subordinance as a marker of hypercortisolism. Some unexpected subtleties. *Ann. N.Y. Acad. Sci.* **771**, 626–639.

Sapolsky, R. M., and Spencer, E. M. (1997). Insulin-like growth factor I is suppressed in socially subordinate male baboons. *Am. J. Physiol.* **273**(4 Pt. 2), R1346–R1351.

Sapolsky, R. M., Alberts, S. C., and Altmann, J. (1997). Hypercortisolism associated with social subordinance or social isolation among wild baboons. *Arch. Gen. Psychiatry* **54**(12), 1137–1143.

Schedlowski, M., Jacobs, R., Stratmann, G., Richter, S., Hadicke, A., Tewes, U., Wagner, T. O., and Schmidt, R. E. (1993). Changes of natural killer cells during acute psychological stress. *J. Clin. Immunol.* **13**(2), 119–126.

Schlesinger, M., and Yodfat, Y. (1991). The impact of stressful life events on natural killer cells. 2nd International Society for the Investigation of Stress Conference: Stress, Immunity and AIDS (1989, Athens, Greece). *Stress Medicine* **7**(1), 53–60.

Schwartz, C. E., Foley, F. W., Rao, S. M., Bernardin, L. J., Lee, H., and Genderson, M. W. (1999). Stress and course of disease in multiple sclerosis. *Behav. Med.* **25**(3), 110–116.

Seeman, T. E., Berkman, L. F., Blazer, D., and Rowe, J. W. (1994). Social ties and support and neuroendocrine function: The MacArthur studies of successful aging. *Ann. Behav. Med.* **16**(2), 95–106.

Sephton, S. E., Sapolsky, R. M., Kraemer, H., and Spiegel, D. (2000). Diurnal cortisol rhythm as a predictor of breast cancer survival. *J. Nat. Cancer Inst.* **92**(12), 994–1000.

Shavit, Y., Terman, G. W., Martin, F. C., Lewis, J. W., Liebeskind, J. C., and Gale, R. P. (1985). Stress, opioid peptides, the immune system, and cancer. *J. Immunol.* **135**, 834s–837s.

Sheridan, J. F., Dobbs, C., Jung, J., Chu, X., Konstantinos, A., Padgett, D., and Glaser, R. (1998). Stress-induced neuroendocrine modulation of viral pathogenesis and immunity. *Ann. N.Y. Acad. Sci.* **840**, 803–808.

Sherman, B. F., Bonanno, G. A., Wiener, L. S., and Battles, H. B. (2000). When children tell their friends they have AIDS: Possible consequences for psychological well-being and disease progression. *Psychosom. Med.* **62**(2), 238–247.

Smyth, J. M., Stone, A.A., Hurewitz, A., and Kaell, A. (1999). Effects of writing about stressful experiences on symptom reduction in patients with asthma or rheumatoid arthritis: A randomised trial. *JAMA* **281**, 1304–1309.

Snyder, B. K., Roghmann, K. J., and Sigal, L. H. (1993). Stress and psychosocial factors: Effects on primary cellular immune response. *J. Behav. Med.* **16**(2), 143–161.

Solano, L., Costa, M., Salvati, S., Coda, R., Aiuti, F., Mezzaroma, I., and Bertini, I. (1993). Psychosocial factors and clinical evolution in HIV-1 infection: A longitudinal study. *J. Psychosom. Res.* **37**(1), 39–51.

Spiegel, D. (1992). Effects of psychosocial support on patients with metastatic breast cancer. *J. Psychosoc. Oncol.* **10**(2), 113–120.

Spiegel, D. (1993). Psychosocial intervention in cancer. *J. Nat. Cancer Inst.* **85**(5), 1198–1205.

Spiegel, D. (1995). How do you feel about cancer now?—Survival and psychosocial support. *Public Health Rep.* **110**(3), 298–300.

Spiegel, D. (1996). Cancer and depression. *Br. J. Psychiatry* **168**(Suppl. 30), 109–116.

Spiegel, D. (1999). Healing words. Emotional expression and disease outcome. *JAMA* **281**(14), 1328–1329.

Spiegel, D. (2001). Mind matters. Coping and cancer progression. *J. Psychosom. Res.* **50**(5), 287–290.

Spiegel, D., Bloom, J. R., Kraemer, H. C., and Gottheil, E. (1989). Effect of psychosocial treatment on survival of patients with metastatic breast cancer. *Lancet* **2**, 888–891.

Spiegel, D., Sephton, S. E., Terr, A. I., and Stites, D. P. (1998). Effects of psychosocial treatment in prolonging cancer survival may be mediated by neuroimmune pathways. *Ann. N.Y. Acad. Sci.* **840**, 674–683.

Stein, M., Miller, A. H., and Trestman, R. L. (1991). Depression, the immune system, and health and illness. Findings in search of meaning. *Arch. Gen. Psychiatry* **48**(2), 171–177.

Taylor, S. E. (1983). Adjustment to threatening events: A theory of cognitive adaptation. *Am. Psychol.* **38**, 1161–1173.

Taylor, S. E., Kemeny, M. E., Reed, G. M., Bower, J. E., and Gruenewald, T. L. (2000). Psychological resources, positive illusions, and health. *Am. Psychol.* **55**(1), 99–109.

Temoshok, L., Heller, B. W., Sagebiel, R. W., Blois, M. S., Sweet, D. M., DiClemente, R. J., and Gold, M. L. (1985). The relationship of psychosocial factors to prognostic indicators in cutaneous malignant melanoma. *J. Psychosom. Res.* **29**(2), 139–153.

Theorell, T., Orth-Gomer, K., and Eneroth, P. (1990). Slow-reacting immunoglobulin in relation to social support and changes in job strain: A preliminary note. *Psychosom. Med.* **52**(5), 511–516.

Theorell, T., Blomkvist, V., Jonsson, H., Schulman, S., Berntorp, E., and Stigendal, L. (1995). Social support and the development of immune function in human immunodeficiency virus infection. *Psychosom. Med.* **57**, 32–36.

Thomas, P. D., Goodwin, J. M., and Goodwin, J. S. (1985). Effect of social support on stress-related changes in cholesterol level, uric acid level, and immune function in an elderly sample. *Am. J. Psychiatry* **142**, 735–737.

Thornton, S., Troop, M., Burgess, A. P., Button, J., Goodall, R., Flynn, R., Gazzard, B. G., Catalan, J., and Easterbrook, P. J. (2000). The relationship of psychological variables and disease progression among long-term HIV-infected men. *Int. J. STD and AIDS* **11**(11), 734–742.

Tschuschke, V., Hertenstein, B., Arnold, R., Bunjes, D., Denzinger, R., and Kaechele, H. (2001). Associations between coping and survival time of adult leukemia patients receiving allogeneic bone marrow transplantation: Results of a prospective study. *J. Psychosom. Res.* **50**(5), 277–285.

Turner-Cobb, J. M., and Steptoe, A. (1996). Psychosocial stress and susceptibility to upper respiratory tract illness in an adult population sample. *Psychosom. Med.* **58**, 404–412.

Turner-Cobb, J. M., Steptoe, A., Perry, L., and Axford, J. (1998). Adjustment in patients with rheumatoid arthritis and their children. *J. Rheumatol.* **25**, 565–571.

Turner-Cobb, J. M., Sephton, S. E., Koopman, C., Blake-Mortimer, J., and Spiegel, D. (2000). Social support and salivary cortisol in women with metastatic breast cancer. *Psychosom. Med.* **62**(3), 337–345.

Uchino, B. N., Cacioppo, J. T., and Kiecolt-Glaser, J. K. (1996). The relationship between social support and physiological processes: A review with emphasis on underlying mechanisms and implications for health. *Psychol. Bull.* **119**(3), 488–531.

Uvnas-Moberg, K. (1997). Physiological and endocrine effects of social contact. *Ann. N.Y. Acad. Sci.* **807**, 146–163.

van der Pompe, G., Duivenvoorden, H. J., Antoni, M. H., Visser, A., and Heijnen, C. J. (1997). Effectiveness of a short-term group psychotherapy program on endocrine and immune function in breast cancer patients: An exploratory study. *J. Psychosom. Res.* **42**(5), 453–466.

Van Rood, Y. R., Bogaards, M., Goulmy, E., and Houwelingen, H. C. (1993). The effects of stress and relaxation on the *in vitro* immune response in man: A meta-analytic study. *J. Behav. Med.* **16**(2), 163–181.

Vassend, O., and Eskild, A. (1998). Psychological distress, coping, and disease progression in HIV-positive homosexual men. *J. Health Psychol.* **3**(2), 243–257.

Vedhara, K., Nott, K. H., Bradbeer, C. S., Davidson, E. A., Ong, E. L., Snow, M. H., Palmer, D., and Nayagam, A. T. (1997). Greater emotional distress is associated with accelerated $CD4^{+}$ cell decline in HIV infection. *J. Psychosom. Res.* **42**(4), 379–390.

Vedhara, K., Schifitto, G., and McDermott, M. (1999). Disease progression in HIV-positive women with moderate to severe immunosuppression: The role of depression. Dana Consortium on Therapy for HIV Dementia and Related Cognitive Disorders. *Behav. Med.* **25**(1), 43–47.

Wadhwa, P. D., Dunkel-Schetter, C., Chicz-DeMet, A., Porto, M., and Sandman, C. A. (1996). Prenatal psychosocial factors and the neuroendocrine axis in human pregnancy. *Psychosom. Med.* **58**(5), 432–446.

Watson, M., Haviland, J. S., Greer, S., Davidson, J., and Bliss, J. M. (1999). Influence of psychological response on survival in breast cancer: A population-based cohort study. *Lancet* **354**(9187), 1331–1336.

Waxler-Morrison, N., Hislop, T. G., Mears, B., and Kan, L. (1991). Effects of social relationships on survival for women with breast cancer: A prospective study. *Soc. Sci. Med.* **33**, 177–183.

Wilder, R. L. (1998). Hormones, pregnancy and autoimmune diseases. *Ann. N.Y. Acad. Sci.* **840**, 45–50.

Zautra, A. J., Hoffman, J., Potter, P., Matt, K. S., Yocum, D., and Castro, L. (1997). Examination of changes in interpersonal stress as a factor in disease exacerbations among women with rheumatoid arthritis. *Ann. Behav. Med.* **19**(3), 279–286.

Zhao, X. Y., Malloy, P. J., Krishnan, A. V., Swami, S., Navone, N. M., Peehl, D. M., and Feldman, D. (2000). Glucocorticoids can promote androgen-independent growth of prostate cancer cells through a mutated androgen receptor. *Nature Med.* **6**(6), 703–706.

THE ROLE OF PSYCHOLOGICAL INTERVENTION IN MODULATING ASPECTS OF IMMUNE FUNCTION IN RELATION TO HEALTH AND WELL-BEING

J. H. Gruzelier

Department of Cognitive Neuroscience and Behavior
Imperial College London
London W6 8RF, United Kingdom

I. Introduction

An optimistic basis for the emergent field of psychoneuroimmunology (PNI) has stemmed from the wealth of evidence of the importance of psychological influences on immunity in addition to well being and health. As this is perhaps best exemplified by the compromise of immunity due to anxiety, depression, and stress (Ader *et al.*, 2001; O'Leary, 1990; Bennett Herbert and Cohen, 1993; Leonard and Miller, 1995; Evans *et al.*, 2000), it should follow that benefits for immunity and health will accrue from psychological interventions that reduce stress and anxiety and alleviate depression. However, acknowledgment has been tempered, especially in neuroscience, through the complexity of brain–immune interactions which makes it difficult to predict with any accuracy patterns of immune up- and downregulation across immune parameters. Accordingly if psychological approaches are to win acceptance for favorably biasing immunity, evidence of improvements in health, and not simply well-being, is essential for validation. Such evidence may in turn help to unravel the poorly understood complexities of immune function, though as this review will show, these principles alone are unable to reconcile the complexity of a highly redundant, dynamic system full of checks and balances, redistribution, and migration.

This is still an early, data gathering stage of investigation, and, in the absence of models allowing predictions to be made which encompass the functional complexity of the immune system, the hypothesized outcomes of psychological interventions on immune parameters have been unitary and unidirectional, that is, predicting upregulation of the single or multiple measures of immunity assayed. Measures typically have included numerative measures of white blood cells, including neutrophils, lymphocyte subsets, and natural killer (NK) cells, and functional assessment of NK cell activity, by target cell killing assay and of lymphocyte competence through blasto-genesis in response to mitogens. Cortisol from saliva or plasma has been obtained as a widely used measure of the stress–neuroendocrine system indexing stimulation of the hypothalamic-pituitary-adrenal (HPA) system. Secretory immunoglobulin A (S-IgA), measured in saliva, has provided an indication of mucosal immune regulation, while attempts at assessment of immune cytokine activity is beginning to shed light on the functional balance of the immune system.

Here a review is provided of the effects of psychological interventions on immunity. The focus has been given to approaches that explicitly or implic-itly relate to neurophysiology such as those that involve relaxation with and without imagery training, sometimes accompanied by neurophysiological monitoring. Excluded are mixed intervention packages where the emphasis has been predominantly on cognitive stress management, those solely using cognitive approaches such as disclosure of worries, and conditioning studies.

II. Relaxation Training

A. SINGLE INTERVENTIONS

1. *Healthy Subjects*

Considering first single-intervention approaches, Kiecolt-Glaser *et al.* (1985) compared the effect of progressive muscle relaxation with social support or no-contact on elderly participants living alone in a residential facility. The intervention groups were seen three times a week for a month. An improvement in self-reports of distress was found in the relaxation group and this was accompanied by an increase in NK cell activity and a decrease in antibody levels to the herpes simplex virus (HSV), implying reduced activity of latent virus. Importantly, despite a lapse in relaxation practice the decrease in antibody levels persisted 1 month later. In healthy subjects McGrady *et al.* (1987a,b) found that both biofeedback training and flotation in a tank with restricted environmental stimulation (REST) was successful in lowering levels of plasma cortisol, while Green *et al.* (1988) examined the effect of daily relaxation practice for 3 weeks on immunoglobulins and found that relaxation increased secretion rate of S-IgA as well as serum levels of IgA, IgG, and IgM in comparison with controls.

2. *Patients*

S-IgA was examined in a mixed group of eight oncology patients in the second of a two-phase meditative music relaxation and music improvisation program (Burns *et al.*, 2001). Controlling for volume changes S-IgA rose during the relaxation session, though there was no change in salivary cortisol. Improvisation had no influences on the immune measures, and, while there was a significant reduction in tension with music relaxation, this and other mood ratings did not correlate with the immune changes.

B. MIXED INTERVENTIONS

1. *Healthy Subjects*

Turning to mixed interventions, approaches of relatively short duration were examined in two studies where healthy subjects were evaluated for levels of stress. The first selected highly stressed subjects on the basis of both the social readjustment scale of Holmes and Rahe (1967) and low neutrophil activation (Peavey *et al.*, 1985). Despite a relatively brief intervention consisting of 2 weeks of daily home practice with physical and mental relaxation tape recordings, together with electromyography (EMG) and skin temperature biofeedback learned to criterion, the phagocytic activity of neutrophils increased when compared with high-stress control participants, while there were improvements in levels of anxiety and coping ability. White cell blood counts were unaffected.

The second study included medical students selected for high or low psychometrically measured stress and who were examined during the summer vacation to avoid the pressures of adaptation to medical school (McGrady *et al.*, 1992). The training schedule consisted of only 4 sessions of EMG and skin temperature biofeedback which were combined with 4 weekly group sessions of relaxed breathing, progressive muscle relaxation, and autogenic/ imagery training without home practice with audiocassettes. There was evidence of increased lymphocyte response to stimulation by mitogens, indicative of enhanced immune competence compared with a control group. There was also a reduction in neutrophils, which was responsible for lowering the total white blood cell count, as lymphocyte number was invariant. Both changes were positively associated with a reduction in forehead muscle tension. No effect was found for cortisol. A post hoc analysis of individual differences in level of stress revealed that the changes were more pronounced in low-stress subjects, which indicates that the four-session intervention schedule was not long enough to have impact on the highly stressed subjects. Notwithstanding, the reduction in neutrophils was interpreted as compatible with stress reduction, because neutrophil levels increase with bacterial infection, inflammatory processes, and stress.

2. *Patients*

Mixed intervention approaches have also been studied in patients. Patients with stage 1 breast cancer following radial mastectomy were assigned to either an immediate or a delayed intervention group, who began treatment after 25 weeks (Gruber *et al.*, 1993). The intervention package over 18 months consisted of EMG biofeedback, progressive muscle relaxation, letting-go relaxation, and guided immune imagery including twice-daily home practice. There was a favorable outcome for immunity on a range of measures including NK cell activity, peripheral blood lymphocyte count, concanavalin A, and mixed lymphocyte responsiveness, but immunoglobulin measures on the whole were not altered. The wide-ranging benefits for immunity did not translate into psychological benefits, with only anxiety showing a reduction among a range of questionnaire measures. In a second study (Taylor, 1995) the patients were men with HIV infection who were asymptomatic and had CD4 counts below $400/\mu l$ indicating a degree of immunosuppression; of all immune parameters the CD4 lymphocyte subset best mirrors symptomatic progression to AIDS. For 20 weeks patients were given biweekly sessions of biofeedback/progressive muscle relaxation, hypnosis, and meditation and were compared with a nonintervention control group. Compared with controls the intervention was successful in elevating CD4 counts and this was accompanied by improvement in anxiety, mood, and self-esteem.

C. LABORATORY STRESSOR

One attempt has been made to validate the effects of relaxation training in moderating the effects of a laboratory stressor. Johnson *et al.* (1996)

examined 24 healthy participants for their response to a laboratory stressor after randomization to relaxation training or to a control group collecting a range of immunological and psychological parameters along with blood pressure before and following 3 weeks of daily practice with tapes of progressive muscle relaxation alternating with hypnotic relaxation. Initial assessment included hypnotic susceptibility with the Stanford Clinical hypnotic induction scale followed by the Creative Imagination Scale (CIS) which is strongly correlated with hypnotic susceptibility. On the second occasion the hypnotic induction was repeated for the relaxation group and both groups were then exposed to a role-play stressor, after which the various measures were repeated.

Mood assessment with the Hospital Anxiety and Depression Scale (HADS) showed that the relaxation group showed a reduction in anxiety while the control group showed an increase in depression. After training, the Profile of Mood States (POMS) disclosed that the relaxation group was more composed, clearheaded, elated, confident, and energetic and they were more relaxed prior to the role play. While changes in mood clearly advantaged the relaxation group a mixed pattern of results was observed among the immune parameters assayed from serum and urine. Consistent with the changes in mood, prior to the stressor both groups showed a decline in mitogen response to phytohaemagglutinin (PHA), an effect which was stronger in the relaxation group. Inconsistently in the relaxation group IgA levels increased and cytokine IL-1 levels decreased while only in the control group did NK cell activity decline. More congruent with mood changes was the response to the stressor. The relaxation group showed increased responsiveness to PHA whereas responsiveness declined in the controls; the relaxation group showed an increase in IL-1; diastolic blood pressure also fell in the relaxation group. Hypnotizability assessed by the CIS moderated the results by correlating with the increase in IgA following relaxation practice, while there were differential correlations with hypnotizability in the experimental and control groups in the poststress change in IL-1—positive in the relaxation group and negative in the control group.

D. SUMMARY

There has on the whole been a successful outcome on immunity and well-being from a diverse range of relaxation interventions, whether applied singly or as part of a mixed program, perhaps with the exception of some of the prestressor effects on immunity of Johnston *et al.* (1999). A diversity of subjects has been involved including vulnerable groups such as geriatrics, hypertensives, and oncology patients, as well as healthy participants evaluated or exposed to stress. Typically where more than single measures of immune function and well-being have been assessed a patterned response has been obtained with some measures "improving" and others remaining unchanged, or as in the case of Johnston *et al.* (1999) some showing down-regulation. No studies have examined beneficial influences on health.

III. Guided Imagery of the Immune System

A. EFFECTS ON IMMUNITY AND WELL-BEING

Training in guided imagery about strengthening immune function has been a popular technique incorporated with relaxation schedules and with the narrative often adapted to the nature of the patient's illness. Metaphorical images of a healthy immune system in a state of combat with invading viruses and germs may be invoked once the participant has been placed in a state of deep relaxation. In children the dynamics have been expressed through a puppet play (Olness *et al.,* 1989). The procedure was inspired largely on pragmatic grounds by work with oncology patients (Simonton *et al.,* 1978). Inherent in the approach will be anxiety desensitization based on classical Pavlovian conditioning principles, for by virtue of evoking illness-related images in a state of deep relaxation there is a process of desensitization to the stress of the illness and its impact on the body. Furthermore, the active cognitive engagement that is necessarily invoked by eliciting targeted imagery about immune system dynamics, will involve different neurophysiological processes than relaxation per se, with implications for brain–immune pathways. Conceivably the more active cognitive attitude that is encouraged may also contribute to differential effects on mood compared with the more passive imagery that accompanies relaxation training, a thesis developed later. As will be seen the outcome has on the whole been promising.

1. *Children*

The most well-known early study is that of Olness *et al.* (1989) who demonstrated the importance of the specificity of immune imagery in increasing S-IgA in children. Children were randomized to three groups comparing a single session of self-hypnosis relaxation combined with nonspecific immune-related suggestions, with self-hypnosis which included specific suggestions about increasing salivary immunoglobulins and with a control group engaged in conversation. All had an initial orientation session with parents including a general relaxation tape to practice at home, a puppet play depicting basic immune system components, a saliva sample, and a hypnotic susceptibility assessment with the Stanford Children's Hypnotic

Susceptibility Scales. They returned 2 weeks later when further samples were taken at the beginning and end of a 25-min session. S-IgA levels remained stable between sessions but increased significantly in the group given specific imagery while remaining unchanged both in the group with nonspecific imagery and in the control group. Serum IgG levels were not altered. Hypnotic susceptibility was not associated with the effect, nor were ratings of interest and alertness. An advantage for imagery training was replicated by Hewson-Bower and Drummond (1996) with 45 healthy children and 45 children with recurrent upper respiratory tract infections. While there was an increase in concentration of S-IgA after both relaxation conditions, S-IgA/albumin ratio, a measure of local mucosal immunity, increased only with immune imagery. In both the healthy children and those with compromised health the increases in the two measures were associated with degree of rated relaxation.

2. *Healthy Adults*

In healthy adults Rider and Achterberg (1989) boldly set out to demonstrate the putative importance of relations between specificity of immune imagery and specificity of changes in immune functions. Imagery was employed that differentially targeted two different immune parameters, either total lymphocyte count or neutrophil count, with imagery training targeting either one. Healthy participants were first given training in progressive muscle relaxation and immune imagery, which they were encouraged to draw. They were then given 20-min tape recordings which began with relaxation instructions followed by the appropriate immune imagery and entrainment music and were instructed to practice at home several times a week for 3 weeks. Predicted imagery-specific changes in immune function were disclosed, but were in the direction of a reduction in lymphocytes and neutrophils. The direction of change was hypothesized to reflect perhaps a systemic migration of leukocytes from the bloodstream to tissues and lymphoid structures. This issue aside it is unclear what the mechanism for mind–body–immune specificity could be, aside from some process of an altogether different order, such as suggestion, the neurophysiology of which is unknown.

In another study upregulation of S-IgA was reported (Rider *et al.*, 1990). Subjects were randomly assigned to a control group or to two groups who listened for 17 min to imagery-inducing music or to the same music preceded by instructions about imagery of the immune system; the immune imagery group had been lectured both on the immune system and on the production of S-IgA. Participants were instructed to practice at home every other day for 6 weeks. S-IgA was assessed at the beginning and end of the session before training, and similarly at Weeks 3 and 6. There were advantages for both immunity and well-being for both relaxation approaches when compared with the controls. Antibody production was higher and there were reductions in fatigue, tension, and somatic symptoms of anxiety. There were also benefits for immune imagery over music shown by higher S-IgA at Weeks 3 and 6, while the immune imagery group rated themselves as less confused/bewildered. The added advantages for immune imagery could not be attributed to group differences in vividness of imagery ratings.

Zachariae *et al.* (1994) reported a wholly negative outcome in two largely uncontrolled studies, but their healthy subjects had a much shorter training period of only 3 weeks compared with the 6-week schedule of Rider *et al.* (1990). Training consisted of either muscle relaxation assisted by music or relaxation immune imagery assisted by music, involving 30 min in the laboratory weekly for 3 weeks together with daily home practice with an audiocassette. In one study the dependent variables included monocyte chemotaxis (MC) and lymphocyte proliferative responses (LPR) to three mitogens and in the other study NK cell activity, all measured at the beginning and end of the sessions. They were unable to find any sustained systematic effects across 3 weeks, nor at the 3-week follow-up, which was included in the second study. No relations were found between perceived stress and immune parameters. However, within-session effects were observed in all 3 weekly sessions. In one study there was a reliable reduction in LPR in both intervention groups, a change not seen in a control group included for 1 week only. In the second study NK cell activity decreased within sessions for all three groups, followed by a return to baseline 1 h later, and with a slight increase following a mental stressor given at follow-up.

A more positive outcome was reported by Ruzyla-Smith *et al.* (1995), despite a rather short intervention schedule. They compared hypnosis with REST flotation and a nonintervention control group. Hypnosis was presented in two group sessions 1 week apart during which participants practiced a rapid alert hypnotic induction twice a day. This was compared with a group having two 1-h sessions of REST flotation. Venipuncture took place at baseline and 1 h after the two group-hypnosis and the two REST sessions and was accompanied by an electrodermal assessment of sympathetic autonomic activity. All subjects watched a film about the immune system and were assessed for hypnotic susceptibility with the selection of only those subjects with high and low susceptibility. Hypnosis led to higher B-lymphocyte cell counts than flotation, whereas T-cell counts increased with both interventions, but only in high-hypnotic-susceptibility subjects. T-cell subset analysis indicated that the beneficial effects of hypnosis were due to the CD4 helper T cells.

The effect on salivary cortisol of a longer training programme of thirteen weeks was examined by McKinney *et al.* (1997) and combined with a six week follow-up. When compared with waiting list controls, six weeks of training led to reductions in ratings of depression, fatigue and total distress. Beneficial effects in lowering cortisol, which correlated with improvements in mood, were not apparent until the follow-up assessment.

3. Patients

Despite widespread clinical application of imagery training there have been few reports with patients that included immunity. The one formal study involved women with breast cancer and while the outcome was beneficial in terms of well-being it was disappointing from the perspective of immunity (Richardson *et al.*, 1997). Forty-seven women who completed treatment for primary breast cancer were assigned to one of three options: standard care, 6 weeks of imagery training, or additional social support. No differential effects on a range of immune parameters were found, though there was a differential impact on mood and coping. Imagery training reduced stress and increased vigor and social and functional quality of life in comparison with the other groups. The psychosocial support group also disclosed advantages in overall coping and acceptance of death, while both interventions when compared with standard care shared improved coping skills and perceived social support with a tendency toward enhanced meaning in life.

4. Examination Stress

Attempts at moderating the influence of stressors have used the stress of exams in university students, a naturally occurring stressor (Kiecolt-Glaser *et al.*, 1986; Whitehouse *et al.*, 1996; Gruzelier *et al.*, 2001a). Though a convenient stressor it occurs against a background of academic stressors and peer pressures which, as will be seen, complicate interpretations of outcome. In the first attempt Kiecolt-Glaser *et al.* (1986) examined first-year medical students comparing the effects of relaxation training on mood and immune function with a control group. The intervention consisted of 3 weeks of relaxation training involving five group sessions of self-hypnosis, progressive relaxation, autogenic training, and imagery exercises; a menu from which students could self-select in order to practice at home prior to exams. NK cell activity and peripheral blood T lymphocytes were measured consisting of CD4$^+$ (helper) and CD8$^+$ (cytotoxic) cells and the CD4:CD8 ratio. There were beneficial effects on mood such that the global distress score on a psychopathology symptom inventory and the obsessive–compulsive subscale were elevated at exam time for the control group but not for the intervention group. However, the advantages for the relaxation group did not translate to immune function. This negative result may have occurred for a number of reasons. The type of relaxation procedure was not standardized but was deliberately left to the students' choices, a diversity unlikely to have had homogeneous influences on immune function. There was also evidence of immune suppression for the group as a whole, for the declines in NK cell activity, CD4% of total lymphocytes and CD4:CD8 ratio were from a baseline that the authors noted was at the bottom of the normal range. Notwithstanding, while the relaxation intervention did not buffer the decline in helper cells and NK cell activity, the amount of home practice did correlate positively with the percentage of CD4 cells. However, while this advantage may be attributed to the relaxation training, it may be the result of motivation, or both factors, because frequency of practice was uncontrolled.

Whitehouse et al. (1996) examined self-hypnosis/relaxation with 21 first-year medical students over 19 weeks from the beginning of the academic year. There were four evaluations—during orientation, in late semester, in the examination period, and concluding with a postsemester recovery assessment. Students were given 90 min, weekly, group training sessions aimed at practicing relaxation and discussing their experiences. These included home practice, which they were requested to do daily for 15 min. From the Harvard Group Scale of Hypnotic Susceptibility and the Inventory of Self-Hypnosis, administered in the first two group sessions, participants were found to be mainly in the medium- and high-susceptibility ranges. As in the Kiecolt-Glaser study while relaxation training preferentially lowered anxiety and distress there were no differences in immune function. Unlike Kiecolt-Glaser's study the exam period was accompanied by upregulation of immune function including increases in B-lymphocyte counts, in activated T lymphocytes, in NK cell cytotoxicity, and in PHA- and pokeweed mitogen (PWM)-induced blastogenesis. The exam period also saw higher ratings of total mood disturbance, fatigue, loss of vigor, hostility, depression, and obsessive-compulsive symptoms. Importantly ratings of the quality of the self-hypnosis relaxation exercises predicted NK cell activity and cell number, though frequency of practice was unrelated to immune changes. However, practice positively correlated with global measures of symptom severity during the exam and preexam assessments, interpreted as evidence that self-hypnosis/relaxation was being practiced with the intention of stress reduction.

The use of first-year students may have reduced the chances of more demonstrable benefits for the psychological intervention. There was evidence on some scales of distress peaking at orientation (tension/anxiety, confusion/bewilderment) or being as high at orientation as at exam time (two global distress scales and anxiety), or remaining high throughout the semester until falling during the postsemester recovery (depression/dejection). This problem may also have compromised Kiecolt-Glaser's study and can be inferred from an assessment of immune function and mood in British medical students (Baker et al., 1985). Baker examined self-report anxiety, CD4 counts, and salivary cortisol, comparing changes in first-year clinical students on their second day in a medical school new to them, with measures 4 months later, and with changes in second-year clinical students assessed at the same time points. Changes in opposite directions on all measures were found between first- and second-year students. Measures were significantly higher in the first assessment in the first-year students than in both their second assessment and in the first assessment of second-year students. Second-year students showed nonsignificant increases over the same period as exams approached.

A more successful outcome for immunity in medical students at exam time was found in a study by the author and colleagues with self-hypnosis training involving instructions including immune imagery, physical relaxation, and cognitive alertness which was compared with a nonintervention control group (Gruzelier *et al.*, 1998, 2001a). There were 31 students of both sexes of whom the majority (21) were second-year students, preferential recruitment aimed to reduce the likely immune suppression accompanying adaptation to medical school life in first-year students. Immune and mood parameters were obtained 4 weeks before the exams prior to training and were repeated during the exams. As a requirement for participation volunteers agreed to practice three times a week to control for frequency of practice. This was recorded in diaries and checked telephonically. Frequency of practice was controlled to avoid the confound that the correlation with number of practice sessions could simply relate to personality differences associated with enthusiasm for practicing self-hypnosis. The hypnosis group contained subgroups of high and low hypnotically susceptible participants. Lymphocyte subpopulations measured were defined as $CD4^+$ T helper and $CD8^+$ T cytotoxic cells, $CD3^+$ total T cells, $CD19^+$ B cells and $CD16^+$ NK cells. Plasma cortisol was also assayed, and a lifestyle questionnaire documented exercise; sleep; consumption of alcohol, coffee, tea, and cigarettes; medication; and major life changes, including health. Here there were too few reported illnesses to make evaluation of the effects on illness possible. Mood was assessed by the Speilberger state anxiety scale and by the Thayer activation–deactivation checklists consisting of calmness, energy, tiredness, and tension, while personality was assessed with activated and withdrawn scales developed out of research on personality and hemispheric functional specialization, elaborated later in Section V. In keeping with lateralized influences on the immune system, reviewed in the final section, we predicted that cognitively activated students would respond advantageously to self-hypnosis training while withdrawn individuals would be less responsive.

As has been well documented, NK cell counts and CD8 counts were found to decline in controls with the stress of exams. Importantly, this decline did not occur in the hypnosis group indicating that hypnosis was successful in buffering both the decline in NK cell counts and the decline in CD8 cells that was observed in the control group and their proportion relative to CD4 cells. All these effects were independent of lifestyle changes, and, as they were significantly correlated, suggested an integrated pattern and underlying process. They proved not to be influenced by hypnotic susceptibility. In the face of the stress of exams, which increased anxiety and tension for participants as a whole, there were additional benefits associated with hypnosis. These included higher energy ratings at exam time than in the control group, while increased calmness after hypnosis training correlated positively with increased CD4 counts. Exam counts of NK cells, which were buffered by hypnosis correlated positively with energy and negatively with anxiety. Furthermore an increase in calmness with hypnosis was associated with an increase in CD4 cells and to a lesser extent with CD3 cells. On the other hand, the B lymphocyte CD19 showed positive relations with the increase in anxiety with exams and with exam levels.

Interestingly, cortisol increased with hypnosis whereas there was no change in the control group. Furthermore, there were positive correlations between changes in NK cells and changes both in CD8 cells and cortisol, suggesting that the cell and humoral-mediated changes may be to some extent an integrated pattern of change. The mood ratings and measures of cortisol shed further light on the findings. In line with negative effects of cortisol in relation to stress in the control group, cortisol correlated positively with tiredness at exam time and the increase in cortisol in controls at exam time correlated with their increase in anxiety. Neither of these relations were found following the hypnotic intervention. Apparently hypnosis may bring about a dissociation of the negative effects of cortisol on mood, an impression replicated in a subsequent study (discussed later).

Turning to individual differences in personality, for the group as a whole the activated personality scale predicted increases in all the lymphocytes as well as their levels at exam time. These relations could be largely seen to be a function of the cognitive activation component. This was despite the fact that the hypnotic intervention by elevating lymphocyte numbers would be expected to mask this relation; indeed correlations between cognitive activation and lymphocyte counts in the nonintervention control group were of the higher significance. The correlation between the activated personality score and the change in CD8 counts in controls is shown in Fig. 1. There were no relations between personality and NK cell number or cortisol.

B. Demonstration of Effects on Health

The two studies by the author and colleagues that followed (see Fox *et al.*, 1999; Gruzelier, 2001; Gruzelier *et al.*, 2001b) succeeded in demonstrating

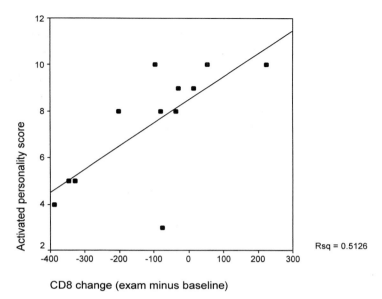

CD8 change (exam minus baseline)

Fig. 1. The correlation between the activated personality score and the change in CD8 T lymphocytes in control medical students (regression coefficient, $t = 3.24$, $p < 0.009$) (Gruzelier *et al.*, 2001a).

improvements in health as a result of self-hypnosis training. While the studies are small in scale they represent an important step for the PNI field by validating the changes in immune function as a result of psychological interventions. One study involved medical students at exam time and took place the winter following our earlier study (outlined on p. 393), when viral infections were more prevalent (Gruzelier *et al.*, 2001b). The other study involved patients with a chronic medical condition herpes simplex virus–2, a form of genital herpes (Fox *et al.*, 1999; Gruzelier, 2001).

1. *Exam Stress*

In the design of the second self-hypnosis study with medical students we set out to elucidate the importance of the type of imagery (Gruzelier *et al.*, 2001b). Whereas one hypnosis group received imagery of the immune system as before, another received self-hypnosis training but with imagery of the immune system replaced with imagery of deep relaxation. Otherwise all aspects of the induction were exactly the same. The same immune parameters were assayed but the design of the study differed in that the order of baseline and treatment conditions were counterbalanced, with the baseline assessment taking place approximately after the exam in half of the students.

First, in the replication study the overall pattern of change in immune parameters for the students as a whole differed from the first study in that there was a highly significant decline in total lymphocyte count whereas in the first study the total lymphocyte count remained stable. Second, whereas NK cell number rose in the first study, in the replication NK cell number declined in second-year students but rose on average in first-year students whose NK cell counts at baseline, as anticipated, were the more suppressed. Considering the results against this background, as before hypnosis buffered the decline in CD8 cells, as well as their percentage of total lymphocytes and the CD8 ratio relative to CD4 cells.

In answer to the question whether the type of imagery training was important, results confirmed those of Olness *et al.* (1989). Crucially, as shown in Fig. 2, it was only the training in imagery of immunity that was successful in buffering the compromise in lymphocytes, with the greater differentiation seen in CD3 and CD4 numbers. Thus relaxation imagery failed to halt the stress-induced decline in total lymphocytes and in the lymphocyte subsets

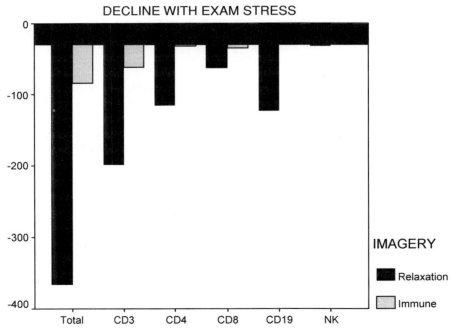

FIG. 2. The changes in lymphocytes and NK cells in medical students following self-hypnosis imagery training with either immune or relaxation imagery (Gruzelier *et al.*, 2001b).

whereas immune imagery buffered decline in all lymphocytes. There were no significant changes in NK cells or cortisol.

Eight students reported illnesses around the exam period. Compared with the remainder, the illness group had a highly significant decline in CD4 counts whereas the decline was not significant in the group that remained well. Importantly, fewer students in the immune imagery group fell ill— 2/11 (18%) compared with 6/9 (67%) of the controls and 5/11 (56%) of those with relaxation imagery. The difference between the immune imagery and control groups was highly significant.

Mood ratings confirmed that students felt less calm at exam time. In view of the dissociation between cortisol and negative mood with hypnosis in the previous study we examined correlations between cortisol and the two Thayer scales of negative affect—tension and tiredness—before and after hypnosis training in the combined hypnosis group. Whereas tiredness correlated with baseline cortisol, there was a negative correlation between the changes in cortisol and tiredness following hypnosis. In other words, here the relation following hypnosis training between cortisol and negative affect went beyond a null relation to a reversal. Consistent with this, increases in tiredness following hypnosis training were associated with decreases not increases in cortisol. To elucidate the individual variation in the effect of exam stress on cortisol in relation to the mood scales, the 10 subjects in the hypnosis group who showed a decline in cortisol were compared with the eight who showed an increase in cortisol. In support of stress raising cortisol, those students characterized by the increase in cortisol had higher levels of tension at baseline compared with those experiencing a decline in cortisol.

2. Herpes Simplex Virus-2

An intervention study on patients who presented with herpes virus chronically and with a high frequency of recurrence is described in detail in this section. HSV-2 is a distressing condition shown by persistently elevated levels of anxiety (Carney *et al.*, 1994), often profound psychosexual morbidity (Mindel, 1993), and compromised psychological well-being and quality of life (Goldmeier and Johnson, 1982). An association between psychological distress and reduced resistance to HSV was first reported by Lycke *et al.* (1974) when depressed patients were found to possess increased antibodies to the latent herpes viruses HSV, Epstein-Barr virus, and cytomegalovirus. Confirmation in subsequent investigations took the form of negative affect in medical students at exam time and in spouses about to divorce, while the reduced resistance to HSV was found in conjunction with a reduction in NK cell activity, CD4 lymphocyte counts, and lymphocyte proliferative responses (see Sheridan *et al.*, 1994). NK cells, macrophages, CD4 and CD8 lymphocytes, INF-α and INF-γ, IL-2, and leukocyte migration inhibitory factor are all significant in protecting against HSV (Rinaldo and Torpey, 1995). In reviewing evidence over a decade Green and Koscis (1997) observed that in many patients the psychological impact of the disorder overshadowed the physical morbidity. This was also true of the patients of Fox *et al.* (1999) in whom the HADS disclosed pathological levels of anxiety at baseline in five patients while two reached the threshold for pathological depression.

The 20 patients were due to discontinue prophylactic antiviral medication for a trial period or were reluctant to take the medication. They were examined at baseline and after a 6-week course of self-hypnosis, which was delayed by 6 weeks for half of the group. After a group hypnosis session they were given self-hypnosis cassette recordings to take home and were recommended to practice a minimum of three times a week, a total of 18 times; the mean was 17 sessions. Hypnosis, as for the student study, involved instructions of relaxation, immune imagery, cognitive alertness, and ego strengthening (Gruzelier *et al.*, 2001a). A wide range of immune measures was incorporated. Immune parameters included numerative measures as before, including plasma cortisol, CD3, CD4, CD8$^+$ T-lymphocyte populations, CD19$^+$ B cells, and CD16$^+$ NK cells. In addition functional NK cell activity was assessed including peripheral blood mononuclear cell (PBMC) nonspecific NK cell cytotoxic activity, HSV-specific NK cell cytotoxic activity of PBMCs infected with HSV-1, HSV-specific cytotoxicity following stimulation with interleukin-2 (termed LAK cell activity) as well as HSV-specific antigen-dependent cellular cytotoxicity (ADCC). Psychometric measures consisted of a Thayer activation–deactivation checklist, the HADS, and the activated and withdrawn personality dimensions.

The intervention had beneficial effects on the occurrence of genital herpes. The number of recurrences fell with hypnosis by a remarkable 40% reaching 48% with the exclusion of two patients who experienced stressful life events in the course of the 6 week intervention. Thirteen patients (65%) showed a reduction in recurrences and were termed responders. The median number of occurrences in the 6 weeks before was 2 (range 0–6) and in the 6 weeks of self-hypnosis was 1 (range 0–3). The results could not be explained away on the basis of demographic or clinical factors, hypnotic susceptibility, or frequency of practice, in fact nonresponders tended to practice more, perhaps due to their visible lack of improvement in herpes recurrence. The beneficial effect on health coincided with an enhancement in well-being. Importantly, this was not restricted to clinical responders for in participants as a whole there was a reduction in HADS Anxiety and Depression, in other words participation in the study had a beneficial effect on anxiety and depression, whether or not there was clinical improvement.

Notwithstanding, relief from anxiety was associated with clinical benefits, and correlational analysis confirmed a significant positive correlation between reduced frequency of recurrence and reduced anxiety.

Along with the clinical benefits there were demonstrable effects on NK cell function in the direction of immune upregulation. In the case of NK cell numbers, whereas nonresponders on average showed a fall, there was a significant increase in responders. Furthermore, functional NK cell activity in responders showed significant changes in HSV-specific NK cell cytotoxicity and HSV-specific cytotoxicity following stimulation with interleukin-2. On the other hand, lymphocytes tended to increase for the group as a whole. Furthermore, there was a tendency for frequency of practice to be correlated positively with immune upregulation, notably with increases in NK cell numbers, CD8 cells, HSV-specific NK cell cytotoxicity, and HSV-specific LAK activity.

Turning to individual difference predictors of particular significance was the finding that cognitive activation correlated positively with clinical improvement. The higher the personality score the fewer the herpes recurrences during the 6 weeks of hypnosis training. In view of the absence of correlations with baseline herpes frequencies, frequency of tape use, and any difference in tape use between responders and nonresponders, this relation could be interpreted as independent of motivation and other extraneous factors and instead as a demonstration that cognitive activation predicted a better clinical response to hypnotherapy.

Cognitive activation also correlated with the change in NK cell parameters. This was seen with NK cell cytotoxicity including a decrease following hypnotherapy in HSV-specific cytotoxicity and in HSV-specific LAK activity, and a positive correlation with posthypnosis levels of HSV-specific cytotoxicity. There was also a positive correlation with some baseline levels namely HSV-specific cytotoxicity, as shown in Fig. 3, and HSV-specific LAK activity. Consistent with the bipolarity of activity–withdrawal, the withdrawn syndrome also correlated negatively with NK cell activity at baseline (i.e., the opposite direction to cognitive activation). Hypnotizability was found to be associated with upregulation of immune function and mood. Immune parameters included changes in NK cell percentage and HSV-specific LAK cell cytotoxicity, the posthypnosis CD8 count, and baseline CD3, CD4, and CD8 lymphocyte counts. Mood parameters included reduction in tiredness and depression.

In conclusion, the clinical outcome was promising and in the year following the study some patients remained off medication and experienced very little recurrence of HSV. While the clinical benefits require confirmation in a controlled clinical trial, advantages of self-hypnosis with directed immune imagery training for immunity, well-being, and health replicate the student

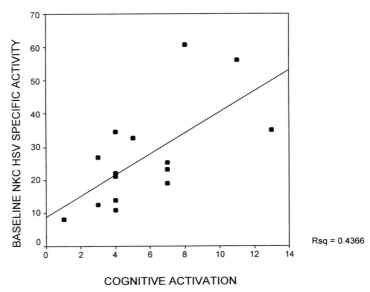

FIG. 3. Baseline HSV-specific cytotoxicity correlated with cognitive activation in HSV-2 patients.

studies (Gruzelier *et al.*, 1998, 2001a,b), as does the predictive ability of the personality factor cognitive activation.

C. SUMMARY OF THE EFFECTS OF IMAGERY TRAINING

1. *Clinical and Psychological Benefits*

To summarize the clinical benefits of imagery training, the marked reduction in recurrence of HSV-2, which was almost halved by 6 weeks of training and which occurred in patients who were chronic and severely ill, attests to the importance of the impact on the immune system of self-hypnosis training with directed immune-related imagery (Fox *et al.*, 1999). There were also psychological benefits in the form of reductions in clinical anxiety and depression. Benefits were shown in a medical student study where imagery training with hypnosis was successful in preventing illnesses at exam time in comparison with hypnotic relaxation training and controls (Gruzelier *et al.*, 2001b).

However, on the whole, the results of the imagery training studies to date which are all too few given the widespread adoption of the practice have largely been carried out with healthy subjects. While these provide a mixed picture in terms of changes in immune function, the most consistent results have clearly been psychological, which suffice it to say are sufficient in themselves to support clinical applications. Psychological benefits included increased coping skills, perceived social support, and enhanced meaning in life in breast cancer patients (Richardson *et al.*, 1997) and reductions in fatigue, tension, somatic symptoms of anxiety, depression, and distress in healthy subjects (Rider *et al.*, 1990; McKinney, 1997). Furthermore in oncology patients these benefits were over and above benefits of social support in reducing stress, increasing vigor, and increasing social and functional quality of life (Richardson *et al.*, 1997), while in healthy subjects immune imagery training resulted in more clear-headedness over and above the benefits of music imagery (Rider *et al.*, 1990).

2. *Immunity*

Effects on immune function per se have provided a mixed picture. Aside from the benefits for patients with HSV-2 this study confirmed that hypnosis can improve NK cell function, and, as shown by others, can reduce antibody titers to HSV (Kiecolt-Glaser *et al.*, 1985; Zachariae and Bjerring, 1990). In support of mediational changes in the immune system, the reduction in herpes recurrences was accompanied by functional enhancement in NK cell activity in the form of HSV-specific cytotoxicity and HSV-specific cytotoxicity following stimulation with IL-2, along with an increase in NK cell number. In the one other clinical study with adults advantages for a range of measures were not forthcoming over 6 weeks in oncology patients (Richardson *et al.*, 1997). However, the null aspect of this result was mitigated somewhat by the fact that it was shared with the psychosocial support intervention. In the study with medical students mediational changes in the immune system were supported by the buffering of the decline in lymphocytes, and these were specific to the immune imagery group rather than to the relaxation imagery group (Gruzelier *et al.*, 2001b).

Otherwise in the other study with healthy subjects the picture is mixed. Rider *et al.* (1990) found that S-IgA levels increased significantly in a group given specific immune imagery while remaining unchanged both in a group with nonspecific imagery and in controls. Serum IgG levels were not altered. Similarly in children S-IgA levels increased significantly with immune specific imagery while remaining unchanged with nonspecific imagery and in nonintervention controls, and serum IgG levels again were not altered (Olness *et al.*, 1989). However, Hewson-Bower and Drummond (1996) could not replicate the effect with S-IgA in children, though they found that IgA/albumin ratio increased with both types of imagery. In contrast, in a 6-week intervention study McKinney *et al.* (1997) found that cortisol was reduced but not until a 6-week follow-up, and while this did correlate with a lasting improvement in mood, the importance for this effect of imagery training per se has not been tested. In another study 3 weeks was insufficient to show advantages in MC, LPR and NK cell activity (Zachariae *et al.*, 1994) and in a further study brought about changes in a direction opposite to hypothesized advantages (Rider and Achterberg, 1989).

It is also the case that hypnosis may interfere with the conventional relation between cortisol and stress and increase cortisol (Gruzelier *et al.*, 2001a,b); after all cortisol is necessary for normal metabolism as evinced with the surge in cortisol level on awakening.

Examination of within-session variation in immune measures (e.g., Zachariae *et al.*, 1994) brought home the plasticity of immune parameters. Aside from questioning the biological significance of up- and downregulation this raises the methodological desirability of replication and retest reliability and validity.

3. *Stressor Validation*

Validation of the efficacy of psychological intervention on immunity by examining the ability to moderate the influence of stress had useful outcomes, in particular, evidence of improvements in health in students around the exam period (Gruzelier *et al.*, 2001b). At the same time difficulties were disclosed with the stressor validation approaches, notably the need to control for background stressors which will impact on the immunity of otherwise convenient samples of mostly young, undergraduate students. This may have been responsible for the fact that Whitehouse *et al.* (1996) found only limited benefits from relaxation training in the form of some reduction in anxiety and distress at exam time, and both Kiecolt-Glaser *et al.* (1986) and Whitehouse *et al.* (1996) found that improvements in mood were not paralleled by immune changes that were distinguishable from controls. However, in support of the efficacy of the psychological intervention the quality of self-hypnosis training was positively associated with immune upregulation (NK cells) (Whitehouse *et al.*, 1996) and frequency of practice was associated with immune upregulation (CD4 cells) (Kiecolt-Glaser *et al.*, 1986). In the Gruzelier studies the value of self-hypnosis was seen in the effect on CD8 counts, extending to NK cell number in the first study and to other lymphocyte populations in the second study. Furthermore, by comparing training in imagery of the immune system versus purely relaxation imagery in the second study, immune imagery was found to be largely responsible for the effects on lymphocytes. Consistency was shown by the fact that despite the exam period being experienced as stressful, energy ratings rose at exam time following hypnosis training compared with controls in both studies, and by the novel finding of a dissociation between the expected negative effects on mood and cortisol increase following hypnosis. The main inconsistency was the failure to replicate the strong buffering effect on the decline in NK cell number evinced in the first study. This may have been due to the inclusion of two streams of students, with the consequence of a greater diversity of prevailing background stressors, as found by Whitehouse *et al.* (1996), or it may have been due to other differences in experimental design such as the counterbalancing of the baseline measure 4 weeks before and after the exams. One other difference in outcome was the relation between hypnotic susceptibility and immune upregulation in CD8 percentage in high hypnotic susceptibles in the second study, whereas there was no relation with susceptibility in the first study.

D. INDIVIDUAL DIFFERENCES

Individual differences in personality also had a significant impact. In the herpes study cognitive activation predicted herpes recurrence together with functional enhancements in a range of NK cell parameters in HSV-2 patients (Gruzelier, 2001). In the student study while there was no relation with NK cell counts, cognitive activation correlated with the full range of T and B lymphocytes. In the clinical study correlations with personality and pathological mood were also found with baseline immune parameters in the form of positive correlations with cognitive activation and negative correlations with withdrawal, especially its affective component, along with clinical depression and anxiety scales. As cognitive activation is part of a behaviorally outward directed personality dimension it will influence recruitment for experimental studies and, as recruitment procedures vary, so will the composition of studies across investigations.

Both hypnotic susceptibility and frequency of practice were also found to have an impact on results; however, improvements in health in students were not restricted to high hypnotic susceptibles or to patients who practiced frequently (Gruzelier *et al.*, 2001b).

IV. Hypnosis and Immunity

Given that self-hypnosis training brought about benefits in health, aside from upregulation of immunity and enhancement in well-being, and these include what may be for the first time a demonstration that hypnotherapy may bring about an improvement in a chronic medical condition, the influence of hypnosis on immunity is worthy of further consideration and research. Historically hypnotherapy has been successful as a stress-reduction technique and there is evidence that it may facilitate mind–body influences on the immune system. It may also shed light on brain–immune pathways involved in the mediation of therapeutic benefits on the immune system, for hypnosis has been shown to produce reproducible influences on brain functional organization in the form of states of dissociation between brain regions which are not everyday occurrences (Gruzelier, 1998, 2000; Oakley, 1999; Rainville *et al.*, 1999).

A. HYPNOTHERAPY AND IMMUNITY

Drawing together the results reviewed here, aside from the three studies of the author and colleagues', hypnosis has been widely used. Self-hypnosis training was included in the mixed approach of Kiecolt-Glaser *et al.* (1985) with the elderly where there was upregulation of NK cell activity and a decrease in antibody levels to HSV with 4 weeks of training. It was one of a menu of options in Kiecolt-Glaser's study of exam stress with medical students (Kiecolt-Glaser *et al.*, 1986), where frequency of home practice over 3 weeks correlated with an increase in the CD4 percentage of total lymphocytes, though nothing conclusive can be stated about hypnosis given the menu approach. In the 19-week study by Whitehouse *et al.* (1996) the quality of self-hypnosis relaxation exercises predicted both NK cell number and functional activity during medical student exams. In children Olness *et al.* (1989) found that hypnotic imagery targeting the immune system increased S-IgA in contrast to nonspecific imagery. Ruzyla-Smith *et al.* (1995) found that a rapid alert hypnotic induction practiced twice a day for 1 week led to higher B-lymphocyte cell counts than two 1-h sessions of REST flotation, and along with REST increased CD4 T lymphocytes.

B. IMAGERY VERSUS RELAXATION

There have been several reports that imagery directed at the immune system was superior to imagery of relaxation or to other relaxation approaches including REST flotation, a particularly deep form of relaxation (Olness *et al.*, 1989; Ruzyla-Smith *et al.*, 1995; Gruzelier *et al.*, 2001b). In other words the critical influence in bringing about the beneficial influences on cell-mediated immunity appears to be the particular imagery and suggestion in the mind of the participant. This exemplifies that hypnosis achieves its effects through a mind–body connection. A particularly graphic example is afforded by a clinical vignette with an asthmatic patient who following instructions of hypnosis underwent a severe asthmatic episode triggered simply by the verbal association of cow bells in a meadow—cows have fur to which the patient was allergic. Furthermore, the asthmatic fit was resolved through the verbal suggestion that the patient was taken up in a helicopter where the oxygen content of the air was enriched, even though this metaphor was false (D. Ewan, personal communication, 1998). It was the mental imagery coupled with the belief in the suggestion that was critical to the mind–body connection.

Accordingly the beneficial influences of self-hypnosis on immune upregulation and the associated clinical benefits cannot be attributed to incidental influences such as relaxation. Hypnosis typically does include relaxation (Williams and Gruzelier, 2000), or at least mental as distinct from physical relaxation, given that participants may be hypnotized while pedaling a bicycle (Cikurel and Gruzelier, 1990). However, we have demonstrated that relaxation procedures, including flotation, while altering neurophysiology do not duplicate the effects of hypnosis (Raab and Gruzelier, 1994; Gruzelier and Brow, 1985).

C. Skin Hypersensitivity

The evidence that hypnosis may facilitate the modulation of immune function has perhaps been most extensively investigated in association with immediate and delayed responses in skin sensitivity to allergens, Mantoux reactions, and various dermatological conditions including warts and congenital ichthyosiform erythrodermia (Hall, 1983; Mason and Black, 1958; Sinclair-Geiben and Chalmers, 1959; Black et al., 1963) and has almost always involved suggestions of imagery of the immune system. While numerous affirmative results have been reported, the phenomenon remains elusive (Locke et al., 1987, 1994) and high hypnotic susceptibility appears necessary (Hall, 1983).

Fry et al. (1964) provided one of the more comprehensive of the early studies. This was a two-part investigation with 47 asthma patients with known skin sensitivity to house dust or pollen, unselected for hypnotizability, and all but one naive to hypnosis. In the first part 18 patients were randomized to hypnosis or control groups and tested 2 weeks apart, the control group without hypnosis, and the hypnosis group first without hypnosis and the second time with an hypnosis session, the last of three training sessions. Hypnosis included instructions to attenuate the skin reaction. Four strengths of allergen were administered bilaterally to the forearms. All were attenuated bilaterally to some extent with hypnosis, reaching significance with the two lower strengths. In the second part the remaining patients were randomized into three groups all of whom were given three sessions of hypnosis over 2 weeks but with different instructions—bilateral nonreactivity, right-arm nonreactivity, or no immune-related suggestion. Compared with the baseline assessment all groups showed attenuated reactions bilaterally. Successful localization to one arm had been reported by Black (1963) the previous year, and subsequently was demonstrated by Zachariae and Bjerring (1990) who also successfully localized nonreaction to one arm through suggestion whereas the contralateral reaction was unaffected. Other successful attempts at attenuating reactions have been reported by Dennis and Philippus (1965) with antigens and histamine and Zachariae *et al.* (1989) with both immediate and delayed flares and wheals of Mantoux reactions. In contrast Smith *et al.* (1992) failed with the group instructed to attenuate reactions, yet found that subjects could enhance reactions when instructed.

Some light on variability of response has been shed by examining daily fluctuations in flare and wheal sizes to histamine in association with fluctuations in mood (Laidlaw *et al.*, 1994, 1996). Liveliness was found to account for 31% of the daily variance of wheal size in the direction of the more lively the smaller the reaction, whereas listlessness showed the opposite effect. In a further study subjects were invited to construct their own imaginative method of decreasing the size of the skin reaction and to visualize this with hypnosis (Laidlaw *et al.*, 1996). Of the participants, 28/32 were able to reduce the size of their wheals to a highly significant degree. Mood ratings in the form of irritability and tension, and a change to higher blood pressure readings, were associated with less success.

V. Research Directions

A. HETEROGENEITY, SPECIFICITY, AND VALIDATION

The review has shown that changes in both cellular and humoral immunity have resulted from a diverse collection of psychological interventions alone or in combination with measures including NK cell number and activation, antibody levels to the herpes simplex virus, cortisol, neutrophil number and activation, lymphocyte number and activation, monocyte chemotaxis, and salivary and serum immunoglobulins. While in general the direction of at least some of the changes was congruent with immune function theory, not all measures changed in consort. Most commonly some were found to remain invariant, and sometimes changes were in a direction consistent with immune compromise, perhaps the outcome of redistribution within and migration from the bloodstream. It is recognized that there are fundamental questions about the functional significance of the direction of change of virtually all immune parameters. This is because of the complexity of interrelations between such tightly integrated systems of checks and balances on immunity. For this reason the impact of a therapeutic intervention on immune parameters by themselves while of interest is made more compelling by concurrent measures of well-being, and even better by measures of health.

Methodological factors will be responsible for some of the heterogeneity of findings. There is heterogeneity of subjects and intervention techniques which cannot be presumed to have identical effects on neurophysiology and immune functions. Heterogeneity extends to training schedules, which ranged from a single session to daily practice for 10 weeks or more, and the nature of the immune assay. The latter may involve the relatively benign collection of saliva or involve a stressful venipuncture. Validation will be assisted by attempts at greater refinement or specificity such as more standardized, and homogeneous intervention techniques rather than mixed intervention packages, sometimes offered as a menu of opportunities. The strategy of varying one feature of an intervention while other components are held constant will assist with clarifying the processes involved (Gruzelier et al., 2001b). Interesting attempts have been made with imagery training to compare the outcome on immune parameters of targeting particular immune functions (Rider and Achterberg, 1989; Gregerson, 1995). Specificity has been successfully demonstrated, though the meaning of the changes in terms of immune function has not been clear (Rider and Achterberg, 1989). Attempts at specificity have also included modification of skin sensitivity to allergens and both attenuation and enhancement of reactions.

Ambiguities about the direction of change in an immune parameter make validation through effects on other domains of measurement essential at this early stage of basic neuroscientific understanding. Most often immune changes have been accompanied by validation through mood ratings with beneficial effects including reductions in ratings of global distress, anxiety, coping ability, fatigue, tension, depression, obsessive–compulsive behavior, and bodily symptoms of anxiety, and increases in energy and calmness; though improvements have tended to be restricted to isolated features of mood rather than improvement overall. Occasionally physiological parameters have provided validation such as EMG, blood pressure, and skin temperature biofeedback learned to criterion (Peavey et al., 1985; McGrady et al., 1989, 1992; Laidlaw et al., 1996). Validation has included the effect of laboratory and real-life stress in the form of medical school examinations.

Now there are the beginnings of more conclusive clinical validation observed both through benefits on health demonstrated by improvement in

a chronic illness, HSV-2 (Fox *et al.*, 1999), and through suggestive evidence
of illness prevention at times of stress in healthy people (Gruzelier *et al.*,
2001b). The validating steps of immune parameters and health are necessary because of the complexity of immune functions and the complexity of
stress, both of which are poorly understood, such that changes in the direction of upregulation or downregulation are ambiguous. One glimpse is
afforded by the seminal experiments of Selye (1955) who from examining
the effects of stress on animals evolved a general adaptation syndrome. This
delineated functional influences of stress on the endocrine system which
were bidirectional, with an initial increase in function characterizing an
alarm stage, followed by a gradual decline in function characterizing a resistance stage, after which a stage of exhaustion was reached. In psychology,
functional bidirectionality has been enshrined in the famous inverted-U
relation between stress and performance, but has seldom been alluded to
in PNI investigations. One implication is that the direction of change in
an immune parameter may differ according to the degree and duration of
stress encountered and the natural history of exposure to stress. Accordingly
stressed subjects may disclose a reduction in a parameter with the psychological intervention through alleviation of Selye's alarm stage, or an increase
in the same parameter in the resistance phase through recovery from
suppression. In nonstressed subjects an increase in an immune parameter
may signify an enhancement of immune competence. It follows that it is
unlikely that immune parameters respond to psychological interventions in
the same way in healthy and vulnerable individuals.

Targeted immune imagery may offer advantages over and above other
approaches whose main aim is relaxation (Olness *et al.*, 1989; Rider *et al.*,
1990; Gregerson *et al.*, 1996; Gruzelier *et al.*, 2001b). Compatible with the
requirement of alert, cognitive involvement in generating immune imagery,
imagery training when compared with other relaxation approaches has been
found to reduce confusion–bewilderment (Rider *et al.*, 1990) but has led to
more anxiety and less calmness (Gregerson *et al.*, 1996). Fluctuations in
mood have had demonstrable effects on the modification through directed
hypnotic imagery of the size of the skin sensitivity reaction (Laidlaw *et al.*,
1992b).

B. INDIVIDUAL DIFFERENCES

1. *Hypnotic Susceptibility*

Heterogeneity confounds will also be reduced by investigation of individual differences. Historically this has been exemplified by the phenomenon
of hypnotic susceptibility which underpins dramatic differences in abilities

to respond to instruction and suggestion and to relax and undergo cognitive changes. Importantly, both cognitive and physiological flexibility has been ascribed to hypnotic susceptibility (Crawford and Gruzelier, 1992). For this reason it has been assessed by many investigators in order to select subjects high in susceptibility or high in absorption which correlates highly with hypnotic susceptibility (Jasnowski and Kugler, 1987), to stratify groups as high or low in hypnotizability (Zachariae *et al.*, 1994; Gregerson *et al.*, 1996; Gruzelier *et al.*, 2001a), or for purposes of correlation (Olness *et al.*, 1989; Whitehouse *et al.*, 1996; Gruzelier *et al.*, 2001b, Gruzelier, 2001). In support of a propensity for dynamic changes Zachariae *et al.* (1994) found that it was high susceptibles who demonstrated within-session changes in NK cell activity and in lymphocyte proliferative responses. Similarly absorption, which is correlated with hypnotic susceptibility, showed advantages with relaxation and alertness procedures (Jasnowski and Kugler, 1987; Gregerson *et al.*, 1996). Gregerson *et al.* (1996) found that high absorbers increased their salivary S-IgA within session more than low absorbers, and absorption was associated with mindfulness–cognitive flexibility ratings. In Ruzyla-Smith's study high susceptibility subjects increased CD4 T cells with both two REST sessions with flotation and 2 weeks of rapid, alert self-hypnosis (Ruzyla-Smith *et al.*, 1995). We found that high-susceptible students increased their CD8 percentage through imagery training more than low susceptibles (Gruzelier *et al.*, 2001b), while hypnotic susceptibility in patients with herpes showed positive correlations with some baseline and post-training lymphocyte counts, increases in NK cell percentage and functional activity, as well as correlation with reductions in tiredness and depression (Gruzelier, 2001).

At the same time hypnotic susceptibility has been found unrelated to the increase in salivary IgA within a single session in children (Olness *et al.*, 1989). Nor was it found relevant to the effects of 10 sessions of hypnosis, 9 sessions of which were practiced at home with a tape recording (Gruzelier *et al.*, 2001a), perhaps because hypnotic susceptibility though possessing trait characteristics is at the same time modifiable with practice, a commonplace clinical occurrence. Future research would benefit by reexamining hypnotic susceptibility at the end of training to determine whether susceptibility has in fact altered.

2. *Cognitive Activation*

Germane to the issue of specificity has been the predictive potential of the activated temperament and an action-oriented personality, particularly the cognitive activation component (Gruzelier *et al.*, 2001a; Gruzelier, 2001). Historically most research has focused on individual differences accompanying immune compromise such as depression, loneliness, and submissiveness

(Scheifer *et al.*, 1983; Ramirez *et al.*, 1989; O'Leary, 1990; Bennett Herbert and Cohen, 1993; Zisook *et al.*, 1994). More recently attention has turned to personality features associated with health and well-being such as hardiness, self-esteem, humor, and expressing emotion rather than suppressing it (Diong and Bishop, 1999; Fernandez-Ballesteros *et al.*, 1998; Johnston *et al.*, 1999; McClelland and Cheriff, 1997; Skevington and White, 1998; Spiegal, 1999; Valdimarsdottir and Bovbjerg, 1997). The results relating to the activated personality, which encompasses aspects of cognitive activation, behavioral activity, and the expression of positive affect, extend the literature on individual differences in psychology relating to immune upregulation, while at the same time they are complimentary with current personality findings and may offer insights about neurophysiological underpinnings.

The construct of the activated personality first grew out of research on schizophrenia, and functional laterality. Active versus withdrawn syndromes were originally delineated by psychophysiological asymmetry parameters and subsequently the syndromes were found to be characterized by a range of lateralized neuropsychological and psychophysiological processes in schizophrenia (Gruzelier, 1999, for review). Later these syndromes were found to apply to personality dimensions in the normal population where they were also associated with opposite cognitive functional asymmetry patterns (Richardson and Gruzelier, 1994; Gruzelier *et al.*, 1995; Gruzelier, 1996; Gruzelier and Doig, 1996). Applicability to immune function follows evidence of lateralized influences on the immune system. This has included evidence in animals. Dependent on the side of unilateral neocortical ablations in rodents, opposite effects have been found on IgG-plaque-forming cells and mitogen-induced splenic T-cell proliferation: enhanced with an intact left hemisphere and compromised with an intact right hemisphere (Renoux *et al.*, 1983a,b; Neveu *et al.*, 1986). In animals NK cell activity has been impaired with left-side ablation (Bardos *et al.*, 1981) while lower NK cell activity has been associated with left-paw preferences (Betancur *et al.*, 1991). In humans a model has been proposed linking left-handers, who as a group have a greater reliance on right-hemispheric processing, with an increased incidence of autoimmune disorders (Geschwind and Behan, 1984; Geschwind and Galaburda, 1985; Lindsay, 1987). Reduced NK cell activity has been found in nurses with a preferential right frontal EEG activation compared with those with the opposite asymmetry (Kang *et al.*, 1991).

The writer proposed a heuristic model incorporating evidence of lateralized influences on the immune system with hemispheric specialization theory including associations with the left hemisphere of approach behavior, positive affect, and immune upregulation and with the right hemisphere of withdrawal, negative affect, and immune downregulation (Gruzelier, 1989). A test of the model was afforded by longitudinal assessment of asymptomatic men with HIV infection (Gruzelier *et al.*, 1996). Confirming the model both EEG and neuropsychological asymmetry patterns assessed at study onset were predictive of CD4 counts 2–3 years later; participants with a left-hemisphere functional preference on first assessment had higher counts than those with a right-hemisphere functional preference, and vice versa. Subsequently Clow and colleagues have reported theoretically consistent asymmetries in salivary IgA and free-cortisol concentration following lateralized transmagnetic stimulation of temporoparietal occipital cortex (Gruzelier *et al.*, 1998; Evans *et al.*, 2000).

The activated temperament, cognitive activation in particular, versus withdrawal are in keeping with left and right hemispheric specializations in the form of approach versus avoidant behavior (and hence the chosen descriptors activated and withdrawn), as well as with hemispheric specialization relating to positive versus negative affect. The particular relation here between immune function and the active syndrome subscale active speech, which describes speaking and thinking quickly, provides compelling support for the association of immune upregulation with left-hemisphere functional preference, given the unambiguous left-hemisphere involvement in speech production. There are clear links between the activated personality dimension and fighting spirit, laughter, and exercise, which have demonstrable advantages for immune function (Greer, 1983; Fernandez-Ballesteros *et al.*, 1998; Skevington and White, 1998; Valdimarsdottir and Bovbjerg, 1997).

Finally, aside from its predictive validity, the activated versus the withdrawn personality dimensions provide insights into the neurophysiological basis of mechanisms mediating psychoneuroimmunological influences. It follows that the activated personality dimension, with its neurophysiological validation may help in clarifying individual differences in psychoneuroimmunological mechanisms. Guidance for individual differences in patient response and compliance may follow the predictive ability of the personality trait. Hypnosis training may succeed better in a depressed patient with the active syndrome trait (showing the potential for modification), in contrast to a depressed patient with a withdrawn personality trait. It also makes sense of why directed imagery of the immune system, with its active, cognitive requirement, when compared with the more passive imagery of relaxation training, appears the more successful form of intervention.

C. Conclusion

Research on clinical interventions in this field of psychoneuroimmunological research is very much in its infancy. As evinced by the nature of the experimental designs of most studies, investigations have been carried out on a limited budget by dedicated pioneers. Typically there has been a small number of training sessions, sometimes just one, and a small number of immune parameters which have been assayed once or twice, seldom more frequently. Nonetheless the conviction of pioneering researchers that psychological interventions may benefit health through influence on the immune system and well-being is being vindicated. Hypnosis, which has shown reproducible effects on the immune system, as an intervention is relatively simple, inexpensive, and of relatively short duration, and through home practice is saving of clinical time, though it must be applied in an informed and responsible manner (Gruzelier, 2000). It provides a scientific approach with which the mind–body connection is amenable not only to experimental investigation, but also to experimental variation and control. The body of evidence reviewed here, when seen in conjunction with the greater accumulation of evidence from studies focusing simply on relations between immune compromise and stress, justifies a larger investment in basic cognitive and immunological science to formally evaluate clinical intervention strategies and to elucidate the mechanisms involved.

References

Ader, R., Felten, D. L., and Cohen, N. (eds.) (2001). "Psychoneuroimmunology." 3rd ed. Academic Press, London.

Baker, G. H. B., Irani, M. S., Byrom, N. A., Nagvekar, N. M., Wood, R. J., Hobbs, J. R., and Brewerton, D. A. (1985). Stress, cortisol concentration and lymphocyte subpopulations. *Br. Med. J.* **290,** 1393.

Banyai, E. I., and Hilgard, E. R. (1976). A comparison of active-alert hypnotic induction with traditional relaxation induction. *J. Abnorm. Psychol.* **85,** 218–224.

Bardos, P., Degenne, D., Lebranchu, Y., Biziere, K., and Renoux, G. (1981). Neocortical lateralisation of NK activity in mice. *Scand. J. Immunol.* **13,** 609–611.

Bennett Herbert, T., and Cohen, S. (1993). Depression and immunity: A meta-analytic review. *Psychol. Bull.* **113,** 472–483.

Betancur, C., Neveu, P. J., Vitiello, S., and Le Moal, M. (1991). Natural killer cell activity is associated with brain asymmetry in male mice. *Brain Behav. Immun.* **5,** 162–169.

Black, S. (1963). Inhibition of immediate-type hypersensitivity response by direct suggestion under hypnosis. *Br. Med. J.* **6,** 925–929.

Black, S., Humphrey, J. H., and Niven, J. S. (1963). Inhibition of Mantoux reaction by direct suggestion under hypnosis. *Br. Med. J.* **6,** 1649–1652.

Burns, S. J., Harbuz, M. S., Hucklebridge, F., and Bunt, L. (2001). A pilot study into the therapeutic effects of music therapy at a cancer help center. *Altern. Ther.* **7,** 48–56.

Carney, O., Ross, E., Bunker, C., Ikkos, G., and Mindel, A. (1994). A prospective study of the psychological impact on patients with a first episode of genital herpes. *Genitourin-Med.* **70,** 40–45.

Cikurel, K., and Gruzelier, J. H. (1990). The effect of an active-alert hypnotic induction on lateral asymmetry in haptic processing. *Br. J. Exp. Clin. Hypnosis* **7**, 17–25.

Crawford, H. J., and Gruzelier, J. (1992). A midstream view of the neuropsychophysiology of hypnosis: Recent research and future directions. *In* "Hypnosis: Research Developments and Perspectives" (W. Fromm and M. Nash, eds.), 3rd ed., pp. 227–266. Guilford Press, New York.

Deinzer, R., and Schuller, N. (1998). Dynamics of stress related decrease of salivary immunoglobulin A(S-IgA): Relationship to symptoms of the common cold and studying behavior. *Behav. Med.* **23**, 161–169.

Dennis, M., and Philippus, M. J. (1965). Hypnotic and non-hypnotic suggestion and skin response in atopic patients. *Am. J. Clin. Hypnosis* **7**, 342–345.

Dennis, M. S. (1985). Stress, cortisol concentrations and lymphocyte subpopulations. *Br. Med. J.* **290**, 1393.

Diong, and Bishop (1999). Anger expression, coping styles and well-being. *J. Health Psychol.* **4**, 81–96.

Evans, P., Hucklebridge., F., and Clow, A. (2000). "Mind, Immunity and Health: The Science of Psychoneuroimmunology." Free Associations Books, London.

Fernandez-Ballesteros, R., Ruiz, M. A., and Garde, S. (1998). Emotional expression in healthy women and those with breast cancer. *Br. J. Health Psychol.* **3**, 41–50.

Fox, P. A., Henderson, P. C., Barton, S. E., Champion, A. J., Rollin, M. S. H., Catalan, J., McCormack, S. M. G., and Gruzelier, J. H. (1999). Immunological markers of frequently recurrent genital herpes simplex virus and their response to hypnotherapy: A pilot study. *Int. J. STD AIDS* **10**, 730–734.

Fry, L., Mason., A. A., and Pearson, R. S. B. (1964). Effect of hypnosis on allergic skin responses in asthma and hay-fever. *Br. Med. J.* **1**, 1145–1148.

Geschwind, N., and Behan, P. (1984). Laterality, hormones and immunity. *In* "Cerebral Dominance" (N. Geschwind and N. Galaburda, eds.). Harvard University Press, Cambridge, MA.

Geschwind, N., and Galaburda, A. M. (1985). Cerebral lateralisation: Biological mechanisms, associations and pathology. I. A hypothesis and a program for research. *Arch. Neurol.* **42**, 428–459.

Goldmeier, D., and Johnson. A. (1982). Does psychiatric illness affect the recurrence rate of genital herpes? *Br. J. Venereal Dis.* **58**, 40–43.

Green, J., and Koscis, A. (1997). Psychological factors in recurrent genital herpes. *Genitourin Med.* **73**, 253–258.

Green, M. L., Green, R. G., and Santoro, W. (1988). Daily relaxation modifies serum and salivary immunoglobulins and psychophysiologic symptom severity. *Biofeedback and Self-Regulation* **13**, 187–199.

Greer, S. (1983). Cancer and the mind. *Br. J. Psychiatr.* **143**, 535.

Gregerson, M. B., Roberts, I. M., and Amiri, M. M. (1996). Absorption and imagery locate immune responses in the body. *Biofeedback and Self-Regulation* **21**, 149–165.

Gruber, B. L., Hersh, P., Hall, N. R. S., Waletzky, L. R., Kunz, J. F., Carpenter, J. K., Kverno, K. S., and Weiss, S. M. (1993). Immunological responses of breast cancer patients to behavioral interventions. *Biofeedback and Self-regulation* **18**, 1–22.

Gruzelier, J. H. (1989). Lateralisation and central mechanisms in clinical psychophysiology. *In* "Handbook of Clinical Psychophysiology" (G. Turpin, ed.), pp. 135–174. Wiley, Chichester.

Gruzelier, J. H. (1996). The factorial structure of schizotypy. I. Affinities and contrasts with syndromes of schizophrenia. *Schizophr. Bull.* **22**, 611–620.

Gruzelier, J. (1998). A working model of the neurophysiology of hypnosis: A review of evidence. *Contemporary Hypnosis* **15**, 5–23.

Gruzelier, J. (1999). Functional neuro-psychophysiological asymmetry in schizophrenia: A review and reorientation. *Schizophr. Bull.* **25**, 91–120.

Gruzelier, J. H. (2000). Redefining hypnosis: Theory, methods and integration. *Contemporary Hypnosis* **17**, 51–70.

Gruzelier, J. H. (2001). A review of the impact of hypnosis, relaxation, guided imagery and individual differences on aspects of immunity and health. *Stress* **5**, 147–163.

Gruzelier, J. H., and Brow, D. (1985). Psychophysiological evidence for a state theory of hypnosis and susceptibility. *J. Psychosomat. Res.* **29**, 287–302.

Gruzelier, J., and Doig, A. (1996). The factorial structure of schizotypy. II. Patterns of cognitive asymmetry, arousal, handedness and gender. *Schizophr. Bull.* **22**, 621–634.

Gruzelier, J., Burgess, A., Stygall, J., Irving, G., and Raine, A. (1995). Patterns of cerebral asymmetry and syndromes of schizotypal personality. *Psychiatry Res.* **56**, 71–79.

Gruzelier, J., Burgess, A., Baldeweg, T., Riccdio, M., Hawkins, D., Stygall, J., Catt, S., Irving, G., and Catalan, J. (1996). Prospective associations between lateralised brain function and immune status in HIV infection: Analysis of EEG, cognition and mood over 30 months. *Int. J. Psychophysiol.* **23**, 215–224.

Gruzelier, J., Clow, A., Evans, P., Lazar, I., and Walker, L. (1998). Mind–body influences on immunity: Lateralised control, stress, individual difference predictors and prophylaxis. *Ann. N.Y. Acad. Sci.* **851**, 487–494.

Gruzelier, J., Smith, F., Nagy, A., and Henderson, D. (2001a). Cellular and humoral immunity, mood and exam stress: The influences of self-hypnosis and personality predictors. *Int. J. Psychophysiol.* **42**, 55–71.

Gruzelier, J. H., Levy, J., Williams, J. D., and Henderson, D. (2001b). Effect of self-hypnosis with specific versus nonspecific imagery: Immune function, mood, health and exam stress. *Contemporary Hypnosis* **18**, 97–110.

Hall, H. R. (1983). Hypnosis and the immune system: A review with implications for cancer and the psychology of healing. *Am. J. Clin. Hypnosis* **25**, 92–103.

Hall, H. R. (1989). Research in the area of voluntary immunomodulation: Complexities, consistencies and future research considerations. *Int. J. Neurosci.* **47**, 81–89.

Hall, H. R., Minnes, L., Tosi, M., and Olness, K. (1992a). Voluntary modulation of neutrophil adhesiveness using a cyberphysiologic strategy. *Int. J. Neurosci.* **63**, 287–297.

Hall, H. R., Mumma, G. H., Longo., S., and Dixon, R. (1992b). Voluntary immunomodulation: A pilot study. *Int. J. Neurosci.* **63**, 275–285.

Hall, H. R., Papas, A., Tosi, M., and Olness, K. (1996). Directional changes in neutrophil adherence following passive resting versus active imagery. *Int. J. Neurosci.* **85**, 185–194.

Hewson-Bower, B., and Drummond, P. D. (1996). Secretory immunoglobulin A increases during relaxation in children with and without recurrent upper respiratory tract infections. *J. Dev. Behav. Pediatr.* **17**, 311–316.

Holmes, T. H., and Rahe, R. H. (1967). The social readjustment rating scale. *Journal of Psychosomatic Research* **11**, 213–218.

Jasnowski, M. L., and Kugler, J. (1987). Relaxation, imagery and neuroimmunomodulation. *Ann. N.Y. Acad. Sci.* **496**, 772–730.

Johnson, V. C., Walker, L. G., Heys, S. D., Whiting, P. H., and Eremin, O. (1996). Can relaxation training and hypnotherapy modify the immune response to stress, and is hypnotizability relevant? *Contemp. Hypnosis* **13**, 100–108.

Johnston, M., Earll, L., Giles, M., McClenahan, R., Stevens, D., and Morrison, V. (1999). Mood as a predictor of disability and survival in patients newly diagnosed with ALS/MND. *Br. J. Health Psychol.* **2**, 127–136.

Kang, D. H., Davidson, R., Coe, C., Wheeler, R. E., Tomarken, A. J., and Ershler, W. B. (1991). Frontal brain asymmetry and immune function. *Behav. Neurosci.* **105,** 860–869.

Kiecolt-Glaser, J. K., Garner, W., Speicher, C., Penn, G. M., Holliday, J. E., and Glaser, R. (1985). Psychosocial enhancement of immunocompetence in a geriatric population. *Health Psychol.* **4,** 25–41.

Kiecolt-Glaser, J., Glaser, R., Strain, E. C., Stout, J. C., Tarr, K. L., Holliday, J. E., and Speicher, C. E. (1986). Modulation of cellular immunity in medical student. *J. Behav. Med.* **9,** 5–21.

Laidlaw, T. M., Booth, R. J., and Large, R. G. (1994). The variability of type 1 hypersensitivity reactions: The importance of mood. *J. Psychosom. Res.* **38,** 51–61.

Laidlaw, T. M., Booth, R. J., and Large, R. G. (1996). Reduction in skin reactions to histamine following a hypnotic procedure. *Psychosom. Med.* **58,** 242–248.

Leonard, B. E., and Miller, K. (Eds.), (1995). "Stress, the Immune System and Psychiatry." Wiley, Chichester.

Lindsay, J. (1987). Laterality shift in homosexual men. *Neuropsychologia* **25,** 965–969.

Locke, S. E., Ransil, B. J., Covino, N. A., Toczydlowski, J., Lohse, C. M., Dvorak, H. F., Arndt, K. A., and Frankel, F. H. (1987). Failure of hypnotic suggestion to alter immune response to delayed-type hypersensitivity antigens. *Ann. N.Y. Acad. Sci.* **496,** 745–749.

Locke, S. E., Ransil, B. J., Zachariae, R., Molay, F., Tollins, K., Covino, N. A., and Danforth, D. (1994). Effect of hypnotic suggestion on the delayed-type hypersensitivity response. *JAMA* **272,** 47–52.

Lycke, E., Norrby, R., and Roos, B.-E. (1974). A serological study on mentally ill patients with particular reference to the prevalence of herpes virus infections. *Br. J. Psychiatr.* **124,** 273–279.

Mason, A. A., and Black, S. (1958). Allergic skin responses abolished under treatment of asthma and hay fever by hypnosis. *Lancet* **1,** 877–880.

McClelland, D. C., and Cheriff, A. D. (1997). The immunological effects of humor on secretory IgA and resistance to respiractory infection. *Psychol. Health* **12,** 329–344.

McGrady, A., Turner, J. W., Jr., Fine, T. H., and Higgins, J. T., Jr. (1987a). Effects of bio-behaviorally-assisted relaxation training on blood pressure, plasma renin, cortisol, and aldosterone levels in borderline essential hypertension. *Clin. Biofeed. Health* **10,** 16–25.

McGrady, A., Woener, M., Bernal, G. A. A., and Higgins, J. T., Jr. (1987b). Effect of biofeedback-assisted relaxation on blood pressure and cortisol levels in normotensives and hypertensives. *J. Behav. Med.* **10,** 301–310.

McGrady, A., Conran, P., Dickey, D., Garman, D., Farris, E., and Schurmann-Brzezinski, C. (1992). *J. Behav. Med.* **15,** 343–354.

McKinney, C. H., Antoni, M. H., Kumar, M., and Times, F. C., *et al.* (1997). Effects of guided imagery and music (GIM) therapy on mood and cortisol in healthy adults. *Health Psychol.* **16**(4), 390–400.

Mindel, A. (1993). Long-term clinical and psychological management of genital herpes. *J. Med. Virol.* **Suppl. 1,** 39–44.

Neveu, P. J. (1993). Brain lateralisation and immuno-modulation. *Int. J. Neurosci.* **70,** 1917–1923.

Oakley, D. A. (1999). Hypnosis and conversion hysteria: A unifying model. *Cogn. Neuropsychiatr.* **4,** 243–265.

O'Leary, A. (1990). Stress, emotion and human immune function. *Psychol. Bull.* **108,** 363–382.

Olness, K., Culbert, T., and Den, D. (1989). Self-regulation of salivary immunoglobulin A by children. *Pediatrics* **83,** 66–71.

Peavey, B. S., Lawlis, G. F., and Goven, A. (1985). Biofeedback-assisted relaxation: Effects on phagocytic capacity. *Biofeedback and Self-Regulation* **10,** 33–47.

Raab, J., and Gruzelier, J. H. (1994). A controlled investigation of right hemispheric processing enhancement after restricted environmental stimulation (REST) with flotation. *Psychol. Med.* **24,** 457–462.

Rainville, P., Hofbauer, R. K., Paus, T., Duncan, G. H., Bushnell, M. C., and Price, D. D. (1999). Cerebral mechanisms of hypnotic induction and suggestion. *Journal of Cognitive Neuroscience* **11**(1), 110–125.

Ramirez, Craig *et al.* (1989). Stress and relapse of breast cancer. *Br. Med. J.* **298,** 291–293.

Renouz, G., Biziere, K., Renouz, M., Guillaumin, J., and Degenne, D. (1983a). A balanced brain asymmetry modulates T-cell mediated events. *J. Neuroimmunol.* **5,** 227–238.

Renoux, G., Biziere, K., Renoux, M., and Guilluamin, J. (1983b). The production of T-cell inducing factors in mice is controlled by the brain neortex. *Scand. J. Immunol.* **17,** 45–50.

Richardson, A., and Gruzelier, J. H. (1994). Visual processing, lateralisation and syndromes of schizotypy. *Int. J. Psychophysiol.* **18,** 227–240.

Richardson, M. A., Post-White, J., Grimm, E. A., Moye, L. A., Singletary, S. A., and Justice, B. (1997). Coping, life attitudes, and immune responses to imagery and group support after breast cancer treatment. *Altern. Ther.* **3,** 62–70.

Rider, M. S., and Achterberg, J. (1989). Effect of music-assisted imagery on neutrophils and lymphocytes. *Biofeedback and Self-Regulation* **14,** 247–257.

Rider, M. S., Achterberg, J., Lawlis, G. F., Goven, A., Toledo, R., and Butler, J. R. (1990). Effect of immune system imagery on Secretory IgA. *Biofeedback and Self-Regulation* **15,** 317–333.

Rinaldo, C. R., Jr., and Torpey, D. J. (1995). Cell-mediated immunity and immunosuppression in herpes simplex virus infection. *Immunodeficiency* **5,** 33–90.

Ruzyla-Smith, P., Barabasz, A., Barabasz, M., and Warner, D. (1995). Effects of hypnosis on the immune response: B-cells, T-cells, helper and suppressor cells. *Am. J. Clin. Hypn.* **38**(2), 71–79.

Scheifer, S. J., Keller, S. E., Camerino, M., Thornton, J. C., and Stein, M. (1983). Suppression of lymphocyte stimulation following bereavement. *JAMA* **250,** 374.

Selye, H. (1955). Stress and disease. *Science* **122,** 625–631.

Sheridan, J. F., Dobbs, C., Brown, D., and Zwilling, B. (1994). Psychoneuroimmunology: Stress effects on pathogenesis and immunity during infection. *Clin. Microbiol. Rev.* **7,** 200–212.

Shor, R. E., and Orne, E. C. (1962). "Harvard Group Scale of Hypnotic Susceptibility: Form A." Consulting Psychologists Press, Palo Alto, C A.

Simonton, O. C., Matthews-Simonton, S., and Creighton, J. (1978). "Getting Well Again." J. B. Tarcher, Los Angeles.

Sinclair-Geiben, A. H., and Chalmers, D. (1959). Evaluation of treatment of warts by hypnosis. *Lancet* **2,** 480–482.

Skevington, S. M., and White, A. (1998). Is laughter the best medicine? *Psychol. Health.* **13,** 157–169.

Smith, G. R., McKenzie, J. M., Marmer, D. J., and Steele, R. W. (1985). Psychologic modulation of the human immune response to varicellazoster. *Arch. Intern. Med.* **145,** 2110–2112.

Smith, G. R., Conger, C., O'Rourke, D. F., Steele, R. W., Charlton, R. K., and Smith, S. S. (1992). Psychological modulation of the delayed-type hypersensitivity skin test. *Psychosomatics* **33**(4), 444–451.

Spiegal, D. (1999). Healing words: Emotional expression and disease outcome. *JAMA* **281,** 1328–1329.

Taylor, D. N. (1995). Effects of a behavioral stress-management program on anxiety, mood, self-esteem, and T-cell count in HIV positive men. *Psychol. Res.* **76,** 451–457.

Thayer, R. E. (1967). Measurement of activation through self-report. *Psychol. Res.* **20,** 663–678.

Valdimarsdottir, H. B., and Bovbjerg, D. H. (1997). Positive and negative mood: Association with natural killer cell activity. *Psychol. Health* **12,** 319.

Whitehouse, W. G., Dinges, D. F., Orne, E. C., Keller, S. B., Bates, B. L., Bauer, N. K., Morahan, P., Haupt, B. A., Carlin, M. M., Bloom, P. B., Zaugg, L., and Orne, M. T. (1996). Psychosocial and immune effects of self-hypnosis training for stress management throughout the first semester of medical school. *Psychosom. Med.* **58,** 249–263.

Williams, J. D., and Gruzelier, J. H. (2001). Differentiation of hypnosis and relaxation by analysis of narrow band theta and alpha frequencies. *Int. J. Clin. Exp. Hypnosis* **49,** 185–286.

Zachariae, R., and Bjerring, P. (1990). The effect of hypnotically induced analgesia on flare reaction of the cutaneous histamine prick test. *Arch. Dermatol. Res.* **282,** 539–543.

Zachariae, R., Bjerring, P., and Arendt-Nielsen, L. (1989). Modulation of Type 1 immediate and Type IV delayed immunoreactivity using direct suggestion and guided imagery during hypnosis. *Allergy* **44,** 537–542.

Zachariae, R., Hansen, J. B., Andersen, M., Jinquan, T., Petersen, K. S., Simonsen, C., Zachariae, C., and Thestrup-Pedersen, K. (1994). Changes in cellular immune function after immune specific guided imagery and relaxation in high and low hypnotizable healthy subjects. *Psychother. Psychosom.* **61,** 74–92.

Zisook, S., Sucter, S. R., Irwin, M., Darko, C. F., Sledge, P., and Resovsky, K. (1994). Bereavement, depression and immune function. *Psychiatry Res.* **52,** 1–10.

INDEX

reflex IgA secretion, 203
schizophrenia, 289–290
Hypothalamic regulation
inflammation, 141–143
Hypothalamic suprachiasmatic
nucleus, 9
Hypothalamus
HPA response, 53

I

IDDM, 9, 277
IFN, 105–111
schizophrenia, 286–287
IFN-γ, 26, 27
IgA. *See* Immunoglobulin
A (IgA)
IgE, 29
IGF, 80
IgG1, 29
IL, 105–111
IL-1
activation of the HPA axis,
52–56
changes in body temperature,
50–52
cytokines signal the brain
mechanisms, 49–58
effects on behavior, 56–58
schizophrenia, 284
IL-2
schizophrenia, 284–286
sleep deprivation, 115
IL-3
schizophrenia, 287
IL-4, 25, 26
IL-6, 115, 117
HPA axis, 55
schizophrenia, 286
IL-10
glucocorticoids, 167
IL-12, 25, 26
HIV, 172
IL-18
schizophrenia, 287
IL-1α, 49
IL-1β, 49, 52, 110
IL-1R, 20
Imagery
vs. relaxation, 404–405

Immune activity
circadian features, 112
Immune cells
diagram of the relationship, 48
function
norepinephrine, 18–19
Immune conditioning
early studies, 326–327
Immune CRH expression
mechanisms which control, 68
Immune defenses, 2
Immune function
psychological intervention, 383–412
research directions, 406–412
heterogeneity, specificity, validation,
406–408
individual differences, 408–411
Immune POMC expression
regulated, 74
Immune reactivity
behavioral lateralization, 309–317
Immune responses
conditioning
rediscovered, 328
health and disease
CRH and related peptides, 68–70
immunoneuropeptides, 80–81
modulators, 67–83
neuropeptides, 67–83
opioid peptides, 75–79
pro-opiomelanocortin (POMC)
peptides, 70–75
trophic peptides, 79–80
Immune system
behavioral conditioning, 325–348
cerebral lateralization, 303–318
dysregulation
depression, 256–259
guided imagery, 388–394, 394–403,
401–402
hypnosis, 403–406
hypnotherapy, 404
left-cortical ablations, 305–309
and neuroendocrine system relationship,
262–263
origin of dysregulations, 263
role of cortisol, 262
role of CRF, 262
right-cortical ablations, 305–309
role, 2

CONTENTS OF RECENT VOLUMES